Pharmaceutical Powder Compaction Technology

DRUGS AND THE PHARMACEUTICAL SCIENCES

Series Executive Editor

Recent Titles in Series

210. Pharmaceutical Stress Testing: Predicting Drug Degradation, Second Edition; Steven W. Baertschi, Karen M. Alsante, Robert A. Reed, ISBN 9781439801796, 2011
209. Pharmaceutical Process Scale-Up, Third Edition; Michael Levin, ISBN 9781616310011, 2011
208. Sterile Drug Products: Formulations, Packaging, Manufacturing and Quality; Michael K. Akers, ISBN 9780849339966, 2010
207. Advanced Aseptic Processing Technology; James Agalloco, James Akers, ISBN 9781439825433, 2010
206. Freeze-Drying/Lyophilization of Pharmaceutical & Biological Products, Third Edition; Louis Rey, Joan May, ISBN 9781439825754, 2010
205. Active Pharmaceutical Ingredients; Development, Manufacturing, and Regulation, Second Edition; Stanley Nusim, ISBN 9781439803363, 2009
204. Generic Drug Product Development: Specialty Dosage Forms; Leon Shargel, Isadore Kanfer, ISBN 9780849377860, 2010

Pharmaceutical Powder Compaction Technology

Second Edition

Metin Çelik PhD
President, Pharmaceutical Technologies International, Inc.
Belle Mead, New Jersey, U.S.A.

Research Professor of Pharmaceutical Technology
College of Pharmacy, Near East University
Lefkosia, Turkish Republic of Northern Cyprus

informa
healthcare

Previous edition published in 1995 by Marcel Dekker, Inc., New York, New York.
This edition published in 2011 by Informa Healthcare, Telephone House, 69–77 Paul Street, London EC2A 4LQ, UK

Simultaneously published in the USA by Informa Healthcare, 52 Vanderbilt Avenue, 7th Floor, New York, NY 10017, USA.

Informa Healthcare is a trading division of Informa UK Ltd. Registered Office: 37–41 Mortimer Street, London W1T 3JH, UK. Registered in England and Wales number 1072954.

A CIP record for this book is available from the British Library.

ISBN-13: 978-1-4200-8917-2
ISSN (print): 0360-2583
ISSN (online): 2154-5219

Orders may be sent to: Informa Healthcare, Sheepen Place, Colchester, Essex CO3 3LP, UK
Telephone: +44 (0)20 7017 6682
Email: Books@Informa.com
Website: http://informahealthcarebooks.com
Printed and bound in the United Kingdom

For corporate sales please contact: CorporateBooksIHC@informa.com
For foreign rights please contact: RightsIHC@informa.com
For reprint permissions please contact: PermissionsIHC@informa.com

Typeset by MPS Limited, a Macmillan Company
Printed and bound in the United Kingdom

Contents

Contributors

Carl T. Allenspach Bristol-Myers Squibb, New Brunswick, New Jersey, U.S.A.

Chris Balducci Novartis Pharmaceuticals Corporation, East Hanover, New Jersey, U.S.A.

Gerad K. Bolhuis Department of Pharmaceutical Technology and Biopharmacy, University of Groningen, Groningen, The Netherlands

Jean-Daniel Bonny Acino Pharma AG, Basel, Switzerland

Gary Bubb Specialty Measurements Incorporated, Lebanon, New Jersey, U.S.A.

Graham Buckton School of Pharmacy, University of London, London and Pharmaterials Ltd., Reading, U.K.

Metin Çelik Pharmaceutical Technologies International, Inc., Belle Mead, New Jersey, U.S.A.; and Faculty of Pharmacy, Near East University, Lefkosia, Turkish Republic of Northern Cyprus

James Cheney Celgene Corporation, Summit, New Jersey, U.S.A.

Olivia Darmuzey Novartis Pharma Stein AG, Stein, Switzerland

Hans de Waard Department of Pharmaceutical Technology and Biopharmacy, University of Groningen, Groningen, The Netherlands

Arne W. Hölzer Mölndal, Sweden

Hans Leuenberger Ifiip LLC, Institute for Innovation in Industrial Pharmacy, and CINCAP LLC, Center for Innovation in Computer-Aided Pharmaceutics, Basel, Switzerland

Michael N. Leuenberger Amificas Inc., American Institute for Innovative Computer-Aided Solutions, Oviedo, Florida, U.S.A.

Ruth Leu-Marseiler Novartis Pharma Ltd., Basel, Switzerland

Rosario LoBrutto TEVA Pharmaceuticals, Pamona, New York, U.S.A.

Susan K. Lum Patheon Inc., Whitby Development Center, Ontario, Canada

Eşay Okutgen PharmaCircle, Skillman, New Jersey, U.S.A.

Anant Paradkar Institute of Pharmaceutical Innovation, University of Bradford, Bradford, U.K.

Maxim Puchkov CINCAP LLC, Center for Innovation in Computer-Aided Pharmaceutics, Basel, Switzerland

Alan Royce Novartis Pharmaceuticals Corporation, East Hanover, New Jersey, U.S.A.

Colleen E. Ruegger Novartis Pharmaceuticals Corporation, East Hanover, New Jersey, U.S.A.

Hedinn Valthorsson Novartis Pharma Stein AG, Stein, Switzerland

Dong Xiang Novartis Pharmaceuticals Corporation, East Hanover, New Jersey, U.S.A.

Peter York Institute of Pharmaceutical Innovation, University of Bradford, Bradford, U.K.

Erika A. Zannou Novartis Pharmaceuticals Corporation, East Hanover, New Jersey, U.S.A.

Preface to the first edition

Oral administration is the dominant method of delivering drugs to the human systemic blood circulation because of its safety and simplicity. Thus, great interest has been focused within pharmaceutical science on the design of oral dosage forms with optimal therapeutic properties. The prevailing oral dosage form today is the tablet due to its elegance. Tablets of various types and biopharmaceutical properties—from conventional, disintegrating tablets, to advanced modified release systems—exist, but their common denominator is the way in which they are formed, that is, powder compaction. Physical and technological aspects of this process, from a pharmaceutical point of view, are the theme of this book.

The complexity of the compaction process—what at first sight seems to be a simple mechanical operation—was recognized early. Problems still exist in large-scale production of tablets, such as low tablet strength, capping, limited use of direct compression, and sensitivity to batch variability of starting materials. Moreover, the use of basic physical data in formulation work in order to predict tabletting behavior of particles such as compressibility (ability to reduce in volume) and compactibility (ability to cohere into compacts) is limited. Thus, tablet formulation must still be based to a large extent on empirical knowledge rather than on scientific theory.

An improved theoretical understanding of the compaction process will enable a more rational approach to the formulation of tablets. However, the investments in research on the physics of the compaction process have, in relative terms, been limited in universities and the pharmaceutical industry. In spite of this, a large number of publications on the theme of the formation of tablets by compaction exist today in the pharmaceutical literature. This literature can be broadly classified into three categories: (1) reports on specific formulations and their compactibility and on formulation solutions to compaction-related problems, (2) studies on mechanisms of and theories for the compression and the compaction of pharmaceutical powders (such studies also include articles dealing with the development and evaluation of methods for theoretical studies), and (3) evaluation, with recognized methods and theories, of the compression and compaction behavior of pharmaceutical tabletting excipients.

In the older literature, publications were focused mainly on the practical aspects of the preparation of tablets. However, since the late 1940s, articles focused on the theoretical aspects of the compaction process have been presented in the pharmaceutical scientific literature. As a consequence of the growing interest in directly compactable formulations, new excipients with improved tabletting performance have been developed and the compaction characteristics of these have been the object of scientific studies. Despite this growing literature on the physics and technology of powder compaction, the interest in bringing together the accumulated knowledge in the form of comprehensive reference works has hitherto been limited. It is thus a great pleasure for the editors of this volume to present a book on theoretical and practical aspects of the process of forming compacts by powder compression. This is, to our knowledge, the first book devoted entirely to this theme. It has been made possible by the contribution of chapters from researchers throughout Europe and North America. To achieve the high level needed, only recognized scientists, representing academia or the pharmaceutical industry, have been involved, and each contributor has been encouraged to focus on his or her field of expertise. The role of the editors has been to primarily select topics and authors for the contributions and to find a suitable structure for the book. The consequence of this is that different concepts and beliefs in the field of powder compaction are presented and discussed in the book, and we have not attempted to hide this diversity. This diversity reflects the complexity of studying and establishing theories for the handling and processing of "real" materials. Moreover, there are also different traditions with respect to the nomenclature used in the discussion on powder compaction, and this inconsistency among researchers in this respect is also reflected in this book. The editors allowed each author to use terms in accordance with his or her tradition. However, to improve the stringency in the use of the nomenclature for the future, a short list of definitions follows this preface.

During the preparation of this book, some topics within the area of pharmaceutical powder compaction have not been dealt with as separate chapters, as they are not covered extensively in the literature. Examples of such topics are energy aspects of the formation of tablets, physical instability in compacts during storage, and mathematical expressions for the tensile strength of compacts. However, these topics are discussed and references are given in some of the chapters of this book.

Although great progress in the theoretical understanding of the compaction process has been made since the late 1940s, the need for further research is obvious. It is our hope that this volume can contribute to and stimulate such intellectually challenging research.

We are very grateful to Marcel Dekker, Inc., for taking the initiative to prepare a book on pharmaceutical powder compaction technology. We express our sincere appreciation especially to Sandra Beberman and Ted Allen for pleasant cooperation during the preparation of this book, for their qualified contributions, and for their support and patience with us in our role as editors.

We are also very grateful to all contributors to this volume, for their positive attitude to share their expertise in the field of powder compaction and for the time and effort taken to write articles of high quality. Without their collaboration and contributions, the writing of this book would never have been accomplished.

Finally, we would like to thank Mrs. Eva Nises-Ahlgren for qualified administrative work in connection with the preparation of this book.

Goran Alderborn
Christer Nystrom

Preface to the second edition

In 1996, the editors of the first edition of this text, Alderborn and Nystrom,[1] were fully aware of the complexity of the compaction process and the limited, at that time, scientific theories applicable to the process. In their preface, they recognized the need for an improved theoretical understanding of compaction and compression in order to enable a more rational approach to the formulation of the solid dosage forms described as "tablets." At the same time, Alderborn and Nystrom realized the need to include information on the practical and technological aspects of compaction and compression and they largely succeeded in these endeavors, as judged by the wide acceptance of the multiauthored first edition that covered a range of theoretical and practical areas.

Even so, Alderborn and Nystrom were aware that certain topics within the area of pharmaceutical powder compaction were not dealt with as extensively as desired, due in large part to a lack of extensive coverage in 1996 when the first edition was published. It is the aim of the second edition to remedy that situation, while at the same time building on and expanding the scope of the first edition.

In Chapter 1 (*Intermolecular bonding forces: Where materials and process come together*), Buckton revisits intermolecular bonding forces, the point where materials and process come together. The stages involve an initial compression to push the particles closer together and a subsequent phase where closely packed particles no longer reduce the packing volume by simple movement around each other, but deform either by brittle fracture or by plastic flow. It is these processes of plastic flow and brittle fracture, and most usually a combination of the two, that give rise to compact formation rather than a reversible compression/decompression of the mass.

Lum shows in Chapter 2 (*Viscoelastic models*) that viscoelasticity has a tremendous effect on powder compaction and in order to produce compacts with desired properties an understanding of mechanical properties and constituent materials is needed. Problems that arise during tablet production are a consequence of viscoelasticity combined with poor interparticle bonding. The degree of interparticle bonding and bond strength is assumed to be largely governed by the magnitude of the true interparticle contact area. The amount of elastic recovery of the particles depends on the release of elastic strain during decompression and tends to disrupt interparticle bonds. This stress relaxation is a function of the viscoelasticity of the material that in turn is influenced by the speed of compression and decompression. At high compaction speeds, the internal stress after compaction is high and the propensity of the material for elastic recovery is considerably higher than at low compaction speeds.

In Chapter 3 (*Application of percolation theory and fractal geometry to tablet compaction*), Leuenberger and coauthors have updated their work in the area of percolation theory, fractal geometry, and the use of virtual R&D reality so as to allow new insights into the physics of tablet compaction and the properties of the tablets. The results attained so far are promising and should stimulate further research in this field.

Bubb addresses postcompaction data analysis techniques in Chapter 4 (*Postcompaction data analysis techniques*) using data collected from an instrumented rotary tablet press. The

[1]*Pharmaceutical Powder Compaction Technology*, Alderborn G and Nystrom G, eds, Marcel Dekker, Inc., New York, 1996.

discussion of an individual pulse analysis is followed by an in-depth analysis of compaction and ejection profiles. Bubb is also the author of Chapter 5, titled *Tablet press instrumentation in the research and development environment*, where he makes the point that an instrumented tablet press in an R&D environment is a necessity if one wishes to have a deeper understanding of compaction principles, shorten development time, and enable easier transition from R&D machines into production equipment.

In Chapter 6 (*Advanced compaction research equipment: Compaction simulators*), compaction simulators are discussed by Ruegger and Çelik. As they point out, with the increasing use of quality by design during drug product development, the standardization of functionality tests is more important than ever to determine the design space for tablet compaction. Although less sophisticated types of equipment may be used to determine the compaction design space, the use of a compaction simulator will increase the likelihood of success for the development of a robust tablet formulation while minimizing the amount of material needed for drug product development and characterization of the formulation and the individual components.

The current state of compactibility functionality tests is reviewed in Chapter 7 (*Compactibility functionality test*) by Okutgen. As this author points out, probably the most important aspect of pharmaceutical formulation development is the selection of suitable excipients since these can profoundly influence the properties of the dosage form, especially when the drug concentration is small. This type of information and the existence of standard test methods to measure such properties are important to assure consistent quality and functioning of different excipients as well as one excipient from lot-to-lot. When the dosage form is a tablet, it is necessary to consider the assessment and comparison of the compaction behavior of the inactive ingredients, including diluents, binders, antiadherents, glidants, and lubricants. Okutgen shows that, even today, the formulator lacks both a single source of reference and the existence of a standard test method to assess and compare the compaction behavior of excipients that play a critical role in the compactibility of a powder composition.

Chapter 8 (*Compaction properties of directly compressible materials*) deals comprehensively with the compaction properties of directly compressible materials. The authors, Bolhuis and de Waard, have both updated and expanded the senior author's chapter in the first edition. They point out that in spite of enormous improvements in wet granulation techniques (high shear granulation, fluid bed granulation, extrusion granulation, continuous granulation, and all-in-one granulation), tablet production by direct compaction has increased steadily over the years because it offers economic advantages through its elimination of the wet granulation and drying steps. As a result, it requires fewer unit operations in production, which means less equipment and space, lower labor costs, less processing time, and lower energy consumption. Additionally, the elimination of the wet granulation step increases the stability of drugs that can degrade by moisture and/or heat. Another advantage of direct compaction is that the tablets generally disintegrate into primary particles rather than into granules. The increased surface area for dissolution may result in a fast drug release for some drugs and some drug products.

Lubrication issues in direct compaction is the title of Chapter 9 by Bolhuis and Hölzer. According to these authors, lubricants are commonly included in tablet formulations in order to reduce die wall friction during both compaction and ejection of the tablet. Their presence, however, may cause undesirable changes in tablet properties, such as decreased diametrical breaking strength and tablet softening.

Paradkar and York discuss crystal engineering and particle design in Chapter 10 (*Crystal engineering and particle design for the powder compaction process*). These authors note that for several decades, in addition to investing huge resources in the drug discovery process for identifying new chemical entities, pharmaceutical interests have been engaged in the engineering of drug substances to overcome challenges in drug delivery linked to their pharmaceutical properties. Particle size, shape, crystal habit, crystal form, density, and porosity are primary properties of solids; whereas flowability, compressibility, compactibility, consolidation, dust generation, and air entrapment during processing are secondary ones. Their chapter shows that it is critically important to employ efficient and effective particle design techniques with the ability to produce small, uniform particles with desired primary and secondary properties for the development of pharmaceutical products. In the area of particle engineering the approaches are focused on alteration of primary and secondary particle properties by a single step process using techniques such as spherical crystallization and crystallo-co-agglomeration. Green processes like supercritical fluid technology may well be preferred over solvent-based crystallization methods considering stringent regulatory controls and the desired control of primary and secondary drug particle characteristics.

The pharmaceutical industry, whether brand name or generic, has over the past few years been increasing its focus on developing combination products, which have been a key component of very successful lifecycle management strategies mainly due to significant advantages in both therapeutic and commercial aspects, including extension of patent coverage. Thus, Chapter 11 (*Compaction of combination products* by Allenspach and Zannou) focuses mainly on formulation options for compaction products and their impact on compaction behavior when a monolayer tablet is not feasible due to compaction properties, intended biopharmaceutical performance (e.g., dissolution/bioavailability and/or stability), and/or line extension or market differentiation. A major topic is multilayer tablet compression (bilayer being the most common) including tablet strength and layer adhesion strength measurement and prediction, formulation considerations to optimize compaction properties, and manufacturing considerations during the compression unit operation.

Balducci and a number of coauthors observe in Chapter 12 (*Quality by design and compression*) that the goal of quality by design is to define a drug product manufacturing process that is flexible enough to allow mitigation of inherent variability in the process that may arise from raw materials or equipment fluctuations. In contrast, traditional manufacturing processes are considered too rigid to account for inherent variability of the process and provide no opportunity for continual improvement. They show that when quality by design is fully implemented into the development strategy, the critical sources of process and material attribute variability can successfully be identified, measured, and understood so that they can be controlled with the appropriate control strategy. Statistical design provides an economical use of resources, especially when many process parameters exist. Also, statistical design facilitates an in-depth understanding of the process and provides strong assurances to regulatory agencies regarding superior process quality.

The final chapter (Chapter 13, *Expert systems and their use in pharmaceutical applications*) is a review that seeks to describe the current status of expert systems applicable to the development of pharmaceutical formulations and processes. Although the literature on expert systems has expanded rapidly since the time first edition was published, the author, Çelik, opines that very few expert systems have been successfully applied to pharmaceutical systems, with even fewer applied to powder compaction. Hopefully, this deficiency will be remedied in the future as more companies realize the value in developing and applying their own expert systems. Stay tuned for the third edition!

James Swarbrick
Executive Editor, Drugs and the Pharmaceutical Sciences
PharmaceuTech Inc.
Pinehurst, North Carolina, U.S.A.

1 Intermolecular bonding forces: Where materials and process come together

Graham Buckton

INTRODUCTION

The application of a compaction force to a powder will require a number of mechanisms to allow the powder bed to convert to a compressed compact—a tablet. The stages involve an initial compression to push the particles closer together and a subsequent phase where closely packed particles no longer reduce the packing volume by simple movement around each other but deform either by brittle fracture or by plastic flow. It is these processes of plastic flow and brittle fracture, and most usually a combination of the two, that give rise to compact formation, rather than a reversible compression/decompression of the mass. These concepts and the process of elastic recovery will be discussed in more detail elsewhere in this book. The processes of brittle facture and plastic flow give rise to surface contact and bond formation. The bond strength will be linked to the surface area over which bonding can occur and the type of bonding that happens (1–7).

THE COMPRESSION STAGE

Prior to the process of forming a compact, the powder bed must be compressed to consolidate the powder. The extent of volume reduction during compression will be related to a number of factors and will ultimately set up the process of compact formation (through suitable bonding). The starting properties of the powder(s) to be compacted and the way that these properties change during the early phase of the compaction process will be significant. Major contributors to this process can be expected to be particle size and shape with high surface area and surface roughness being regarded as favorable to set up the compaction process (8). That being said, the process becomes dominated by the deformation mechanism with highly fragmenting brittle materials exposing a large increase in surface area for compact formation, which will most probably override any effect of the size, surface area, and roughness of the starting materials.

Inevitably, the physical form of a material, in terms of whether it is amorphous or crystalline, and if crystalline the properties of that crystal, will affect the compression and compaction processes. Different crystal forms include salts, polymorphs, hydrates, and solvates, each of which can have different external habits (needles or prisms), melting points, and mechanical properties. While amorphous materials have a tendency to plastic deformation and crystalline to brittle fracture, it is far from being as simple as that with many materials having both brittle and plastic behaviors, and some changing their behavior as a function of strain rate (tablet machine speed). While these points will not be discussed here in any detail, it is vital to remember that the understanding of the interplay between the materials properties and the process is the cornerstone of Quality by Design. The principles of Quality by Design are that critical sources of variability in a product are identified and explained, variability is controlled by understanding the materials and the process, and product quality attributes can be accurately and reliably predicted using a design space relating to materials properties, the process, and the environmental/other factors that may be significant. In this short chapter, the nature of materials is touched upon and their interaction with atmospheric humidity mentioned in one example. In no way will the coverage be comprehensive, but it will hopefully be indicative.

COMPACTION

Bonding surface area can be regarded as the effective surface area that is involved in the interaction between particles. It is not possible to be certain or to have a practical measurement of the actual surface area that is available during tabletting. The surface area will alter because of fracture and flow and will be influenced by the porosity of the materials. Rather than having a direct measurement, surrogate secondary properties are often correlated to tablet strength; these would include particle size, particle shape, and particle surface energy (9) along with an understanding of compaction mechanism and bond strength (8,10).

BONDING MECHANISMS

Rumpf (11) described bonding mechanisms as being of five types (Box 1).

Box 1 The Types of Bonding That Can Give Rise to Tablet Formation

1. **Solid bridges**
2. Moveable liquids
3. Non–freely moveable binders
4. **Attraction between solid particles**
5. **Mechanical interlocking**

The three that are regarded as dominant are shown in bold.

Solid bridges are formed between two particles by processes such as crystallization of amorphous portions of solid, sintering, or chemical reactions (such as salt formation between adjacent particles).

Bonding between moveable liquids is related to surface tension forces. As compaction is assisted by the presence of some moisture, this may well play a role in the process for tablets. Non–freely moveable binders would include the binders used in wet granulation processing, such as polyvinylpyrrolidone, which make powders into better compact formers.

Attractions between solid particles will be discussed further below. Mechanical interlocking can be envisaged during tabletting, by both plastic flow and brittle fracture. Further, it is probable that this assists with increasing surface area for interparticulate bonding as much as physical interlocking.

While all these binding mechanisms can be regarded as of significance, it is probable that solid bridge formation, intermolecular forces, and mechanical interlocking are dominant (12).

SOLID BRIDGES

Pilpel and York proposed that particle contact points experience such high local stresses during the compaction process that asperity melting occurs. This does not necessarily mean that the melting point of the material is reached across the sample, but rather that at extreme local pressure, the combination of local heating and pressure is sufficient to cause molecular movement, and as the pressure is released, perhaps during the compaction process as a slip of fragmentation occurs or during release of the pressure as the punch is removed, the adjacent particles return to the crystalline state and the newly formed crystalline region bridges more than one particle, most probably forming a network structure, giving the tablet great strength. Consequently, the formation of solid bridges can inevitably be viewed as a strong bonding mechanism as the bonds will have the strength of the intermolecular forces that are holding the particles together. The strength of bonds within crystals will vary from material to material depending on the forces that hold the crystal together. This is what gives rise to changes in melting points, enthalpies of fusion, and surface energies.

The crystallization of amorphous, or partially amorphous, materials to give rise to solid bridges (13–16) is very common, by both intentional use of crystallizing excipients and uncontrolled use of partially amorphous materials. A good example of intentional use is that of spray-dried lactose, which is a commercially available tabletting excipient with an amorphous content of around 15% to 20%. The amorphous content causes improved tabletting properties compared with crystalline material alone. This is due to the amorphous material adding a plastic deformation to the usual brittle fracture of crystalline lactose and the fact that the amorphous material will crystallize to form solid bridges during compaction. Sebhatu et al. (17) studied the effect of storage at 57% relative humidity on the compression properties of spray-dried lactose (Fig. 1) and found that the tensile strength of the tablets increased during the first four hours of storage. During this period the water content increased as water was absorbed into the amorphous lactose, resulting in a plasticizing effect and an increase in the mobility of the amorphous phase. The mobile amorphous material was able to flow under compression and produce stronger tablets. Storage for longer times resulted in the amorphous material crystallizing, and as such, the benefit that amorphous material can add to the compression process was lost. As crystallization had occurred, there was no change in tablet properties for any increase in storage time at 57% RH. Subsequent to this, the compacts stored for zero to four hours at 57% RH prior to compaction were exposed to 57% RH post compaction (Fig. 2) and were seen to produce stronger tablets over time, because of the amorphous material crystallizing to form solid bridges. The tablets that were produced from lactose that was crystalline prior to compaction (those stored for more than four hours at 57% RH prior to compaction) did not show postcompaction changes in properties (Fig. 2).

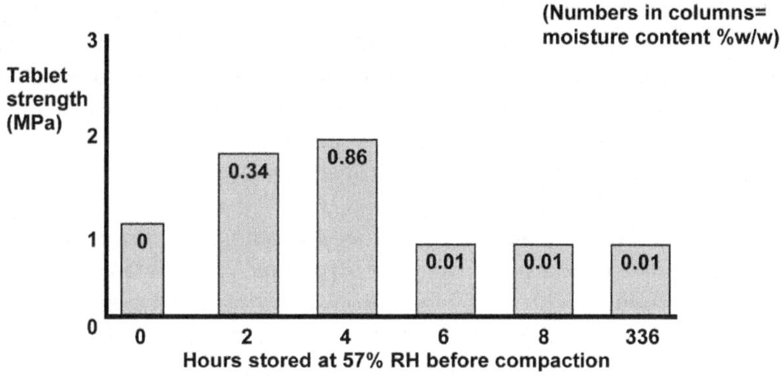

Figure 1 Effect of storage at 57% humidity.

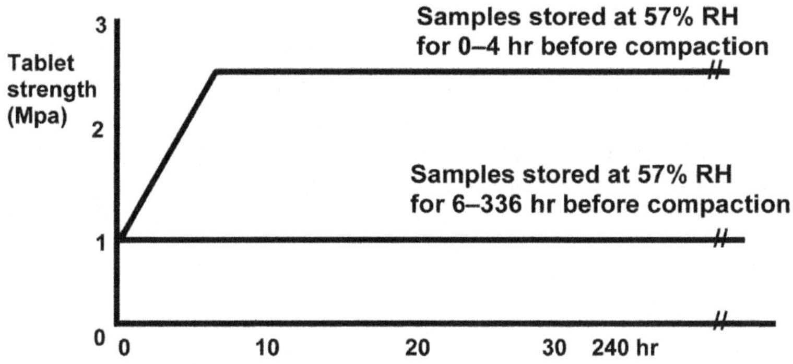

Figure 2 Schematic representation. From Ref. 17.

These data indicate that the deformation process is altered by having the mix of crystalline and amorphous (brittle and plastic materials, respectively) and that the solid bridge formation continues post compaction as the amorphous lactose converts to crystalline material, with improved tablet strength. The comments above show that it is important to understand the state of physically unstable materials prior to tabletting, for example, if amorphous lactose has crystallized, the outcome will be different and the product properties will change. Depending on drug dose, the properties of the active can also affect the tabletting behavior, and it is relatively common for processed active to be partially amorphous. For high-dose drugs, this could well be significant for the tablet strength.

The discussion around solid bridges highlights that new crystalline regions can be created during tabletting. With this in mind, it is important to realize that (to a greater or lesser extent) the polymorphic form of the drug present in the final compressed tablet can change as a consequence of compaction, either by the crystallization from the amorphous state or by solid-state transitions induced by the pressure or combination of local heat and pressure that occur. These solid-state transitions have potential significance not only for compaction properties but also for dissolution and bioavailability of poorly soluble drugs and for intellectual property infringement for compounds that have protected physical forms.

INTERFACIAL FORCES

Tablets will, to a large extent, be held together by interfacial forces, also known as long-range attractive forces. These long-range forces, which are lower energy than covalent bonding forces, are due to physical bonding on different types, all with energies usually less than around 40 kJ/mol. Long-range forces are electrical in nature and can be considered as electrostatic (which can be both attractive and repulsive depending on whether the two materials carry the opposite or the same charge, respectively), van der Waals, and hydrogen bonding. While electrostatic forces will often be induced on powders during mixing and other dry processing events, it is van der Waals and hydrogen bonding forces that are most likely to make the major contribution during tabletting.

Van der Waals forces consist of dipole, induced dipole, and dispersion forces. Many molecules have an imbalance of charge across their structure. Water is an example of such an asymmetric charge distribution, with an oxygen atom carrying a net negative charge, being balanced by two hydrogen atoms, each of which has a slight net positive charge. Most drugs are ionizable and have such an asymmetric charge distribution. Certain molecules exist with a permanent charge imbalance and can simplistically be regarded as small bar magnets. Such materials are said to have *permanent dipoles,* and interactive forces are due to attraction between the negative pole of one molecule when in reasonably close contact with the positive pole of another. These interactive forces are also termed *orientational* or *Keesome* interactions, and the molecules that take part in such interactions are regarded as polar molecules. Hydrogen-bonding interactions are a specific sort of this type of bonding, resulting from the fact that hydrogen consists of only one proton and one electron, making it very strongly electronegative. When hydrogen bonds, its electron is lost, leaving an exposed proton. This unique situation causes a strong attraction between the proton and an electronegative region from another atom. The strength of the hydrogen bond results in drastically different properties of interaction, exemplified by the fact that water has such a surprisingly high melting and boiling point (in comparison with non-hydrogen-bonded materials). Strengths of hydrogen bonds fall in the range 10 to 40 kJ/mol (18). Many excipients and active pharmaceutical ingredients are able to hydrogen bond. Examples of excipients with hydrogen bonding potential would include sugars, celluloses, and starches.

A bond between carbon and oxygen would be expected to be dipolar; however, if the molecule of carbon dioxide (not of significance for tabletting, but the easiest molecule with which to demonstrate the concept) is considered (O=C=O), it can be seen that the molecule is, in fact, totally symmetrical, the dipole on each end of the linear molecule being in perfect balance with that on the other end. Even though these molecules do not carry a permanent dipole, if they are placed in the presence of a polarized material, a dipole will be induced on the (normally symmetrical) molecule, such that interaction can occur. Such interactions are common and are termed *dipole–induced dipole* (or Debye) interactions.

London–van der Waals forces are termed dispersion forces. These are interactions between molecules that do not have a charge imbalance and the ability to have an induced dipole either. Essentially, these are interactions between nonpolar materials. These dispersion forces occur between all materials, and thus, even though the interaction forces are weak, they make a very significant contribution to the overall interaction between two molecules. Dispersive forces are extremely complex to describe but can be considered in a simplistic fashion by considering the fact that the electrons that spin around two neighboring nonpolarized atoms will inevitably not remain equally spaced and consequently result in local imbalances in charge, which leads to transient induced dipoles. These induced dipoles and the forces that result from them will constantly be changing, and obviously, the magnitude of these interactions is small compared with the permanent and induced dipole situations described above. As stated above, dispersion forces are complicated interactions; examples of their properties include their long range (in the order of 10 nm, which is significantly longer than the bond length), nonadditivity (because neighboring molecules exert an influence on two interacting molecules), and the fact that the interactions can obviously be attractive or repulsive (but in practice are considered to be attractive over long distances). A typical strength for van der Waals bonding would be 1 kJ/mol.

It follows then that the total van der Waals force acting between two molecules can be considered as the sum of the three contributing types of forces, that is, dipole-dipole (orientation, or Keesome), dipole–induced dipole (Debye or induced), and induced dipole–induced dipole (London dispersion) forces (19–22).

THE EFFECT OF LUBRICANT ON TABLET BONDING

As essentially every compressed tablet has a lubricant as part of the formulation, and as it is known that the presence of lubricant can affect tablet strength (as well as dissolution rate, etc.), it is worth considering the effect of the lubricant. Magnesium stearate is the most commonly used and investigated lubricant for tablets, but its complexity presents problems in gaining a full understanding of its influence. The British Pharmacopoeia describes magnesium stearate as "a mixture of magnesium salts of different fatty acids," so there is an understanding that this is not a chemically pure material. It is known that different polymorphic forms, different hydrate levels, and different sizes and shapes of the particles all have an effect on the properties of magnesium stearate (23). It is therefore not surprising that materials from different vendors and even lots from the same vendor can have different behaviors in products. That said, there are a number of publications on magnesium stearate and its effects in tablets that do highlight general trends. It is clear that magnesium stearate will be used as small particle size material and will be mixed (for critical duration with respect to the effect) just prior to tabletting, such that magnesium stearate will be positioned between particles and between the particles and the die wall. It is therefore to be expected that the interfacial bonding that has been described above will be altered by the presence of this excipient between two other particles. Indeed, it has been shown (3,24,25) that increasing the amount of magnesium stearate in a powder mix results in the remaining strength of a compact falling to essentially a plateau level (3). It has been argued that the plateau level that is obtained provides information on the contribution of the solid bridge formation in a compaction process on the basis that the process of fracture (or presumably plastic flow, as a number of the materials studied were not brittle) will penetrate the lubricant covering of the particles and override their contribution to the bonding process.

CHEMICAL IMAGING IN TABLETS

Advances in chemical imaging have provided a way to better understand the compaction process. A number of methods now exist including Terahertz spectroscopy, which is good for studies around density differences in tablets, showing defects and potential for lamination. More regular use comes from near infrared (NIR) and/or Raman spectroscopy. These methods can be complimentary as NIR provides information about water and its distribution and binding energy, often to some extent masking other effects, and Raman does not detect a signal for water; hence, together, they can provide a lot of information about what is in a tablet and where it is distributed, which in turn can shed light on how the compaction process may have progressed. A chemical image is constructed by measuring spectra in each defined region (often 1 μm x 1 μm) of a sample and then repeating until an entire area is mapped. By knowing the characteristic peaks for each component in the formulation, it is possible to do analysis of each μm square and see which component(s) is (are) present in that region. In Figure 3, a map of a sample is shown, and the upper plot shows the distribution of the active, and the bottom plot, one of the excipients. It can be seen that the regions where the active has high intensity (black), the excipient is absent, and vice versa. This method can be used to show how different batches of the same product differ and explain what has happened during the processing to cause this (mixing failure, compaction differences, etc.). Equally, as different polymorphic forms and amorphous forms have different spectra, it is possible to use this approach to see what changes have happened to materials during the tabletting process and to see if these are the same in each batch. Understanding this helps control batch-to-batch variability in tablet properties, stability, and dissolution; it also helps to ensure that physical-form patents are not being infringed. As the spectra are taken on each individual small area, the resolution and detection sensitivity is very high indeed—such that individual particles can be indentified and their properties (polymorphic form/amorphous content) studied in the formulation.

CONCLUSIONS

Tablets are held together predominantly by solid bridges and interfacial forces of attraction, which can be hydrogen bonding or the range of van der Waals interactions. Solid bridges are caused by the material properties and the compaction process, and will be prone to give different tablet strengths as factors such as the amorphous content change (equally likely to be true with changes in polymorphic form, hydrates and particle size, surface area, and roughness). The interfacial forces can act over ranges much longer than a simple bond, and the strength of these will be related to the nature of the materials that are compressed and the surface area over which bonding can take place, which is also linked to the materials properties and the processing used.

Materials can and will change their form during compaction, and these form changes will alter tablet bonding and the subsequent tablet properties—chemical stability and dissolution included. There are techniques available that allow tablets to be studied and the properties to be understood.

The interplay between input materials properties, which must be very well understood, and critical parameters controlled, the processes used to make the tablets and the measured properties of the tablets that are produced, especially by techniques such as chemical imaging,

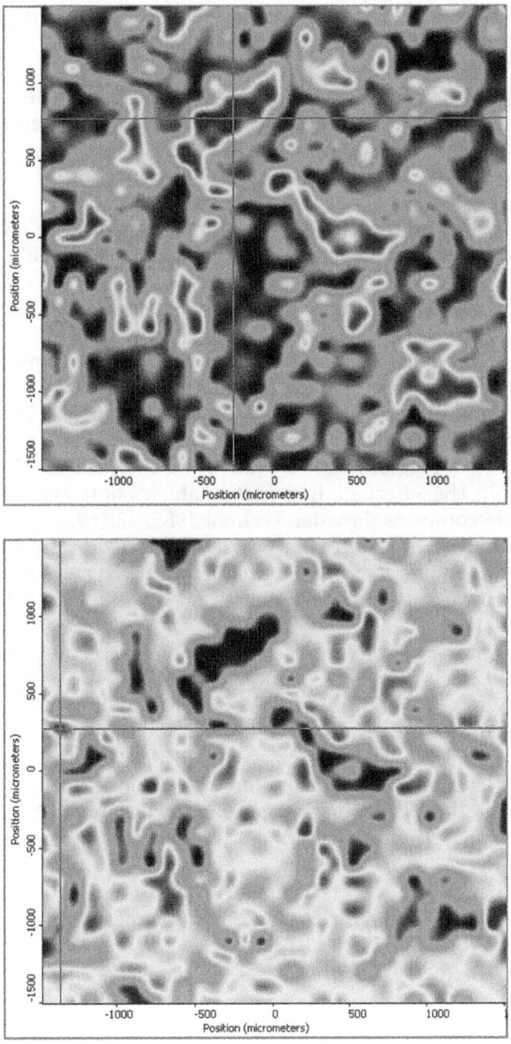

Figure 3 Raman maps of a tablet showing the distribution of the active (*top*) and excipient (*bottom*). Black, high intensity; white, low intensity.

provides the basis for Quality by Design and the guarantee that uniformly high-quality product can be produced with an established design space.

REFERENCES

1. Nyström C, Karehill PG. 1. Studies on direct compression of tablets XVI. The use of surface area measurements for the evaluation of bonding surface area in compressed powders. Powder Technol 1986; 47:201.
2. Chowhan ZT, Chow YP. Int J Pharm Technol Prod Manuf 1981; 2:29.
3. Karehill PG, Börjesson E, Glazer M, et al. Bonding surface area and bonding mechanisms–two important factors for the understanding of powder compactability. Drug Dev Ind Pharm 1993; 19:2143.
4. Karehill PG, Nyström C. Studies on direct compression tablets XXI. Investigation of bonding mechanisms of some directly compressed materials by strength characterization in media with different dielectric constants (relative permittivity). Int J Pharm 1990; 61:251.

5. Karehill PG, Nyström C. Studies on direct compression tablets XXII. Investigation of strength increase upon ageing and bonding mechanisms for some plastically deforming materials. Int J Pharm 1990; 64:27.
6. Karehill PG, Glazer M, Nyström C. Studies on direct compression of tablets XXIII. The importance of surface roughness for compactability of some directly compressible materials with different bonding and volume reduction properties. Int J Pharm 1990; 64:35.
7. Benbow JJ. In: Stanley-Wood NG, ed. Enlargement and Compaction of Particulate Solids. London: Butterworths, 1983:171.
8. Duberg M, Nyström C. Studies on direct compression tablets XII. The consolidation and bonding properties of some pharmaceutical compounds and their mixtures with Acivel 105. Int J Pharm Technol Prod Manuf 1985; 6:17.
9. El Grindy NA, Samaha MW. Tensile strength of some pharmaceutical compacts and their relation to surface free energy. Int J Pharm 1983; 13:35–46.
10. Stanley-Wood NG, Shubair MS. The variation of surface topography of granules under compression with degree of binder addition. Powder Technol 1980; 25:57.
11. Rumpf H. Basic principles and methods of granulation I. II. Chem Eng Tech 1958; 30:144.
12. Führer C. Lab Pharma Probl Technol 1977; 269:759.
13. Rumpf H. In: Knepper WA, ed. Agglomeration. New York: Interscience, 1962:379.
14. Down GRB, McMullen JN. The effect of interparticulate friction and moisture on the crushing strength of sodium chloride compacts. Powder Technol 1985; 42:169.
15. Mitchell AG, Down GRB. Recrystallization after powder compaction. Int J Pharm 1984; 22:337.
16. Ahlneck C, Alderborn G. Moisture adsorption and tableting II. The effect on tensile strength and air permeability of the relative humidity during storage of tablets of 3 crystalline materials. Int J Pharm 1989; 56:143.
17. Sebhatu et al. Effect of moisture sorption on tableting characteristics of spray dried (15% amorphous) lactose. Pharm Res 1994; 9:1233.
18. Joesten MD, Schaad LJ. Hydrogen Bonding. New York: Dekker, 1974.
19. Israelachvili JN. Intermolecular and Surface Forces. London: Academic Press, 1985:21.
20. Derjaguin BV. Sci Am 1960; 203:47.
21. Derjaguin BV, Abrikosova II, Lifshitz EM. Quart Rev Chem Soc 1956; 10:295.
22. Israelachvili JN, Tabor D. van der Waals forces theory and experiment. Prog Suf Membr Sci 1973; 7:1.
23. Rowe RC, Sheskey PJ, Quinn ME (eds.). Handbook of Pharmaceutical Excipients. Pharmaceutical Press, London, 2009.
24. de Boer AH, Bolhuis GK, Lerk CF. Powder Technol 1978; 25:75.
25. Bolhuis GK, Lerk CF, Zijlstra HT, et al. Film formation by magnesium stearate during mixing and its effect on tabletting. Pharm Weekblad 1975; 110:317.

2 Viscoelastic models

Susan K. Lum

INTRODUCTION

S.G. the director of production at XYZ Pharmaceuticals looked more harassed than usual today. He had been raked over the coals for allowing 9 out of 17 1200 kg scale lots from a recent commercial campaign fail acceptable quality level (AQL) inspection due to fractured tablets with the majority of defects being capping, lamination or hairline cracks. The coating operator and production managers were reporting in again, "sorry but it's the same problem as before, sometimes it works and sometimes it doesn't, this pan has loads of split tablets." How did it get to this point? The press was overhauled and many mechanical issues that could have led to tablet breakage, like the bottom punch seals, were fixed. The tablet pan speed in coating was slowed to a crawl to reduce stress on the tablets. All operator and engineering shifts were trained and retrained to make sure the tablets were compressed to meet specification. There were only so many engineering controls to be done—it was filed this way.

Pharmaceutical formulation development difficulties often arise during scale-up since significant increase in production and local strain rates occur. Such cases are commonly encountered in technical transfer since there is limited understanding regarding the effect of process variables on material properties at either the pilot or commercial scale. Technical difficulties can sometimes be managed with extreme engineering controls, but often the root cause is the formulation. A tablet is only as robust as the components that form it. Often these materials form differently under different compression speeds. This strain rate sensitivity has lead to other well publicized manufacturing cases like the FDA-triggered recall in 2005 of all lots of a controlled-release tablet made by a large multinational (1). It was found that the tablets could split apart. This deficiency could cause patients to receive a portion of the tablets that lack any active ingredient or, alternatively, a portion that contained the active without the intended controlled-release effect. Clearly, the materials chosen, the formulation, and processing together defined the boundary of failure. How could these known risks have been better managed?

Solving such formulation problems begins with an understanding of the factors contributing to tablet quality. As it will be described in depth in subsequent chapters of this text, the compaction of powders is a complex process. Powder densification consists of sequential and concurrent processes where particles are brought into intimate contact, interparticulate bonds are formed, and dimensional changes occur as a result of stress redistribution. Time-dependent phenomena inherent in these processes govern the ultimate quality of the compact produced. Permanent densification occurs when stress exceeds the elastic strain limit of a given material. The extent of this nonrecoverable deformation depends on the time in which the applied stress exceeds this material yield stress. Thus, intrinsic tablet strength is a sum of the following:

1. Permanent interparticulate bonds formed intrinsically during compaction. High plasticity or irreversible deformation drives the formation of permanent particle to particle contact zones.
2. Reversible or recoverable, elastic deformation. Highly elastic material stores mechanical energy during compaction but will release it again once the stress is removed. This bounce-back is often seen as cracking, capping, or lamination during the unloading phase of the compression cycle or upon ejection.

3. Fracture toughness determines the extent to which interparticulate contact regions fracture or are crushed during compaction with time.
4. Time-dependent recovery.

It is viscoelastic theory that serves to quantify this time dependence of material properties. This provides a description of the relationship between the strain rate or the time component of the material undergoing viscoelastic deformation, the stress or the force component required to produce that strain rate, and the effect.

Empirical evidence has shown that tablet strength and durability in downstream processing are a function of both the materials as well as the processing of these materials. The rate at which compacts are formed determines the final tablet strength and viability (2–14). With the advent of modern processes, the high speed of compaction accentuates the time-dependent component of material behavior. This complexity is reflected in large-scale manufacturing of tablets where problems of low tablet strength, capping, and sensitivity to material batch variability exist and can be quite costly. Improving the theoretical understanding of the compaction process would enable a more rational approach to the formulation of tablets (15).

Problems that arise during tablet production are a consequence of viscoelasticity combined with poor interparticle bonding. The degree of interparticle bonding and bond strength is assumed to be largely governed by the magnitude of the true interparticle contact area. The amount of elastic recovery of the particles depends on the release of elastic strain during decompression and tends to disrupt interparticle bonds. This stress relaxation is a function of the viscoelasticity of the material, which in turn is influenced by the speed of compression and decompression. At high compaction speeds, the internal stress after compaction is high and the propensity of the material for elastic recovery is considerably higher than at low compaction speeds.

The concept and much of the theory for time-dependent materials properties originates in engineering sciences with the study of materials such as metals, soils, and ceramics. By applying information from these other areas to the field of solid pharmaceutics, the compaction process may be distilled into mechanical models with mathematical terms. Applied viscoelastic theory predicts the behavior of these models under stress. The behavior may be simplified by considering the bulk powder mass as a continuum or by considering a large number of individual particles in an aggregate blend on the basis of the behavior of an average single particle.

VISCOELASTIC RHEOLOGY

Viscoelastic substances are materials whose behavior lies between that of perfectly elastic solids and that of perfectly viscous fluids. Macromolecular pharmaceuticals, polymeric materials, glasses, and even concrete are considered to be viscoelastic materials; the increased use of these materials has brought with it the need to ascertain their mechanical behavior under a variety of conditions of stress, temperature, and flow (16).

The behavior of a viscoelastic material can often be described by a mechanical model whose behavior, under a specific regimen of tests, closely resembles that of the material in question. This model may then be used to predict the behavior of the material under different test conditions. Such models have traditionally been visualized as systems of springs and dashpots; the springs provide purely elastic behavior, and the dashpots purely viscous behavior in the Newtonian sense (Fig. 1). Such basic elements of springs and dashpots are useful rheological analogs for stress and strain.

Stress, σ, as shear, tensile, or compressive, describes surface forces acting perpendicularly while strain, ε, describes the local deformation in relation to original dimensions. Under

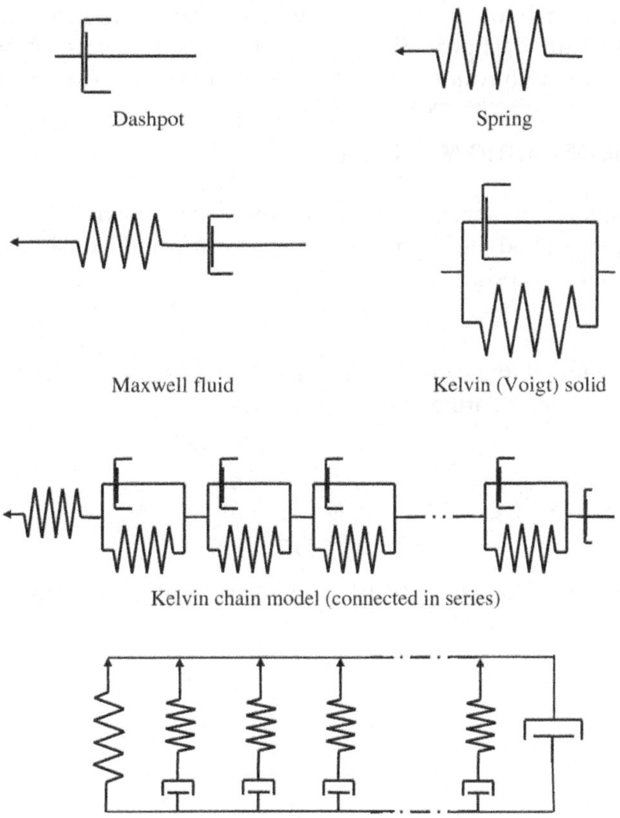

Figure 1 Elements in viscoelastic models.

uniaxial loading, the behavior of materials commonly used in tabletting is history dependent, and past compressions can affect present densification.

Elasticity

A spring represents linear elastic behavior. An ideal helicoidal spring element would be perfectly linear and massless. When a force is applied, the length of the spring increases to a certain dimension, and when the force is removed, the spring returns to its original state. Hooke's law states that the stress is proportional to the strain.

$$\sigma(t) = E\varepsilon(t) \tag{1}$$

where E is the modulus of elasticity with dimension (force/length^2). The spring has a creep function $H(t-\tau)/E$ and a relaxation function $EH(t-\tau)$. Powders, in compression, are constrained by the die wall upon dilation once a load is removed.

Viscosity

A dashpot or piston moving in a cylinder filled with a viscous fluid represents viscous behavior. The dashpot as an ideal viscous element extends at a rate proportional to the force applied according to Newton's law. Stress is proportional to the strain rate.

$$\dot{\varepsilon}\,(t) = \frac{\sigma(t)}{\eta} \tag{2}$$

where $\dot{\varepsilon} = \partial\varepsilon/\partial t$ is the rate of strain and η is the viscosity coefficient with dimension (force $\times$ time $\times$ length^{-2}). With dashpots, the greater the speed or rate of strain, the greater the resistance; since there is no restoring force, a dashpot remains extended upon load removal.

COMPOSITE VISCOELASTIC MODELS

Maxwell Model

Different combinations of springs and dashpots afford flexibility in portraying different responses. The Maxwell body is a combination of a spring and a dashpot in series. In series, the strain in each element is additive.

$$\varepsilon = \varepsilon_s + \varepsilon_d \tag{3}$$

where subscript s refers to the elastic spring and d, the viscous dashpot. Differentiating allows the insertion of the differentiating elastic equation (1) to give

$$\dot{\varepsilon} = \dot{\varepsilon}_s + \dot{\varepsilon}_d = \frac{\dot{\sigma}}{\kappa} + \frac{\dot{\sigma}}{\eta} \tag{4}$$

$$\sigma + p_1\dot{\sigma} = q_1\dot{\varepsilon} \tag{5}$$

where κ is Young's modulus of elasticity, η, viscosity, and the hybrid constitutive parameters correspond to

$$p_1 = \frac{\eta}{\kappa} q_1 = \eta \tag{6}$$

Kelvin–Voigt Model

The Kelvin body is a combination of a spring and a dashpot in parallel. With elements in parallel, the stress of each element is additive.

$$\sigma = \sigma_s + \sigma_d \tag{7}$$

and

$$\sigma = \kappa\varepsilon + \eta\dot{\varepsilon} = q_0\varepsilon + q_1\dot{\varepsilon} \tag{8}$$

The stress in the Kelvin element is related both to the elongation of the spring and to the rate of deformation of the dashpot.

Generalized Models

Linear Viscoelasticity

Many materials, notably polymers, exhibit time-dependent behavior in their relationships between stress and strain. The common features of a three-parameter or standard solid viscoelastic behavior include an initial elastic response to an applied or eliminated stress, a delayed elastic response, and often a permanent strain that is acquired through the action of creep.

A summary of some basic arrays of these viscous and elastic elements that make physical sense is depicted in Table 1 (17). Simple two-parameter models such as the Maxwell or Kelvin–Voigt elements, as mentioned above, are composed of either serial or parallel connections, respectively. When in series, the entire elongation of the model is equal to the sum of the elongations of its components. When in parallel, the elongation of both elemental parts is equal at any time. Stress is then proportional to both the elongation of the spring and the rate of deformation of the dashpot. These simple models illustrate the principle central to the study of

Table 1 Rheological Models with Variable Parameters Used to Describe Viscoelastic Material Behavior

Designation	Elements	Differential equation	Creep function	Relaxation function
1 parameter (Hooke)		$\sigma = E\varepsilon$	$\frac{1}{E}$	E
2 parameter (Maxwell)	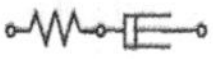	$\frac{\dot{\sigma}}{E} + \frac{\sigma}{\eta} = \dot{\varepsilon}$	$\frac{1}{E} + \frac{t}{\eta}$	$Ee^{-\frac{tE}{\eta}}$
2 parameter (Kelvin–Voigt)	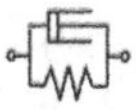	$\sigma = E\varepsilon + \eta\dot{\varepsilon}$	$\frac{1}{E}\left(1 - e^{-\frac{tE}{\eta}}\right)$	$E + \eta\delta(t)$
3 parameter (standard)	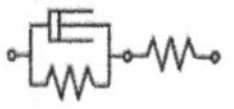	$\sigma(E_1 + E_2) + \dot{\sigma}\eta$ $= E_1E_2\varepsilon + E_1\eta\dot{\varepsilon}$	$\frac{1}{E_1} + \frac{1}{E_2}\left(1 - e^{-\frac{tE_2}{\eta}}\right)$	$E_1e^{-\frac{t(E_1+E_2)}{\eta}} +$ $\frac{E_1E_2}{E_1 + E_2}\left(1 - e^{-\frac{t(E_1+E_2)}{\eta}}\right)$
4-parameter solid	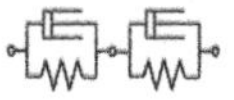	$\sigma(E_1 + E_2) + \dot{\sigma}\eta_1\eta_2$ $= E_1E_2\varepsilon + \eta_1\eta_2\ddot{\varepsilon}$ $+ (E_1\eta_2 + E_2\eta_1)\dot{\varepsilon}$	$\frac{1}{E_1}\left(1 - e^{-\frac{tE_1}{\eta_1}}\right) +$ $\frac{1}{E_2}\left(1 - e^{-\frac{tE_2}{\eta_2}}\right)$	$\frac{E_1E_2}{E_1 + E_2} + \frac{\eta_1\eta_2}{\eta_1 + \eta_2}\delta(t) -$ $\left(\frac{E_1E_2}{E_1+E_2} - \frac{E_1\eta_2+E_2\eta_1}{\eta_1+\eta_2} + \left(\frac{1}{E_1+E_2}\right)\frac{\eta_1\eta_2}{(\eta_1+\eta_2)^2}\right)e^{-\frac{t(E_1+E_2)}{\eta_1+\eta_2}}$

Source: From Ref. 17.

linear viscoelasticity. Each model is described by a linear differential equation with constant coefficients of the type (16)

$$\sigma + p_1\dot{\sigma} + p_2\ddot{\sigma} + \cdots = q_0\varepsilon + q_1\dot{\varepsilon} + \cdots \tag{9}$$

or

$$\left(\sum_0^m p_k \frac{d^k}{dt^k}\right)\sigma = \left(\sum_0^n q_k \frac{d^k}{dt^k}\right)\varepsilon \tag{10}$$

Since the equation can be divided by a constant without changing its meaning, p_0 is always set to unity. In the viscoelastic regime, elastic deformation depends directly on stress, whereas viscous internal stress depends on the rate of deformation.

Solutions to the above constitutive equation for a linear viscoelastic material form an important part of the theory of viscoelasticity. The nonzero coefficients dictate the type of behavior displayed by the model. For instance, if $q_0 = 0$, the model displays fluid flow; otherwise, it behaves as a solid. If $n = m$, the model shows an initial elastic response to an applied stress; if $n = m + 1$, this response is absent.

Although the behavior of some solids approaches Hooke's law at infinitesimal strains and the behavior of some liquids approach Newton's law at infinitesimal strain rates, these are mere ideals. Most materials show characteristics between the two. Viscoelastic behavior is hallmarked by creep, stress relaxation, and stress-strain hysteresis. Viscoelastic materials dissipate energy during deformation and consequently recover some of the deformation it has undergone. Although Kelvin–Voigt and Maxwell models are often used to describe viscoelastic behavior, these models are extremely simplified. The Kelvin–Voigt model may represent creep of a viscoelastic material but fails to represent the stress relaxation. The Maxwell model may present the stress relaxation of a viscoelastic material but fails to register the creep behavior. A response closer to the behavior of a real viscoelastic material with time dependence comes from more complex combinations.

Simple Maxwell or Kelvin models are limited and cannot describe material properties with high strain rates or over long periods of time. More complex combinations in series or in parallel have been used in generalized models to overcome these limitations. It is possible to generalize spring and dashpot models into more complicated forms using an arbitrary number of Maxwell units in parallel or Voigt units in series. The physical microconstants, which are assigned to each elastic or viscous element in the array, form the basis for the hybrid coefficients, p_k and q_k, in the constitutive equation for the mechanical model.

Whereas simple models are inadequate representations of real viscoelastic behavior, a three-parameter, standard, model consisting of two springs and a dashpot has successfully portrayed real material behavior (18,19). A standard model may be thus employed in the extension of the description of particle contact to the viscoelastic case. Linear viscoelastic relationships are valid for small strains and assume the principle of superposition holds (19) a stress history.

$$\sigma(\tau) = \sigma_1(\tau) + \sigma_2(\tau), \tau \ni (\tau_0, t) \tag{11}$$

corresponds to a strain history of

$$\varepsilon(\tau) = \varepsilon_1(\tau) + \varepsilon_2(\tau) \tag{12}$$

Materials often exhibit linear behavior at low stresses and nonlinear viscoelastic behavior at high stresses. The stress-strain relations for a linear viscoelastic material are commonly expressed as a relaxation function, $\Psi(t)$. This function, as deduced above from a standard spring-dashpot model, expresses the stress response to a step change in strain.

In operational form, this standard model (Fig. 2) appears as

$$E_1\eta\dot{\varepsilon} + E_1E_2\varepsilon = \sigma(E_1 + E_2) + \eta\dot{\sigma} \tag{13}$$

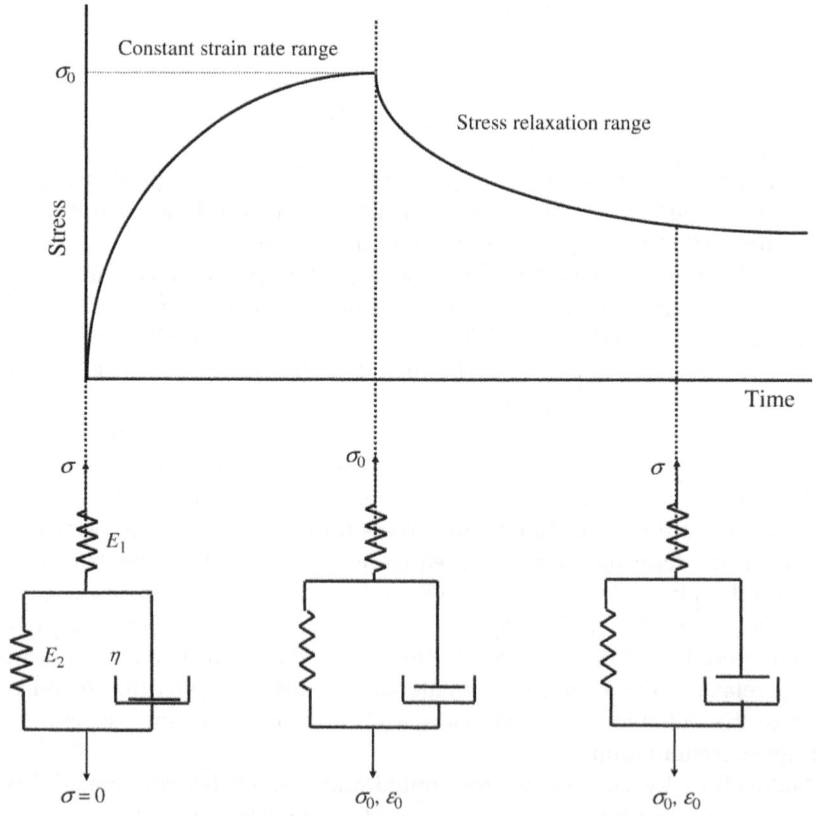

Figure 2 Illustration of the stress-time relation in association with a standard three-parameter rheological model.

By integrating over a strain history $\varepsilon(t)$

$$\therefore \sigma(t) = \varepsilon(t)\frac{E_1E_2}{E_1+E_2} + \frac{E_1^2}{E_1+E_2}\int_0^t e^{-\frac{(t-\tau)(E_1+E_2)}{\eta}}\dot{\varepsilon}(\tau)d\tau \tag{14}$$

Wherein the creep function is

$$\phi(t) = \frac{1}{E_1} + \frac{1}{E_2}\left(1 - e^{-\frac{tE_2}{\eta}}\right) \tag{15}$$

and the relaxation function

$$\Psi(t) = E_1 e^{-\frac{t(E_1+E_2)}{\eta}} + \frac{E_1E_2}{E_1+E_2}\left(1 - e^{-\frac{t(E_1+E_2)}{\eta}}\right) \tag{16}$$

Simplified three-dimensional form for viscoelastic constitutive description. When a problem has more than one stress component, then a generalization of the viscoelastic law to the three-dimensional form may be needed (16). If a viscoelastic material is isotropic, a hydrostatic stress must produce a dilatation and no distortion. The stress and strain are related by the differential equation

$$\sigma + p_1\dot{\sigma} + p_2\ddot{\sigma} + \cdots = q_0\varepsilon + q_1\dot{\varepsilon} + \cdots \tag{9}$$

$$\left(\sum_0^m p_k \frac{d^k}{dt^k}\right)\sigma = \left(\sum_0^n q_k \frac{d^k}{dt^k}\right)\varepsilon \tag{10}$$

or shorter by

$$P''\sigma = Q''\varepsilon \tag{17}$$

On the other hand, if shear is present, then the shear strain coupled with isotropy of the material requires that

$$\left(\sum_0^{m'} p'_k \frac{d^k}{dt^k}\right)\hat{\sigma} = \left(\sum_0^{n'} q'_k \frac{d^k}{dt^k}\right)\hat{\varepsilon} \tag{18}$$

or

$$P'\hat{\sigma} = Q'\hat{\varepsilon} \tag{19}$$

where $\hat{\sigma}$ and $\hat{\varepsilon}$ correspond to the components of the stress and strain deviators.

The operator pairs P'' and Q'', and P' and Q', which describe the viscoelastic material, are entirely independent of each other. To each pair, the relation stated in equation (15) or (16) is applicable for a standard three element model. These operator pairs for different rheological models are also summarized in Table 2.

UNIAXIAL CONFIGURATION

In uniaxial tension, there is only one stress, σ_x and strain has three components, the axial strain ε_x and the lateral contractions $\varepsilon_y = \varepsilon_z$. As the operators are assumed linear and time invariant, the commutative properties yield

$$(P''Q' + 2Q''P')\sigma_x = 3Q''Q'\varepsilon_x \tag{20}$$

$$(P''Q' - Q''P')\sigma_x = 3Q''Q'\varepsilon_y \tag{21}$$

Table 2 Viscoelastic Models with Variable Parameters: Generalized

Model	Name	Differential equation inequalities	Creep compliance $\Phi(t)$	Relaxation modulus $\Psi(t)$	Complex compliance, real $G_1'(\omega)$	Complex compliance, imaginary $G_2''(\omega)$
	Elastic solid	$\sigma = q_0\varepsilon$	$\frac{1}{q_0}$	q_0	$\frac{1}{q_0}$	0
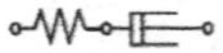	Viscous fluid	$\sigma = q_1\dot{\varepsilon}$	$\frac{t}{q_1}$	$q_1\delta(t)$	0	$-\frac{1}{q_1\omega}$
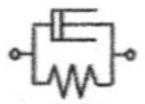	Kelvin solid	$\sigma = q_0\varepsilon + q_1\dot{\varepsilon}$ $\sigma = E\varepsilon + \eta\dot{\varepsilon}$	$\frac{1}{q_0}(1 - e^{-\lambda t}), \lambda = \frac{q_0}{q_1}$	$q_0 + q_1\delta(t)$	$\frac{q_0}{q_0^2+q_1^2\omega^2}$	$-\frac{q_1\omega}{q_0^2+q_1^2\omega^2}$
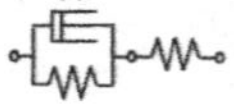	3-parameter solid	$\sigma + p_1\dot{\sigma} = q_0\varepsilon + q_1\dot{\varepsilon}$ $q_1 > p_1 q_0$ $\sigma(E_1 + E_2) + \dot{\sigma}\eta = E_1 E_2\varepsilon + E_1\eta\dot{\varepsilon}$	$\frac{p_1}{q_1} e^{-\lambda t} + \frac{1}{q_0}(1 - e^{-\lambda t})$ $\lambda = \frac{q_0}{q_1}$	$\frac{q_1}{p_1} e^{-\frac{t}{p_1}} + q_0\left(1 - e^{-\frac{t}{p_1}}\right)$	$\frac{q_0+p_1q_1\omega^2}{q_0^2+q_1^2\omega^2}$	$-\frac{(q_1-q_0p_1)\omega}{q_0^2+q_1^2\omega^2}$

Source: From Ref. 16.

for applied uniaxial stress. By analogy,

$$(Q''P' + 2P''Q')\varepsilon_x = 3P''P'\sigma_x \tag{22}$$

$$(Q''P' - P''Q')\varepsilon_x = 3P''P'\sigma_y \tag{23}$$

follows for applied uniaxial strain (in stress relaxation experiments).

Equations (20) to (23) serve as the viscoelastic equivalent of a complete statement of Hooke's law for uniaxial tension. Interpretation requires special choices for the operators. While shear deformation may be rather large, the change of volume measured by e is always very limited. It seems therefore, reasonable to neglect the latter completely and to assume $e = 0$. This corresponds to $P'' = 0$ and $Q'' = 1$. The constitutive equations for uniaxial stress in tension are then

$$2P'\sigma_x = 3Q'\varepsilon_x \quad \text{and} \quad -P'\sigma_x = 3Q'\varepsilon_y \tag{24}$$

and for applied uniaxial strain,

$$2Q'\varepsilon_x = 3P'\sigma_x \quad \text{and} \quad -Q'\varepsilon_x = 3P'\sigma_y \tag{25}$$

Elastic Parameters as a Limiting Case

In a viscoelastic body, all of the stresses, strains, and displacements occurring under load are time dependent. The elastic solid is then a limiting case of a viscoelastic material. The moduli of the springs of the rheological model do not correspond simply to Young's modulus as time, temperature and stress dependence are overt considerations. If, for example, one considers a material tested in tension, the plot of the compliance would show that the ratio $\varepsilon_x/-\varepsilon_y$ varies with time. This time dependence indicates that the concept of Poisson ratios, etc., is not very meaningful for a viscoelastic material (16, p. 178).

The linear elastic law (Hooke's law) would render the four operators (eqs. 24 and 25) as multiplicative constants and the formulation of the stress deviation as

$$\sigma_x = \frac{E}{(1+\nu)(1-2\nu)}[(1-\nu)\varepsilon_x + \nu(\varepsilon_y + \varepsilon_z)] \tag{26}$$

The differential operators P and Q can be replaced by polynomials $\wp(s)$ and $\vartheta(s)$ in considering an elastic body; hence,

$$\wp''(s)\bar{\sigma} = \vartheta''(s)\bar{\varepsilon} \tag{27}$$

$$\wp'(s)\bar{\tilde{\sigma}} = \vartheta'(s)\bar{\tilde{\varepsilon}} \tag{28}$$

where $\bar{\sigma}$ and $\bar{\varepsilon}$ are the Laplace transforms of the time-dependent stresses and strains.

These algebraic relations define the limits of their elastic counterparts if

$$3K \to \frac{\vartheta''(s)}{\wp''(s)}, \quad 2G \to \frac{\vartheta'(s)}{\wp'(s)} \tag{29}$$

$$E \to \frac{3\vartheta'\vartheta''}{2\wp'\vartheta'' + \vartheta'\wp''}, \quad \nu \to \frac{\wp'\vartheta'' - \vartheta'\wp''}{2\wp'\vartheta'' + \vartheta'\wp''} \tag{30}$$

For a standard (three-parameter material), a reduced or effective modulus $E_R(t, \tau)$ offers estimates of an instantaneous $E_R(\tau_0, \tau_0)$ and an asymptotic modulus $E_R(\infty, \tau)$ (18).

$$E(\tau_0, \tau_0) = \frac{1}{E_1}, \quad E(\infty, \tau_0) = \frac{E_1 E_2}{E_1 + E_2} \tag{31}$$

STRESS RELAXATION AND CREEP COMPLIANCE

As delineated in Table 2, the differential equations for the standard models are often used in materials description. The primary advantage in these models lies in the relatively few material constants needed to describe many experimental curves; they confer the ability to easily predict the results of one mechanical test from the results of another without explicit knowledge of the constitutive equations from continuum mechanics or structure.

Two standard tests applied to viscoelastic materials are that of creep and stress relaxation. Ideally, in a creep experiment, an instantaneous step increase in stress is applied to the material. Strain is then time dependent and proportional to σ_0. Likewise, ideally in a stress relaxation experiment an instantaneous step increase in strain is imposed on a sample. This means applying at $t = 0$ whatever stress is needed to achieve the desired ε_0, holding the strain constant, and monitoring the stress. The resultant experimental data may be linearized according to equation (1) (18).

$$\sigma(t) = \sigma_0 \Psi(t) \tag{32}$$

where $\Psi(t)$ is the relaxation modulus that decreases monotonically to a horizontal asymptote. To solve for $\Psi(t)$, one lets $\varepsilon = H(t)$, the Heaviside function, so $\bar{\varepsilon}(s) = 1/s$; solution then involves the LaPlace transform. The relaxation modulus is the product of an elastic response and a relaxation function expressing the time-dependence of mechanical response.

PRACTICAL APPLICATION

The application of viscoelastic modeling in pharmaceutics relies on these concepts developed originally in the fields of materials engineering and metallurgy. Building on the rheological models designed to mimic materials behavior is a step in the process to quantify observed time-dependent materials behavior.

By assuming that the energy input or work of compression correlates with the tablet forming properties of materials, force displacement measurements have been used to examine the deformation process (44). Force displacement curves are obtained from the accurate measurements of both the upper and lower punch force and displacements.

The greater area under the force-displacement curve may indicate a greater amount of energy throughput in the tabletting process. A larger energy consumption is not necessarily indicative of a better, stronger tablet. The inability to specifically allocate energy use amongst competing processes mutes the usefulness of this measurement. The resolution of the energy usage during compaction is again, as alluded above, quite difficult. The energy used may be attributed to a number of concurrent processes including internal friction, die wall friction, fracture, bonding, heating, bonding, melting, as well as plastic or viscoelastic deformation.

Upadrashta et al. (20), for example, examined the compaction characteristic of various viscosity grades of ethylcellulose using force-displacement, ejected Heckel analysis and work calculations from instrumented press test data, after accounting for machine deformation. The investigation served to provide a qualitative rank ordering of the grades of the polymer with the degree of compactibility. The better material tablet properties depended on the molecular weight of the polymer and were attributed to the elastic nature of the polymer and the elastic component of the performed work.

STRESS RELAXATION METHODS

The ability of a material to relieve stress under pressure has also been used to characterize the consolidation mechanism of materials. One of the first observations recognizing the role of viscoelasticity in tablet compaction was reported by Rees and Rue (21). Their empirical experiments with starches and celluloses displayed stress relaxation; time-dependent change of the load occurred when the tabletting machine was halted at its lowest point. David and

Augsberger (2) also loaded tablets in a static manner using constant stress or strain. By applying rudimentary mechanical models to the load-displacement profiles for the entire compact, they showed that various agents and excipients were neither ideally elastic, nor ideally viscous, or plastic materials.

Rippie and coworkers (9,22–24) advanced these observations with important macromolecular tabletting materials on a rotary machine. A time shift between the load maxima and displacement maxima during dynamic compression was attributed to the properties of the compacting substance. In the decompression and the post compression phases of rotary tabletting, the load versus time curves were found from radial and axial load measurements while the displacement versus time functions were deduced from the rotary geometry. By assuming that the expansion volume upon decompression was purely an elastic process and the deformation itself, purely viscoelastic, three dimensional viscoelastic models with many elements in series were adjusted to conform to the compression profile of the entire tablet. Arithmetic evaluation of the mechanical model fit resolved the stress strain relation in standard form. The resultant hybrid parameters, p_k and q_k, thus determined on the whole compact, were found to relate to the maximum compression load.

In attempting to correlate the calculated hybrid parameters with the observed behavior of the directly compactible powders examined, it was noted that the elastic parameters were indicative of the degree of interparticulate bonding. Lamination was more likely with the presence of high residual die stresses and large negative terminal elastic parameters.

The aforementioned negative viscoelastic parameters elude a physical explanation (17). While it is easier to calculate these hybrid parameters rather than the microconstants associated with each element of the mechanical model, neither should be negative in a real solution. In a four-parameter model, for example, up to four solutions for a microconstant may be obtained from regression decomposition; their physical validity must be tested to determine the real solution. The numerical stability of the regression fitting method should be addressed.

Moreover, the direct application of rheological models to the entire compact inherently assumes the tablet to be a continuum. The mathematical development of the mechanical models in materials engineering is valid for fully dense, homogeneous, isotropic materials. Rippie and coworkers (24) have acknowledged that their calculated parameters cannot be considered true material constants. The "negativity" of the elastic constants was then suggested to be a consequence of the internal structure of the compact; the internal stresses arising from internal fracturing or bond breaking during expansion processes are transmitted to the die and "overshadows the punch stress reduction from the particulate strain" (24, p. 711).

Consideration of the powder as a continuum also means ignoring the voids between the particles. It would be difficult to envision when these interparticulate pores are not important. At the beginning of compaction, these pores contribute to particle and internal structure rearrangements. In the latter stages of compaction, the porous network of the tablet is progressively established and beyond a critical point, the pore network bears load.

The physical basis for their calculated parameters must be deduced by quasi-inference. The viscous parameters calculated, for example, must be interpreted in the context of "extent of flow within the compact, rather than viscosity" (24, p. 711). The predictive capability of this model form is hampered by the absence of physical analogy.

CREEP COMPLIANCE METHODS

As an alternative to stress relaxation experiments, creep testing may be performed where the change in strain is monitored as a function of step changes in stress. Staniforth and Patel (25) tested both starch and microcrystalline cellulose in a tensile tester at 14 kN/min held at three constant loads. Like Rippie and coworkers, they attempted to apply a rheological model directly to the powder compaction data. The presence of pores and void spaces within the compact, as discussed previously, precludes the extracted parameters from representing physically real material properties. In this case, no attempt was made to calculate hybrid parameters from the assumed eight-parameter Voigt–Maxwell fluid model; instead, the compliance curve was examined and dissected qualitatively into portions that were interpreted as representing just the elastic or the plastic contributions. The mathematical description of the creep function in an eight-parameter model of the form

$$\Phi(t) = \frac{1}{E_1} + \sum_{i=2}^{k} \left(\frac{1}{E_i} \left(1 - e^{-\left(\frac{E_i}{\eta_i}\right) t} \right) \right) + \frac{t}{\eta_1} \tag{33}$$

shows that the intercept of the compliance curve is related to the elastic contribution. This elastic term, however, is a complex function of each of the four springs comprising this rheological model. The simple subtraction of the intercept from the compliance curve cannot provide a quantitative measure of the plastic contribution alone to deformation. The time-dependent elastic recovery of the material is a factor that is neglected by this analysis.

Malamataris et al. (26,27) also used creep experiments performed on a tensile tester for starch, Emcompress and paracetamol to ascertain a more fundamental measure of tabletability. In attempting to apply a Maxwell rheological form directly to the entire compact, viscoplastic deformation was assumed. The subsequent decomposition of the creep compliance curve as interpreted by the authors has several contradictions. While the reciprocal slope of the compliance curve for a Maxwell fluid should provide an apparent viscosity, the shape of the experimental curve is clearly parabolic. The assumption of a Maxwell form is inappropriate; a higher order rheological model would be more suited to the curvature of the observed data. With a three-parameter model, the reciprocal of the slope of the linear portion of the curve would be a measure of the complex modulus (defined in Table 1 as $E_1E_2/E_1 + E_2$). Apparent viscosity, η, cannot be used interchangeably as a measure of the extent of the elastic retardation as assumed in their analysis. Their derived values quantifying the elastic recovery to plastic compression are therefore in error.

The primary advance in the work of Malamataris and coworkers (26–28) is in recognizing the effect of pore structure on the derived apparent material properties. Tablets were made at different packing fractions, the rheological model applied, and the material properties calculated from curves extrapolated to describe a fully dense or zero porosity compact. Nevertheless, rheological model concepts were formulated for homogeneous constituent materials; the direct application to porous whole tablets has not been validated.

Celik and Aulton (29) quantified time-dependent deformation by creep analysis of microindentation data. Test data in the linear viscoelastic region were deconvolved to derive viscoelastic parameters of compacts for higher order rheological models, consisting of a Maxwell unit in series with several Voigt units. Elastic compliance, contributions of the retarded elastic region to the total compliance and residual shear viscosities (in the region of nonrecoverable viscous flow) were explicitly derived from creep compliance spherical indentation experiments.

PARTICLE-BASED MODELS

Numerical simulation is commonly used in soil mechanics and metallurgy for the determination of stress distribution under complex loadings. Typical studies include that by Williamson et al. (30), which describes the densification of tin cylindrical particles in uniaxial hot pressing. A general particle level continuum model based on input of elastic properties from long and shear wave velocity test data and tensile yield data from Swinkels et al. (31) was used to provide full stress and strain fields within the particles. This model assumed an unequal sized particle distribution, elastic perfectly plastic behavior, a von Mises yield condition and power law creep without boundary diffusion. Mesh refinement along the surface of the particle served to smooth the predicted curves. The reports by Lu and Shi (32), Nolan and Kavanagh (33), or Yen and Chaki (34) exemplifies dynamic numerical Monte Carlo simulations for the isostatic pressing of a random packed green compact.

The constitutive models supplied by Wang (35) and Sinka (36) provide other examples of a finite element simulation of uniaxial compression under closed die conditions. In this case, densification conditions were confined to frictionless interaction and negligible elastic recovery due to the lateral constraints of the die walls. Likewise, Mueller (37) created numerical constructs using sequential packing models for monosized spheres with different mean packing densities. Die wall effects were included as periodic boundary conditions but the final outcome was highly dependent on the assumption of the diameter aspect ratio and the assigned coordinate system. In general, a realistic construct consists of explicit relations between stress and strain agreeing with macroscopic observations at various strain rate conditions. Correlations between macroscopic and microscopic parameters are also required.

Continuum models assume that the granular media behaves as a homogeneous medium. Although there are many simulations using continuum models, these models use equivalent flow properties derived experimentally and they cannot represent local effects. Continuum calculations are mathematically complex and yield results that sometimes differ by up to an order of magnitude from experimental values (38).

Finite element simulations are conceptually simpler but rely heavily on computing power in simulating systems. This approach involves the application of Newtonian dynamical equations to a system of impacting particles, which includes keeping track of all forces and moments at each step and integrating the equations to obtain the new state of the system. A statistical distribution is often assumed to determine the state of the particles. This statistical approach eliminates the need for some force calculations and integration routines, but it affords little flexibility. Three dimensional calculations of particle impaction are nontrivial as changes in velocity are nonplanar. The definition of the vertices of the element make up contributes to the wide variability in eventual outcomes.

Micromechanical models are also used in powder metallurgy to determine stress and density distributions during compaction, but these approaches are not so familiar in pharmaceutical powder technology. As a consequence, the few examples of models for compaction have often been directly derived from soil mechanics or metallurgy. Duncan-Hewitt and Weatherly (39), for example, adapted their model from powder metallurgy.

Particle deformation modeling, in contrast to continuum mechanical treatments, describes the response of powder compacts during densification as a framework of linked particles. The linked framework connects the centers of particles through interparticle contacts. The behavior of each link is then based on the unit problem for the interaction between individual spheres; any two contact particles form a link and carries the force that would be transmitted form one particle to another through that contact. By dealing with individual particles, it becomes possible to predict the evolution of different packings from different particle material properties. This approach bridges the gap between unit models on the level of individual particles and the continuum behavior of very large packings of particles.

The physical basis for particle based models has been experimentally verified by a number of investigators. Jagota and Dawson (40,41) validated the Voronoi cell model by comparing particle model predictions with experimental data for the packing of sintered monosized glass spheres. They proposed a truss structure to represent powder morphology based on a micromechanical framework of linked particles with local as well as global descriptions that were suited for high porosity cases. Li and Funckenbusch (42) made actual measurements on contact areas developed on individual particles belonging to randomly dense packed powder aggregates during their consolidation in a closed die and by hot isostatic pressing. The applicability of simple geometric conditions to simulate the deformation behavior of complex powder preforms has further been supported by the findings of McMeeking and coworkers (43,45–47) and Fischmeister (48). McMeeking (49) observed that the lateral flow experienced by porous preforms varied with the relative density of the preform and the geometric condition of their packing. The work of Swinkels et al. (31) examined the hot isostatic pressing of lead, tin and PMMA based on a single sphere indentation model where the primary mechanism of deformation included plastic yielding and power law creep. The experimental results were found to be in good agreement with the model predictions.

The application of particle models established in materials engineering fields for different materials must account for the differences that exist between distinct classes of granular materials. Metallic particles, for example, often pack uniformly and deform plastically while soil particles are usually rigid, densify by rearrangement and are susceptible to shear deformation. Ceramic particles are very small and fine. The van der Waals attraction between such particles invariably overshadows their weight so that the initial relative densities may be as small as 0.2 (50). Pharmaceutical tabletting particles are yet another class of materials that can be subject to rearrangements, shear, elastic, plastic, and viscoelastic deformation and interparticulate attraction and bonding. Specific models for these materials require specific properties like viscoelasticity be described and implemented in the code to determine the mechanical behavior of the powder.

In the adaptation of a particle-based model developed by Artz (51), Duncan-Hewitt and Weatherly (39) modeled ductile pharmaceutical powders as an assembly of monosized spheres packed with a random structure. The local link between the particles assumed a flow condition hinging on static indentation. While the model predictions correlated well with observed behavior in plastic powders, the derivation needs to be expanded to include time-dependent deformation for application to a broader class of granular pharmaceutical powders.

Some approaches that have been undertaken to predict the densification of pharmaceutical powders to gauge the status of advancement in the field; it was not meant to serve as an exhaustive review. Existing approaches to assess powder compaction have mainly tried to quantify compression in an empirical form (52–60). The equations proposed to empirically fit compaction data with parameters such as punch and die wall stresses and tablet porosity offer little insight into the physical basis for particle interaction. Therefore, they remain descriptive rather than predictive of powder densification.

Studies have ascertained that it is the material undergoing compaction that determines the nature of the final compact (61–71). The material properties are then essential and fundamental properties of a tablet mass that delimits tablet success. The inherent time dependence of the materials behavior affects and would be affected by the time periods for the different parts of the compaction cycle, an interaction that is reflected in the mechanical strength of the final compact.

This aspect of viscoelasticity needs to be addressed in the development of a predictive model for tablet densification. It is an inherent property of the material powder; the constituent particles of the tablet mass. Present predictive models in pharmaceutics fail to address this time dependence explicitly, or in a form that avails itself of the material constants.

SUMMARY

Viscoelasticity has a tremendous effect on powder compaction; to produce compacts with desired properties, an understanding of mechanical properties and constituent materials is needed. Despite this link between time-dependent material properties and the sensitivity of consolidation mechanism to changes in the rate of application of the compressing force in solids, there have been few studies relating the frequency dependence of viscoelastic behavior of tablets during compaction. Powder compaction is a bulk phenomenon, which cannot be fully described by simple compression or tensile models. Since powder compaction phenomena occurs within a die, the importance of studying tablet manufacture using consolidation speeds that are related or are similar to those used in practice are paramount. Mathematical translation of the viscoelastic properties of powders may be by differential equations and combinations and permutations of unit spring or dashpot elements. Hence, the behavior or powder blends under stress-strain may be calculated as a function of depth within a die and different loading and compression conditions. Building on basic relaxation and creep tests, the increasing density within a compact at different depths and hardening of these blends arising from creep and deformation responses to applied stress may be predicted in terms of numerical simulation. Developing a working mathematical model of the mechanical and viscoelastic behavior of powders that occurs during the compaction of a tablet would allow for rational design of complex compacts including multi-layer and controlled-release products.

LIST OF SYMBOLS

a, $a(t)$	particle contact radius (radius of circular contact region) varying with time
$a(R'')$, $a(\rho)$	normalized average contact area defined in terms of R'', which in turn is defined by relative density
d	distance
E	elastic modulus
E_1, E_2, η	elastic moduli of the springs and viscosity coefficient of the dashpot of a standard rheological model
f	average local contact force
F	uniaxial applied force
F_{att}	molar attraction function ($J^{1/2} \cdot cm^{3/2}/mol$)
G	shear modulus (MPa)
h	sample thickness (m)
H_k	hardness number (MPa)
H_v	Vickers hardness
K, G	bulk, and shear modulus
$k_1 + k_2$	elastic constant term $= (1 - v^2)/\pi E$ upon substitution
l	length
p	pressure (MPa)
P	load (N)
$P(t)$	applied compressive force acting normal to the surface
p, $p(r,t)$	normal contact pressure that varies with contact radius and time
p, p_{eff}	applied far field stresses
p_h	hydrostatic component
Pi	probability
q, q_0	concentrated pressure, maximum pressure
r	distance from sphere center, x coordinate of point (r,z) of indenting sphere, $r << R$
R	radius of the spherical particle
σ	stress
ε	strain
T	absolute temperature
τ	shear stress
t_f	time at which stress relaxation begins, total strain is held constant

T_g	glass transition temperature
v	the Poisson's ratio
x, y, z	Cartesian coordinates
Z, Z_0	coordination number, initial contact number
z_1, z_2	ordinate in Cartesian coordinates of deforming surface points, designated 1 and 2
$\dot{r}$	plastic deformation rate
$\dot{\varepsilon}_0$	constant strain rate
$\dot{\varepsilon}$	rate of strain
$\frac{dn}{dc}$	refractive index increment
$\dot{\gamma}$	shear strain rate
$\dot{\tau}$	shear stress rate
ρ	density (g/mL)
ν	frequency of motion
τ	interval within domain t, also designated $\eta/(E_1 + E_2)$
ν	Poisson's ratio
λ	the wavelength of light
ϕ	volume concentration of the spheres
α	a constant
λ and μ	Lamé moduli
σ_∞	limiting stress during stress relaxation
$\Delta\tau$	the excess turbidity of the solution over the pure solvent
$\Phi(t)$	creep function
$\varepsilon(t)$	strain that varies with time
$\sigma(t)$	stress that varies with time
$\Psi(t)$	the relaxation function
ρ, ρ_0	relative density, initial relative density
α, α_1, α_2	total distance of approach of two spheres; of each sphere designated 1 and 2
σ_f	stress at time corresponding to beginning of stress relaxation (strain rate = 0)
ΔH_m	the heat of fusion (kJ/mol)
δ_{ij}, ε_{ij}, σ_{ij}	the Kronecker delta, strain and stress in the ith and jth direction, respectively.
Δl	change in distance
ε_t	true strain
σ_x, σ_y, σ_z	punch stress and the die wall stresses
σ_y.	yield stress

REFERENCES

1. FDA news release #P05-10, March 4, 2005.
2. David ST, Augsberger LL. Plastic flow during compression of directly compressible fillers and its effect on tablet strength. J Pharm Sci 1977; 66(2):155–159.
3. Hoag SW, Rippie RG. Thermodynamic analysis of energy dissipation by pharmaceutical tablets during stress unloading. J Pharm Sci 1994; 83:903–908.
4. Ishino R, Yoshino H, Hirakawa Y, et al. Influence of tabletting speed on compactibility and compressibility of two direct compressible powders under high speed compression. Chem Pharm Bull 1990; 38(7):1987–1992.
5. Armstrong NA, Palfrey LP. The effect of machine speed on the consolidation of four directly compressible tablet diluents. J Pharm Pharmacol 1989; 41:149–151.
6. Maarschalk KV, Zuurman K, Vromans H, et al. Porosity expansion of tablets as a result of bonding and deformation of particulate solids. Int J Pharm 1996; 140(2):185–193.
7. Paronen P, Ilkka J. Porosity-pressure functions. In: Alderborn G, Nystrom C, eds. Pharmaceutical Powder Compaction Technology. Drugs and the Pharmaceutical Sciences Series, 71. New York: Marcel Dekker, Inc., 1996:55–76.

8. Rees JE, Rue PJ. Time-dependent deformation of some direct compression excipients. J Pharm Pharmacol 1978; 30:601–607.
9. Rippie EG, Danielson DW. Viscoelastic stress/strain behaviour of pharmaceutical tablets: analysis during unloading and postcompression periods. J Pharm Sci 1981; 70(5):476–481.
10. Roberts RJ, Rowe RC. The effect of punch velocity on the compaction of a variety of materials. J Pharm Pharmacol 1985; 37:377–384.
11. Roberts RJ, Rowe RC. The effect of the relationship between punch velocity and particle size on the compaction behaviour of materials with varying deformation mechanisms. J Pharm Pharmacol 1986; 38:567–571.
12. Ruegger CE, Celik M. The effect of compression and decompression speed on the mechanical strength of compacts. Pharm Dev Technol 2000; 5(4):485–494.
13. Tye C-K, Sun C, Amidon GE. Evaluation of the effects of tableting speed on the relationships between compaction pressure, tablet tensile strength, and tablet solid fraction. J Pharm Sci 2005; 94(3):465–472.
14. York PJ. A consideration of experimental variables in the analysis of powder compaction behavior. Pharm Pharmacol 1979; 31:244–246.
15. Duncan-Hewitt WC, Papadimitriopoulos E. Deformation kinetics of KBr crystals predict tablet stress relaxation. J Pharm Sci 1994; 83:91–95.
16. Flugge W. Viscoelasticity. 2nd ed. New York: Springer-Verlag, 1975:1–21, 159, 178.
17. Muller F. Viscoelastic models. In: Alderborn G, Nystrom C, eds. Pharmaceutical Powder Compaction Technology. Drugs and the Pharmaceutical Sciences Series, 71. New York: Marcel Dekker, Inc., 1996:99–132.
18. Creus GJ. Viscoelasticity—Basic theory and applications to concrete structures. In: Brebbia CA, Orszag SA, eds. Lecture Notes in Engineering. Vol 16. New York: Springer-Verlag, 1985:18–42.
19. Ferry JD. Viscoelastic properties of polymers. 2nd ed. New York: John Wiley & Sons Inc. 1970:671.
20. Upadrashta SM, Katikaneni PR, Hileman GA, et al. Compressibility and compactibility properties of ethylcellulose. Int J Pharm 1994; 112:173–179.
21. Rees JE, Rue PJ. Elastic and viscoelastic properties of compressed tablets. J Pharm Pharmacol 1977; 29 (7–12):37–43.
22. Moe DV, Rippie EG. Nondestructive viscoelastic analysis of anisotropy in compressed tablets. J Pharm Sci 1997; 86(1):26–32.
23. Morehead WT, Rippie EG. Timing relationships among maxima of punch and die-wall stress and punch displacement during compaction of viscoelastic solids. J Pharm Sci 1990; 79(11):1020–1022.
24. Rippie EG, Morehead WT. Structure evolution of tablets during compression unloading. J Pharm Sci 1994; 83(5):708–715.
25. Staniforth JN, Patel CI. Creep compliance behaviour of direct compression excipients. Powder Technol 1989; 57:83–87.
26. Malamataris S, Rees JE, Hart JP. Influence of loading rate and packing fraction on visco-elastic behaviour of powder compacts. Powder Technol 1992; 69:231–238.
27. Malamataris S, Rees JE. Viscoelastic properties of some pharmaceutical powders compared using creep compliance, extended Heckel analysis and tablet strength measurements. Int J Pharm 1993; 92:123–135.
28. Malamataris S, Hatjichristos Th, Rees JE. Apparent compressive elastic modulus and strength isotropy of compacts formed from binary powder mixes. Int J Pharm 1996; 141:101–108.
29. Celik M, Aulton ME. The viscoelastic deformation of some tableting materials as assessed by indentation rheology. Drug Dev Ind Pharm 1996; 22(1):67–75.
30. Williamson RL, Knibloe JR, Wright RN. Particle-level investigation of densification during uniaxial hot pressing: continuum modeling and experiments. J Eng Mater Technol 1992; 114(January):105–110.
31. Swinkels FB, Wilkinson DS, Artz E, et al. Mechanisms of hot-isostatic pressing. Acta Metall 1983; 31(11):1829–1840.
32. Lu GG, Shi X. Computer simulation of isostatic powder compaction by random packing of monosized particles. J Mater Sci Lett 1994; 13:1709–1711.
33. Nolan GT, Kavanagh PE. Random packing of nonspherical particles. Powder Technol 1995; 84: 199–205.
34. Yen KZY, Chaki TK. A computer simulation of rearrangement and compaction of particles under various internal and external stresses. In: Madan D, Anderson I, Fraizer W, et al., eds. Computational & Numerical Techniques in Powder Metallurgy. Chicago: The Minerals, Metals & Materials Society, 1993:139–164.
35. Wang PT. Thermomechanical deformation of powder-based porous aluminium. Part II: constitutive model including densification hardening. Powder Technol 1991; 66:21–32.

36. Sinka IC, Cunningham JC, Zavaliangos A. Analysis of tablet compaction. II. Finite element analysis of density distributions in convex tablets. J Pharm Sci 2004; 93(8):2040–2053.
37. Mueller GE. Numerical simulation of packed beds with monosized spheres in cylindrical containers. Powder Technol 1997; 92:179–183.
38. Hogue C, Newland D. Efficient computer simulation of moving granular particles. Powder Technol 1994; 78:51–66.
39. Duncan-Hewitt WC, Weatherly GC. Modeling the uniaxial compaction of pharmaceutical powders using the mechanical properties of single crystals I: ductile materials. J Pharm Sci 1990; 79(2):147–152.
40. Jagota A, Dawson PR. Micromechanical modeling of powder compacts—II. Truss formulation of discrete packings. Acta Metall 1985; 36(9):2563–2573.
41. Jagota A, Dawson PR. Micromechanical modeling of powder compacts—I. Unit problems for sintering and traction induced deformation. Acta Metall 1988; 36(9):2551–2561.
42. Li EKH, Funckenbusch PD. Geometrical considerations regarding the densification rates of monosize or bimodal-size particle systems during hot isostatic pressing. In: Clauer AH, deBarbadillo JJ, eds. Solid State Powder Processing. Columbus: The Minerals, Metals & Materials Society, 1990:127–134.
43. Fleck NA, Kuhn LT, McMeeking RM. Yielding of metal powder bonded by isolated contacts. J Mech Phys Solids 1992; 40(5):1139–1162.
44. Heckel RW. Density-pressure relationships in powder compaction. Trans Met Soc AIME 1961; 221:671–675.
45. Kuhn LT, McMeeking RM. Power-law creep of powder bonded by isolation contacts. Int J Mech Sci 1992; 24(7):563–573.
46. Kuhn LT, Xu J, McMeeking R, Constitutive models for the deformation of powder compacts. In: Madan D, Anderson I, Fraizer W, et al., eds. Computational & Numerical Techniques in Powder Metallurgy. Chicago: The Minerals, Metals & Materials Society, 1993:123–138.
47. McMeeking RM, Kuhn LT. A diffusional creep law for powder compacts. Acta Metall Mater 1992; 40 (5):961–969.
48. Fischmeister HF, Arzt E. Densification of powders by particle deformation. Powder Metall 1983; 26 (2):82–88.
49. McMeeking RM, and Kuhn, LT. A diffusional creep law for powder compacts. Acta Metall Mater 1992; 40(5):961–969.
50. Yu AB, Bridgwater J, Burbidge A. On the modelling of the packing of fine particles. Powder Technol 1997; 92:185–194.
51. Adams MJ, McKeown R. Micromechanical analyses of the pressure-volume relationships for powders under confined uniaxial compression. Powd Tech 1996; 88:155–163.
52. Altaf SA, Hoag SW. Deformation of the Stokes B2 rotary tablet press: quantitation and influence on tablet compaction. J Pharm Sci 1995; 84(3):337–343.
53. Artz E. The influence of an increasing particle coordination on the densification of spherical powders. Acta Metall 1982; 30:1883–1890.
54. Blattner D, Kolb M, Leuenberger H. Percolation theory and compactibility of binary powder systems. Pharm Res 1990; 7(2):113–117.
55. Bouvard D, Lange FF. Relation between percolation and particle coordination in binary powder mixtures. Acta Metall Mater 1991; 39(12):3083–3090.
56. Celik M, Okutgen E. A feasibility study for the development of a prospective compaction functionality test and the establishment of a compaction data bank. Drug Development and Industrial Pharmacy 1993; 19(17–18):2309–2334.
57. Coffin-Beach DP, Hollenbeck RG. Determination of the energy of tablet formation during compression of selected pharmaceutical powders. Int J Pharm 1983; 17:313–324.
58. Doelker E, Shotton EJ. The effect of some binding agents on the mechanical properties of granules and their compression characteristics. J Pharm Pharmacol 1977; 29:193–198.
59. Holman E, Leuenberger H. The relationship between solid fraction and mechanical properties of compacts—the percolation theory model approach. Int J Pharm 1988; 46:35–44.
60. Holman LE. The compaction behaviour of particulate materials. An elucidation based on percolation theory. Powder Technol 1991; 66:265–280.
61. Johnson KL. Contact Mechanics. New York: Cambridge University Press, 1985:542.
62. Khossravi D, Morehead WT. Consolidation mechanisms of pharmaceutical solids: a multi-compression cycle approach. Pharm Res 1997; 14(8):1039–1045.
63. Kroschwitz JI, ed. Concise Encyclopedia of Polymer Science and Engineering. New York: Wiley & Sons, 1990:17–21.

64. Lange FF, Atteraas L, Zok F, et al. Deformation consolidation of metal powders containing steel inclusions. Acta Metall Mater 1991; 39(2):209–219.
65. Leuenberger H, Leu R, Bonny J-D. Application of percolation theory and fractal geometry to tablet compaction. In: Alderborn G, Nystrom C, eds. Pharmaceutical Powder Compaction Technology. Drugs and the Pharmaceutical Sciences Series, 71. New York: Marcel Dekker, Inc., 1996:133–164.
66. Leuenberger H. The compressibility and compactiblity of powder systems. Int J Pharm 1982; 12:41–55.
67. Pinto JF, Podczeck F, Newton JM. Investigations of tablets prepared from pellets produced by extrusion and spheronisation. II. Modelling the properties of the tablets produced using regression analysis. Int J Pharm 1997; 152:7–16.
68. Ragnarsson G. Force-displacement and network measurements. In: Alderborn G, Nystrom C, eds. Pharmaceutical Powder Compaction Technology. Drugs and the Pharmaceutical Sciences Series, 71. New York: Marcel Dekker, Inc., 1996:77–98.
69. Roberts RJ, Rowe RC. Brittle/ductile behaviour in pharmaceutical materials used in tabletting. Int J Pharm 1987; 36:205–209.
70. Rue PJ, Rees JE. Limitations of the Heckel relation for predicting powder compaction mechanisms. J Pharm Pharmacol 1978; 30:642–643.
71. Toure P, Puisieux F, Duchene D, et al. Energy terms in tablet compression cycles. Powder. Technol 1980; 26:213–216.

3 Application of percolation theory and fractal geometry to tablet compaction

Hans Leuenberger, Maxim Puchkov, Ruth Leu-Marseiler, Jean-Daniel Bonny, and Michael N. Leuenberger

SHORT INTRODUCTION TO PERCOLATION THEORY AND FRACTAL GEOMETRY

Percolation theory (1) and fractal geometry (2) represent novel powerful concepts that cover a wide range of applications in pharmaceutical technology (3). Both concepts provide new insights into the physics of tablet compaction and the properties of compacts (4–12).

Percolation Theory

Different types of percolation can be distinguished: random-site, random-bond, random-site-bond, correlate chain, etc. Generally, percolation theory deals with the number and properties of clusters (1). A percolation system is considered to consist of sites in an infinitely large real or virtual lattice. Applying the principles of random-site percolation to a particulate system, a cluster may be considered as a single particle or a group of similar particles that occupy bordering sites in the particulate system (Fig. 1A, B). In the case of bond percolation, a group of particles is considered to belong to the same cluster only when bonds are formed between neighboring particles.

In random-bond percolation, the bond probability and bond strength between different components can play an important role. The bond probability ρ_b can assume values between 0 and 1. When $\rho_b = 1$, all possible bonds are formed and the tablet strength is at its maximum; that is, a tablet should show maximal strength at zero porosity when all bonds are formed. To form a stable compact it is necessary that the bonds percolate to form an "infinite" cluster within the ensemble of powder particles filled in a die and put under compressional stress. Tablet formation can be imagined as a combination of site and bond percolation phenomena. It is evident that for a bond percolation process the existence of an infinite cluster of occupied sites in a lattice is a prerequisite. Figure 2 shows the phase diagram of a site-bond percolation phenomenon (13).

Example for percolation on a square lattice for $\rho = 0.6$ (1). The infinite cluster is marked by lines.

Site percolation is an important model of binary mixture consisting of two different materials. In the three-dimensional case, two percolation thresholds, ρ_c, can be defined: a lower threshold, ρ_{c1}, where one of the components just begins to percolate, and a second, upper percolation threshold, ρ_{c2}, where the other component ceases to have an infinite cluster. Between the two thresholds the two components form two interpenetrating percolating networks. Below the lower or above the upper percolation threshold, the clusters of corresponding components are infinite and isolated. Thus, in site percolation of a binary powder mixture, ρ_c corresponds to a critical concentration ratio of the two components. From emulsion systems these concentrations are well known where oil-in-water or water-in-oil emulsions can be prepared exclusively.

Table 1 shows critical volume-to-volume ratios for well-defined geometrical packing of monosized spherical particles. The critical volume-to-volume ratios depend on the type of percolation and the type of lattice. In the case of real powder systems the geometrical packing is a function of the particle size, the particle size distribution, and the shape of the particles.

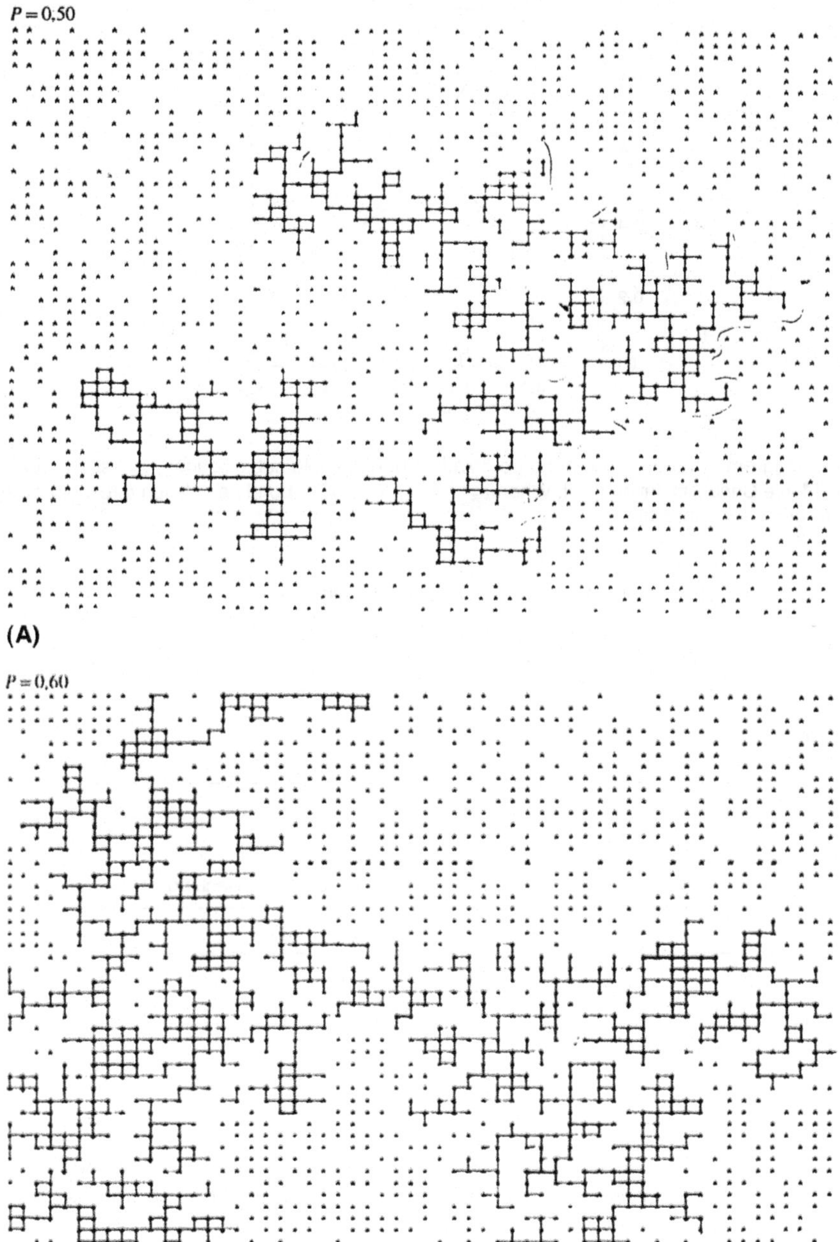

Figure 1 (**A**) Example for percolation on a square lattice for $P = 0.50$. Occupied sites are shown with asterisk; empty sites are ignored. Two clusters are marked by lines. (**B**) Example for percolation on a square lattice for $P = 0.60$. Occupied sites are shown with asterisk; empty sites are ignored. The first formation of an infinite cluster is marked by lines. *Source*: From Ref. 1.

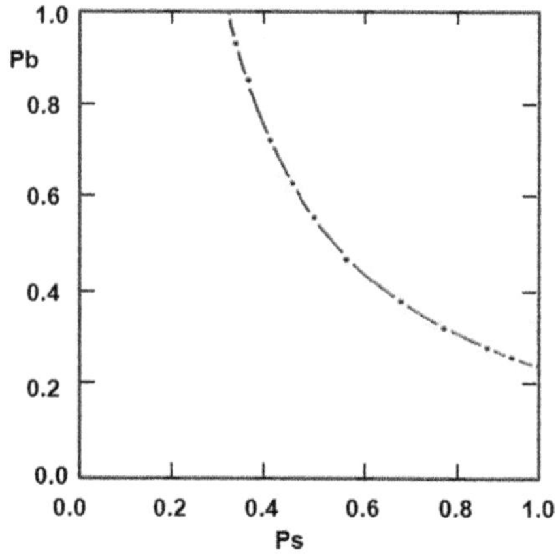

Figure 2 Phase diagram of random site bond percolation in the simple cubic lattice: ρ_s = site probability, ρ_b = bond probability [Monte Carlo simulation (13)]. When $\rho_s = 1$, $\rho_b = 0.249$ = bond percolation threshold; when $\rho_b = 1$, $\rho_s = 0.312$ = site percolation threshold.

Table 1 Selected Percolation Thresholds for Three-Dimensional Lattices

Lattice type	Site	Bond
Diamond	0.428	0.388
Simple cubic	0.312	0.249
Body-centered cubic	0.245	0.179
Face-centered cubic	0.198	0.119

Table 2 Coordination Numbers of Isometric Spherical Particles for Different Packing Structures

Lattice type	Coordination number	Porosity
Diamond	4	0.66
Simple cubic	6	0.48
Body-centered cubic	8	0.32
Face-centered cubic	12	0.26

As different types of packing of monosized spherical particles show different porosities, a powder system that has porosity ε can be represented in an idealized manner as an ensemble of monosized spheres having hypothetical mean diameter x and a mean coordination number z corresponding to hypothetical (idealized) geometrical packing. Table 2 shows the coordination number z of isometric spherical particles of different packing structures.

Using the simplified model of powder systems mentioned, the following equation was developed (14):

$$z = \frac{\pi}{\varepsilon} \tag{1}$$

for porosities in the range $0.25 < \varepsilon < 0.5$. This equation is a rough estimate and does not hold for compacts, where usually $\varepsilon < 0.25$.

At a percolation threshold some property of a system may change abruptly or may suddenly become evident. Such an effect starts to occur close to ρ_c, and is usually called a critical phenomenon. As an example, the electrical conductivity of a tablet consisting of copper powder mixed with Al_2O_3 powder may be cited. The tablet conducts electricity only if the copper particles from an infinite cluster within the tablet, spanning the tablet in all three dimensions.

In case of a pharmaceutical tablet consisting of an active drug substance and excipients, the principle of function is not the electric conductivity and the tablet usually does not consist of a binary powder system compressed. However, often also in case of a complex tablet composition the system can be reduced to a type of binary powder system dividing the drug and excipients involved in two classes of function, such as material that is swelling or is easily dissolved in water. Thus, in case of a mixture KCI-Sta-RX 1500 cornstarch the two percolation thresholds expected are well recognized as a function of the disintegration time of the tablet (Fig. 3) (3). It is evident that the optimal amount of the disintegrant corresponds to the lower percolation threshold. Above the percolation threshold, the formulation of the tablet has problems to disintegrate due to the swelling properties of the disintegrant, which blocks the further uptake of water molecules. This becomes very significant above the upper percolation threshold. The physical model of the disintegration process is described in Figure 4 as a function of the lower percolation threshold. A detailed study concerning the rational estimation of the optimum amount of nonfibrous disintegrant applying percolation theory for a binary fast disintegrating formulation can be found in (15). It is important to distinguish two cases of the arrangement of the disintegrant particles in the powder compact (Fig. 5A, B).

Fractal geometry is related to the principle of self-similarity; that is, the geometrical shape is kept identical independent of the scale, magnification, or power of resolution (2). In practice the range of self-similarity may, however, sometimes be limited to only a few orders of magnitude.

Fractal Geometry

A typical case of fractal geometry is the so-called Coastline of Britain Problem (2): the length of the coastline is continuously increasing with increasing power of resolution, that is, with a smaller yardstick to measure the length. Thus a log-log plot of the length of the coastline as a function of the length of the yardstick to perform a polygon approximation yields a straight line with slope $1 - D_1$, where D_1 is equal to the fractal dimension of the coastline. Coastlines with perfect self-similarity can also be constructed mathematically (Fig. 6). The fractal

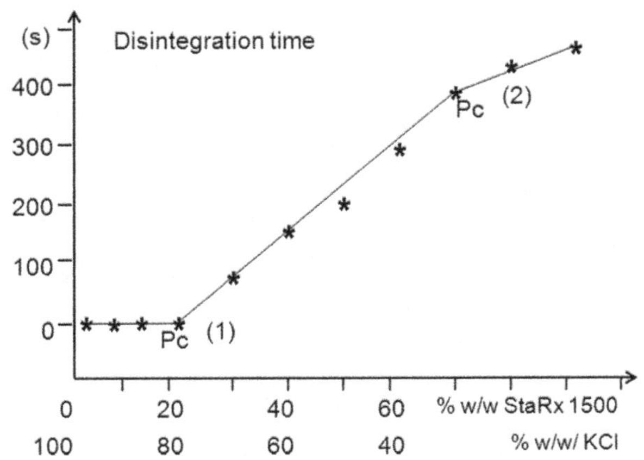

Figure 3 Percolation thresholds for the compacted binary mixture KCl-Sta-Rx 1500; tablet property: disintegration time. *Source*: From Ref. 3.

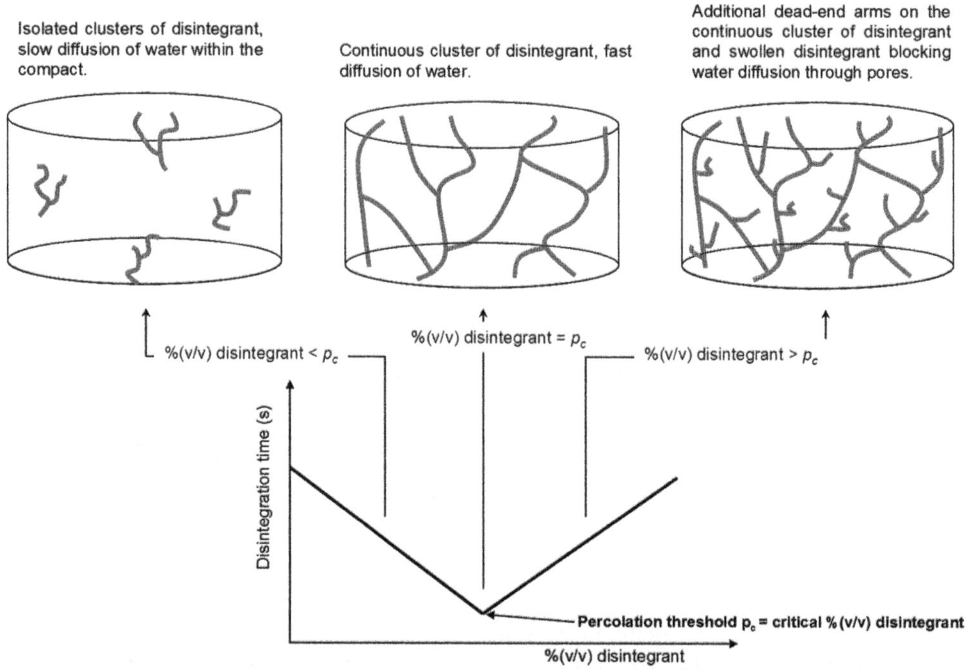

Figure 4 Physical concept of the disintegration process. In this context, two different cases of the diffusion paths of water molecules have to be considered, which initiate the disintegration process of the tablet by the swelling action of the disintegrant. The two cases are linked to the packing of the particles, which are assumed to be spherical with the relative sizes (radius r = disintegrant, radius R = neighboring particle, i.e., drug substance, filler, etc.) and are shown in Figure 5A, B. *Source*: From Ref. 15.

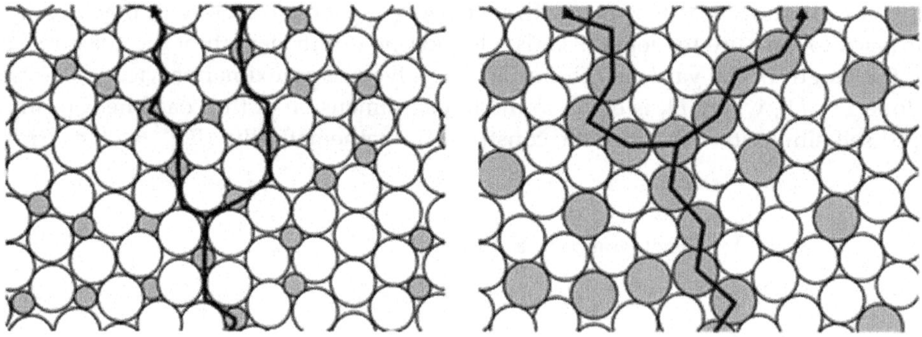

Figure 5 (**A**) The "percolation threshold" and the subsequent "infinite" path depend on the size ratio (radius) of the disintegrant r and the surrounding particles R (filler, drug substance). Case r versus R: $r/R \leq (\sqrt{3}-1)$. (**B**) The "percolation threshold" and the subsequent "infinite" path depend on the size ratio (radius) of the disintegrant r and the surrounding particles R (filler, drug substance). Case r versus R: $r/R > (\sqrt{3}-1)$. *Source*: From Ref. 15.

dimension of a coastline is thus in between the Euclidean dimension 1 for a straight line and 2 for surface.

It is also fruitful to imagine a fractal surface dimension D_s describing the roughness of a surface. Such a description again includes the prerequisite of a self-similar shape independent of the scale. In this respect the introduction of a surface fractal is very advantageous in powder

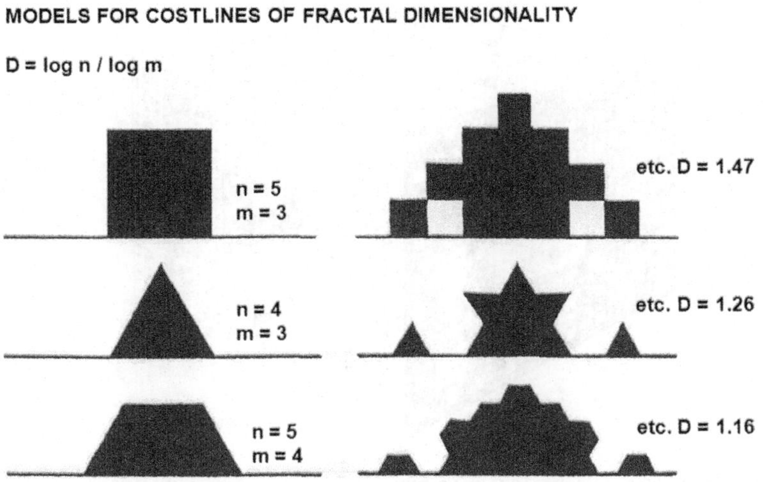

Figure 6 Different self-similar structures as a coastline models; m = theoretical number of equal parts of the unit length projected on a straight line and n = theoretical number of equal parts of unit length describing the coastline structure. *Source*: From Ref. 4.

technology: the result of a measurement of the specific surface or a powder is, as is well known, dependent on the power of resolution of the apparatus (e.g., Blaine, mercury intrusion porosimetry, nitrogen gas adsorption BET method, etc.). Thus the result of a measurement with a chosen method is not able to describe adequately the roughness of the surface. However, if this roughness shows at least within a certain range an approximate self-similarity, it is possible to describe the surface by indicating a value for the specific surface and a value for the surface fractal D_s. Consequently, it is possible to know the specific surface data for different powers of resolution applying a log-log plot of the specific surface as a function of the yardstick length, describing the power of resolution, where the slope of the resulting straight line is equal to $2 - D_s$. As the slope is negative, that is, the surface is larger for a smaller yardstick length—the value for D_s between 2 and 3.

In case of porous material it is also possible to define a volume fractal. This concept is based on the fact that as a function of the power of resolution to detect a pore volume or pore size the void volume is increased. It is evident that in a practical case the porosity of a material attains a limiting value; that is, the self-similarity principle is only valid within a limited range. On the basis of a mathematical self-similar model of pores, that is, a Menger sponge (Fig. 7), the relationship between the accessible void space (sum of pore volumes) and the power of resolution of the pore size was established (9). For this purpose the mercury intrusion porosimetry is the method of choice as the pore volume; that is, the void space of a tablet is filled with mercury as a function of the mercury intrusion pressure, which is related to the accessible pore size. It is, however, necessary to keep in mind that as a consequence of percolation theory not all of the pores are accessible in the same way and some of them are not accessible at all; that is, there may be closed pores present and pores of increased size may be hidden behind pores of smaller size, a fact that is responsible for the hysteresis loop between filling up and draining off the mercury from the void space within the tablet. These reservations have to be taken into account when the volume fractal dimension D_v of a porous network is determined.

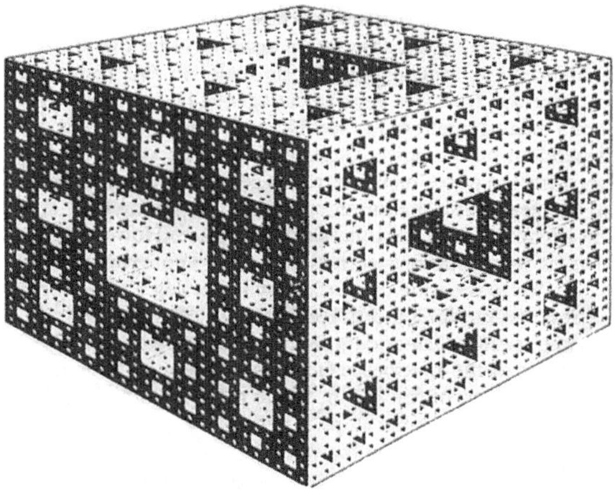

Figure 7 Menger sponge with fractal dimension of 2.727 (idealized three-dimensional network of a pore system). *Source*: From Ref. 2.

In case of a porous network the solid fraction, that is, relative density $\rho_r = 1 - \varepsilon$ of the tablet (determined according to the pore volume fraction $\varepsilon(d)$ filled up by mercury)—as a function of the pore diameter d is related to the volume fractal D_v as follows:

$$\log \rho_r = (3 - D_v) \log d + c \tag{2}$$

with c = constant and $(1 - D_v)$ = slope of the above linear equation.

It is a unique property of the Menger sponge that its surface and volume fractals D_s and D_v are identical and equal to 2.727 = log 20/log 3.

On the other hand, in case of an agglomerate or aggregate of the size L consisting of identical primary particles of diameter δ the following relationship holds:

$$\log \rho_r = (3 - D_v) \log\left(\frac{\delta}{L}\right) + c \tag{3}$$

with c = constant.

A volume fraction Φ can be attributed to such an aggregate. The above equation plays an important role in the gelification of, for example, silica particles, which is by nature a percolation process. It has to be taken into account, however, that secondary aggregates of volume fraction Φ (fractal blobs) and size L and not the primary individual silica particles of diameter δ are the percolating units. Such frail structures may have fractal dimensions below $D_v = 2$. In case of Aerosil 200 aggregates a fractal dimension $D_v = 1.77$ was determined (16). Thus, depending on the structure of an aggregate, the range of D_v overlaps the range of linear and surface fractals, introduced in a first step to trigger the imagination. This is not a contradiction, as the definitions of linear, surface, and volume fractals are arbitrarily related to the Euclidean dimensions to which we are better accustomed. In fact, the electron micrograph of an Aerosil aggregate shows a chainlike structure leading to a fractal dimension of 1.77 as mentioned. Other types may have a fractal dimension close to 2 or even 3. Well known by the work of Mandelbrot (2), the concept of fractal geometry has numerous applications in other fields. In this chapter the concept of fractal geometry is treated only in respect to the physics of tablet compaction and the resulting tablet properties.

THE FORMATION OF A TABLET

Filling of the die: For simplicity, it is imagined that the volume of the die is spanned by a three-dimensional lattice. The lattice spacing is assumed to be of the order of a molecular diameter. Thus granules represent clusters of primary particles and primary particles are considered as clusters of molecules. After pouring the particles/granules to be compacted into the die, the lattice sites are either empty forming pores or occupied by molecules forming clusters with site occupation probability ρ_s. This site occupation probability is equal to the relative density $\rho_r = 1 - \varepsilon$, with ε = porosity (10).

Loose Powder Compacts

To fill hard gelatine capsules, one of the principles consists in performing a loose powder compact as a unit dosage form. As only relatively low compressional force is applied, no brittle fracture or plastic flow is expected. However, at the relative density ρ_r in the range of $\rho_p \leq \rho_r \leq \rho_t$ (ρ_p = poured, ρ_t = tapped relative density) bonds are already formed at the contact points throughout the powder bed. This process can be considered as a bond percolation problem. For a sufficient strength of the powder bed only a weak compressional stress σ_c is needed.

In this simplified model, it is assumed that the particle size of the powdered substances is sufficiently small to form a cohesive compact under a weak compressional stress. In practice the residual moisture content and the individual substance specific capacity to form weaker or stronger bonds have to be taken into account. Thus, a bond percolation threshold ρ_{rc} is expected in the range $\rho_p \leq \rho_{rc} \leq \rho_t$.

Dense Powder Compacts

Pharmaceutical Tablets

Tablets represent the majority of solid dosage forms on the pharmaceutical market. This position is due to its elegance and convenience in application. Thus, a tablet usually has, among other properties, smooth surfaces, low friability, and sufficient strength (e.g., tensile strength or deformation hardness). For the production of such tablets the compressional stress σ_c needs to be important enough to induce plastic flow and/or brittle fracture of the primary granules or particles, that is, to produce simultaneously new surfaces and bonds in this dense powder compact.

Process of Uniaxial Compression

The usual tabletting machines work according to the principle of uniaxial compression. Thus the upper and lower surface areas of the tablet remain constant during the compression process; the thickness of the tablet is reduced with application of the compressional stress σ_z in the z-direction. Because of the initial high porosity of the powder bed, the radial transmission σ_r of the main stress cannot be calculated easily.

In the following simplified model of uniaxial compression, the radial stress σ_r need not be specified explicitly. However, a lateral displacement or rearrangement of particles occupying former pore sites is allowed. The compression process is now studied starting from a loose powder compact with a relative density ρ_p and a moving upper punch. Again a three-dimensional lattice with lattice spacing of molecular diameter spanning the die volume is imagined. During the compression the number of sites to be occupied is constantly reduced and the material (particles, granules, i.e., cluster of molecules) is available to occupy remaining sites. According to the principle of uniaxial compression the mean particle-particle separation distance is more reduced in the z-direction than in the lateral directions. Thus it can be assumed that in the beginning a one-dimensional bond percolation (i.e., a chain of molecules) is responsible for the stress transmission from the upper to the lower punch due to the

Table 3 Physical Characterization of Aerosil 200 and Different Types of Carbon Black

Material	BET surface (m^2/g)	Density (g/cm^3)			Fractal dimension	
		True	Poured	Tapped		
Aerosil 200	202	2.2	0.016	0.022	1.77	± 0.05
Types of carbon black						
Noir d'acétylène	80	1.87	0.027	0.041	1.99	± 0.10
TB number 4500	57	1.83	0.043	0.066	1.76	± 0.12
TB number 5500	206	1.84	0.025	0.043	1.83	± 0.13
Sterling FT	15	1.85	0.27	0.40	3.0	

Source: From Ref. 16.

repulsion forces of the electron shell (Born repulsion forces). After the rearrangement of the particles/granules at a relative density $\rho_r = \rho_r$* an important buildup of stress occurs as particles/granules can no longer be displaced easily. At this relative density ρ_r* the compact can be considered as the "first" dense tablet.

On a molecular scale the molecules react in a first approximation as a hard-core spheres model and span as an infinite cluster the die. This situation is typical for a site percolation process. Above the percolation threshold $\rho_c = \rho_r$* still-empty lattice sites can be occupied due to brittle fracture and/or plastic flow of particles. Thus at higher compressional stress one can imagine that new bonds are formed with a certain bond formation probability ρ_b and sites are occupied with a site occupation probability ρ_s typical for a site-bond percolation phenomenon. Figure 2 represents the phase diagram of a site-bond percolation process. Thus, due to the complex situation during the formation of a tablet, no sharp percolation threshold is expected.

Stress Transmission in the Die

It is well known that the compressional stress is transmitted from the upper punch to the lower punch by means of particle-particle contact in the powder bed. Thus, one may expect that the stress is conducted similar to electric current. As a consequence it is of interest to measure at the same time the stress and the electric current transmitted. This experimental work was realized by Ehrburger et al. (16), using as a conducting material different types of carbon black and for comparison the electrical insulating silica particles Aerosil 200. The physical characterization of the material tested is compiled in Table 3.

Silica was chosen because this material is often used to study the gelation process, which can be adequately described by percolation theory. From nonlinear regression analysis the parameters of the following power laws were determined:

$$C = C_o(\rho - \rho_c)^{\mu} \tag{4}$$

$$\sigma_c = \sigma_o(\rho - \rho_o)^{T} \tag{5}$$

where

C = conductivity $[(\Omega\ \mathrm{cm})^{-1}]$,
C_o = scaling factor,
σ_c = compressional stress transmitted,
σ_o = scaling factor,
ρ_c, ρ_o = percolation thresholds, and
t, T = experimentally determined scaling exponents, expected to equal the electrical conductivity coefficient $\mu = 2$.

Table 4 Stress Transmission and Conductivity of Aerosil 200 and Different Types of Carbon Black

	Stress transmission			Conductivity		
Material	ρ_o	τ	ρ_{max}	ρ_c	t	ρ_{max}
Aerosil 200	0.025	1.5 ± 0.1	0.06	–	–	–
Types of carbon black						
Noir d'acétyléne	0.032	2.9 ± 0.1	>0.24	0.024	1.9 ± 0.1	≥0.24
TB number 4500	0.050	2.2 ± 0.1	0.15	0.040	1.8 ± 0.1	≥0.27
TB number 5500	0.033	2.1 ± 0.1	0.10	0.019	1.8 ± 0.1	>0.2
Sterling FT	0.27	3.9 ± 0.2	>0.56	0.27	3.4 ± 0.2	>0.56

Source: From Ref. 16.

The results of these investigations are compiled in Table 4. The authors (16) concluded that both the stress transmission and the conductivity follow the power laws of percolation. The value of ρ_{max} should indicate the range of relative densities ρ_r: $\rho_o < \rho_r < \rho_{max}$ and $\rho_c < \rho_r < \rho_{max}$ where this power law is still valid. Table 4 shows rather low values for the percolation thresholds ρ_o and ρ_c. This fact can be explained that in the process of percolation secondary agglomerates (aggregates of size L) consisting of primary carbon black or silica particles (of size δ) are responsible for the stress transmission. Taking into account the fractal geometry of these aggregates, that is, the volume fractal D_v, Ehrburger et al. (16) obtained a good estimate for the ratio L/δ, using the following equation based on the percolation threshold ρ_o:

$$\rho_0 = 0.17\left(\frac{L}{\delta}\right)^{D_v^3} \tag{6}$$

The values calculated for L/δ were in good agreement with estimates obtained from independent experiments (BET and porosimetry measurements).

TABLET PROPERTIES

A number of tablet properties are directly or indirectly related to the relative density ρ_r of the compact. According to the percolation theory the following relationship holds for the tablet property X close to the percolation threshold ρ_c:

$$X = S(\rho - \rho_c)^q \tag{7}$$

where

X = tablet property,
ρ = percolation probability,
ρ_c = percolation threshold,
S = scaling factor, and
q = critical exponent.

In case of a tablet property X, the values of S and q are not known a priori. In addition, the meaning of ρ and ρ_c has to be identified individually for each property X. In the case of site percolation the percolation probability ρ is identical to the relative density ρ_r as mentioned earlier. For obvious reasons one can expect that the tensile strength σ_t and the deformation hardness of P are related to the relative density ρ_r in accordance with the percolation law (eq. 7).

Unfortunately, in a practical case only the experimental values of σ_t, P, and ρ_r of the tablet are known. In this respect, it is important to take into account the properties of the power law (eq. 7) derived from percolation theory: (*i*) the relationship holds close to the percolation threshold but it is unknown in general how close; (*ii*) the a priori unknown percolation threshold ρ_c and the critical exponent q are related. Thus there is a flip-flop effect between ρ_c and q, a low ρ_c value is related to a high q value, and vice versa. As a consequence, the data evaluation based on nonlinear regression analysis to determine S, ρ_c, and q may become very tedious or even impossible.

In selected cases such as percolation in a Bethe lattice, the percolation exponent q is equal to unity. A list of selected exponents for the dimensions $D = 2, 3$ and infinite dimensions (i.e., Bethe lattice, valid for $D > 6$) known from first principles is compiled in Table 5 and cited from Stauffer (1). Details should be read there concerning the relevant equation for the property of the system described.

In case of the tablet properties such as tensile strength σ_t and deformation hardness P no meaningful results can be obtained without additional expertise. It was a rewarding endeavor (10) to combine the following two equations derived earlier (17) with the well-known Heckel equation (18):

$$\sigma_t = \sigma_{t\max}(1 - \exp(-\gamma_t \sigma_c)) \tag{8}$$

$$P = P_{\max}(1 - \exp(-\gamma \sigma_c)) \tag{9}$$

where

σ_{tmax} = maximum tensile strength at $\varepsilon \to 0$,
P_{max} = maximum deformation hardness at $\varepsilon \to 0$,
σ_c = compressional stress, and
γ, γ_t = compression susceptibility.

$$\ln(1/(1 - \rho_r)) = a + b\sigma_c \tag{10}$$

where a, b = constants specific to the particulate material compressed and σ_c = compressional stress. The derivation takes into account that for $\sigma_c = 0$ in equation (10), ρ_r is equal to ρ_c. The combination of equations (8) and (10), respectively, equations (9) and (10) yields the following general relationships (eqs. 11 and 12). This derivation takes into account that for $\sigma_c = 0$ in equation (10), ρ_r is equal to ρ_c.

$$\sigma_t = \frac{\sigma_{t\max}(\rho_r - \rho_c)}{(1 - \rho_c)} \tag{11}$$

$$P = \frac{P_{\max}(\rho_r - \rho_c)}{(1 - \rho_c)} \tag{12}$$

with ρ_c = critical relative density (percolation threshold).

Table 5 Percolation Exponents for Two Dimensions, Three Dimensions, and in the Bethe Lattice and the Corresponding Quantity

	Dimension			
Exponent	2	3	Bethe	Quantity/property
a	−2/3	−0.6	−1	Total number of clusters
B	5/36	0.4	1	Strength of an infinite cluster
Γ	43/18	1.8	1	Mean size of finite clusters
V	4/3	.9	1/2	Correlation length
M	1.3	2.0	3	Conductivity

It is evident that equations (11) and (12) are formally identical with the fundamental law of percolation theory (eq. 7).

$$\sigma_t = S(\rho_r - \rho_c)^q \text{ with } S = \frac{\sigma_{t\max}}{(1-\rho_c)} \quad \text{and} \quad q = 1 \tag{13}$$

$$P = S^1(\rho_r - \rho_c)^q \text{ with } S^1 = \frac{P_{\max}}{(1-\rho_c)} \quad \text{and} \quad q = 1 \tag{14}$$

The exponent $q = 1$ corresponds to a percolation in a Bethe lattice (1).

The general relationships (11), (12) can be specified, on the one hand, for the formation of loose compacts and, on the other hand, for the formation of dense compacts.

For loose compacts, that is, at a low-pressure range, ρ_c equals the bond percolation threshold ρ_0. $P_{\max}$ and $\sigma_{t\max}$ of the low-pressure range do not correspond to the maximal possible deformation hardness or tensile strength, respectively, at the relative density $\rho_r = 1$, but describe the strength of the substance specific particle-particle interaction at low relative densities, where the primary particles have not yet lost their identity. The scaling factors S and S^1 are a measure of the strength of this interaction. In this range the compact can still be separated into the original particles.

It has to be kept in mind, that the Heckel equation (eq. 10) is not correct in the low pressure regime. For this case the modified Heckel equation (19) has been derived, which takes into account the initial curvature in the classical Heckel plot and defines the critical relative density ρ_c for such loose compacts, when the stress transmission starts at $\sigma_c \geq 0$:

$$\sigma_c = \frac{1}{C}\left\{\rho_c - \rho_r - (1-\rho_c)\ln\left[\frac{(1-\rho_r)}{(1-\rho_c)}\right]\right\} \tag{15}$$

It is evident, that for $\sigma_c = 0$, ρ_r is equal to ρ_c. Equation (15) takes care of the whole range of the Heckel plot and there is no need to select arbitrarily the data for the linear section of the classical Heckel plot.

For the formation of dense compacts, that is, at a median pressure range, ρ_c equals the site percolation threshold ρ_r*. In this case $P_{\max}$ and $\sigma_{t\max}$ correspond to the maximum deformation hardness or tensile strength, respectively, of the substance at $\rho_r \to 1$. The value of ρ_r* can be best calculated on the basis of the classical Heckel equation (10) for $\sigma_c = 0$, that is,

$$\ln\left(\frac{1}{(1-\rho_r*)}\right) = a \tag{16}$$

At ρ_r* it can be assumed, that the primary particles start to agglomerate and to lose the original shape and identity. This can happen by plastic flow and/or brittle fracture due to local stress concentrations. This concept of ρ_r* is in complete agreement with the classical interpretation of a corresponding to the yield pressure, when the powder system starts to undergo, for example, a plastic flow.

At higher relative densities of the tablet the pore network may no longer form an infinite cluster. Thus another percolation threshold p_c has to be expected for $\rho_r = \rho_\pi$. It is evident that this threshold is important, for example, the disintegration time. Because of the complexity of the tablet formation and the limited size of a tablet again no sharp percolation threshold is expected at ρ_π.

Experimental methods and the data evaluation are described in detail in Ref. (10). For the physical characterization of the materials used see Table 6. The experimentally determined Heckel plot is approximated by two linear sections: a linear section for low and another for median pressures. The percolation thresholds ρ_o and ρ_r* are calculated on the basis of the intercepts of these two linear sections of the classical Heckel plot. Thus there is not an absolute

Table 6 Physical Characterization of the Substances Used

Substances studied	Densities					Mean particle size (μm)
	True (g/mL)	Poured (g/mL)	ρ_p	Tapped (g/mL)	ρ_t	
Avicel PH 102 FMC 2843	1.58	0.325	0.206	0.439	0.278	97.0
Caffeine anhydrous Sandoz 88828	1.45	0.323	0.223	0.417	0.288	55.0
Emcompress $CaHPO_4.2H_2O$ Ed. Mendell & Co. E27B2	2.77	0.714	0.258	0.870	0.314	106.0
Lactose *a*-monohydrate DMV 171780	1.54	0.562	0.365	0.735	0.477	53.2
PEG 10000 Hoechst 605331	1.23	0.568	0.462	0.719	0.585	135.8
Sta-RX 1500 Sandoz 86823	1.50	0.606	0.404	0.741	0.494	68.1

necessity to evaluate the modified Heckel equation according to equation (15), which needs a nonlinear regression analysis of the experimental data.

It is evident that the Heckel equation is an approximation of the pressure-density profile and the application of two linear sections leads to a better fit of the relationship found in reality (Figs. 8–10). According to equations (13) and (14) σ_t and P are plotted against the relative density and linearized by two regression lines for the same ranges as in the Heckel plot (Figs. 11–16). In Table 7 the experimentally determined percolation thresholds from the Heckel plot (eq. 10), from the plot of tensile strength against relative density (eq. 13), and from the plot of deformation hardness against relative density (eq. 14) are summarized. Table 8 shows the values of S, S', and the resulting values for σ_{tmax}, P_{max} at the median pressure range. The values of σ_{tmax}, P_{max}, γ, and γ_t, according to equations (8) and (9), are compiled in Table 9 and the corresponding plots are shown in Figures 17 and 18. The comparison of the results and estimated standard deviations indicates that equations (13) and (14) offer in general more reliable estimates for the σ_{tmax} and P_{max} values than equations (8) and (9). However, the squared correlation coefficients of the

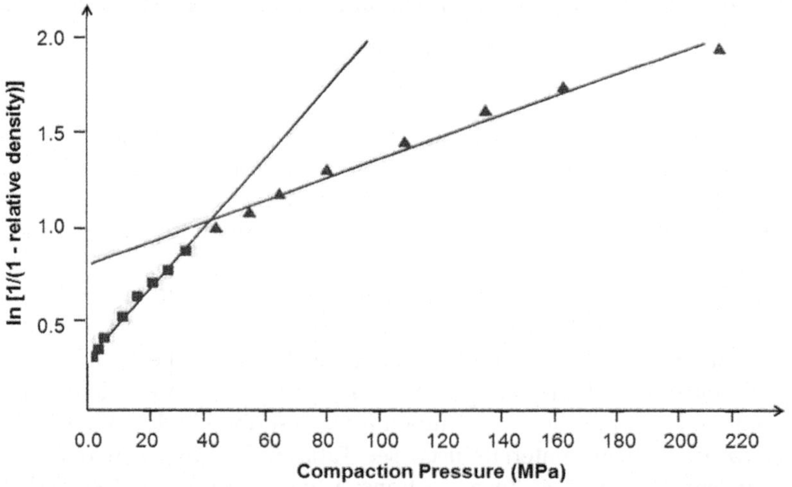

Figure 8 Heckel plot of microcrystalline cellulose (Avicel) to determine the percolation thresholds for loose and dense compacts (ρ_o and ρ_r*).

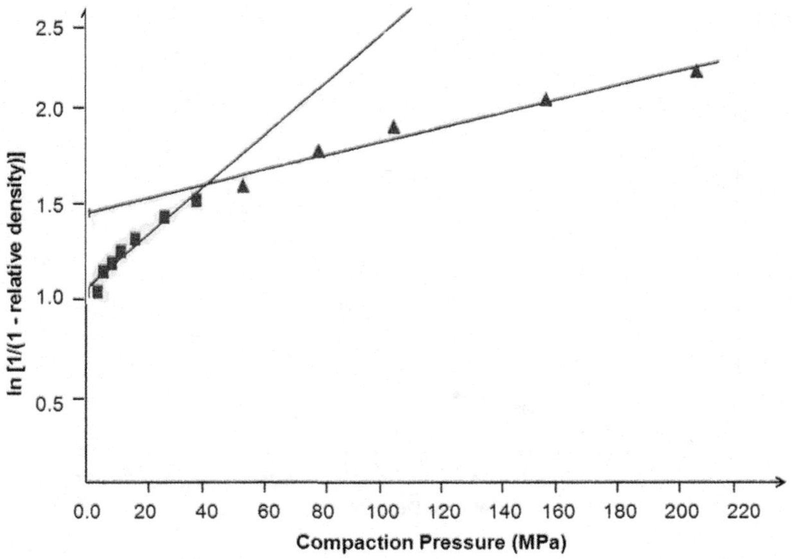

Figure 9 Heckel plot of lactose to determine the percolation thresholds for loose and dense compacts (ρ_o and $\rho_r{}^*$).

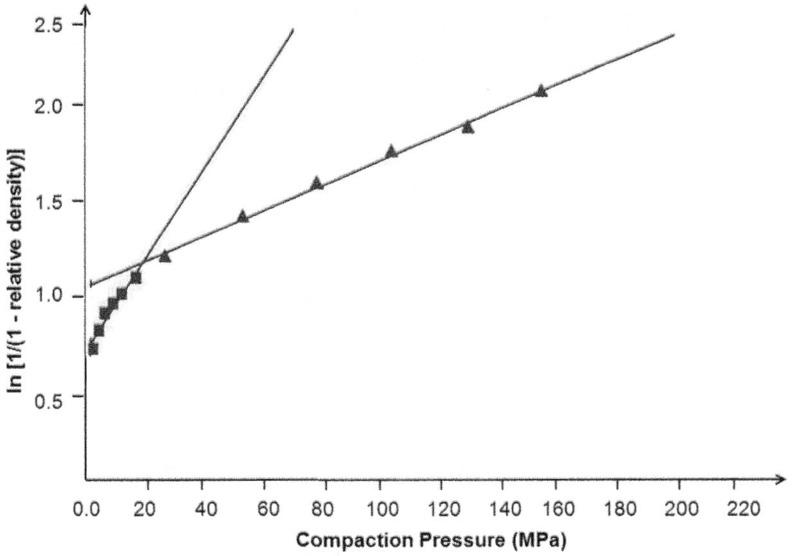

Figure 10 Heckel plot of caffeine to determine the percolation thresholds for loose and dense compacts (ρ_o and $\rho_r{}^*$).

evaluation according to equations (13) and (14) are lower than according to equations (8) and (9). This is due to the fact that the linear model is not as flexible as the exponential model parameters with large standard errors in spite of the good squared correlation coefficients. In the linear model the data are fitted by a straight line, which results in poorer fits but at the same time more reasonable values for the parameters P_{max} and σ_{tmax}.

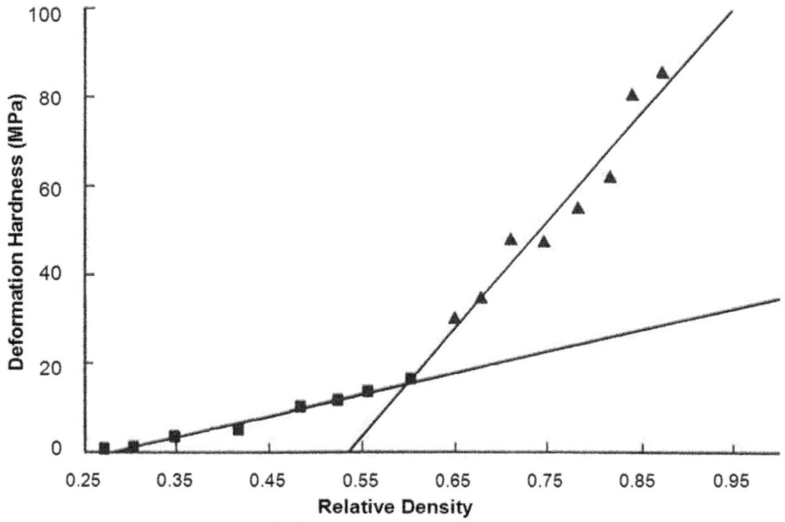

Figure 11 Tablet property: deformation hardness of Avicel compacts as a function of relative density according to equation (14).

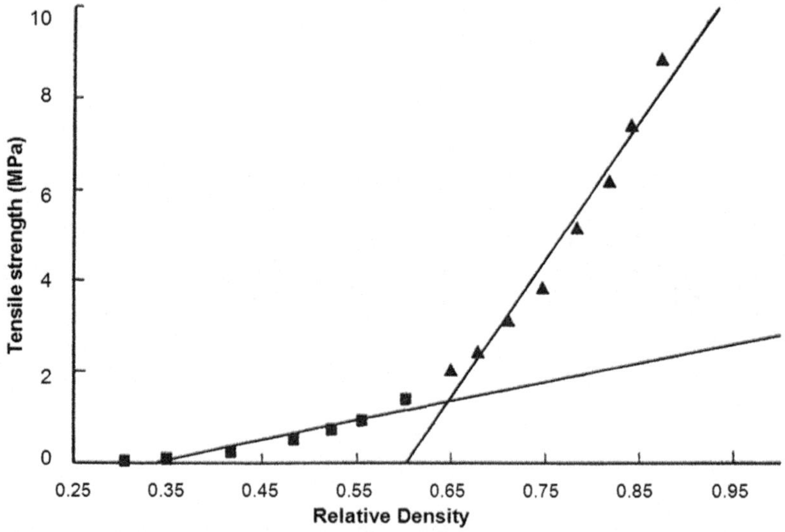

Figure 12 Tablet property: tensile strength of Avicel compacts as a function of relative density according to equation (13).

The description of the other tablet properties such as disintegration time (3), dissolution rate, etc., according to percolation theory are limited as long as the theoretical models are not established to allow more to be known about the respective percolation exponents. However, in the special case of a matrix-type slow-release system (8,9) it is possible to apply simultaneously the concept of percolation theory and fractal geometry. This is the topic of the next section.

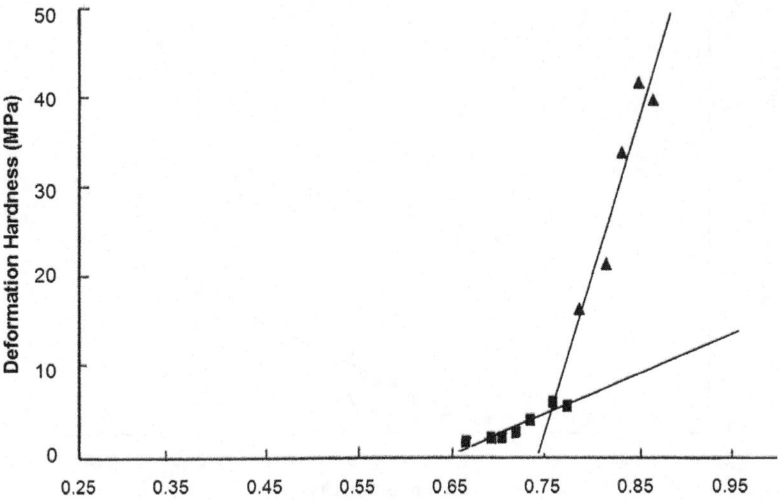

Figure 13 Tablet property: deformation hardness of lactose compacts as a function of relative density according to equation (14).

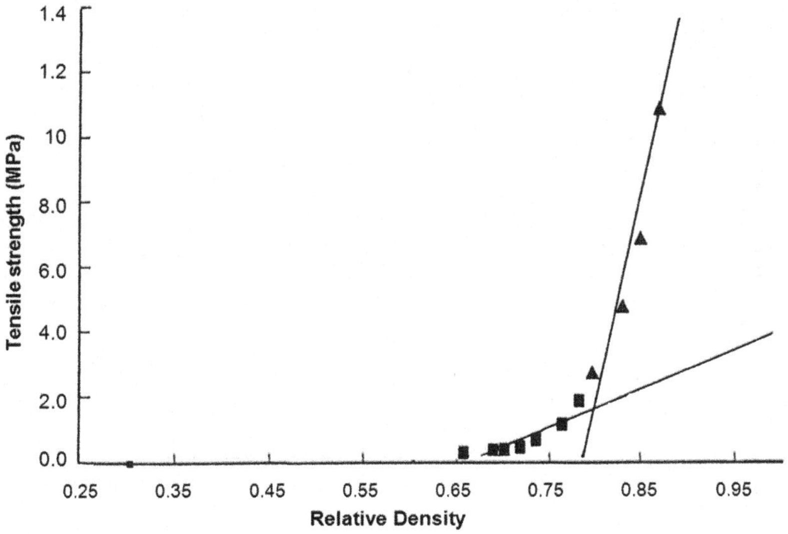

Figure 14 Tablet property: tensile strength of lactose compacts as a function of relative density according to equation (13).

DRUG DISSOLUTION FROM A MATRIX-TYPE CONTROLLED-RELEASE SYSTEM

Ants in a Labyrinth and Drug Dissolution Kinetics

Molecules of an active substance, which are enclosed in a matrix-type controlled-release system, may be called ants in a labyrinth (1) trying to escape from an ordered or disordered network of connected pores. For site occupation probabilities p_s far above the percolation

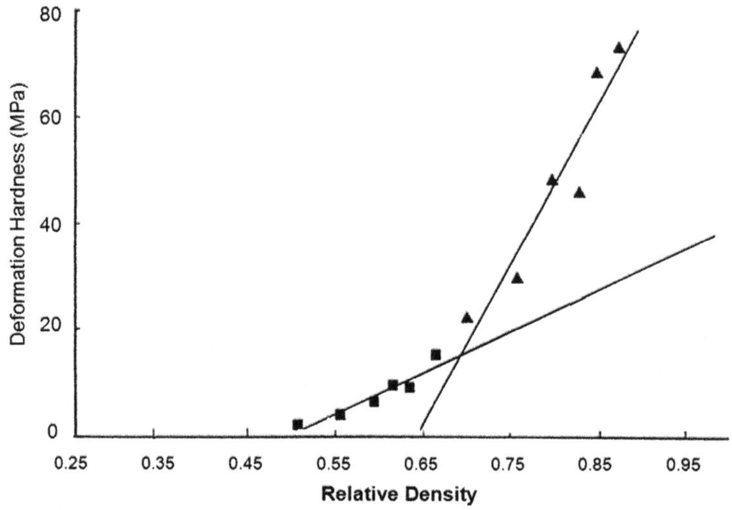

Figure 15 Tablet property: deformation hardness of caffeine compacts as a function of relative density according to equation (14).

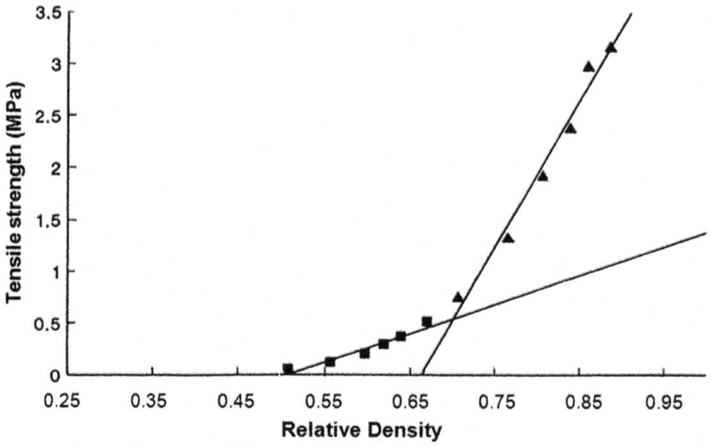

Figure 16 Tablet property: tensile strength of caffeine compacts as a function of relative density according to equation (13).

threshold p_c, the random walk distance R of such an ant is related to time t as follows: $R^2 = Dt$, where D is the diffusivity. In a more general form this diffusion law can be expressed as $R \propto t^k$ with $k = 0.5$ for p_s above p_c. For p_s far below p_c, R approaches a constant for large times, that is, $k = 0$. Right at the critical point the value of k ranges between these two extremes and is about 0.2 in three dimensions. This process is called anomalous diffusion.

From these first principles one can conclude that there is at least one percolation threshold, that is, p_{c1}, where the active drug is completely encapsulated by the water-insoluble matrix substance and where the usual square root of time law for the dissolution kinetics is no

Table 7 Comparison of Values for ρ_c Experimentally Determined According to Equations (10), (13), and (14) for (a) Loose and (b) Dense Compacts

Substances studied	(a) Loose compacts (i.e., low-pressure range $\rho_c = \rho_o$					
	Heckel plot Equation (10)		Tensile strength Equation (13)		Deformation hardness Equation (14)	
	$\rho_o \pm s$	r^2	$\rho_o \pm s$	r^2	$\rho_o \pm s$	r^2
Avicel	0.275 ± 0.013	0.922	0.333 ± 0.023	0.911	0.278 ± 0.011	0.980
Caffeine	0.521 ± 0.028	0.939	0.464 ± 0.016	0.946	0.500 ± 0.010	0.952
Emcompress	0.429 ± 0.025	0.924	0.454 ± 0.016	0.854	0.448 ± 0.020	0.801
Lactose	0.660 ± 0.026	0.945	0.659 ± 0.013	0.820	0.646 ± 0.015	0.841
PEG 10000	0.591 ± 0.028	0.966	0.644 ± 0.016	0.897	0.576 ± 0.008	0.978
Sta-RX 1500	0.506 ± 0.015	0.976	0.521 ± 0.010	0.917	0.499 ± 0.012	0.927
	(b) Dense compacts (i.e., median-pressure range $\rho_c = \rho_r^*$)					
Substances studied	$\rho_r^* \pm s$	r^2	$\rho_r^* \pm s$	r^2	$\rho_r^* \pm s$	r^2
Avicel	0.572 ± 0.039	0.982	0.602 ± 0.013	0.968	0.532 ± 0.025	0.941
Caffeine	0.657 ± 0.023	0.996	0.703 ± 0.019	0.914	0.650 ± 0.028	0.901
Emcompress	0.566 ± 0.006	0.996	0.581 ± 0.008	0.965	0.586 ± 0.018	0.808
Lactose	0.765 ± 0.052	0.966	0.775 ± 0.010	0.930	0.757 ± 0.011	0.934
PEG 10000	0.807 ± 0.147	0.943	0.813 ± 0.011	0.969	0.823 ± 0.033	0.745
Sta-RX 1500	0.619 ± 0.067	0.982	0.688 ± 0.014	0.976	0.595 ± 0.070	0.820

Table 8 S and S' Values for Low- and Median-Pressure Ranges and P_{max} and σ_{tmax} Values Resulting from the S and S' Values for Median Pressures According to Equations (13) and (14)

	Tensile strength (eq. 13)		
	Low-pressure range	Median-pressure range	
Substances studied	$S' \pm s$ (MPa)	$S' \pm s$ (MPa)	$\sigma_{tmax} \pm s$ (MPa)
Avicel	4.199 ± 0.586	30.232 ± 2.261	12.023 ± 0.564
Caffeine	18.924 ± 2.253	2.443 ± 0.376	0.726 ± 0.075
Emcompress	0.467 ± 0.086	3.451 ± 0.379	1.446 ± 0.135
Lactose	0.705 ± 0.165	10.364 ± 1.421	2.332 ± 0.240
PEG 10000	1.621 ± 0.316	14.322 ± 1.475	2.688 ± 0.141
Sta-RX 1500	1.142 ± 0.172	12.167 ± 1.359	3.807 ± 0.282
	Deformation hardness (eq. 14)		
	Low-pressure range	Median-pressure range	
Substances studied	$S \pm s$ (MPa)	$S \pm s$ (MPa)	$P_{max} \pm s$ (MPa)
Avicel	48.17 ± 2.77	239.22 ± 24.55	111.85 ± 6.12
Caffeine	99.16 ± 11.19	309.34 ± 51.33	108.39 ± 10.19
Emcompress	21.60 ± 4.82	262.17 ± 73.77	108.63 ± 26.20
Lactose	40.72 ± 8.85	333.41 ± 44.34	81.17 ± 7.50
PEG 10000	18.45 ± 1.25	139.80 ± 47.18	24.74 ± 4.50
Sta-RX 1500	32.03 ± 4.49	74.69 ± 24.71	30.25 ± 5.11

Table 9 Evaluation of the Data According to Equations (8) and (9) over the Whole Pressure Range

Substances studied	Tensile strength (eq. 8)		
	$\sigma_{tmax} \pm s$ (MPa)	γ_t x$10^{-3} \pm s$ (MPa)	r^2
Avicel	14.44 ± 0.59	5.22 ± 0.30	0.999
Caffeine	4.32 ± 0.34	9.61 ± 1.26	0.995
Emcompress	5.89 ± 12.95	0.57 ± 1.29	0.995
Lactose	494.53 ± 28,849	0.01 ± 0.80	0.994
PEG 10000	3.34 ± 0.39	11.93 ± 2.15	0.992
Sta-RX 1500	5.26 ± 0.70	3.98 ± 0.79	0.990

Substances studied	Deformation hardness (eq. 9)		
	$P_{max} \pm s$ (MPa)	γ_t x $10^{-3} \pm s$ (MPa)	r^2
Avicel	93.22 ± 5.65	12.34 ± 1.44	0.984
Caffeine	94.66 ± 15.64	10.99 ± 3.22	0.968
Emcompress	725.14 ± 10,187	0.32 ± 4.60	0.944
Lactose	58.56 ± 12.15	7.45 ± 2.51	0.958
PEG 10000	37.06 ± 17.73	8.85 ± 5.95	0.924
Sta-RX 1500	20.38 ± 1.44	19.93 ± 4.20	0.945

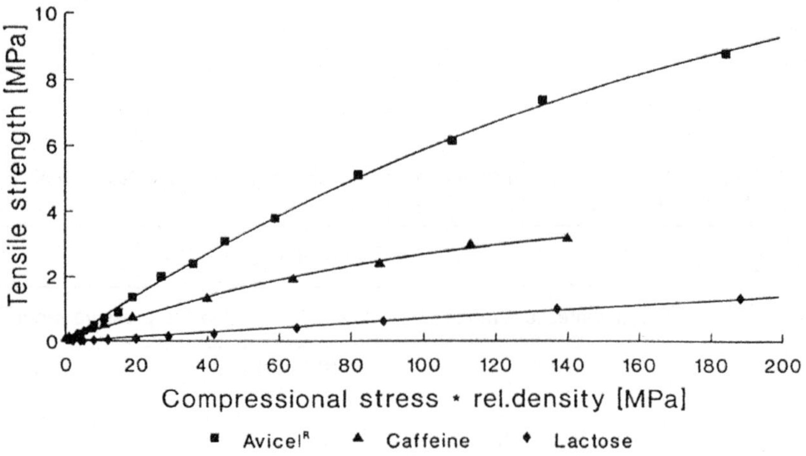

Figure 17 Plot of the tensile strength σ_t versus the relative density ρ_r according to equation (8) for Avicel, caffeine, and lactose.

longer valid. As in general, in a three-dimensional system two percolation thresholds can be expected, an experiment was set up to elucidate this phenomenon. For this purpose, a highly water-soluble model drug (caffeine anhydrous) and a plastic water-insoluble matrix substance (ethyl cellulose) were chosen. The materials and methods used are described in detail by Bonny and Leuenberger (12). Here, only the theoretical background and conclusions are summarized. In this experiment the drug content was varied from 10% to 100% (w/w), and the drug dissolution from one flat side of the tablets was studied. To measure this intrinsic dissolution rate the tablet was fixed into a paraffin matrix to leave only one side accessible for the dissolution medium (distilled water).

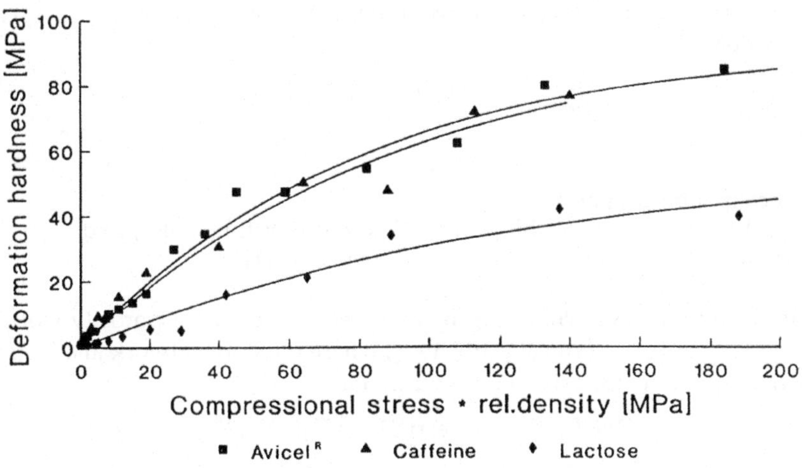

Figure 18 Plot of the deformation hardness P_{max} versus the relative density ρ_r according to equation (9) for Avicel, caffeine, and lactose.

For low drug concentrations, that is, low porosity of the matrix, most of the drug is encapsulated by the plastic matrix and the release is incomplete. At the lower percolation threshold p_{c1} the drug particles begin to form a connective network within the matrix and according to the theoretical considerations the diffusion should be anomalous. At the upper percolation threshold p_{c2}, the particles that should form a matrix start to get isolated within the drug particles and the tablet disintegrates.

The concentration of the drug particles within the matrix can be expressed as site occupation probability p_s. The amount of drug $Q(t)$ released from one tablet surface after the time t is proportional to t^k and the exponent k depends on the percolation probability p_s.

Case 1: $p_s < p_{c1}$. Only the few particles connected to the tablet surface can be dissolved and $Q(t)$ reaches a constant value.
Case 2: $p_s \sim p_{c1}$. Anomalous diffusion (1) with $k \sim 0.2$ in three dimensions.
Case 3: $p_{c1} < p_s < p_{c2}$. Normal matrix-controlled diffusion with $k = 0.5$.
Case 4: $p_{c2} < p_s$. Zero-ordered kinetics with $k = 1$.

Between the two percolation thresholds p_{c1} and p_{c2} the particles of drug and matrix substance form a bicoherent system; that is, the drug release matches the well-known square root of time law of Higuchi (19) for porous matrices.

$$Q(t) = \sqrt{DC_s(2A - (\varepsilon C_s)t)} \text{ with } D = D_0\left(\frac{\varepsilon}{\tau}\right) \quad (17)$$

where

Q = cumulative amount of drug released per unit exposed area,
D_o = diffusion coefficient of drug in permeating fluid,
ε = total porosity of empty matrix,
τ = tortuosity of matrix,
C_s = solubility of drug in permeating fluid,
A = concentration of dispersed drug in tablet, and
D = apparent or observed diffusion coefficient.

Close to the percolation threshold the observed diffusion coefficient obeys a scaling law, which can be written as [1]

$$D \propto D_0(p_s - p_{c1})^{\mu} \tag{18}$$

where

p_s = site occupation probability,
p_{c1} = critical percolation probability (lower percolation threshold), and
μ = conductivity exponent = 2.0 in three dimensions [1].

In the case of a porous matrix p_s can be expressed by the total porosity ε of the empty matrix and p_{c1} corresponds to a critical porosity ε_c, where the pore network just begins to span the whole matrix. Equation (18) can then be written as

$$D \propto D_0(\varepsilon - \varepsilon_c)^{\mu} \text{ resp. } D = \kappa D_0(\varepsilon - \varepsilon_c)^{\mu} \tag{19}$$

with κD_o = scaling factor.

In the cases where the dissolution kinetics are in agreement with equation (17), the dissolution data can be linearized by plotting $Q(t)$ versus t giving a regression line with the slope b,

$$b = \sqrt{DC_s(2A - (\varepsilon C_s))} \tag{20}$$

which leads to

$$D = \frac{b^2}{C_s(2A - \varepsilon C_s)} \tag{21}$$

Combining equations (19) and (21) for D and assuming μ = 2.0 results in

$$\kappa D_o(\varepsilon - \varepsilon_c)^2 = \frac{b^2}{C_s(2A - \varepsilon C_s)} \tag{22}$$

After taking the square root and rearranging the equation (22), a new tablet property β is defined, which depends linearly on $(\varepsilon - \varepsilon_c)$.

$$\beta = \sqrt{(\kappa D_0 C_s)}(\varepsilon - \varepsilon_c) = \frac{b}{\sqrt{(2A - C_s)}} \in \tag{23}$$

that is, the tablet property β is determined by a linear relationship of ε,

$$\beta = c(\varepsilon - \varepsilon_c) \tag{24}$$

where the constant c equals to $\sqrt{\kappa(D_0 C_s)}$.

By the help of equation (24) ε_c can easily be calculated by using a nonlinear, or even a linear, regression analysis, giving a slope of c and an intercept of $-c\varepsilon_c$.

The results of the intrinsic dissolution test of the tablets with the different caffeine loadings are plotted in Figure 19. A change in dissolution kinetics can be assumed between 70% and 80% of caffeine loading. To test this assumption the release data are evaluated according to the model $Q(t) = a + b\sqrt{t}$ by a simple linear regression to clarify for which loadings the square root of time law is fulfilled. To analyze the diffusion mechanism the data are also evaluated according to $Q(t) = a' + b't^k$ by a nonlinear least square fit. The results are compiled in Table 10.

Comparing the squared correlation coefficients of the model evaluation of $Q(t) = a + b\sqrt{t}$, there is a clear decrease in the grade of correlation for caffeine loadings higher than 70%; that

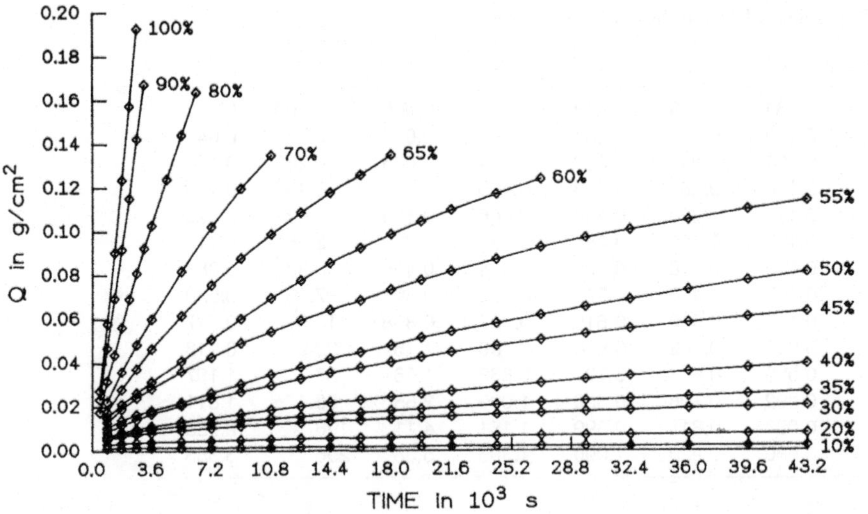

Figure 19 Cumulative amount $Q(t)$ of caffeine released per unit area as a function of time for tablets with caffeine loadings between 10% (w/w) and 100% (w/w).

Table 10 Evaluation of Dissolution Data and Percolation Thresholds

Drug content	$Q(t) = a + b\sqrt{t}$		$Q(t) = a' + b't^k$	
	b	r^2	K	r^2
10	0.006	0.9858	0.17	0.9615
20	0.027	0.9975	0.27	0.9932
Lower percolation threshold expected				
30	0.085	0.9989	0.37	0.9983
35	0.116	0.9994	0.41	0.9973
40	0.189	0.9954	0.52	0.9963
45	0.320	0.9987	0.54	0.9992
50	0.415	0.9929	0.61	0.9980
55	0.593	0.9989	0.56	0.9979
60	0.858	0.9936	0.67	0.9979
65	1.15	0.9983	0.66	0.9991
70	1.56	0.9941	0.74	0.9996
Upper percolation threshold expected				
80	2.55	0.9863	0.84	0.9998
90	4.11	0.9752	1.02	0.9991
100	5.38	0.9784	1.09	>0.9999

is, the drug loadings from 80% to 100% caffeine are no longer in good agreement with the $\sqrt{t}$ law. For low drug concentrations only a small amount of drug is released and the dissolution curve runs nearly parallel to the abscissa. In these cases the correlation coefficient cannot be used as an indicator for compliance with the model.

The estimation of k according to the model $Q(t) = a' + b't^k$ yields values for k between 0.17 and 1.09. For 35% to 55% of caffeine the exponent k ranges between 0.41 and 0.61, which is in good agreement with the $\sqrt{t}$ kinetics with $k = 0.5$. For higher loadings there is a clear change from $\sqrt{t}$ kinetics to zero-order kinetics with $k = 1$. Both evaluations show that the upper percolation threshold p_{c2} lies between 70% and 80% of caffeine.

Table 11 Calculation of D and the Tablet Property β

Drug content	ε_o	ε_d	ε	A	b	D	β
10	0.134	0.078	0.212	0.110	0.006	0.0045	0.013
20	0.128	0.158	0.286	0.225	0.027	0.0437	0.041
30	0.121	0.242	0.363	0.344	0.085	0.282	0.104
35	0.118	0.285	0.403	0.405	0.116	0.446	0.130
40	0.116	0.328	0.444	0.466	0.189	1.03	0.198
45	0.109	0.375	0.484	0.532	0.320	2.58	0.313
50	0.110	0.418	0.528	0.594	0.415	3.88	0.384
55	0.106	0.465	0.571	0.660	0.593	7.13	0.520
60	0.103	0.512	0.615	0.727	0.858	13.5	0.717
65	0.098	0.562	0.660	0.798	1.15	22.2	0.918
70	0.099	0.608	0.707	0.863	1.56	37.7	1.197
80	0.092	0.708	0.800	1.006	2.55	86.4	1.811
90	0.088	0.811	0.899	1.151	4.11	196	2.729
100	0.080	0.920	1.000	1.306	5.38	296	3.353

For the quantitative determination of the lower percolation threshold p_{c1}, that is, the critical porosity ε_c, equations (23) and (24) are used. The needed data are summarized in Table 11.

The initial porosity ε_o before leaching is calculated from the apparent volume V_{tot} and the true volume V_t of the tablet constituents:

$$\varepsilon_o = \frac{(V_{tot} - V_t)}{V_{tot}} \tag{25}$$

ε_d is the porosity corresponding to the volume occupied by the drug substance in the matrix and is calculated as follows:

$$\varepsilon_d \frac{V_{td}}{V_{tot}} = \frac{m_d}{(\rho_d V_{tot})} \tag{26}$$

with $V_{td} = m_d/\rho_d$ = true volume of the drug substance, m_d = total amount of drug present in the tablet, and ρ_d = true density of drug. It is important to keep in mind that the porosity ε in equation (24) corresponds to the total porosity of the empty matrix.

$$\varepsilon = \varepsilon_0 + \varepsilon_d \tag{27}$$

D is calculated according to equation (21), and the tablet property β according to equation (23).

For estimating ε_c with the help of equation (24), the data for ε and β for 35% to 55% of caffeine are used, because in this range there is the best agreement with the normal diffusion law, that is, the $\sqrt{t}$ kinetics with k = 0.5 (Table 10). The nonlinear regression yields $\varepsilon_c = 0.35 \pm 0.01$ and $c = (2.30 \pm 0.15).\ 10^{-3}\ g^{1/2}cm^{-1/2}\ sec^{1/2}$. A linear regression analysis leads to the same result. The critical porosity of 0.35 corresponds to a caffeine content of about 28% (w/w). Figure 20 shows the plot of β versus ε where the point of intersection with the abscissa just indicates ε_c.

In Table 1 selected percolation thresholds (1) for three-dimensional lattices (as volume-to-volume ratios) and in Table 2 the corresponding coordination numbers (20) for isometric spherical particles are compiled. Comparing the experimentally determined percolation threshold of 0.35 (volume-to-volume ratio) with the theoretical values shows good agreement with the simple cubic lattice, which has a site percolation threshold of 0.312. In the tablet a brittle and a plastic substance of different grain size are compacted together so that the

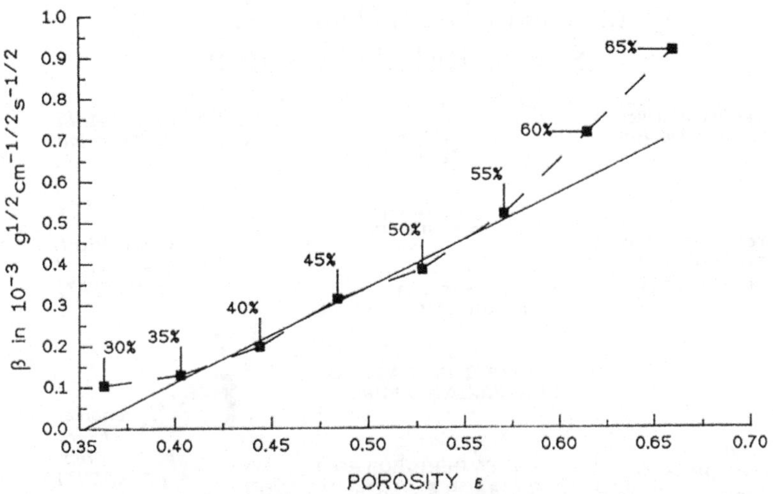

Figure 20 Tablet property β as a function of porosity ε (power law) according to percolation theory.

postulation of a lattice composed of isometric spheres is only a rough estimate, but it points out that the magnitude of ε_c is reasonable.

Fractal Dimension of the Pore System of a Matrix-Type Slow-Release System

If the matrix-type slow-release dosage form of the preceding chapter is removed from the dissolution medium after, for example, a maximum of 60% (w/w) of drug dissolved to guarantee the physical stability of the remaining carcass, the open pore system left can be analyzed by mercury intrusion porosimetry to determine the fractal dimension. It is evident that the pore structure depends on the particle size distribution of the brittle model drug caffeine anhydrous originally embedded in the plastic ethyl cellulose matrix. The volume fractals (9) of leached tablets with an initial caffeine content of 50% (w/w) range between 2.67 and 2.84, depending on the particle size distribution of the water-soluble drug. The system that contained caffeine with a broad particle size distribution (125–355 μm) yielding a fractal volume dimension of 2.734 is closest to the dimension of the Menger sponge (D_v = 2.727).

Fractal Dimension of the Porous Network of a Fast-Disintegrating Tablet

Luy (21) states in his thesis that tablets were produced from granules that showed a very high porosity and internal surface. Those granules were obtained with a novel process technology: vacuum fluidized-bed spray granulation (22,23). Independent of the granule size distribution the resulting tablets released 100% of the soluble drug (solubility in water $< 0.01\%$) within five minutes. The disintegration time was smaller than one minute. The fractal dimension of the porous network ranged from 2.82 to 2.88.

QUALITY BY DESIGN AND PERCOLATION THEORY

How to Achieve a Six Sigma Quality?

An excellent manufacturing process produces a very small amount of defectives. The performance of a manufacturing process can be quantified with its σ *value*. The σ value is based on the Gaussian probability distribution and does not depend on the type of product manufactured. The champion is the Chip-Industry, which is able to produce chips with a σ value = ca 6. This value corresponds to an amount of defectives of ca 0.0003%! It is generally

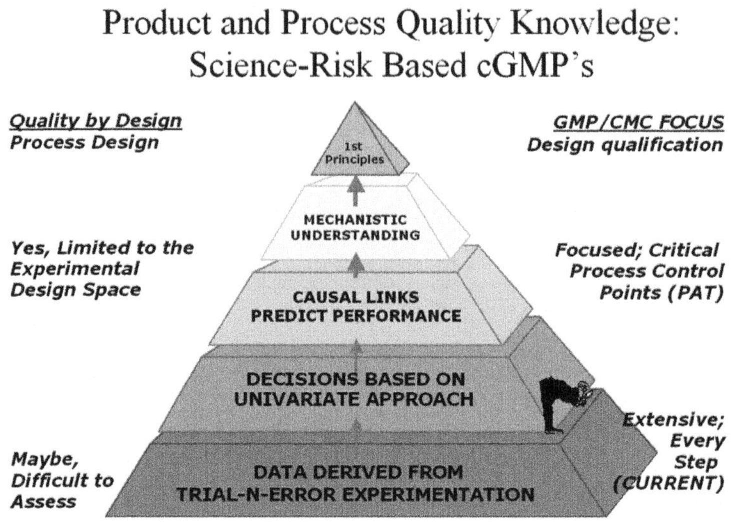

Figure 21 Knowledge pyramid. *Source*: Courtesy of Dr. Ajaz Hussain.

accepted that the σ value in case of the pharmaceutical industry is ca 2, which means in the best case (snapshot view) ca 4.5% of defectives. This low performance prompted the Process Analytical Technology (PAT) and Quality by Design (QbD) initiative of FDA (24). QbD means not to "test in," but to "build in" the required quality. This is only possible thanks to a better process understanding, which needs additional efforts, as schematically shown in Figure 21.

The PAT initiative boosted the installation of additional in-process control units in the manufacturing departments for optimizing the quality. Several pharmaceutical companies in Switzerland and Germany have introduced at-line, on-line or in-line near infrared (NIR) spectroscopy control tools for nearly all process steps such as raw material identification, blending, drying and tabletting (25). Besides NIR spectroscopy, sophisticated tools such as terahertz pulsed imaging have been tested for suitability to monitor the coating process of pharmaceutical tablets (26). Interestingly the PAT initiative did not affect with the same visibility the pharmaceutical R&D departments with their task to use the methods of QbD according to ICH Q8 (27). The implementation of QbD is not free of charge and needs resources. Is it possible to *increase the quality* of a product *and to reduce simultaneously the costs* in pharmaceutical development? PricewaterhouseCoopers (PwC) suggests in the publication series PHARMA 2020 to focus on e-development, virtual R&D (28,29) to shorten time to market and to reduce development costs.

Actually the marketed dosage form such as a tablet is in general still designed at the end of the clinical phase II or at the begin of phase III. This is an important problem, as for the first clinical trials in phase I and II usually a preliminary "service dosage form" is used, which does not has the optimal quality. It is evident, that a change of the formulation needs a bioequivalence test to guarantee the same bioavailability. For the implementation of QbD it is important to start as early as possible in the development process. Therefore it would be ideal to manufacture the samples for the first and all subsequent clinical trials in the quality of the final marketed dosage form. In this context, the question arises, if this task can be resolved by a "virtual R&D" support. The goal would be to calculate "in-silico" tablet formulations, and to predict accurately tablet properties such as the drug dissolution rate, percolation thresholds, etc. Thus, expensive laboratory trials should be replaced by a computer-aided design of

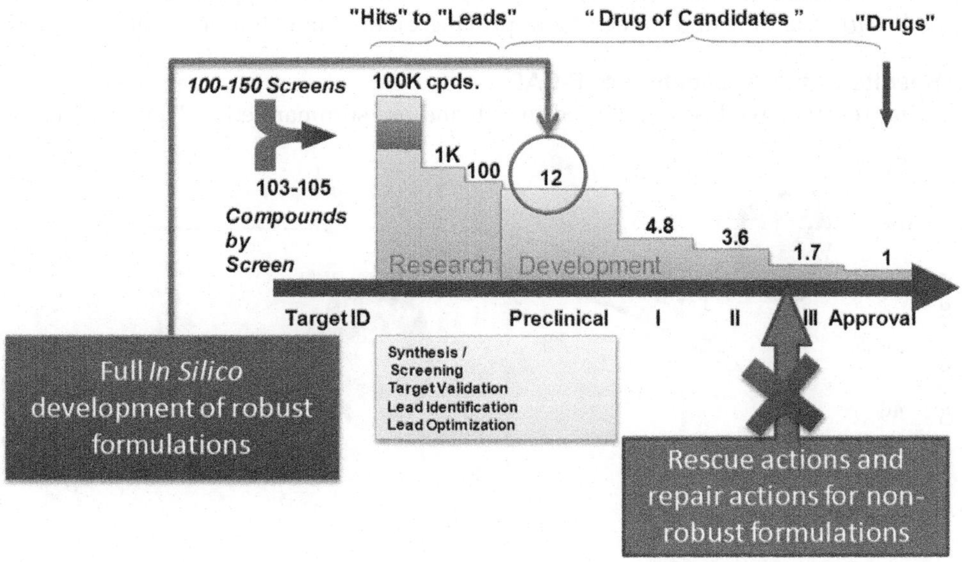

Figure 22 Typical number of drug substances in the pipeline of a pharmaceutical company as a function of the development phase (figure adapted based on a presentation by Dr. Ajaz Hussain), taking into account the capabilities of F-CAD for an early "in-silico" development. *Source*: From Ref. 30.

formulations. This can be realized by applying an appropriate virtual R&D approach (30,31). In such a case optimization and repair actions in a late clinical phase II/III could be avoided (Fig. 22).

Proof of Concept of Virtual R&D Reality for Tablet Formulations

It is important to realize, that percolation thresholds in the three-dimensional space can be only calculated by numerical calculations, using an appropriate software such as Monte Carlo methods. Ideally such a software is based on the use of cellular automata, which can be described as a "first principle" approach (Fig. 21). Such a unique software, Formulation Computer-Aided Design (F-CAD), has been developed by CINCAP LLC (30) as a tool for a reliable and fast development of tablet or capsule formulations ready for market already for clinical phase I. The core module of F-CAD is taking into an account the shape of the dosage form. In fact, the special core module is capable of taking care of the different boundary conditions, which would be extremely difficult by solving classically sets of partial differential equations. This is the advantage, that F-CAD follows *first principles*. The capabilities of F-CAD can be summarized as follows: (*i*) Search for an optimal and robust formulation for the market with a minimum amount of drug substance, that is, already in an early phase of development; (*ii*) explore the whole design space in a minimum of time; (*iii*) define the necessary specifications of the drug substance and the excipients on a scientific base such as particle size distribution search for root cause in case of out-of-specification (OOS) problems; (*iv*) check for equivalence in case of the exchange of an excipient by the same type having, for example, a different size distribution; (*v*) generate a feasibility study for the project including a sensitivity analysis to assure the robustness of the formulation; (*vi*) establish intellectual property rights by providing results for systems before experimental confirmation; (*vii*) effectuate only a minimum of laboratory experiments to confirm the results of the suggested formulations by F-CAD; and (*viii*) save time and money by improving quality and shortening time to market.

Typical Results of the Application of F-CAD

The following results have been published in (30) and are summarized in Figures 23 to 25.

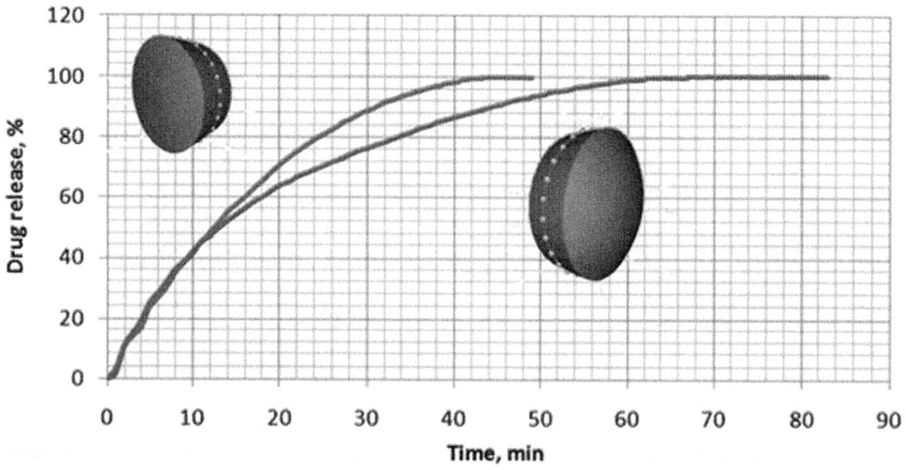

Figure 23 F-CAD calculated dissolution profiles as a function of the shape of a tablet having the same volume and the same composition. *Source*: From Ref. 30.

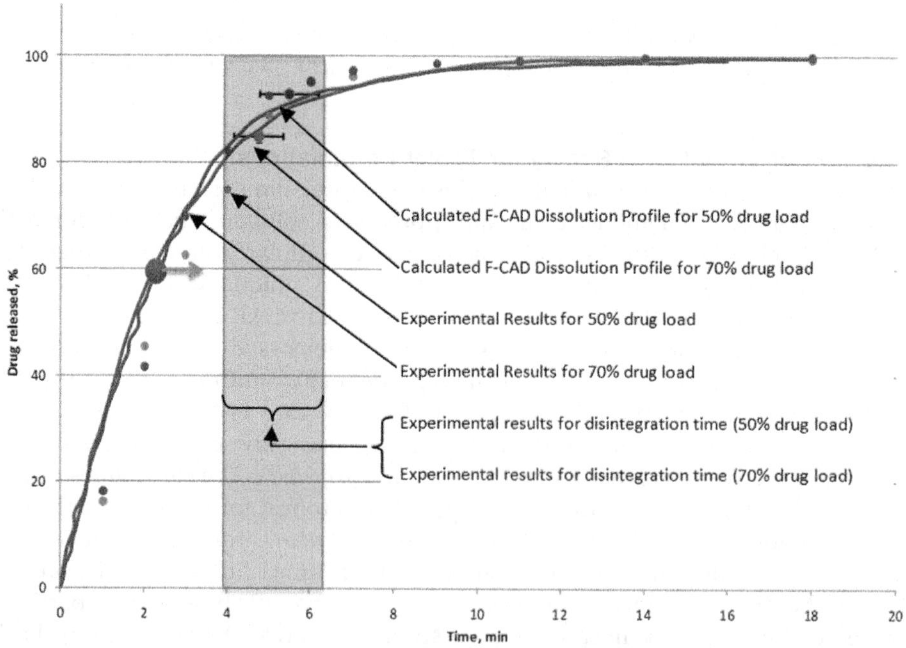

Figure 24 F-CAD calculated dissolution rate profiles for a 50% (w/w) and 70% (w/w) drug load of a robust tablet formulation and laboratory data. The drug substance has a poor wettability, which did not allow a capsule formulation as an early "service dosage form." F-CAD marks in addition the specific time point (*filled black circle*) when the water molecules have reached the geometric center point of the tablet volume. Experimental disintegration times have been observed between ca four and six minutes. *Source*: From Refs. 15 and 30.

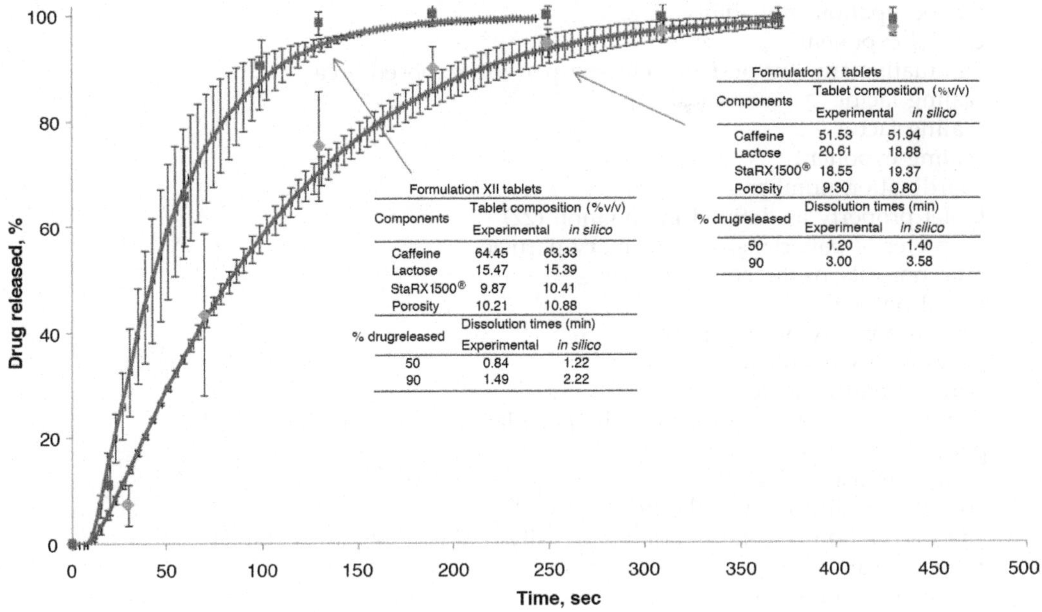

Figure 25 F-CAD calculated dissolution profile including error bars and the corresponding result of the laboratory experiment [PhD thesis, Etienne Krausbauer (15)]. The calculated error bars by F-CAD originate from replicated calculations, taking into account that there is a random mixture of the drug particles within the tablet, that is, that the drug particles do not occupy the identical site in each tablet. *Source*: From Ref. 30.

CONCLUSIONS

The concept of percolation theory, fractal geometry, and the use of virtual R&D reality allow new insights into the physics of tablet compaction and the properties of the tablets. The results attained so far are promising and should stimulate further research in this field.

LIST OF SYMBOLS

A concentration of the dispersed drug in the tablet
b slope of the regression line in the plot of $Q(t)$ versus $\sqrt{t}$
C_s solubility of the drug in the permeating fluid
D diffusivity or apparent diffusion coefficient
D_0 diffusion coefficient of the drug in the permeating fluid
D_1 linear fractal dimension
D_s surface fractal dimension
D_v volume fractal dimension
k exponent (dissolution kinetics)
L size of agglomerate or aggregate
μ critical exponent for conductivity
p percolation probability
p_b bond formation probability
p_c percolation threshold
p_{c1} lower percolation threshold
p_{c2} upper percolation threshold
P deformation hardness (Brinell hardness)
P_{max} maximum deformation hardness at $\varepsilon \rightarrow 0$

p_s	site occupation probability
q	critical exponent
Q	cumulative amount of drug released per unit exposed area
S	scaling factor
S'	scaling factor
T	scaling exponent
z	coordination number
β	tablet property as defined in equation (22)
δ	diameter of particle; size of primary particle
ε	total porosity of the empty matrix
ε_o	initial porosity
ε_c	critical porosity of the matrix
μ	conductivity exponent
ρ_c	percolation threshold
ρ_{max}	maximal relative density where the percolation power law still is valid
ρ_p	poured relative density
ρ_r	relative density
ρ_π	percolation threshold of the pore network
ρ_r^*	site percolation threshold, relative density of the "first" dense tablet
ρ_t	tapped relative density
ρ_o	percolation threshold
σ_t	tensile strength
σ_{tmax}	maximum tensile strength at $\varepsilon \rightarrow 0$
τ	tortuosity of the matrix

REFERENCES

1. Stauffer D, Aharony A. Introduction to Percolation Theory. Revised 2nd ed. London: Taylor and Francis, 1994.
2. Mandelbrot BB. The Fractal Geometry of Nature. San Francisco: Freeman, 1982.
3. Leuenberger H, Rohera BD, Haas C. Percolation theory—a novel approach to solid dosage form design. Int J Pharm 1987; 38:109–115.
4. Leuenberger H, Holman L, Usteri M, et al. Percolation theory, fractal geometry and dosage form design. Pharm Acta Helvetiae 1989; 64:34–39.
5. Leuenberger H, The application of percolation theory in powder technology (Invited review). Adv Powder Technol 1999; 10:323–353.
6. Blattner D, Kolb M, Leuenberger H. Percolation theory and compactibility of binary powder systems. Pharm Res 1990; 7:113–117.
7. Holman LE, Leuenberger H. Powder Technol 1991; 64:233.
8. Leuenberger H, Bonny JD, Usteri M. Proc. Second World Congress Particle Technology, Kyoto, Japan, 1990.
9. Usteri M, Bonny JD, Leuenberger H. Fractal dimension of porous solid dosage forms. Pharm Acta Helv 1990; 65(2):55–61.
10. Leuenberger H, Leu R. Formation of a tablet: a site and bond percolation phenomenon. J Pharm Sci 1992; 81:976–982.
11. Bonny JD, Leuenberger H. Percolation Effects in Controlled Release Matrices. In Proc Intern Symp on Control Rel of Bioact Mater,18 (1991), Controlled Release Society, Inc., 407–408.
12. Bonny JD, Leuenberger H. Matrix type controlled release systems: I. Effect of percolation on drug dissolution kinetics. Pharm Acta Helv 1991; 66:160–164.
13. Stauffer D, Coniglio A, Adam M. Gelation and critical phenomena. Adv Pol Sci 1982; 44:103–155.
14. Smith WO, Foote PD, Busang PF. Packing of Homogeneous Spheres. Phys Rev 1929; 34:1272–1274.
15. Krausbauer E, Puchkov M, Betz G, et al. Rational estimation of the optimum amount of non-fibrous disintegrant applying percolation theory for binary fast disintegrating formulation. J Pharm Sci 2008; 97:529–541 and Krausbauer E. Contributions to a science based expert system for solid dosage form design, PhD thesis, 2007, University of Basel, Faculty of Science.

16. Ehrburger F, Misono S, Lahaye J. Conducteurs granulaires, theories, caractéristiques et perspectives, Journée d'études Oct. 10, Paris, Textes de communication, 1990:197–204.
17. Leuenberger H. The compressibility and compactibility of powder systems. Int J Pharm 1982; 12:41–55.
18. Heckel RW. Density pressure relationship in powder compaction. Trans Metall Soc AIME 1961; 221:671–675.
19. Kuntz M, Leuenberger H. Pressure susceptibility of polymer tablets as a critical property: a modified Heckel equation. J Pharm Sci 88(2):174–179.
20. Higuchi T. Mechanism of sustained action medication: theoretical analysis of rate of release of solid drugs dispersed in solid matrices. J Pharm Sci 1963; 52:1145–1149.
21. Sherrington PJ, Oliver R. Granulation. London: Heyden, 1981:34.
22. Luy B. Vakuum-Wirbelschicht, PhD thesis, University of Basel, Basel, 1991.
23. Leuenberger H, Luy B, Hirschfeld P. Experiences with a novel fluidized bed system operating under vacuum conditions, Proc. Preworld Congress Particle Technology, Gifu, Japan, 1990:113–122.
24. Leuenberger H, Lanz M. Pharmaceutical powder technology—from art to science: The Challenge of FDA's PAT Initiative. Adv Powder Technol 2005; 16:3–25.
25. Maurer L, Leuenberger H. Application of near infrared spectroscopy in the full-scale manufacturing of pharmaceutical solid dosage forms. Pharm Ind 2009; 71:672–678.
26. Maurer L, Leuenberger H. Terahertz pulsed imaging and near infrared imaging to monitor the coating process of pharmaceutical tablets. Int J Pharm 2009; 370:8–16.
27. ICH Q8 pharmaceutical development. Available at: http://www.emea.europa.eu/pdfs/human/ich/16706804en.pdf.
28. Pricewaterhouse Coopers PHARMA 2020 virtual R&D. Available at: http:www.pwc.com/extweb/industry.nsf/docid/705B658C95033AE8852575680022FC75.
29. Hotz D. Swiss Pharma Science Day 2008, Bern, Keynote Lecture "PHARMA 2020: Virtual R&D—which path will you take?" SWISS PHARMA 2008; 30(10):7–8.
30. Leuenberger H, Leuenberger MN, Puchkov M. Implementing virtual R&D reality in industry: n-silico design and testing of solid dosage forms. SWISS PHARMA 2009; 31(7/8): 18–24.
31. Leuenberger H, Leuenberger MN, Puchkov M. Right first time: Computer-aided scale-up for manufacturing solid dosage forms with a shorter time to market. SWISS PHARMA 2010; 32(7/8): 3–13.

4 Postcompaction data analysis techniques

Gary Bubb

There are 10 different types of people, those that speak in binary and those who do not.

INTRODUCTION

Material presented in this chapter covers data collected from an instrumented rotary tablet press. Analysis of data obtained from a single station tablet press or that from a mechanical device designed to either simulate a compression event or produce a specific waveform has or will have been covered in another chapter. There may be some overlap in the first portion of this chapter as individual compression and ejection events are discussed in detail but will be limited to characteristics of the event such as rise, fall, and dwell times. Following the discussion of an individual pulse analysis will be an in depth analysis of compaction and ejection profiles. A compaction profile requires an appreciation of the mechanical strength of the compact so different approaches to determine the tablet strength will be covered. Ejection profiles are extremely useful but often not performed. The choice of a lubricant and the amount used in a formulation has significant affect on both the compact strength and characteristics such as friability, disintegration, and dissolution. Future formulations may be lubricated with an external spray solution injected onto the tooling to either eliminate or minimize internal lubrication requirements. Instrumentation, as critical as it is now will be required for this emerging technology.

SINGLE COMPRESSION EVENT ANALYSIS

The detailed analysis of a compression or ejection event can only be performed provided certain criteria as met. For a compression event analysis the force transducer must be located such that the action line of the force is coincident with the centerline of the transducer. The transducer cannot be remotely located such as on eye bolts, tie rods, or other components. The sample rate to digitize the analog signal from the force transducer must be sufficiently high to capture all of the information, 5000 samples per second or higher is suggested, twice that is recommended. Transducers not meeting the criteria stated can be used for other purposes such as compaction profiles, but not for the type of analysis to follow.

Figure 1 illustrates a single compression event. The three horizontal lines representing 10%, 50%, and 90% of the peak force are shown to visually assist in the following definitions. Note that these definitions are mathematically derived on the basis of the compression trace and, in the instance of dwell times, represent a calculated approximation of a mechanical parameter. Dwell time is the actual time during which the punch tips are at their maximum penetration into the die and is controlled by the flat on the head of the punch as the flat transverses the compression roll. It has been shown that the mathematically calculated values agrees quite closely to the physical values based on tooling head flat, pitch circle diameters, and turret speed.

COMPRESSION WAVEFORM DEFINITIONS

Rise time: time required to go from 10% to 90% of peak force
Dwell time: time from 90% of peak on the rise to 90% of peak on the decay
Fall time: time from 90% to 10% of the curve past peak
Pulse width: time from 50% of the rise to 50% of the decay
Contact time: time from 10% of the rise to 10% of the decay

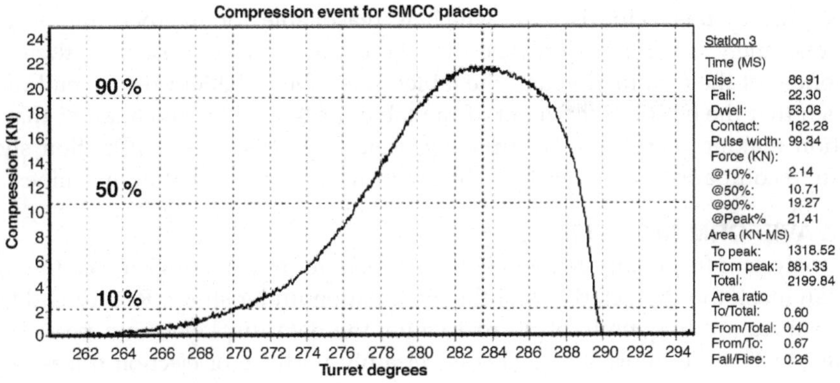

Figure 1 A single compression event.

Station 3	
Time (MS)	
Rise:	86.91
Fall:	22.30
Dwell:	53.08
Contact:	162.28
Pulse Width:	99.34
Force (KN):	
@10%:	2.14
@50%:	10.71
@90%:	19.27
@Peak:	21.41
Area (KN-MS)	
To Peak:	1318.52
From Peak:	881.33
Total:	2199.84
Area Ratio:	
To/Total:	0.60
From/Total:	0.40
From/To:	0.67
Fall/Rise:	0.26

Figure 2 Tabular statistical data.

Most data acquisition systems will automatically calculate these values for you and present them in a tabular form such as shown in Figure 2. Note that the table in Figure 2 presents some other information such as areas to and from peak and certain ratios. These ratios are intended to indicate the degree of brittle verses elastic behavior of the formulation during the compaction. For example, if a material were to be 100% brittle, there would be no elastic re-expansion and the compression curve would immediately drop to zero after the dwell time. Likewise, if a material were to be 100% elastic, the curve would be totally symmetrical on both sides of the peak force. In case of 100% brittle, the area ratio of area from the peak divided by the area to the peak would be zero and the 100% elastic would be 1.0. The value of the ratio of the area from the peak force divided by the area to the peak force will immediately provide some insight into the degree of elasticity of the formulation based on a scale of 0 to 1. For example in Figure 2, this value is 0.67, which is reasonable for microcrystalline cellulose (MCC). It is extremely important to realize that these values are representing the combined elastic component of the formulation and the tablet press. Older tablet presses tend to be very compliant such that the elastic component of the tablet press will be much larger than that of the formulation making these ratios completely useless. At best this information should be used for relative comparison only on data obtained on the same tabletting machine.

EJECTION WAVEFORM ANALYSIS

A detailed analysis of a compression event is open to much criticism as the machine compliance can mask the characteristics of the formulation under study. Real-time displays of compression curves are frequently of no real value, the data displayed in a peak bar graph display being much more informative. This is certainly not true of ejection traces. Real-time displays of ejection are extremely important and can inform the press operator of impending problems before they occur. Figure 3 is a free-body diagram of the forces on a compact within the die during ejection where

$F_{ejection}$ = the measured force required to push the compact out of the die,
F_n = the normal or residual radial die wall force, and
F_f = the frictional force resisting the ejection of the compact.

Basic physics will show that $F_f = F_n\ (\mu)$ where μ is the coefficient of friction (1). It is immediately apparent that the ejection force is a function of the residual radial die force and the coefficient of friction. Residual die wall force is a formulation property controlled by the plasticity of the material while the coefficient of friction is controlled by the lubricant and surface conditions of the die. Lowering either one of these will decrease the ejection force. The initial portion of the trace is the breakaway force, the force required to start the tablet moving. The subsequent portion represents the force required to keep the tablet moving. In addition to the residual radial die wall force, the breakaway force is a function of the static coefficient of friction while the dynamic coefficient resists the continued movement. Figure 4 shows a real ejection profile for a press with only two stations tooled. Based on the high breakaway force relative to the dynamic force there is a red flag even though the actual force values are low. Figure 5 is the same run only several minutes later showing how things have deteriorated based on the unacceptable high ejection forces. The ejection forces have gone from around 300 N to 2000 N. Figure 6 is representative of an ejection trace for a properly lubricated formulation. Note that the breakaway force is much more reasonable.

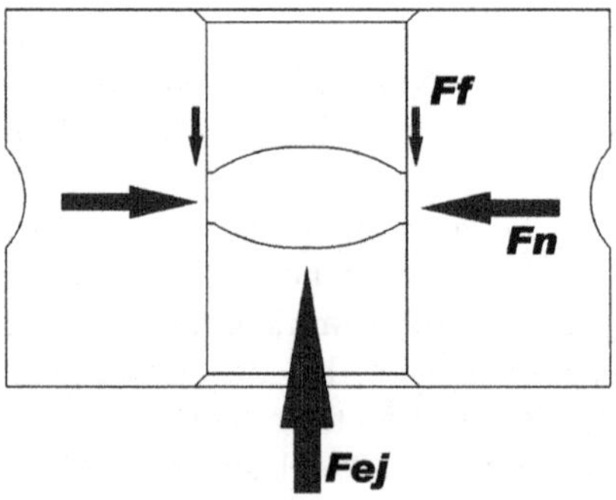

Figure 3 A free-body diagram of the forces on a compact within the die during ejection.

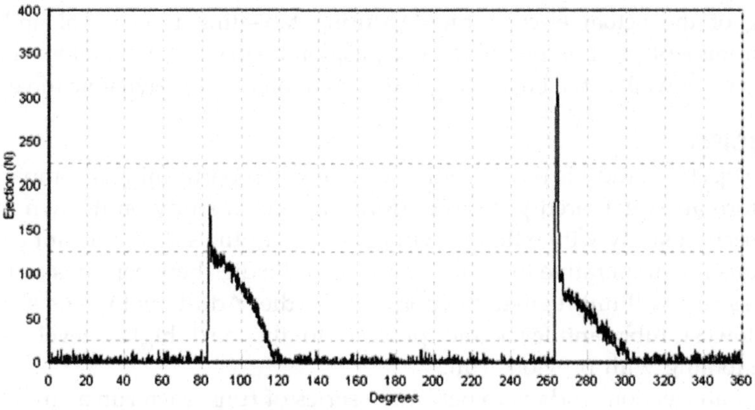

Figure 4 An example of the initial ejection profile.

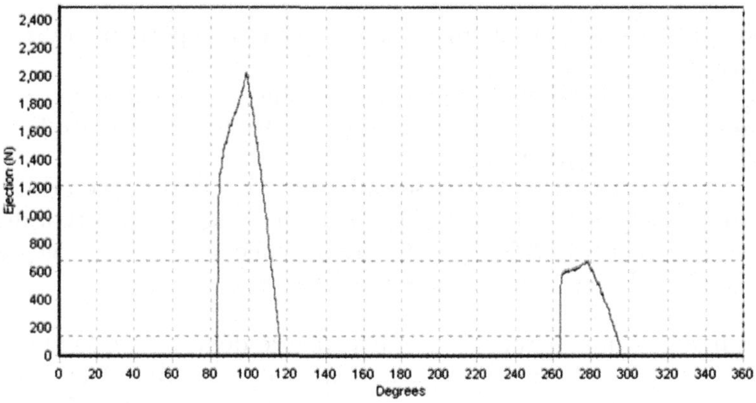

Figure 5 An example ejection profile of a poorly lubricated formulation.

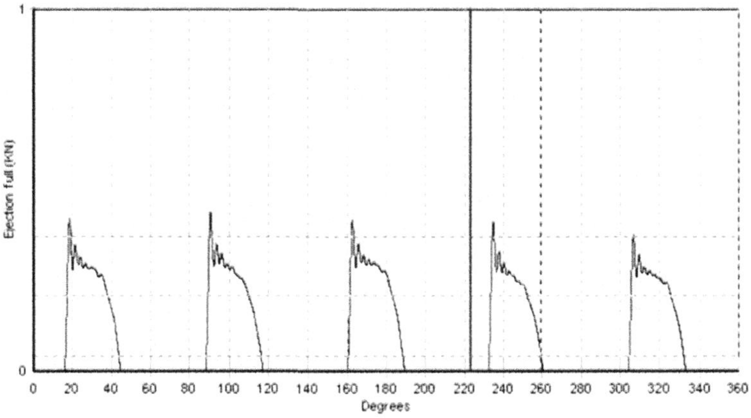

Figure 6 An example ejection profile of a properly lubricated formulation.

Monitoring of the actual ejection profile during tabletting is essential and should be displayed in a prominent position of any data acquisition display. The operator will be looking for the actual values as well as the breakaway forces compared to the dynamic forces of ejection.

EJECTION STUDIES

The beauty of an ejection study is that is comes for free when performing a compaction profile although the information gathered justifies performing such a study on its own merit. When performing an ejection study with either varying amounts of the same lubricant or comparing different lubricants it is imperative to remove and clean the dies between runs as the lubricate from the previous run will impregnate the pores of the die. Additionally, one should always start with the lowest lubricant level and proceed to the next higher level so as to not contaminate the tooling with previous run.

The basis or an ejection study is to perform a series of runs, each run at an incrementally higher force. Based on the resulting graph and desired compression force the ejection force can be determined. Figure 7 shows the results of such a graph. In this example, the number in the parenthesis represents the percent of magnesium stearate used in the formulation. There are several items of interest in this data.

1. Unlubricated MCC displayed erratic ejection forces independent of the compression force level.
2. MCC with just a trace of lubrication, 0.02%, reduced the ejection force to near zero.
3. Dicalcium phosphate with 0.5% lubrication showed increasing ejection force with increasing compression force.
4. The lactose data is really two data sets. The first part of the graph is the result of a container charge of 60% during the blending while the second part of the graph is a container charge of only 20%. It is clear that a smaller container charge dispersed the lubricant better, lowering the ejection force values.

The information shown in Figure 7 is not intended to demonstrate anything other than the usefulness of this type of analysis and as mentioned before the data is automatically obtained during a compaction profile provided the press is instrumented for ejection.

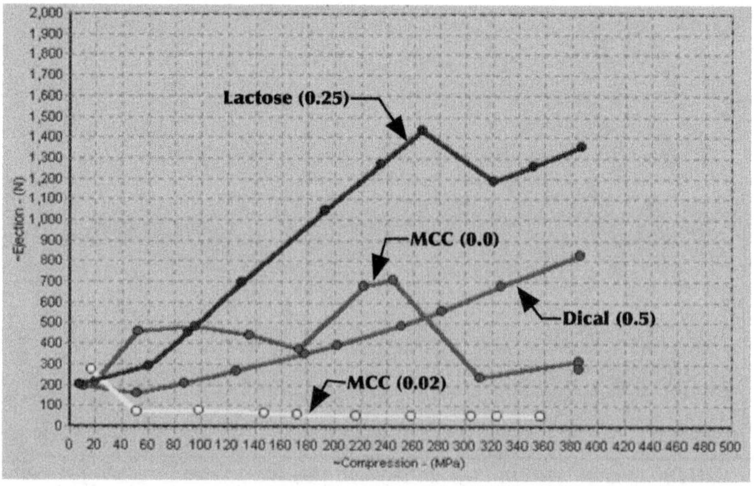

Figure 7 Ejection profiles of compacts made from lactose, microcrystalline cellulose (MCC), and dicalcium phosphate lubricated with magnesium stearate (lubrication concentration values are given in parentheses).

MECHANICAL STRENGTH OF TABLETS

The logical progression for this chapter would be a discussion of compaction profiles, however to do so requires an in depth discussion of the compact strength and ways to test for that strength. Permit me to make a few bold statements at this point. A compaction profile graph must be performed to properly understand the physical characteristics of the compact. The ordinate of that graph should always be in units of pressure. The abscissa should always be tensile strength whenever possible. Data presented as compression force versus hardness is specific to one tooling set and tablet geometric and cannot be related or compared to another compact. Normalizing the data into units of strength and pressure allows comparisons among different tablet sizes and geometries.

Tablets are formed by applying a compression force to a powder and consolidating that power into a compact within a die of a press. This results in a compact that is stronger in compression than it is in tension; therefore, tension is the mode generally tested for. There are several approaches to testing a compact.

- Direct tensile test
- Flexural bending test
- Diametrical compression test
- Work of failure
- Fracture mechanics
- Indentation hardness

DIRECT TENSILE TEST

There are two typical ways to perform a direct tensile test, neither of which is practical for typical pharmaceutical tablets due to the problems of gripping the compact. The first involves machining a dog bone-shaped specimen from the compact shown in Figure 8. Pair of grips attaches to each end of the large portion of the dog bone and the specimen is pulled apart producing a tensile failure. The axial force and displacement are recorded during the test. There are many problems with this approach.

- The dog bone geometry must be machined from the compact resulting in a very small specimen size.
- It is time consuming and expensive to manufacture the specimen making in-process test all but impossible.
- Brittle materials such as pharmaceutical tablets typically fail at the grip attachment or the grip cannot be tightened sufficiently to secure the specimen during the test without slipping.
- Most tensile testing machines are meant for testing larger specimens and are not suited for the small sizes and force involved for this type of application.
- The dog bone geometry that is machined from a compressed compact may not accurately represent the strength of the original compact. Machining can induce residual stress into the dog bone resulting in errors. Additionally, the compact is not homogeneous with regard to the tablet strength.

The second method requires manufacturing a specific set of grips that conform to the tablet geometry and gluing the grips to the compact, leaving only a small portion of the compact

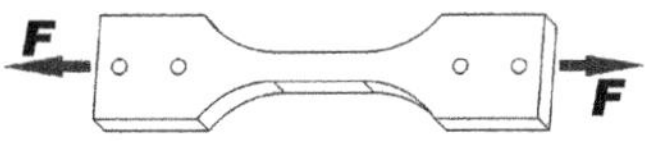

Figure 8 Schematic diagram of direct tensile test involving the dog bone geometry of the specimen.

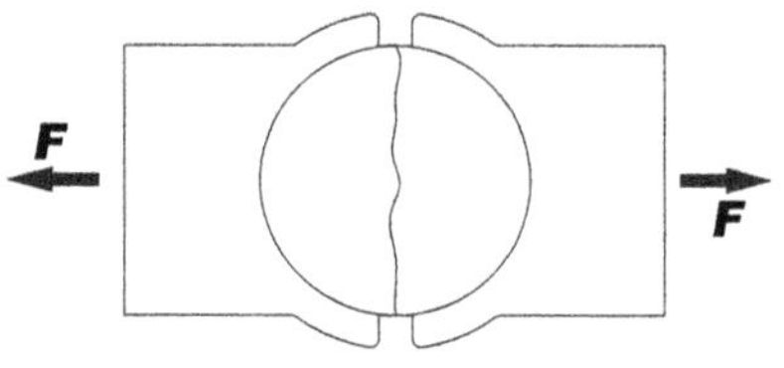

Figure 9 Schematic diagram of direct tensile test involving the gluing of the grips to the specimen.

exposed (Fig. 9). Although this method eliminates some of the problems with machining, the dog bone specimen and grip attachments it still is subject to criticism. The glue line may and probably does reinforce the compact and is still an expensive and time consuming process.

The direct tensile test does offer, however, a simple calculation for the tensile strength of the compact.

The tensile strength is simply the force divided by the cross-sectional area of the specimen.

$$\sigma = \frac{\text{force}}{\text{area}} \tag{1}$$

FLEXURE BENDING TEST

In general, a specimen is machined from the compact in the shape of a rectangular beam. The beam is supported by two rods, one on each end that permits the specimen to rotate freely during the test. A load is then applied to the center of the beam causing a tensile failure opposite the applied force. There are two generally accepted loading conditions; one, a single concentrated load in the center of the beam, the other two equal concentrated loads symmetrically placed at the beam center (2). The single-point load is referred to as a three-point flexure test (Fig. 10) while the two equal concentrated loads are termed a four-point flexure test (Fig. 11). The shear and moment diagram for each loading condition is shown under the loading condition to make visually apparent the advantageous features of the four-point test. At the center of the beam for the four-point loading, the shear is zero and the moment is constant. The disadvantage of the four-point loading is the need for a larger specimen size.

As with the direct tensile test the disadvantages are the requirement to machine a specimen from the tablet and the practicality of performing in-process testing. The precision of the loading fixture and end effects from the support rods will negatively affect the accuracy and reproducibility.

The equations for the three- and four-point flexure test are presented for a rectangular beam specimen. It is conceivable that certain tablet shapes could be used without machining such as rods or even certain capsule designs. The moment of inertia would need to be determined for shapes other than the rectangular geometry presented below.

THREE- AND FOUR-POINT FLEXURE TEST

$$\sigma = \frac{Mc}{I} \tag{2}$$

where

σ = tensile stress,
M = bending moment,
c = distance from the neutral axis to the outer most part of the beam, and
I = second moment of inertia or the area moment of inertia.

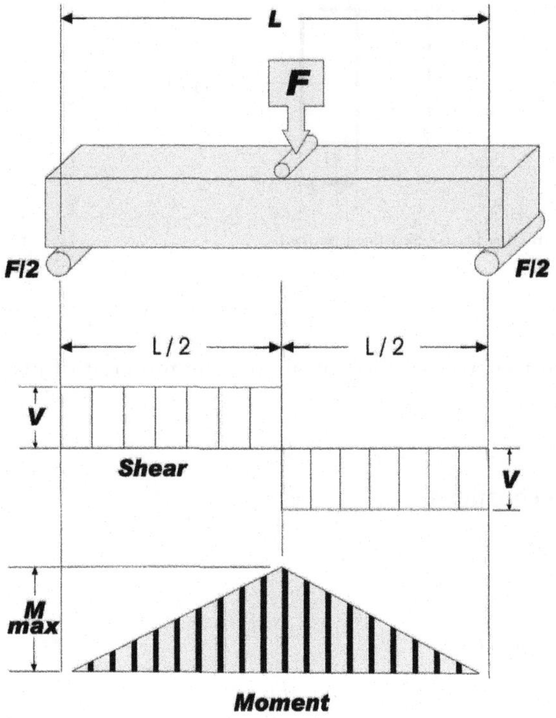

Figure 10 Schematic diagram of a three-point flexure test.

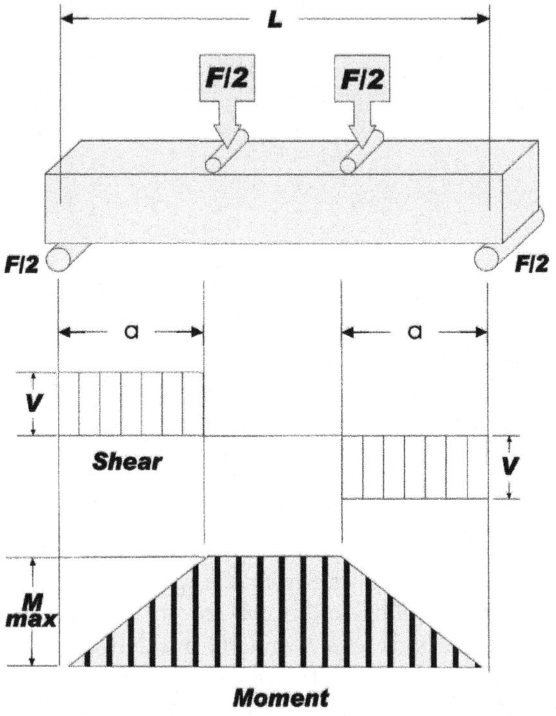

Figure 11 Schematic diagram of a four-point flexure test.

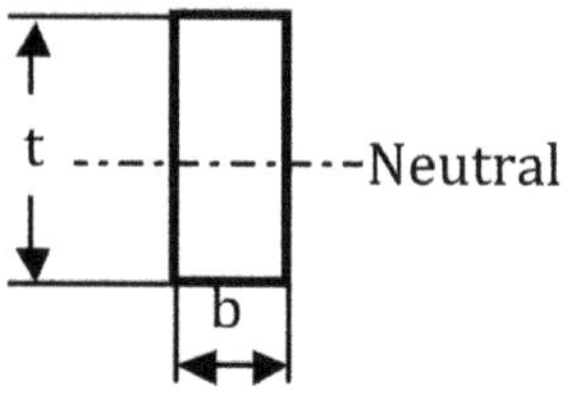

Figure 12 Schematic diagram of a four-point flexure test.

For a rectangle with the cross-section as shown the moment of inertia is (Fig. 12),

$$I = \frac{bt^2}{12} \quad (3)$$

Combining the two equations,

$$\sigma = 6\frac{M}{bt^2} \quad (4)$$

For the Three-Point Flexure Test

$$M = \frac{FL}{4} \quad (5)$$

Therefore,

$$\sigma = \frac{3Fl}{2bt^2} \quad (6)$$

For the Four-Point Flexure Test

$$M = Fa \quad (7)$$

Therefore,

$$\sigma = \frac{3Fa}{bt^2} \quad (8)$$

DIAMETRICAL COMPRESSION TEST

The flexure bending test is generally used with a rectangular form. Circular shapes, which represent a large percentage of tablets produced, represent an unsuitable geometry for either a three -or four-point flexure test. The diametrical compression test has exactly the opposite issue. It generally works very well for circular shapes but is more complicated for other geometries.

There are numerous individuals that can be given credit for determining an analytical solution for the induced stress state resulting from the radial loading of a flat faced circular disk as shown in Figure 13. Barcellos and Carneiro in Brazil and Akazawa in Japan independently developed the equations (eqs. 9 and 10) in the early 1950s.

The equation was also derived by the Russian mathematician Timoshenko and subsequently evolved into an American Society for Testing and Materials Standard, ASTM C-496 primarily used for testing concrete cores.

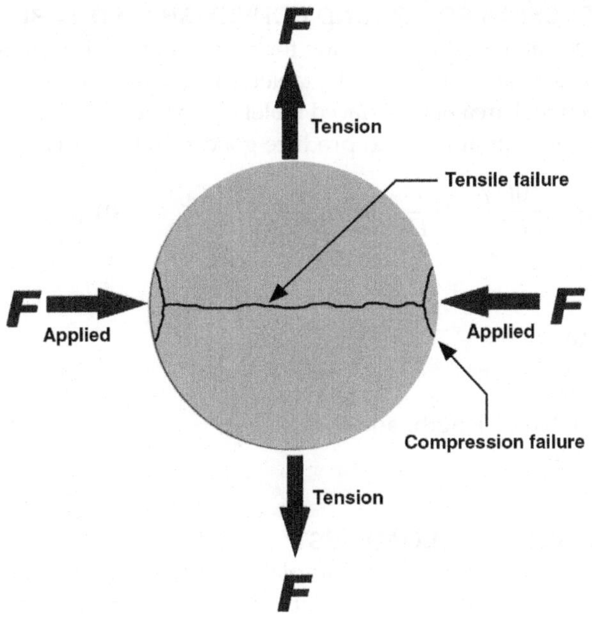

Figure 13 Diametrical compression test.

The equations for the tensile and compressive stress based on the loading as shown in Figure 13 are

$$\text{Tensile} \quad \sigma = \frac{2F}{\pi \phi t} \tag{9}$$

$$\text{Compression} \quad \sigma = \frac{6F}{\pi \phi t} \tag{10}$$

It can be seen from the equations above for a tensile failure to occur the compact must be at least three times stronger in compression than in tension. Therefore, the solution for the tensile strength can only be used provided that the failure is actually a tensile failure as shown in Figure 13. Localized compressive failure at the loading plates is acceptable provided that it represents a small percentage of the specimen diameter. A value of less than 10% is generally considered a reasonable amount of localized compressive failure for the equation to still apply. It is also common for a localized shear failure to occur at the loading point as shown in Figure 14.

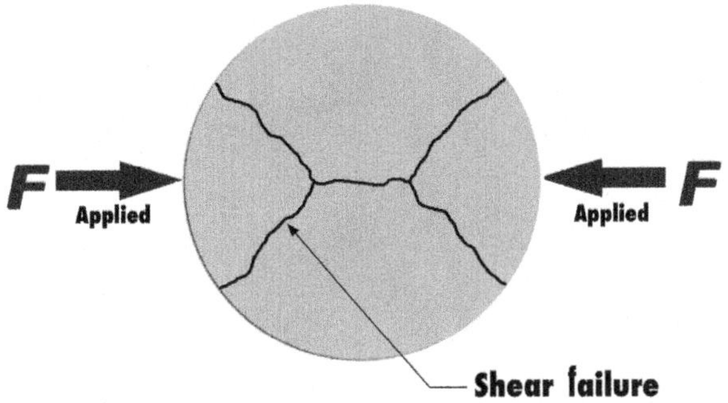

Figure 14 Diametrical compression test: shear failure.

DIAMETRICAL COMPRESSION FOR ROUND CONVEX-SHAPED TABLETS

Newton and Feld produced an empirical equation for round convex-shaped tablets with the help of photoelasticity (3). The central thickness of the tablet is measured and the equation corrects for the difference in cross-sectional area of a flat faced tablet compared to convex shape. The equation has no basis in elastic theory but appears to produce good results and is well accepted.

$$\sigma_t = \frac{10F}{\pi D^2}\left(2.84\frac{t}{D} - 0.126\frac{t}{C_L} + 3.15\frac{C_L}{D} + 0.001\right)^{-1} \tag{11}$$

where

σ_t = tensile strength,
F = breaking force,
D = tablet diameter,
C_L = cylinder (bellyband) length, and
T = tablet thickness.

ISSUES WITH THE DIAMETRICAL LOAD TEST

- The shape and material of the loading platens is critically important to the outcome of the result for the diametrical compression test. Using a softer material for the platens or creating a concave semicircular platen has been shown to decrease the test results variability, probably due to reducing the amount of localizing compression and shear failures at the platen. The disadvantage is the need for custom platens for each tablet diameter.
- The loading rate must also be kept constant to obtain consistent results. It has been shown that some formulations benefit from a faster loading rate while others suffer. The main point is to be consistent.
- The compliance of hardness testers vary from manufacture to manufacture and will produce different results. Some models have a platen that is part of a cantilever beam with strain gages attached to measure the load while others have a load cell as part of the platen. The cantilever beam designs are much more compliant and tend to shatter the compact due to the stored energy from the deflection of the beam that is released when the tablet fails. Figure 15 shows the results of tablets that were manufactured at the same time and tested on three different hardness testers. The tablet on the left was tested on a unit that had a cantilever design and completely destroyed the tablet. The

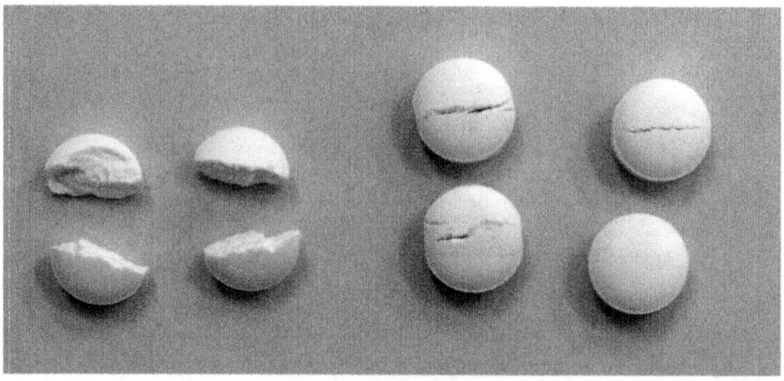

Figure 15 Results of tablets that were manufactured at the same time and tested on three different hardness testers.

tablet on the right barely fractured the top portion of the compact. The numerical readouts also varied by as much as 15% from model to model.

- Work of failure has been used as a useful tool to measure the ductility and toughness of the compact. In this case, the force and axial displacement are both measured; the area under the force-displacement curve being referred to as the work of failure. Compacts with a high work of failure were considered to deform plastically under load while brittle materials required only a small amount of axial loading to induce failure. The problem is that the compliance of the hardness tester itself can overwhelm the material properties of the compact. Work of failure studies must be performed using a rigid hardness tester and all results should be reported using only that model.
- Tablets age and properties change with time. Therefore, the big question is when to test. Properties of some materials start to change immediately after exiting the die and will continue to do so for many hours. In-process testing is frequently performed immediately by selecting tablets directly from the tabletting machine for geometry, weight, and hardness. Shift changes can easily delay this operation by fractions of an hour or even the follow day. Some scientists argue all tablets should age at least 24 hours prior to testing. If this is true then two sets of acceptance criteria will probably be needed, one for the 24-hour ageing and one for immediate acceptance, as it is not practical to wait 24 hours to determine if acceptable tablets are being made.

INDENTATION HARDNESS

Hardness is one of the most commonly used terms in the pharmaceutical industry; and it is used incorrectly most of the time. Hardness is the resistance of a material to indentation, scratching, abrasion, or cutting (4,5). A hardness test is to determine the relative hardness of materials ranging from rubber to hardened steel and is generally expressed in scales of rockwell, brinell, mohs, shore, or durometer. There are more than 30 different tests that exist for determining a material's hardness, and there will be no attempt to cover even one of them in detail in this chapter. In general, the two most common techniques involve either a scratch or an indentation test (5). In the scratch test, a cutting point of a fixed shape is drawn across a surface while under a constant load and the width of the scratch measured. In the indentation method, a constant force is applied to a round or triangular hardened point and the resulting indentation measured. Figure 16 shows a representative test using a spherical ball along with the equation to calculate the hardness (5). This is shown such that the reader has an overall appreciation of hardness and why the term hardness when used in conjunction with a diametrical compression is the wrong use of the term.

FRACTURE MECHANICS

There are many ASTM tests for determining the fracture toughness of a material. Some measure the relative ductility or brittleness of a material, others the propensity for an initial crack or known defect to propagate. The merit of applying these tests to pharmaceutical compacts can be useful in a basic understanding of the types of failures possible, but have little practical use in everyday use. Davies and Newton, Hiestand, as well as Rowe and Roberts have performed excellent experiments for those wishing to investigate this area in detail.

PRACTICAL STRENGTH TESTING AND NORMALIZATION

Data obtained from diametrical compressions tests is probably the single most beneficial information we have readily available today. Given the issues outlined earlier this method offers the most accurate and reproducible results for in-process testing. It has sufficient sensitivity to identify differences among tablets and is quick and easy to perform. It can,

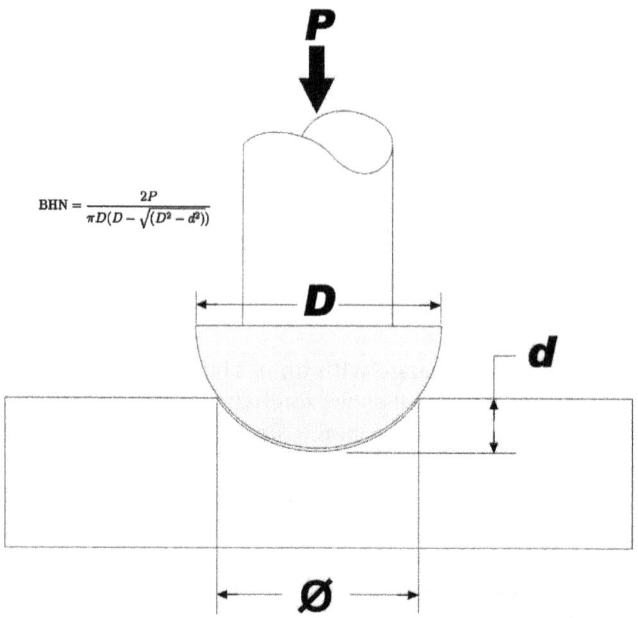

Figure 16 A schematic indentation test using a spherical ball along with the equation to calculate the hardness.

however, be improved upon. Instruments used, aka hardness testers, should be standardized for loading rate, compliance, and platen shapes. A calibration technique that applies a static load to the platen containing the load cell is a good start but not sufficient as the compliance of the instrument is not considered as demonstrated in Figure 15. Consideration should be given to standardized aged compacts that will fail at consistent and known loads. Colored dies could be used to identify the specific breaking force for that color compact.

THE CASE FOR NORMALIZATION

Breaking force data obtained from the tablet hardness tester is generally plotted as a function of the compression force applied to make the tablet. This is beneficial for one size compact only and then with limitations as the compact thickness changes with each compression force change. Changes in tablet thickness and diameter for circular tablets can be accounted for by using tensile strength in place of breaking force. Plotting the results in this fashion is a huge improvement, but true normalization is not realized until the applied compression force is also normalized into compaction pressure. Presentation of the results in this fashion has several advantages.

Tensile strengths become meaningful all by themselves without regard to the tablet size. For example, making the statement that a specific hardness is required to pass a friability test or survive a coating operation is not universally true and applies to one specific size only. Normalizing the data will remove that barrier. If it is determined that a particular tensile strength is required for friability then that value can be a target independent of the tablet size. One need not have multiple design limits for each tablet provided the results are normalized properly. Figures 17, 18, and 19 illustrate this point.

Figure 17 represents three different dose proportional tablet weights; 75, 150, and 300 mg of the same formulation plotted in the traditional method, breaking force versus compression force. This is a useful plot but the results are dependent on the tablet diameter and thickness.

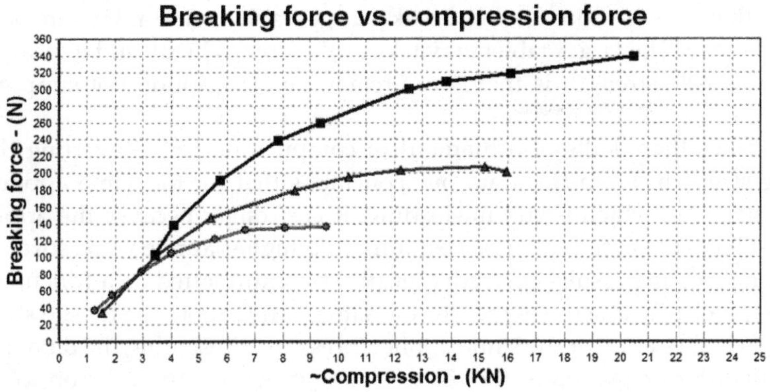

Figure 17 Breaking force versus compression force profiles of three different dose proportional tablet weights: 75, 150, and 300 mg of the same formulation.

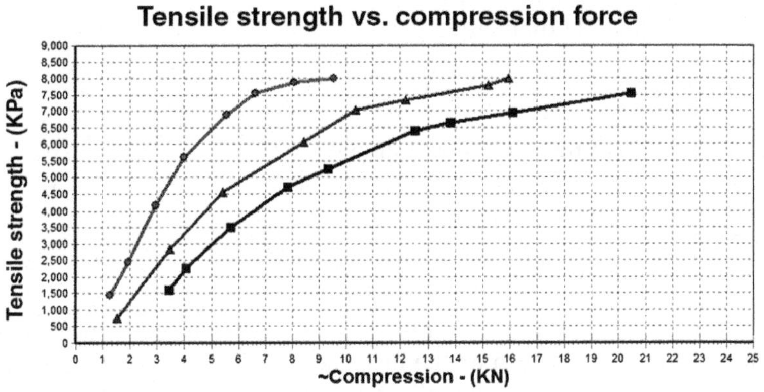

Figure 18 Tensile strength versus compression force profiles of three different dose proportional tablet weights: 75, 150, and 300 mg of the same formulation.

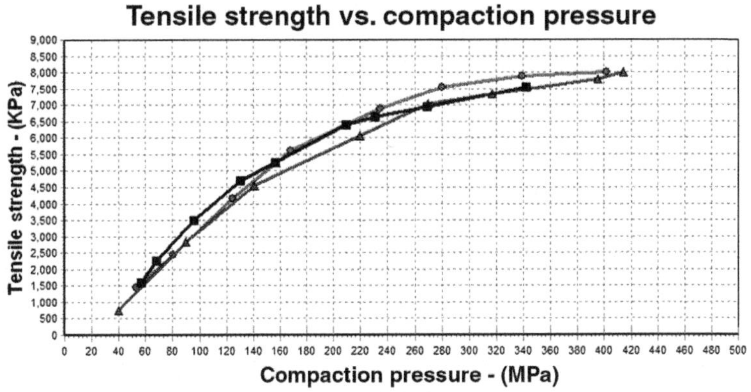

Figure 19 Tensile strength versus compression pressure profiles of three different dose proportional tablet weights: 75, 150, and 300 mg of the same formulation.

The conclusion is that it takes more force to break a larger tablet made at the same compression force than a smaller one. The larger tablet may take more force to break and logically one might conclude that the tensile strength is therefore greater, an assumption not supported by this data presentation.

Figure 18 shows the results of a diametrical compression test converting the breaking force into tensile strength. From this data one concludes that for the same applied force the smaller tablet is stronger, which is still misleading. It is stronger because the applied force is acting over a smaller cross-sectional area, resulting in a higher applied pressure.

Figure 19 presents the results correctly, tensile strength plotted as a function of applied pressure. This graph clearly illustrates that for the same formulation, the tensile strength of the various size tablets is the same for the same applied pressure. Presenting the data in this manor will allow a legitimate comparison of the relative strength of one formulation to another.

STRAIN RATE ANALYSIS

Another valuable post analysis tool is a strain rate study. Almost all formulations show reduced mechanical properties with increased loading rates and shorter dwell times.

Factors affecting the dwell times are as follows:

- Tablet press pitch circle diameter
- Turret speed
- Tooling head flat

Factors affecting the loading rate are as follows:

- Tablet press pitch circle diameter
- Turret speed
- Compression wheel diameter

A strain rate study will keep the compression force constant while increasing the turret speed. This is quite difficult to perform as the increase speed will reduce the amount of time the lower punch is under the feeder potentially causing a less than full die fill; resulting in a lowered compression force. The lower compression force will mean lower tablet mechanical strength. The reduction in mechanical strength can wrongly be attributed to the increased loading rate when in fact it may have been caused by the reduced compression force. For that reason small variations can be compensated for by plotting the ratio of tensile strength/ compaction pressure as a function of the loading rate. Figure 20 shows such a relationship.

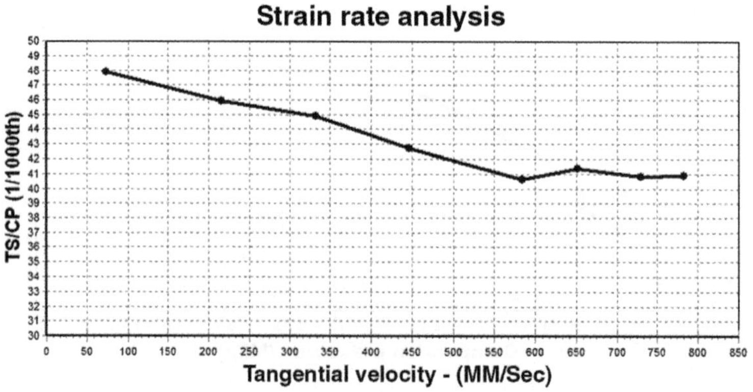

Figure 20 Strain rate analysis.

REFERENCES

1. Shigley JE. Stress Analysis. McGraw Hill Series in Mechanical Engineering. New York: McGraw-Hill Book Company, Inc., 1963:20, 29.
2. Davis PN, Newton MJ. Mechanical Strength. Pharmaceutical Powder Compaction Technology. New York: Marcel Dekker, Inc., 1996:167.
3. Davis PN, Newton MJ. Mechanical Strength. Pharmaceutical Powder Compaction Technology. New York: Marcel Dekker, Inc., 1996:171.
4. Shigley JE. Selection of Materials. McGraw Hill Series in Mechanical Engineering. New York: McGraw-Hill Book Company, Inc., 1963:112.
5. McGregor CW. Mechical Properties of Materials. Standard Hand Book for Mechanical Engineers. 7th ed. New York: McGraw-Hill Book Company, Inc., 1967:5–15.

SUGGESTED READING

Stidel RF. Mechanics of Solids. Standard Hand Book for Mechanical Engineers. 7th ed. New York: McGraw-Hill Book Company, Inc., 1967:3–12.

American Institute of Steel Construction, Inc. Beam Diagrams and Formulas. 6th ed. New York: American Institute of Steel Construction, Inc., 1964:2–122.

5 Tablet press instrumentation in the research and development environment

Gary Bubb

You ask a data acquisition system to tell the truth, the whole truth and nothing but the truth.
Peter Stein

INTRODUCTION

A chapter on tablet press instrumentation could go into tremendous detail on how to design a data acquisition system including selection of the sensors, power supplies, amplifiers, antialiasing filters, analog to digital (A/D) conversion, and the presentation and analysis of the data along with a machine-human interface. Each of the above items could be a unique chapter and in totality a complete text. However, this chapter will not address those topics individually because it is my experience that, unlike in the past when companies were forced to develop in-house systems, today every tablet press manufacture offers a data acquisition system.

The features of a data acquisition system are integral to the mechanical features of a specific model tablet press that purchase decisions are frequently made on the software features.

Tablet press instrumentation discussed in this chapter will be limited to that of force and displacement. Other parameters such as vibration, noise, and temperature can be meaningful but are not commonly used in the research and development arena. The same is true for the measurement of punch pull-up and pull-down forces and tablet press control systems.

OVERVIEW OF A DATA ACQUISITION SYSTEM

Although there are many components that make up an instrumentation system, they will be grouped into five major categories for the purpose of this discussion. Calibration is technically not a component of the system its importance is so significant that it has been included.

1. Sensor types
   a. Piezoelectric
   b. Strain gage
      i. Wheatstone bridge
      ii. Temperature compensation
      iii. Bridge balance
   c. Displacement
2. Signal conditioning
   a. Power supply
   b. Differential amplifier
3. Analog to digital conversion
   a. Resolution
   b. Aliasing filters
4. Representative tablet press sensors for compression, ejection, and takeoff
5. Calibration
   a. Precision, accuracy, and repeatability
6. Analysis software

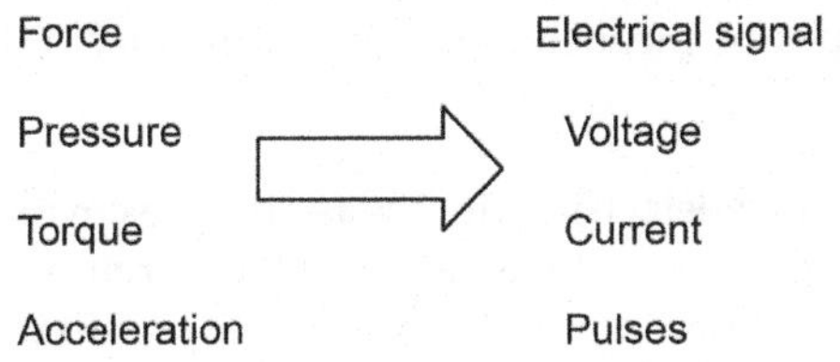

Figure 1 Physical parameter to electrical signal.

SENSOR DEFINITION

In the broad sense a sensor or transducer is a device that transforms one type of energy into another. By this definition, a battery is a transducer (the conversion of chemical energy into electrical). Narrowing the definition to a specific class of transducers, electromechanical, a transducer is a device that converts a physical parameter into an electrical signal that can be measured and or recorded.

Examples of a sensor or transducer are illustrated in Figure 1.

DISCUSSION OF SENSORS FOR FORCE MEASUREMENTS ON A TABLET PRESS

There are two generic types of sensors that have been used for the measurement of compression and ejection forces, piezoelectric and strain gage based. Piezoelectric were the early favorite due to their small size, large self-generating output, and high-frequency response. A drawback to this type of sensor is the low-frequency response allowing its use only in dynamic events. Signal changes due to cable movement and contamination within connectors also are problematic. These could be overcome by carefully routing and anchoring cables but the low-frequency response presents a challenge for calibration. Typically calibrations are performed by gradually applying a force, holding it for several seconds to allow the signal to decay to zero, and then rapidly removing the force. This procedure actually performs a negative force calibration relaying on the belief that a positive and negative calibration were equivalent.

The strain gage–based transducer offers the advantage of a static or DC response. That is to say an applied force will continue to be displayed properly independent of the application time. A piezoelectric sensor will "bleed down" to a zero reading in some seconds, even if the force is still being applied. Additionally a well-designed stain gage–based transducer is an order of magnitude more accurate. For these reasons the strain gage–based transducer has dominated the measurement of forces in the pharmaceutical industry.

PIEZOELECTRIC LOAD CELLS

Piezoelectric force transducers are generally constructed of quartz or piezoceramic elements. The crystal produces an electrical output when experiencing a change in load. The general believe is that they cannot be used for static measurements, their use being limited to dynamic events only. However, this is a misconception. Quartz transducers, paired with appropriate signal conditioners, can offer excellent quasistatic measuring capability (1,2).

Anyone wishing to utilize a piezoelectric force transducer should contact the manufacture of the device for directions. Mounting is extremely important as off center loading can cause large errors. Time constants must be considered. If the load application is slow the peak value will be understated and the return to zero will overshoot the baseline. The signal conditioning must match the sensor impedance (see later in the chapter) and should be tailored to the application. Used properly, piezoelectric force transducers are rugged, accurate devices that are small in size and generally easy to install.

There are two basic types of piezoelectric force transducers, low impedance and high impedance.

- *High impedance*: The piezoelectric effect was first discovered by Pierre and Jacques Curie in 1880. When the element was distorted a current was produced. To relate the current to the deformation a special amplifier is required; a charge amplifier. This system offers the user the most flexibility. Time constants can be made longer allowing easy short-term static calibration. Because they contain no built-in electronics, they have a wider operating temperature range. However, they do come with some significant disadvantages. Because of the high impedance any changes in the resistance or capacitance of the connections between the quartz element and the charge amplifier will likely cause a false signal. Special impedance cables must be used and all connectors need to be free on any contamination. Even the oil from ones fingers is sufficient to cause problems.
- *Low impedance*: Transducers of this type are the same in their construction with the addition of a built in amplifier. This will increase the size of the transducer and limit the temperature range due to the internal electronics but will eliminate the concerns with cable movement and connector contamination. Low-impedance transducers can be used with general purpose cables in environments where high humidity/ contamination could be detrimental to the high insulation resistance required for high-impedance transducers. Also, longer cable lengths between transducer and signal conditioner and compatibility with a wide range of signal display devices are further advantages of low-impedance transducers.

STRAIN GAGE

The strain gage is the basic element in the construction of a strain gage load cell or transducer. There is a common misconception that a quality strain gage load cell is merely installing four strain gages into a Wheatstone bridge and performing a calibration. This is far from the truth. A proper load cell consists of a designed spring element, proper installation of strain gages onto the mechanical spring element, temperature compensation for no load and full load conditions along with a calibration performed after installation into the machine.

Strain gage–based load cells are used by the National Institute of Standards and Technology (NIST) as primary standards for force measurements due to their accuracy, repeatability, and robustness. With today's technology the life expectancy of strain gage–based load cell should approach 25 to 50 years depending on the environment.

There have been many in-house designed instrumentation systems that served the pharmaceutical industry well in the past, some better than others. Because the strain gage–based load cells are by far the dominant sensor on modern tablet presses and the quality of the installations varies widely, there will be a significant discussion on this area.

STRAIN, THE DEFINITION

There are two definitions of strain, true strain, and engineering strain. For all practical purposes in the design of load cells, they are identical as the deformations are so small.

True strain = change in length divided by the current length
Engineering strain = change in length divided by the original length (Fig. 2)

When any item undergoes stress there is a resulting strain, the magnitude varies with the elastic modulus or Young's modulus of elasticity.

- True strain = $\delta L/L$ actual
- Engineering strain = $\delta L/L$ original

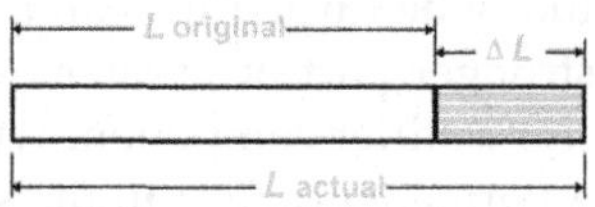

Figure 2 Definition of strain.

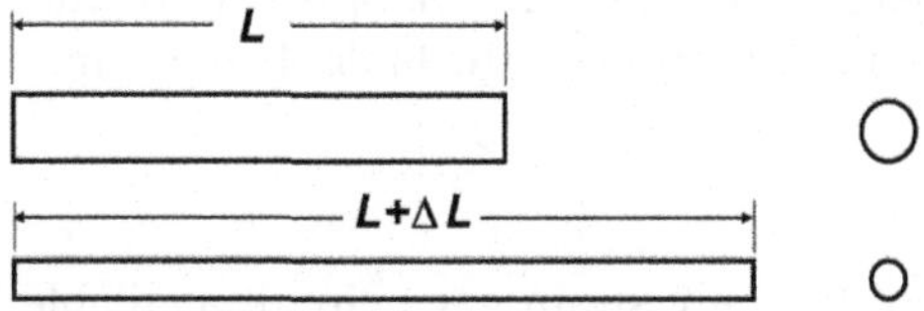

Gage factor = $(\Delta R/R)$ / $(\Delta L/L)$

Picture either a hose or a wire being elongated, either way the resistance will increase as it gets longer and smaller in diameter.

Figure 3 Strain and resistance change.

Picture the image on the left as a length of copper wire. When stretched, the wire becomes longer and smaller in diameter, both contribute to an increase in the resistance of the wire (Fig. 3).

STRAIN GAGES, THE HISTORY

The exact discovery of the strain induced resistance change of electrical wires is not clear; Lord Kelvin did report on the effect in the 1800s. The initial wire strain gage utilized small holes drilled into the part under test at a given distance apart. Small posts were then inserted into the holes and a wire wrapped around the posts. As the part underwent strain the resistance change of the wire was measured and correlated to the strain.

In the 1950s, printed circuit technology gave birth to the bonded foil strain gage. The foil quickly supplanted the wire with better heat dissipation, reduced creep, and much greater design flexibility. Today there are more than 20,000 different patterns using specialized alloys and shapes to assist the strain gage transducer designer.

The two other strain gage types that deserve attention are discussed below.

SPUTTERED OR DEPOSITED METALLIC STRAIN GAGES

Metal films can be vaporized and sprayed onto an electrically insolated surface and used as strain gages. By proper masking the desired strain gage pattern can be deposited directly onto the surface. In this manor, multiple gage patterns can be sprayed at once (3). There are several advantages to this approach; elimination of an organic adhesive and low cost high production rates. The disadvantage at this time is high set up cost and generally lower performance than achievable with rolled alloy foils.

SEMICONDUCTOR STRAIN GAGES

Semiconductor strain gages are generally small silicon chips that have been preferentially cut on a specific silicon crystal axis. Depending on the cut direction the sensitivity can be up to 80 times higher than a typical foil gage. The small size and high sensitivity make them ideal for miniature high-output transducers.

The disadvantages are a high sensitivity to temperature, inability to dissipate heat produced from the excitation voltage, and a reduced linearity, especially at higher strain levels. One of these negative factors can actually be turned into an advantage as designing a spring element for a lower strain means a stronger part or greater overload rating before structural failure would occur. This also makes for a stiffer component with a resulting higher-frequency response. An overload will result in a permanent offset in the strain circuit however not likely to cause structural failure of the component part and possibility taking a machine out of service.

Semiconductor strain gages are ideal for tablet press transducers such as takeoff, scrape off, knockoff, or whatever name you apply to the tablet being removed from the lower punch tip after ejection.

WHEATSTONE BRIDGE

The Wheatstone bridge is not the only strain gage circuit available but is certainly the most commonly accepted for use in industry. It is excellent for use with multiple gage installations and measurements of both static and dynamic events
where

$+E$ is the positive excitation voltage to the circuit,
$-E$ is the negative excitation voltage to the circuit,
$+$ signal is the positive voltage output from the circuit, and
$-$ signal is the negative voltage output from the circuit.

On the basis of Figure 4 and making the initial assumption that all four resistors, wire and wire connections are exactly the same resistance values within each arm or leg of the Wheatstone bridge; the voltage potential at the signal corners would be zero. The beauty of this simple circuit is that even with a large applied excitation voltage the differential voltage at the signal corners is still zero. Therefore, even very small signal changes can be amplified without bias from the excitation voltage. Amplifier gains in excess of 10,000 today show excellent linearity and frequency response, making this circuit extremely sensitivity to minute changes in resistor values.

Let us say that the resisters are strain gages. As pointed out earlier, a wire or foil under a positive strain (tension) will increase in length and decrease in diameter, resulting in an increase in resistance. A compressive force will decrease the wire length, increase the diameter, and lower the resistance. Let us assume for the moment that the strain gage in arm 1 goes into tension resulting in an increase in resistance. The current in the circuit will always take the path of least resistance; therefore, more current will flow through arm 2 and less through arm 1, causing a higher voltage potential at the junction between arms 2 and 3 than the junction of arms 1 and 4. For that reason, the junction between arms 2 and 3 is called the positive signal for

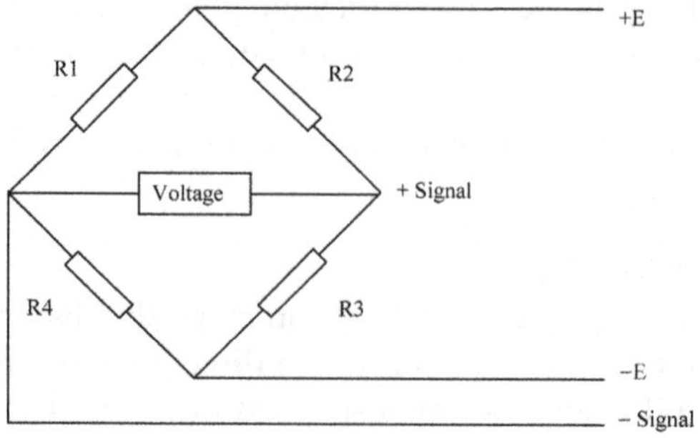

Figure 4 Wheatstone bridge.

this arrangement. Following the same logic if the strain gage in arm 2 went into compression, it would produce the same positive potential as arm 1 going into tension. The same discussion can be offered for arms 3 and 4.

The conclusion to all of this is that an increase in resistance of either arm 1 or 3 will cause a positive output in the circuit while a decrease in resistance in arms 2 and 4 will also cause a positive signal. For this reason, arms 1 and 3 are referred to as the positive arms, while arms 2 and 4 are called the negative arms. The term bridge factor is an expression of the number of equivalent active arms in the circuit. For example, if only arm 1 contained a strain gage that actually saw a strain, the bridge factor would be 1. If the strain gages in arms 1 and 3 saw tension and the strain gages in arms 2 and 4 saw an equal amount of compression, the bridge factor would be 4.

STRAIN GAGE TRANSDUCER CONCEPTS

The well-designed transducer needs to be linear with minimal hysterias, sensitive, exhibit good thermal stability, and have a good return to zero under a no load condition. Additionally, the *transducer should only respond to the force to be measured* and not to any other force or physical parameter. The choice of materials to manufacture the transducer from will be a consideration as well as the design of the spring element and the area where the strain gages will be attached. If the physical design of the transducer is not well thought out the sensor will not perform as hoped. The following simple examples are shown to demonstrate the principle, not an actual design concept (Fig. 5).

The two gages on the top will experience tension as the beam is deflected; therefore, one gage should be installed in arm 1; the other in arm 3 of the Wheatstone bridge. Provided that the other two arms contained only resisters and not strain gages the bridge factor would be 2.0. However, if two additional strain gages were installed on top surface perpendicular to the other two, they would see only Poisson's ratio of the full strain, or 0.3. Therefore, the bridge factor would be 1 + 0.3 + 1 + 0.3 or 2.6. Now if the two strain gages on the bottom that see compression were installed in arms 2 and 4, the bridge factor would be 4. To make a proper transducer the length and thickness of the beam would be designed to provide the desired stress and resulting strain for the material the beam is made of.

There are hundreds of unique transducer concepts that have been utilized for force applications. The roll pin concept for compression force was introduced into the pharmaceutical industry in the early 1980s (4). Prior to that time compression forces on a rotary tablet press were measured with strain gages installed on structural tie rods or eye bolts. Wheatstone

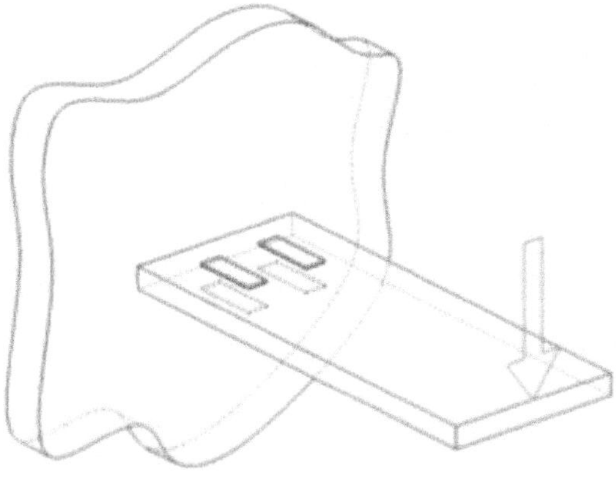

Figure 5 Cantilever beam.

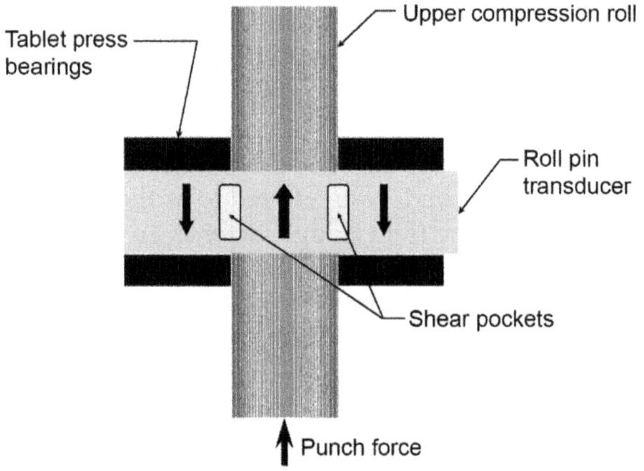

Figure 6 Roll pin shear load cell.

bridges were applied but no additional consideration was given to the spring element design or temperature compensation. To this day many transducers manufactured for the Pharmaceutical Industry are not properly temperature compensated. The load cell roll pin is a good example of a proper design. The sensor is physically close to the force to be measured, the action line of the force is coincident with the load cell, the bridge factor is 4, and it can easily be temperature compensated.

The roll pin load cell replaces the existing roll pin in this application while keeping all of the original functionality, including lubrication. Shown in Figure 6 is a representation of an upper roll load cell. The upper punch is exerting a force on the compression wheel that is being transferred to the center of the roll pin. The pin then transfers the force through the shear pockets to the ends of the pin and finally into the structural support of the machine. In this instance, a compression force is converted into a shear force for the purpose of making a transducer. The shear pocket geometry is conceived to produce the desired sensitivity for the anticipated forces (Fig. 7).

The shear pocket on the left is not under load. The shear pocket on the right is an exaggerated picture of how the real distortion would look. With the strain gage mounted at a 45° angle the strain is positive in this pocket. By carefully choosing the correct strain gage orientation for each of the four pockets, a bridge factor of 4 is obtained and the roll pin responds only to the desired force. One must be careful here as there are three possibilities on how the gages are positioned and only one is correct.

1. The load pin reacts only to the compression force.
2. The load pin reacts only to the torque in the pin from the compression wheel turning.
3. The load pin reacts to both the torque and compression force.

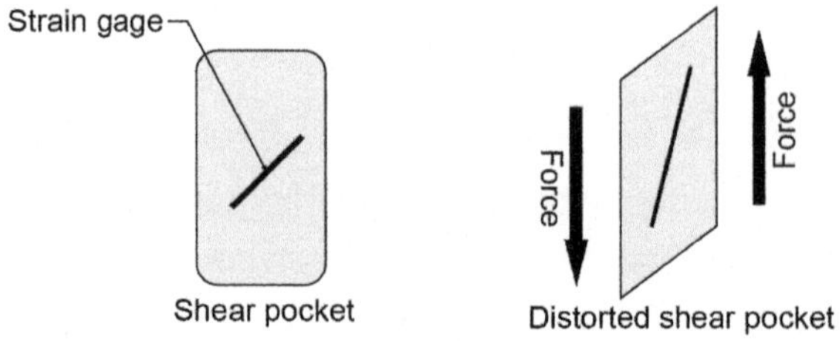

Figure 7 Strain in roll pin transducer.

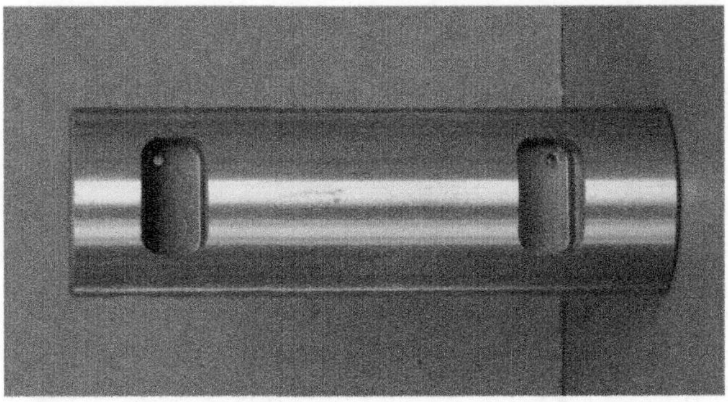

Figure 8 Ungauged roll pin.

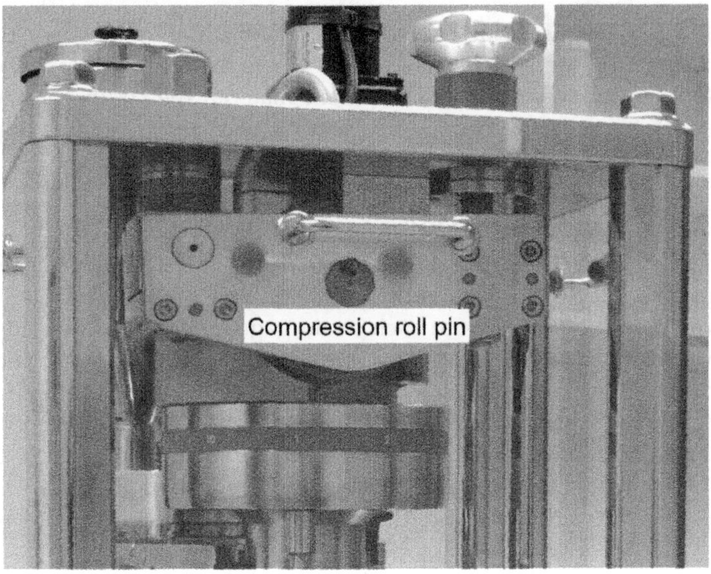

Figure 9 Roll pin transducer in tablet press.

Number three is the most insidious as it will not show up during a calibration with only an axial load applied; however, will yield incorrect information during operation due to the tensional component. A check is to try to rotate the compression quickly without applying an upward force and see if the load cell produces any output. Remember that the torsion affect will be much greater under a compressive force so any output observed no matter how small is a good indication of an improperly installed or wired set of strain gages (Figs. 8 and 9).

TEMPERATURE COMPENSATION

The basic strain gage and Wheatstone bridge circuit is generally adequate for low-accuracy do-it-yourself transducers. These types of systems have in fact served the pharmaceutical industry very well over the past several decades and much benefit has come from these homegrown systems. Even today, some companies promoting themselves as experts are in reality offering transducers only at this quality. This level of thermal compensation, however, is not nearly adequate for a large class of commercial transducers available over the last 20 years.

There are two thermal considerations to account for the following:

1. Zero shift with change in temperature
2. Span or sensitivity change with change in temperature

ZERO SHIFT

There are four orders of temperature compensation for zero shifts that can be achieved on a strain-gaged load cell.

1. Select the proper alloy coefficient of expansion.
2. Use strain gages from the same manufacturing lot for a load cell.
3. Perform an oven temperature test and make corrections.
4. Install active circuitry to correct imperfections from step 3.

ALLOY SELF-TEMPERATURE–COMPENSATING COEFFICIENT

The strain gage manufacture can supply strain gages where the thermal expansion of the alloy closely matches the thermal expansion of the parent material the strain gage is adhered to. Strain output due to a temperature change under no load is referred to as apparent strain. Strain that is apparently there but not the result of a load change.

STRAIN GAGES FROM THE SAME MANUFACTURING LOT

Residual apparent strain from a proper alloy selection can be reduced by using four strain gages from the same manufacturing lot and the use of a full Wheatstone bridge. Provided that an identical apparent strain resulted from each strain gage installation the undesired output from each gage would be the same and the positive and negative arms of the Wheatstone bridge would correct the problem. There would be two negative apparent strains and two positive values, the sum of which would be zero leaving only the desired signal due to force. The problem is the strain gages do not react perfectly alike. There may be slight differences in the alloy or adhesive thickness under the gage, resulting in a change in signal with no change in loading. The telltale sign here is a nonreturn to a zero signal when there is no longer any applied load.

The technology in most strain gage applications include the above two methods of temperature compensation, but that may not be sufficient for more demanding applications. A tablet press used in research may only be run for short durations at a time and not see any appreciable change in temperature near the load cell. Machines that are run for extended periods of time do get warmer and require additional temperature compensation to maintain their reputed accuracy.

WHEATSTONE BRIDGE THIRD-ORDER CORRECTIONS

Now the professionals step in. This is the step that separates the home grown systems from the professional manufacturer. A system should not be promoted as temperature compensated until this step is completed. Two additional temperature sensitive foil adjustable resisters are installed in each adjacent arm of a Wheatstone bridge. The load cell is slowly heated in a controlled oven to observe the apparent strain of the load cell under a no load but increasing temperature environment. The results are recorded and a calculation performed to determine which resister needs to be adjusted and to what value. This extra step is time consuming but necessary as it will improve the zero stability by an order of magnitude. In addition, it serves as a quality control check.

ACTIVE CIRCUITRY

This degree of temperature compensation is required only if extreme accuracy or unusual temperatures are to be encountered. They are routinely not performed nor need they be as part of a tablet press operation. Basically, an accurate temperature sensor is attached as part of the strain gage installation and correction made to the data accordingly.

SPAN OR SENSITIVITY CHANGE WITH TEMPERATURE

The normalized output of a transducer, referred to as mV/V at full scale, will change with temperature. This fact is ignored by the do it yourself crowd but not by commercial manufactures of quality load cells. If this is important or trivial for the pharmaceutical industry is questionable. The change occurs because both the gage factor (sensitivity) of the strain gages and the modulus of elasticity of the spring element are functions of temperature. As an example, for a typical installation, at an increase in temperature of say 50°F (38°C), the increase in the sensitivity of the strain gages is about 0.25% while the decrease in modulus of steel is approximately 0.75%, a 1% total error if left uncorrected.

Span shifts with temperature can be corrected by a inserting a temperature sensitive resistor in the bridge excitation supply line. With a resistor of the proper value and temperature sensitivity the voltage to the Wheatstone bridge will vary to offset the span error. In other words, as the full-scale sensitivity of the bridge increases with temperature, the temperature sensitive resistor will also increase in value, lowering the voltage to the bridge, thereby reducing its output. If done correctly, the net result is a zero change in full-scale output.

The proof that span shift compensation has been done correctly is difficult as the transducer must be calibrated at two different temperatures. The nominal value of a selected temperature sensitive resistor, however, can easily be calculated that will be proper for the material of the spring element. Doing so is not perfect but will reduce the span error by an order of magnitude, making a 1% error discussed in the preceding text a 0.1% error, one that can easily be ignored for use with a tablet press even in a production environment.

WHEATSTONE BRIDGE BALANCE

Bridge balance means zero output when there is no applied load to the transducer. Installation of four strain gages into a Wheatstone bridge will need some method of making the output read zero at zero load. This can be accomplished with external signal conditioning or within the bridge itself. Some external techniques distort the geometry of the Wheatstone and introduce system errors so it is beneficial to perform this task within the confines of the bridge. This is easily accomplished by installing two adjustable, small but identical value, nontemperature-sensitive resistors, one in each adjacent leg of the bridge. By adjusting the proper resistor, the output of the bridge can easily be made to zero.

SUMMARY OF THE WHEATSTONE BRIDGE

The simple circuit shown in Figure 4 has now taken on a different appearance. Installation of additional resistors, both temperature sensitive and nontemperature sensitive for bridge balance, zero shift with temperature, and span change with temperature makes the Wheatstone appear like Figure 10.

DISPLACEMENT SENSOR

There are sensors to measure angular (rotational) and linear position.

Linear displacements sensors are widely used in tablet presses. Single station tablet presses use them to determine the position of the upper and lower punches and to correct for tooling and machine compliance. Production tablet presses use displacement sensors to define,

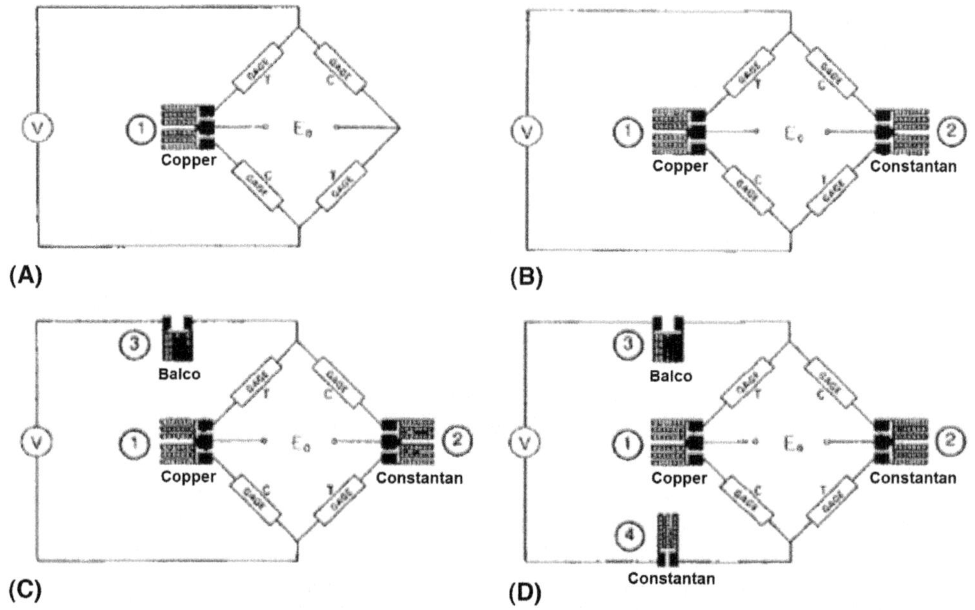

Figure 10 Summary of the Wheatstone bridge. (**A**) High-TCR copper resistor (#1) inserted in corner of bridge circuit and adjusted to maintain bridge balance over the opening temperature range. (**B**) Low-TCR constantan resistor (#2) inserted in second corner of bridge circuit and adjusted for initial zero balance. (**C**) High-TCR Balco resistor (#3) inserted in bridge excitation supply line and adjusted to maintain essentially constant transducer sensitivity (span) over the operating temperature range. (**D**) Low-TCR constantan resistor (#4) inserted in bridge power supply line and adjusted to set the initial span at the desired calibration level.

control, or limit the position of weight cams and roll positions. These types of sensors are available in many forms, from strain gage, linear variable differential transformers (LVDT), to magnetic and optical (3–6).

An LVDT displacement transducer comprises three coils, a primary and two secondary coils. The transfer of current between the primary and the secondary coils of the LVDT displacement transducer is controlled by the position of a magnetic core called an armature. At the center of the position measurement stroke, the two secondary voltages of the displacement transducer are equal but because they are connected in opposition the resulting output from the sensor is zero. As the LVDT's armature moves away from center, the result is an increase in one of the position sensor secondary and a decrease in the other. This results in an output from the measurement sensor. With LVDT's, the phase of the output (compared with the excitation phase) enables the electronics to know which half of the coil the armature is in. The strength of the LVDT sensor's principle is that there is no electrical or mechanical contact across the transducer position sensing element which for the user of the sensor means clean data, infinite resolution and a very long life. There is a slight variation of this concept that is called a gaging head whereby a mechanical spring extends the armature to the fully extended position to come in contact with the moving part to be measured without a mechanical connection as with the free style armature. Some designs also contain electronics so that only a DC voltage needs to be applied from a power supply.

LVDT sensors are very robust with nonlinearity from 0.1% to 1% depending on the model. Measurement ranges are generally from 0.5 mm full scale to 40 plus mm full scale. Frequency response is generally greater than 100 Hz, which is more than adequate for even high-speed tablet press or compaction simulator applications.

Noncontact displacement sensors are rarely used as the range is typically limited to less than 5 mm. One application is to determine if a part is in place for safety considerations.

Rotary displacement sensors are being used more on rotary tablet press today than in the past to accurately define the exact angular position of the turret on a rotary tablet press. Resolvers or their digital counterpart, rotary encoders can resolve an angular change as small as 0.006 degrees. This is useful to determine the exact punch location relative to a compression roll and the resulting force to evaluate the compact relaxation under the constant strain period known as dwell time.

SIGNAL CONDITIONING

Power Supplies

The power supply is the source of excitation to the sensor. Historically power supplies were notoriously noisy electrically and tended to drift or change their output voltage values. Today they are much more stable and smaller in size. That being said it is still prudent to measure the voltage output from the power supply before sampling the voltage from the sensor. In the case of most sensors, the output is directly proportional to the applied voltage, noise included.

Ratiometric measurements are the most accurate method to assure that the reading of the signal is independent of the applied voltage. The output from the load cell is normalized by dividing the output from the sensor by the applied voltage from the power supply. This is expressed as mV/V or so many millivolts out per applied voltage in. All quality load cells are supplied with calibration certificates in mV/V and a good data acquisition system should do the same by measuring the power supply and dividing the output signal by this value. All in situ calibrations should also be performed in mV/V and not just as a number in the final units.

Power supplies generally produce either a constant voltage or a constant current, and there are advantages to each. Lead wire resistance, for example, is not of concern with a constant current system as it is with a constant voltage where extended lead wire length adds an effective resistance in series with the sensor, reducing the voltage to the sensor. Lead wire lengths are generally minimal around tablet presses but the proper calibration should be performed at the point where the lead wires terminate at the input to an amplifier.

The critical item is that the load cell needs to be matched to the power supply or all of the efforts to temperature compensate the transducer will be incorrect. In the United States the standard is for constant voltage power supplies and load cell manufactures assume that to be true. *If you plan on using a constant current power supply, you must order your load cells accordingly.* They will work fine either way, but they will not be properly temperature compensated.

WHAT EXCITATION VOLTAGE SHOULD I USE?

Typical excitation levels used for powering strain gage circuits range from a high of 15 VDC to a low of 3 VDC. Why the large range and what is appropriate? The answer is it depends on the physical size of the strain gage, the gage resistance, the desired accuracy and what material the gage is bonded to. A strain gage is like a toaster grid. Current flowing through the grid produces heat that must be dissipated into the material that the strain gage is bonded to. A strain gage bonded to copper or aluminum will be capable of dissipating much more heat than one bonded to stainless steel and therefore allow much more excitation voltage. Excessive heating will cause a thermal drift causing a shift in the zero base line of the transducer.

So, if too much voltage is applied the transducer will drift, too little and the output will be too small. For a desired moderate- to high-accuracy transducers with the strain gages bonded to steel, the power dissipation should be kept to 2 $W/in.^2$ (3 kW/m^2). The correct excitation level is easy to calculate. Using basic Ohm's law relationship, the following equation is easily derived (3):

$$E = \sqrt{RAP}$$

where

E = the voltage for the Wheatstone bridge,
R = the resistance of the strain gage,
A = the grid area of the strain gage, and
P = the power dissipation of the strain gage discussed in the preceding text.

A typical strain used in roll pins for pre and main compression is a shear pattern from the Measurements Group J2A-06-SO91K-350. This is a 350-Ω gage resistance with a grid size of 0.125 by 0.105 in. (3.18 by 2.67 mm). Inserting these values into the above equation results in an optimal bridge excitation of 6 V. Some wireless systems apply only 3 V to the bridge; this lower value is in consideration for conserving battery power, not for optimizing performance of the strain gage circuit.

STRAIN GAGE AMPLIFIERS

The small millivolt signals from the strain gage Wheatstone bridge need to be amplified to a higher-level voltage for conversion into a digital signal for subsequent analysis. This is generally performed in two steps, each with a purpose. The first amplifier is called a differential or instrumentation amplifier and may only have a gain of one. A second amplifier will usually perform the actual amplification and may have a programmable gain from 100 to 1000 times.

The purpose of the differential amplifier is to remove electrical noise from the environment carried into the amplifier by the electrical cables. The signal cables are in a sense like an antenna with a resistance at the end (the strain gage bridge). The cable from the strain gages should be shielded, and the wires within the cable twisted and not parallel to each other. Nonshielded exposed wires should be minimized as they will be excellent antennas. The positive signal wire should carry the signal from the Wheatstone bridge; the negative signal should remain at 0 V. If the negative lead were to be attached to an electrical ground this would be referred to as a single-ended input.

For a single-ended input, the positive input to the amplifier would see the signal from the strain gages as well as any electrical noise which in turn would be amplified by the high gain second stage amplifier. Provided that the negative lead is not attached to an electrical ground but to the negative side of the differential amplifier, this is called a differential input. Since both wires (positive and negative signal) are run within the same cable and in fact twisted together both should see the same electrical noise. The purpose of the differential amplifier is to take the difference between the two signal leads, which should eliminate the cable noise and allow only the data through to the high gain amplifier. The common mode rejection (CMR) of an amplifier is a measure of how well this is performed. The higher the CMR, the better the noise canceling and subsequent signal-to-noise ratio.

ANALOG TO DIGITAL CONVERSION

The advent of high-speed, high-resolution analog to digital conversion has enabled large quantities of data to be analyzed and displayed in a meaningful way so that either a person or a feedback control system can respond to the data. The purpose of the A/D converter is to change the incoming analog signal to a series of digital numbers. The rate at which this is performed and the resolution of the conversion will have a lot to do with the overall accuracy of the data acquisition system. Although there are many factors that need to be considered such as amplifier settling time, switching rates, programmable amplifiers, only the major three items will be covered.

- Resolution
- Sample rate
- Aliasing and the need for aliasing filters

RESOLUTION

Resolution is the number of parts that an analog signal is represented by and is described by the number of bits for the conversion process. Mathematically, it is expressed as 2^x, where x is the number of bits. A single bit conversion ($x = 1$) with a 5-V DC input can be thought of as any value between 0 and 2.5 V will be put into one bin and any value between 2.5 and 5 will go into a second bin. The greater the number of bins, the greater the resolution. Table 1 shows the relationship between resolution and bits. The last two columns are based on a bipolar setup that is plus and minus the stated amount. The last column is the resolution for a bipolar signal where full scale is 50 kN. The highlighted rows are commonly available commercially today.

Looking at Table 1, it would appear that the 10- or 12-bit resolution would be more than adequate for the acquisition of data on a rotary tablet press, and that would be the case provided that an amplifier gain was unique for each channel that raised the millivolt signal to the full scale of A/D converter. Typical amplifier gains are fixed however and not optimized, letting the resolution of the A/D converter solve the shortcomings. Let us take two realistic examples.

Example 1

A transducer with a 2.0 mV/V output; excitation voltage of 3 V, a fixed gain amplifier of 64 and a 12-bit A/D. Determine the percent resolution and equivalent number of Newtons with a full scale of 50 kN at 5 V.

Transducer output of 6 mV is amplified to 0.384 V with the fixed gain of 64 amplifier. A 12-bit bipolar A/D can measure 1 part in 2048 out of 5 V or 2.4 mV. 2.4-mV resolution with a 0.384-V signal represents 0.64%. Therefore, what appeared as a resolution of 0.05% quickly became 0.64% or 320 N on a 50-kN transducer.

Table 1 Bits Versus Resolution

Bits	Equation	Resolution (one part in)	Percent of full scale	*N* resolution[a]
1	Resolution $= 2^1$	2	100	50,000
2	Resolution $= 2^2$	4	50	25,000
3	Resolution $= 2^3$	8	25	12,500
4	Resolution $= 2^4$	16	12.5	6,250
5	Resolution $= 2^5$	32	6.25	3,125
6	Resolution $= 2^6$	64	3.125	1,562
7	Resolution $= 2^7$	128	1.56	781
8	Resolution $= 2^8$	256	0.78	391
9	Resolution $= 2^9$	512	0.39	195
10	Resolution $= 2^{10}$	1,024	0.20	98
11	Resolution $= 2^{11}$	2,048	0.10	49
12	Resolution $= 2^{12}$	4,096	0.05	24
13	Resolution $= 2^{13}$	8,192	0.024	12
14	Resolution $= 2^{14}$	16,384	0.012	6
15	Resolution $= 2^{15}$	32,768	0.006	3
16	Resolution $= 2^{16}$	65,536	0.003	1.5

[a]transducer

Example 2

A transducer with a 2.0 mV/V output; excitation voltage of 5 V, a fixed gain amplifier of 64 and a 14-bit A/D.

The transducer output is 10 mV amplified to 640 mV with the amplifier. The 14-bit bipolar A/D can measure 1 part in 8192 out of 5 V or 0.61 mV for a resolution of 0.095% or 47.5 kN on a 50-kN transducer. By using a higher excitation and a 14-bit A/D, the resolution became close to seven times better and more in line with the requirements for a tablet press transducer system.

RESOLUTION SUMMARY

High-resolution A/D converters are commonplace today and at reasonable prices and performance. Common practice in the past was to use adjustable amplifier gains to optimize the transducer full scale to that of the input of the A/D converter. For instance, a 10-mV signal would be amplified with an amplifier gain of 500 to produce a 5-V signal for a 5-V input to the A/D converter. Today programmable gain amplifiers are used that cannot be adjusted so the full-scale input signal to the A/D is less than optimal.

SAMPLE RATE

There is more misunderstanding about frequency response, sampling rate and the Nyquest theory than you can envision. Sample rate is easy; it is the number of times a digital reading is taken over a period of time, usually one second. This is sometimes expressed in hertz. Therefore, a 100-Hz digital sample rate is 100 equally time spaced samples taken for each second.

The confusion is the word hertz. In the analog world, hertz refers to the number of cycles/sec. Therefore, in analog speak, a 1-Hz sine wave or 1 cycle/sec may require 10 samples/sec to represent the sine wave. In digital speak, this is a 10-Hz rate. In other words for this example, it takes a 10-Hz digital sample rate to define a 1-Hz analog signal. Got it? Good.

Now for the next item of confusion. The Nyquest theory states that the frequency content of any analog signal can be determined with a sample rate of only twice that of the analog frequency. Some individuals twisted this to mean you only need to double the analog frequency with the digital sample rate to reproduce the original data. That is not what the Nyquest states and it is very misleading. Nyquest states you can obtain correct frequency information this way but says nothing about reproducing the shape of the data. There is a relationship between the number of samples required to define a cycle and the statistical error of missing the peak value of the cycle. The graphic in Figure 11 clearly shows the problem. The analog sign wave is being sampled at a rate of 5 samples/cycle. The computer would basically connect the dots, making a pseudosquare from this sine wave.

Provided that you wish to limit your peak detection error to 0.25% you must sample digitally 100 times the analog frequency contained within the data. Such high sample rates are generally not used and the user is never aware of what is being missed. For tablet press instrumentation, a digital sample rate (hertz) of at least 10,000 is required to cover all presses and transducers.

ALIASING ERRORS

Nyquest States

If frequencies greater than half the sampling rate are allowed to the input of the A/D converter, the higher frequency *will* erroneously be represented by a lower frequency that *cannot be separated from the real data.*

The only way to eliminate this error is to use an antialiasing filter prior to digitizing the input signals. Therefore if a sample rate of 10,000 Hz is to be used a low pass analog filter of

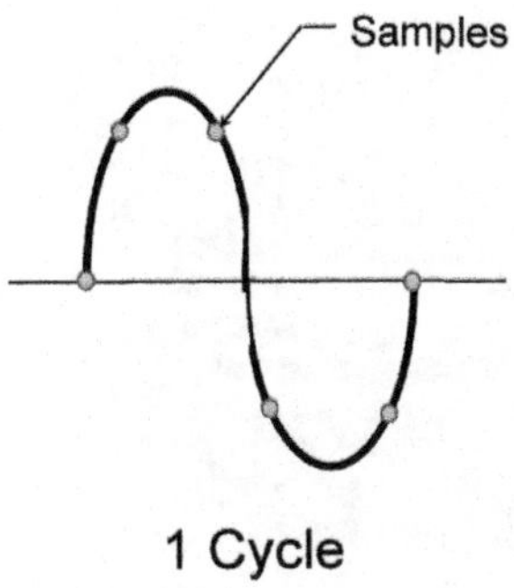

Samples/Cycle	% Error
5	5.00
10	2.50
20	1.25
50	0.50
100	0.25
200	0.10

For example: If the frequency of your data is 100 Hz and you desire a maximum error of 0.25%, you must sample the 100 Hz at 100 samples per cycle or 10,000 samples per second.

Figure 11 Sample rate versus error.

less than 5000 Hz must be used to prevent aliasing errors. This filter will prevent analog frequencies of greater than 5000 Hz from being digitized. Just because the higher frequencies are not present when the system is installed does not mean they will never be present. Changes in equipment in the facility, use of hand-held radios or even new utilities can be the source of high-frequency noise.

Any good data acquisition system must incorporate such protection into the design or the user will someday receive incorrect information and never even know that his system is creating new data to superimpose on the actual data.

A classic example that most of us can relate to is the wagon wheel in a Western movie. The camera is taking pictures at a fixed rate, say 60 frames/sec. If the wagon wheel makes 90% of a rotation between frames the wheel will appear to have rotated backward by 10%. Wrong in both magnitude and direction! The same phenomena will occur will your data acquisition system if it is left unprotected without the use of an antialiasing filter (Fig. 12).

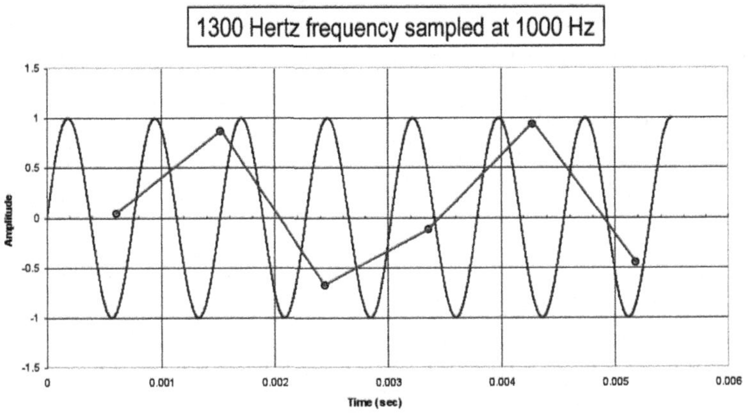

Figure 12 Aliasing error.

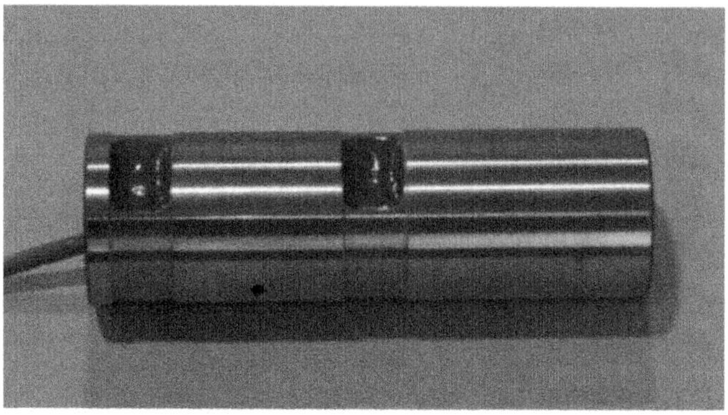

Figure 13 Representative tablet press transducer.

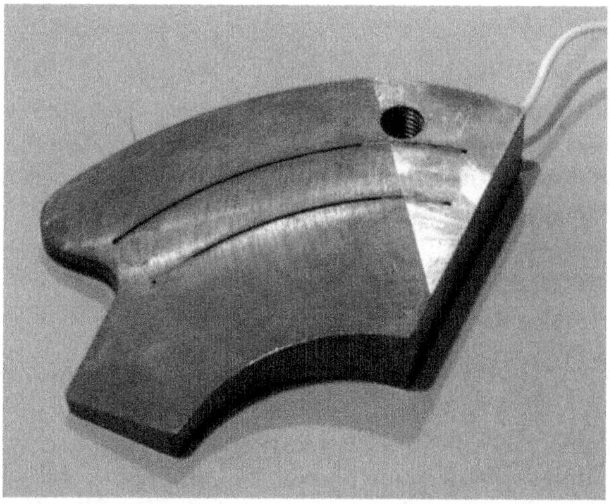

Figure 14 Instrumented ejection ramp.

REPRESENTATIVE TABLET PRESS TRANSDUCERS

Figure 13 shows an instrumented compression roll pin for a Piccola bilayer tablet press.

Figure 14 shows an instrumented ejection ramp for a Riva Piccola tablet press.

Back side of a not yet strain gauged ejection ramp for the Piccola tablet press showing the pockets where the strain gages will be placed. The two spring elements are differential bending beams on each end with a relief in the middle (Fig. 15) (4).

CALIBRATION

Calibration is the comparison of a component or group of components against a known and recognized standard under a specific set of conditions. A system is considered within calibration if it complies or can be adjusted to comply with the acceptable uncertainties.

Validation in the sense of measurement systems is a set of calibrations over the environmental conditions the system must perform within. This implies that if a measurement system is to operate over a specified temperature and humidity range, it must be calibrated over the extremes to be validated.

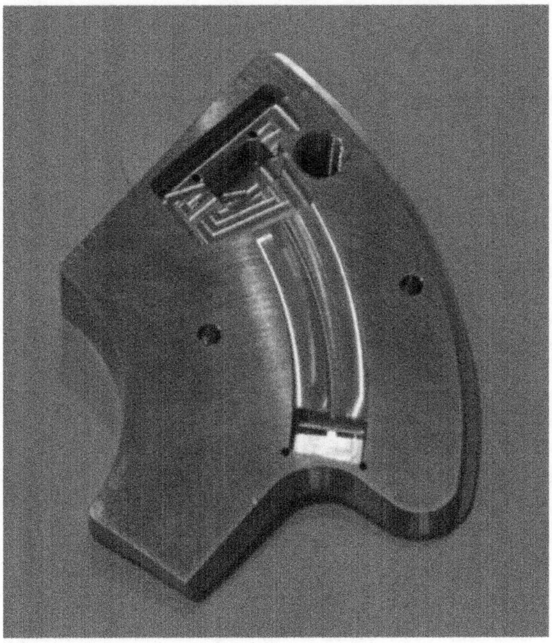

Figure 15 Piccola ejection ramp showing strain gauge pockets.

In the United States, the NIST, maintains standards and is considered the arbiter and ultimate U.S. authority for values of SI units and industrial standards. NIST also provides traceability to its standards by calibration, by which an instrument's accuracy is established by comparing, in an unbroken chain, to the standards maintained by NIST. For each step in the process, the measurement uncertainty is evaluated.

Traceability is the property of a standard whereby it can be related to stated references, usually national or international standards, through an unbroken chain of comparisons, all having stated uncertainties. The level of traceability establishes the level of comparability of the measurement: was the measurement compared to the previous one, a yesterday of a year ago, or to the result of a measurement performed anywhere else in the whole world.

Figure 16 shows the organizational chart for the standards in the United States. It is a federal offense for one to misrepresent their facility and may well result in time spent in jail and a personal meeting in front of the Senate. Most in-house calibration facilities fall into instrument maintenance while companies specializing in calibration services are secondary laboratories. Secondary laboratories rely on a primary laboratory for their internal standard to be calibrated that will in turn rely on a direct NIST calibration for their standards. Therefore, the calibration performed by a process application technician must have an unbroken chain of traceability directly to NIST.

The level of uncertainty increases the longer the chain from NIST. A secondary laboratory will rely on the standards of the primary laboratory to be in compliance with the requirements of the NIST.

CALIBRATION OF TABLET PRESSES

Calibration of a rotary tablet press needs to be done with caution as it is easy to make an incorrect calibration. Calibrated punches can become misaligned, causing excessive friction resulting in a loss of applied force to the machine load cell. The calibrated punches should

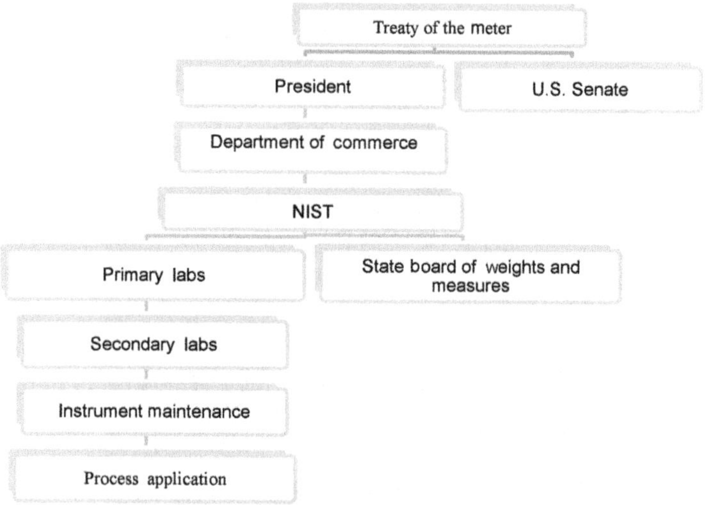

Figure 16 U.S. standards structures.

have at least two standards, one each in the upper and lower punches with procedures to make sure the standards agree with each other before the results can be accepted. One vendor of calibration services uses three standards in line to ensure that none of the applied load is being lost due to friction from misalignment. It is interesting to note that misalignment is not obvious to the eye, and there is no method of knowing that it had occurred if only one reference is used, the resulting calibration will look completely normal, just with incorrect values.

There are two basic methods of performing a static calibration on a rotary tablet press. One is to perfectly align the modified punches between the rolls and apply the load with a hydraulic ram while acquiring data from the standards and the machine load cell. The second method is to install the modified punches prior to the rolls and using the machine hand wheel, roll the punches through the compression cycle. The first method can apply a higher force smoothly and with more control, and is easier to ensure the modified punches are properly aligned. The second method is quicker and does not involve hydraulic rams, pumps, and hoses; however, the load cannot be controlled as well. Both methods produce acceptable results (Figs. 17 and 18).

CALIBRATED PUNCHES

The design of a custom punch to be used as a standard or reference must follow the general rules of transducer design (4).

1. The mechanical design of the punch must be such that it has excellent sensitivity in the direction of the desired force to be measured and low sensitivity to all undesired forces.
2. The placement of the strain gages should be such to electrically cancel any residual stress from all other undesirable forces, such as side loads.
3. Placement of the strain gages within the Wheatstone bridge to cancel unwanted forces and respond only to the desired force.

Let us compare three potential mechanical designs for a 50-kN calibrated punch spring element.

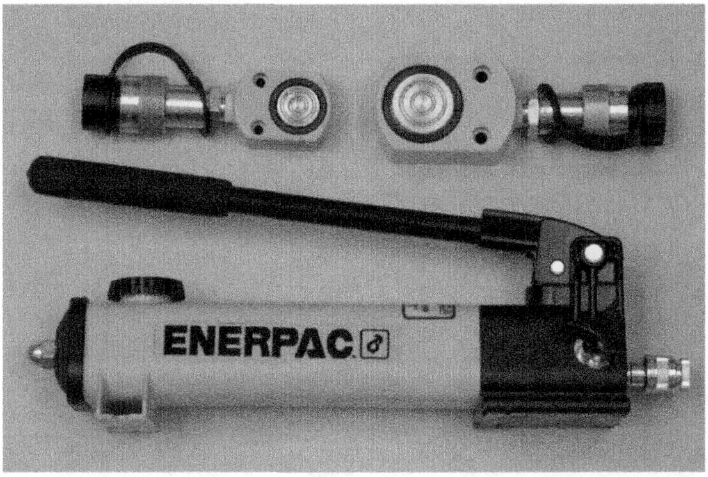

Figure 17 Calibration kit hydraulic pump.

Figure 18 Calibrated punch in tablet press.

Design 1

Machine a smaller diameter on the punch barrel and install a Poisson full bridge set of four strain gages (Fig. 19).

Reducing the outside diameter to 14 mm from the original 19 mm to allow room for the strain gages and yield a correct sensitivity for calibration purposes results in a cross-sectional area of 154 mm^2.

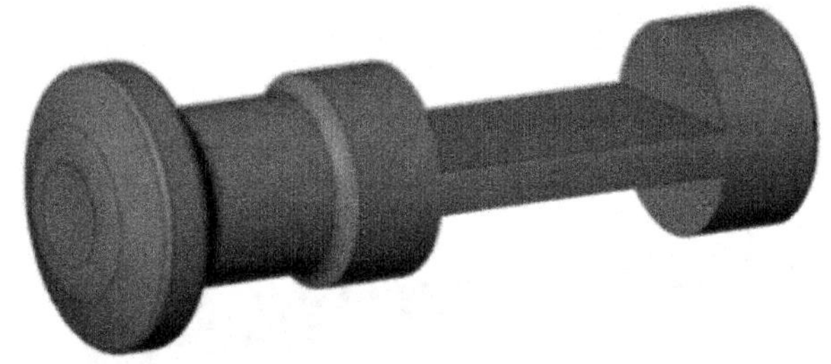

Figure 19 Calibrated punch rectangular.

The axial stress on the reduced area is

$$\text{Stress} = \frac{\text{Force}}{\text{Area}}$$

The equation for bending due to an offset load such as when the punch contacts the roll is

$$\text{Stress} = \frac{mc}{I}$$

where
c is the distance from the punch centerline to the position of the strain gages and m is the bending moment.

I is the moment of inertia, which is

$$\pi d\frac{4}{64}$$

for a circular cross section.

Using the above equations and geometry, the axial and transverse sensitivity can be computed.

Design 2
Machine flats on the punch barrel to install strain gages (Fig. 20).

Design 3
Machine pockets in the punch to install the strain gages. This results in a cross-sectional area resembling a structural member used in building and bridge construction called an I beam. As expected this design offers many advantages. In fact this design is five times more resistant to undesirable bending forces than the other two (Figs. 21 and 22).

Figure 20 Calibrated punch reduced cross section in design.

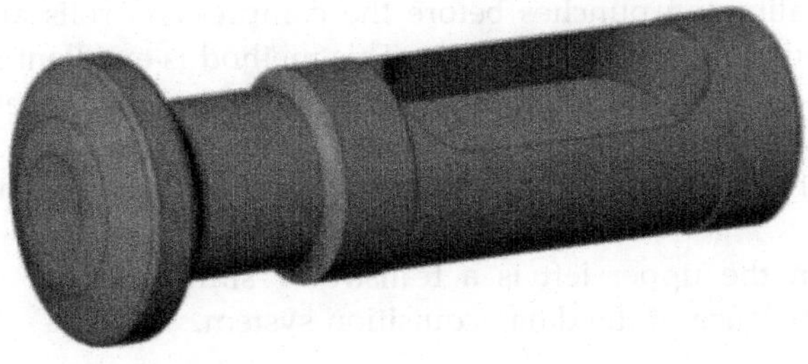

Figure 21 Calibrated punch pocket design.

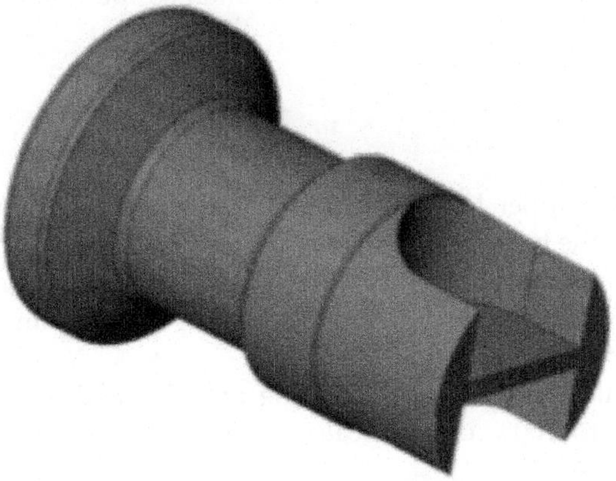

Figure 22 Cross section of pocket design.

USING THE CALIBRATED PUNCH

The strained gaged punch must be calibrated against a recognized standard to be used as a calibration standard. It must be calibrated on a regular interval as dictated by company standard operating procedure (SOP). The SOP at Specialty Measurements Inc. (SMI) is that the punch must be calibrated against a standard every three months and the standard must be sent to an independent agency for certification within the last 12 months. This policy prevents in-house propagation of errors. Another part of the SIM procedure is that one set of strain gages will be installed in each of three pockets, one in the upper punch and two in the lower punch, in essence making three standards in use during a calibration. These three standards must agree within established criteria before the calibration is acceptable.

APPLICATION OF THE FORCE

The force is generally applied in one of three ways:

1. Insert a hydraulic jack in line with the calibrated punches and use a hand pump to apply pressure to the piston. The punches are generally prealigned between the rolls. The load is applied gradually and many points can be obtained from zero to full scale. At SMI over 1000 points are obtained and a regression analysis is performed to obtain the stated sensitivity and errors.
2. Align the calibrated punches between the rolls as before and use the hydraulic system of the tablet press to produce a load in place of the in-line jack.

3. Position the calibrated punches before the compression rolls and rotate the turret manually through a compression cycle. This method is excellent for a quick check of the force measurement system at a limited number of force levels.

The calibration kit shown in Figure 23 shows some of the components used for method one above.

The instrument in the upper left is a transducer simulator and is used to apply a calibrated input to the balance of the data acquisition system.

THE BALANCE OF THE SYSTEM REQUIRES CALIBRATION ALSO!

The emphasis to date in this chapter has been on the actual force transducer installed within the machine. It is, however, only one link in the chain. Other components, collectively referred to as signal conditioning must be calibrated as well, such as the following:

Power supplies
Amplifiers
A/D converters

The instrument in the upper left of Figure 23 is a transducer simulator and is used to apply a calibrated input to the balance of the data acquisition system. It is this instrument that is used to input a traceable ratiometric mV/V signal into the signal conditioning. The transducer is temporary disconnected from the signal conditioning and the transducer simulator installed in its place.

The transducer simulator inputs an ascending and descending signal to the system in 10% increments from 0% to 100% of full scale. All recorded data points are regressed to

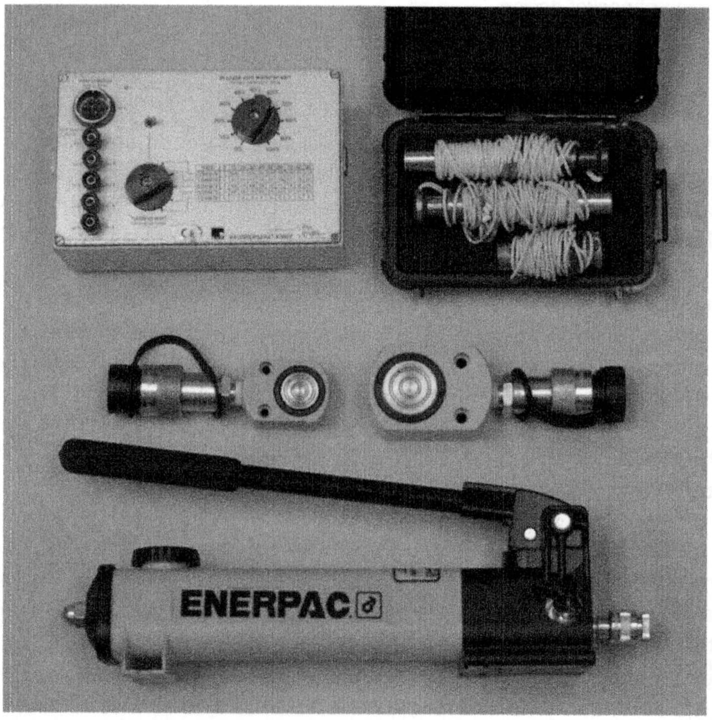

Figure 23 Calibration kit.

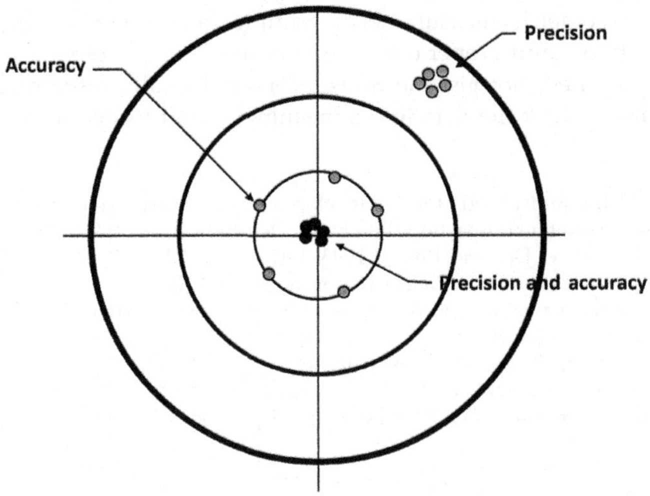

Figure 24 Accuracy versus precision.

determine accuracy and linearity. Power supplies, amplifiers, and A/D convertors are so accurate today that a typical overall error is less than 0.05% of full scale with a rejection tolerance of 0.1% (4).

"It is much better to be approximately accurate than precisely wrong" (7).

Two terms that are frequently interchanged are accuracy and precision. They do not mean the same as illustrated in the example of the target, Figure 24. Precision is the tight grouping of bullets (data) in a location not necessarily where desired. If you were a deer hunter, every shot could be precisely in the same spot, all way over the top or short of the desired target. Making an adjustment in your rifle sights (instrumentation) could correct this problem. Accuracy is a random grouping within a specified tolerance of the target center. A tight accuracy tolerance would lead to precision at the target center.

SUMMARY

An instrumented tablet press in an R&D environment is not a luxury today; it is a necessity if one wishes to practice good science and have a deeper understanding of compaction principles. It is possible to design an in-house system and many have been built and put to good use. Today there are several commercial options that should be considered first to see if they fit into the company needs as thousands of man-hours have been invested into their design by the manufactures. Whatever the path, do instrument or purchase an instrumented tablet press. It will shorten development time; enable easier transition from R&D machines into production models resulting in a quick return on the initial investment.

A properly designed data acquisition system needs to be based on sound mechanical and electrical principles. "You ask a measurements system for the truth, the whole truth, and nothing but the truth, not its opinion." Incorrect components are perfectly willing to moonlight providing more information than you wanted. Some force transducers produce a nice signal when exposed to a strong light source, others from temperature and still others due to improper mounting. This is not acceptable. There are many who purport to being "Instrumentation Experts," do not be duped into believing a fancy software program makes for a well-designed instrumentation system. The transducers must fit the application; power supplies must match the transducer requirements of either constant voltage or constant current, the resolution of the analog to digital conversion must be appropriate for the application and use ratiometric measurements. Sample rates must be determined for the required frequency response and proper use of antialiasing filters employed. The entire system must be able to be calibrated, not just the transducers and finally there must be a software system that can condense all of the data into a meaningful and usable format.

REFERENCES

1. Celik M. Oktugen E. A feasibility study for the development of a prospective compaction functionality test and the establishment of a compaction data bank. Dev Ind Pharm 1993; 19(17–18):2309–2334.
2. Cocolas HG, Lordi NG. Drug Dev Ind Pharm 1993; 19(17–18):2473–2497.
3. Hoag S. Tablet compaction issue. Eur Pharm Rev Issue 2005; 2:104–111.
4. Kistler Instrument Corp., Amherst, NY. Available at: http://www.kistler.com/do.content.us.en-us?content=90_Support_Download. Accessed September 2007.
5. MarshalK. KMA Associates, Brick, NJ, Conversation.
6. Microstrain, Williston, VT. Available at: http://www.microstrain.com. Accessed September 2007.
7. PCB Piezotronics, Inc., Depew, NY. Available at: http://www.pcb.com/techsupport. Accessed September 2007.

SUGGESTED READING

RDP Electrosense, Pottstown, PA. Available at: http://www.rdpe.com/us/mendisp.htm. Accessed September 2007.

Specialty Measurements Inc., Lebanon, NJ: Internal Publications.

Vishay Micro Measurements, Wendell NC. Available at: http://www.andruss-peskin.com. Accessed September 2007.

6 Advanced compaction research equipment: Compaction simulators

Colleen E. Ruegger and Metin Çelik

INTRODUCTION

Throughout this book the importance of understanding the compaction behavior of pharmaceutical materials has been emphasized as well as the critical parameters that should be monitored and evaluated during a compaction study. The quality and reliability of the data that is generated is dependent on the equipment used for the compaction study and the quality of the instrumentation used during the data collection process. Since the instrumentation of the first press was first reported in 1951 by E. F. Brake (1), the sophistication of the compaction equipment as well as the instrumentation has increased significantly.

The work for the development of a standard functionality test is still in progress as addressed in chapter 7.

With the increasing use of quality by design during drug product development, the standardization of functionality tests is more important than ever to determine the design space for tablet compaction. Although less sophisticated types of equipment may be used to determine the compaction design space, the use of a compaction simulator will increase the likelihood of success for the development of a robust tablet formulation while minimizing the amount of material needed for drug product development and characterization of the formulation and the individual components.

This chapter will give an overview of compaction simulators and the different types available as well as how to best utilize them for development of pharmaceutical products.

HISTORY

Since the first patent for a machine to manufacture compacts was granted to Brockedon in 1843 (2), the tablet presses available for making tablets and compacts have evolved into high performing, high productivity, computer controlled machines.

Today, there are three general types of compaction equipment available: single station presses, rotary tablet presses, and universal testing machines (including the Carver press, Instron-type machines, and compaction simulators—hydraulic and mechanical).

Single station tablet presses run at a relatively slow speed and operate with the lower punch in a stationary position. The upper punch is attached to an eccentric arm that causes the upper punch to move with a stamping motion. The advent of tablet press instrumentation has helped to increase the understanding of the compaction process. The instrumentation of a single punch press was first reported in 1951 by E. F. Brake (1). Brake attached strain gauges to the upper punch of a Colton 4B tablet press to measure the upper punch force exerted during compaction.

In 1951, Higuchi and coworkers (3) began a series of reports on the physics of tablet compaction. They instrumented an isolated punch and die set for use with a mechanical-lever machine to determine compression forces and to measure punch displacement. In the third paper of this series, Higuchi et al. (4) instrumented a Stokes Model A-3 single punch tablet press. Strain gauges were bonded to the machine frame to measure forces applied by the upper punch. A load cell was placed below the lower punch and a linear variable differential transducer (LVDT) was attached to determine the movement of the upper punch. Ejection force was also determined. Several others built on Brake's work (5–7).

Rotary tablet presses have more than one set of punches and, generally, operate at higher speeds than single station presses. Compaction is a continuous process occurring on a rotating turret that moves the upper and lower punches between two pressure rollers, producing a squeezing action. In 1963, the instrumentation of these types of presses began with Shotton et al. (8) using stain gauges and a radiotelemetry device to measure force on a Manesty D3 press. Knoechel et al. (9) instrumented a rotary tablet press that could be used in production situations and Wray (10) instrumented a rotary tablet press to obtain compression forces as well as individual ejection forces through the use of a sectioned ejection cam. The instrumentation of rotary tablet presses allowed researchers to increase their understanding of the physics of tablet compaction and to begin to characterize material behavior and aid in formulation development.

However, due to the single-ended compaction of the single station tablet press and the difficulty in instrumenting a rotary tablet press, the data obtained on these types of research machines were not always predictive of the high-speed, production conditions. In addition, rotary tablet presses require relatively large quantities of material that may be costly or may not be readily available during the early stages of drug product development.

Universal testing machines (UTMs) are devices originally used by mechanical engineers and material science scientists for standardized stress testing of materials. These devices usually consist of one or two movable actuators, which are linked to various mechanisms allowing for operation of the device by compression or tension. Punch and die assemblies have been attached to the actuators by pharmaceutical scientists to use this equipment to study tablet compaction.

Shlanta and Milosovich (11) used a Carver press to study the behavior of powder beds under constant strain. The piston on the Carver press was driven by compressed air in place of using a hydraulic loading system. Varsano and Lachman (12) evaluated the stress-strain characteristics of materials using selected loads applied at varying rates and patterns using the Instron, model TM-M, UTM.

In 1972, Rees et al. described a system designed to simulate the double-acting compression effect of a rotary tablet press (13). The device was designed primarily for use with a UTM, but could be used with different types of mechanical or hydraulic equipment. In theory, the system achieved double-ended compaction similar to that of a rotary tablet press, by movement of the die (downward) and the upper punch. The system, however, was unable to follow unusual waveforms or multiple compression and ejection cycles and the authors described its uses to be primarily in the area of preformulation studies. One of the main advantages of such devices is that they can be run under load control as well as displacement control whereas single station and rotary tablet presses follow a cam track that controls the punch displacement. Çelik et al. (14) used a Mayes Universal Testing Press under load control to look at strain changes under constant stress for several pharmaceutical powders, including microcrystalline cellulose, pregelatinized starch, dextrates, and dicalcium phosphate.

The first mention of a true simulation device in the literature was by Hunter et al. (15). The authors described a simulator that they had developed, which was capable of reproducing the multiple compressions and ejection cycle of any pharmaceutical compressor at its normal operating speed. The device consisted of a frame supporting two servo controlled hydraulic actuators connected to a high pressure hydraulic power pack with an electronic control system to operate the servo valve. Standard B or F tooling could be attached to the actuators and the simulator could reproduce any given displacement or force/time profile at the same speeds at which a rotary press would operate.

DESCRIPTION AND TYPES OF COMPACTION SIMULATORS

Historically, a compaction simulator has been defined as a device that is capable of mimicking the exact cycle of any tablet press in real time and recording parameters, such as force and displacement, which are important to the analysis of the compaction event (16). Compaction simulators are also known by several other names including compression simulator, tablet press simulator, integrated compaction research system (ICRS), formulation development tool, and tablet press replicators. The modern-day compaction simulator evolved from UTMs that were developed for the stress testing of materials.

Since 1976, a number of simulators have been developed and described (17–23). Bateman reviewed compaction simulators and briefly discussed some of the work done up to 1988 (24).

Pharmaceutical scientists and engineers have used simulators for a variety of purposes since 1976 including, using milligram quantities, to characterize raw material deformation properties and evaluate various compaction parameters (punch force, ejection force, displacement, speed, etc.) for excipients, drug substances and formulations, formulation development and optimization, scale-up prediction, and troubleshooting of production problems. Utilization of compaction simulators is described in detail in section "Practical Applications of Compaction Simulators."

The modern-day compaction simulator has evolved into a multifunctional machine capable of assisting in all phases of drug product development. There are currently three main types of compaction simulators: (*i*) hydraulic compaction simulators (i.e., Mand and ESH), (*ii*) mechanical linear compaction simulators (i.e., Presster^{TM}), and (*iii*) mechanical rotary cam compaction simulators (i.e., Stylcam^{TM}).

The following section describes the features of each type of compaction simulator.

Hydraulic Compaction Simulators

The full description of the hydraulic compaction simulator is given by Çelik and Marshall (16) and will be summarized here (Fig. 1).

Machine Unit

The heavy machine frame consists of a lower and an upper crosshead, with an intermediate die table, linked by rigid steel columns. A high-performance hydraulic actuator is located in the upper and lower crossheads to drive the punches via force measuring load cells. High-performance servo-valves and hydraulic pressure accumulators control the movement of the hydraulic rams, which move at up to 2 m/sec with forces up to 50 or 100 kN. The intermediate die table carries the die holder, the various powder hoppers, tablet wipe-off arm, and the tablet collection device. The small range punch position measuring transducers are mounted to the table. Any standard or customized punch and die sets can be used with these types of systems, and special instrumented or temperature controlled punches and dies can be used.

Hydraulic Unit

The hydraulic unit provides the power to move the actuators at high pressure for each actuator. Chilled water or fans can be used to maintain the hydraulic fluid at a constant temperature. It is also possible to physically combine the hydraulic and the machine units into one unit.

Control Unit

This electronic unit contains several modules that are used to control and monitor the load and displacement of the upper and lower punches as a function of time. Additionally, a computer

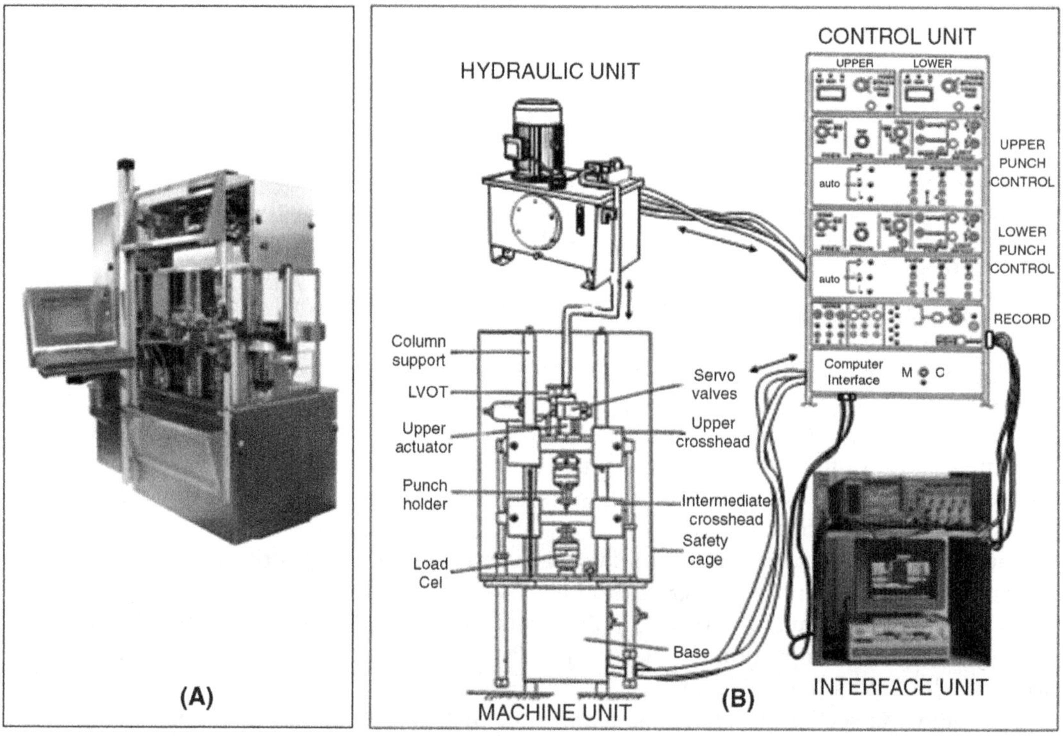

Figure 1 (**A**) A picture of an ESH hydraulic compaction simulator. (**B**) A typical schematic of a hydraulic compaction simulator.

unit is integrated to the control console to run the compaction simulator in a programmable manner. The profiles followed by the hydraulic compaction simulators are discussed in section "Tablet Press Simulation." Load cell amplifiers provide complete conditioning at several ranges for the upper and lower load cells. Punch position transducers are typically LVDT type with special amplifiers for maximum resolution over small displacements. Longer range LVDT's control the actuator movements outside the compaction region. The output signals from the electronic units associated with the force and displacement transducers can be retrieved by the operator via internal or external recording devices such as analog or digital oscilloscopes.

Mechanical Linear Compaction Simulator

The mechanical linear compaction simulator (Presster) basically resembles a single station rotary tablet press with all its individual sections arranged in a straight line rather than in a circle (Fig. 2). As described by Thorsten Neuhaus (2007) (22), in contrast to any rotary tablet press, the Presster operates with just one pair of punches and one single die. The tablet tooling is installed in a turret-analogue carriage that is driven by a drive belt on a horizontal line through the whole machine passing successively the dosing cam, the upper and lower rollers of the pre- and main compaction stations, and finally the ejection cam and the takeoff bar. As the punches on the Presster are guided by punch cams similar to the ones used on rotary tablet presses, the geometrical path of the punches on both types of machines is quite close. Standard punch and die sets for TSM or EU B or D tooling can be used. Compaction rollers of different

(A)

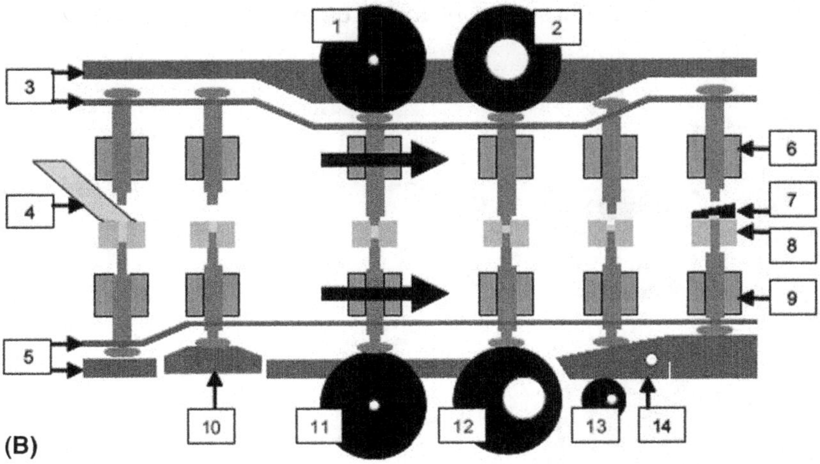

(B)

Figure 2 (**A**) A picture of a Presster^{TM} mechanical linear compaction simulator. (**B**) A typical schematic of a Presster mechanical linear compaction simulator. 1, upper precompaction roller; 2, upper main compaction roller; 3, upper punch cam; 4, hopper; 5, lower punch cam; 6, upper punch bushing; 7, takeoff bar; 8, die; 9, lower punch bushing; 10, dosing cam; 11, lower precompaction roller; 12, lower main compaction roller; 13, adjustment for ejection angle; 14, ejection cam. *Source*: From Ref. 22.

diameters from 7.5 ($\sim$190 mm) to 15.4 ($\sim$390 mm) can be used. The compaction roller equivalent to the press to be simulated should be used.

Compression forces are monitored by strain gauge instrumented compression roller pins, which are fixed on one side to the machine frame. Ejection forces are measured using strain gauge instrumented bolts. The ejection cam itself is not directly instrumented. Linear variable differential transducers (LVDTs), which are attached to each punch with a bracket, are used to measure the vertical punch movements during compaction.

Mechanical Rotary Cam Compaction Simulator

The mechanical rotary cam compaction simulator (Stylcam) (Fig. 3) is set up as a single station eccentric tablet press where both punches follow a programmable electronic cam designed to simulate a rotary tablet press. The machine can be run at a fixed speed or the speed of the cycle can be modulated during the run. This flexibility allows for the simulation of different rotary tablet presses. The tablet press profiles are programmed into the software and no parts are required to be changed to simulate different presses. Adjustments can be made to the filling height, ejection height, precompression force, main compression force and speed. The tablet takeoff force can also be monitored via a transducer placed on the feeding shoe, which detects the force required to remove the tablet from the punch surface. Additional functionality is provided by its ability to manufacture small batches of tablets in an automated mode.

The press is a stand-alone machine the size of some single punch eccentric presses such as Korsch EK-0, approximately 2 ft x 3 ft x 6 ft (w x d x h) and consists of the press, an electrical cabinet containing the programmable drive controller (which can be located remotely), and a PC for data collection and analysis. The upper part of the press is separated into a powder handling zone where the feeder, punches and tablet chute are located and the user is protect by clear, solid panels. The powder handling zone can be supplied for enhanced containment with wash-in-place (WIP) capability.

Each punch is driven by a cam that determines its vertical movements during the cycle. The system has mechanical inserts on the cam to mimic the action of punch movement during the precompression event. The adjustment of main compression force or tablet thickness is

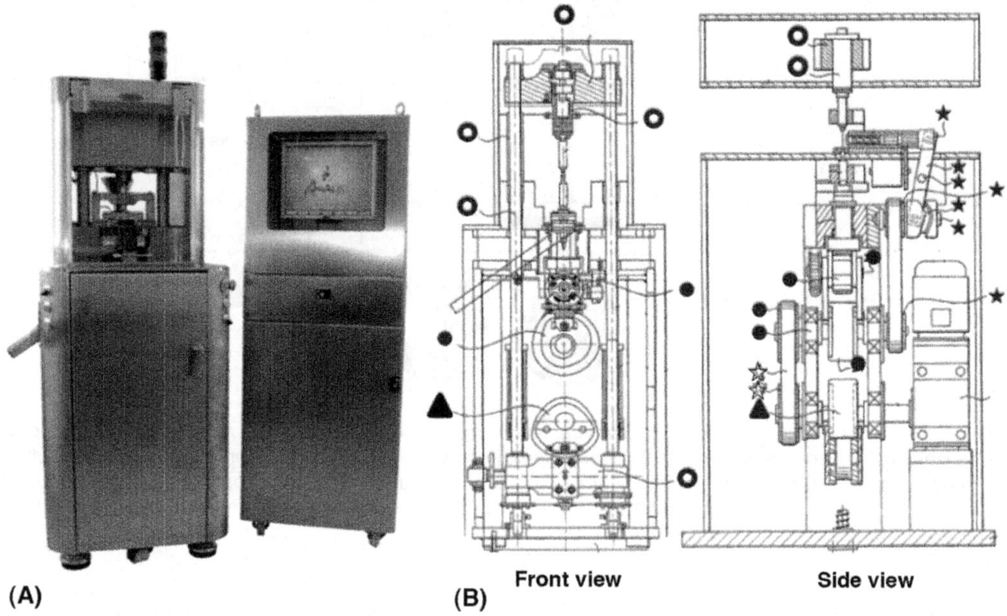

Figure 3 A picture (**A**) and typical schematic (**B**) of a Stylcam mechanical compaction simulator. **O**, System for upper punch movement; •, System for upper punch movement; ⋆, upper and lower cam synchronization ▲, Cam for upper punch movement.

done by moving the upper and lower cams closer or further away similarly to the rolls on a rotary press.

The two cams are synchronized mechanically and driven by a programmable motor that has the ability to accelerate or stop rapidly to adapt the movement of the punches and simulate the compression timings of any targeted rotary press (including relaxation timing and, for most presses, ejection timing). Although the system can adjust quickly there is still significant mechanical inertia resulting in less flexibility in the compression profile than a hydraulic press.

The pressure on the upper and lower punches is monitored by strain gauges located in the upper and lower punch holders. Two displacement captors located on both the upper and lower punch holders facilitate the measurement of the punch movement.

The electric cabinet contains an automat and a programmable drive in addition to the other electrical hardware (fuses, safety, etc.).

COMPARISON OF COMPACTION RESEARCH MACHINES

The type of tablet press or compaction simulator that will be purchased will ultimately depend on the specific needs of the end user. Table 1 compares important features of different types of compaction research equipment. Table 2 compares the features of different types of compaction simulators.

Some of the compaction simulators can also be fitted with additional accessories, a few of which are listed below.

- Instrumented dies
- Instrumented punches
- Temperature controlled tooling
- Instrumented takeoff transducers
- Roller compaction tooling
- Different hopper designs and sizes or multiple hoppers for making multilayer tablets

Table 1 Comparison of Single Station Presses, Tablet Presses, and Compaction Simulators

Attribute	SSP	MSP	Hydraulic	Mechanical linear	Mechanical cam
Easy to operate	Yes	Yes	Yes	Yes	Yes
Small amount of material required	Yes	No	Yes	Yes	Yes
Different compaction profiles	No	No	Yes	Limited	Limited
Rotary press simulation	No	Yes	Yes +/–	Yes +++	Yes ++
Easy to set up	Yes	Depends	Moderate	Yes	Yes
Easy to instrument	Yes	No	Yes	Yes	Yes
Data analysis	Poor to very good	Poor to very good	Very good to excellent	Very good	Very good
Space requirements	Small	Small to moderate	Moderate to large	Moderate	Small
Multilayer capability	Yes	Yes for some models	Yes	May be	May be
Roller compaction simulation option	No	No	Yes	Yes	No
Cost	Low to moderate	Moderate to high	Moderate to high	Moderate	Moderate

Source: Adapted from Ref. 16.

Table 2 Comparison of Compaction Simulators

	Hydraulic	Mechanical linear	Mechanical cam
Dwell time range	Unlimited	5.8–230 msec (12.7-mm punch head diameter)	3 msec minimum
Maximum vertical punch velocity (per punch)	1 m/sec/punch		
Maximum production cycle	10 tablets/min	4 tablets/min	50 tablets/min
Feeding mechanism	Manual filling Automatic feedshoe Automatic paddle filling	Manual filling Automatic feedshoe	Manual filling Automatic feedshoe Automatic paddle filling
Precompression capability	Yes	Yes	Yes
Multiple compaction events	Yes	No	No
Force range (kN)			
Pre-	0–50 kN	0–10 kN	0–50 kN
Main	0–50 kN	0–50 kN	0–50 kN
Profiles (types)	Sawtooth Sinewave Tablet press theoretical cam profile Versatile customization	Tablet press cam profile	Approximate sawtooth Sinewave Tablet press theoretical cam profile Limited customization
Stress/strain studies	Both	Strain	Strain
Load control/ displacement control	Both	Displacement only	Displacement only
Bridging force capability	Yes	No	No

There are different machine design features to consider when deciding what type of compaction simulator to purchase as described in this section. There are also other practical consideration that will mainly be driven by the needs of the customer when performing compaction experiments and general maintenance of the equipment. Cost, complexity of the studies to be performed, ease of use, space requirements, and portability and customer service are all important considerations. We will only offer an opinion on the complexity of the studies to be performed since the other considerations are more subjective.

Some common advantages for all compaction simulators are as follows:

- Compaction simulators can be used to evaluate the following:
  - The effect of tooling variation
  - Scale-up parameters
  - Build-up effects such as adhesion problems (much easier with mechanical rotary machine due to number of tablets produced per hour)
  - The effect of process variables (speed, etc.)
  - Basic compaction mechanisms
  - Tablet properties (strength, disintegration, dissolution) under identical manufacturing conditions (since the compaction history of each individual tablet is known)
- Milligram quantities of material are required.
- Fingerprinting of actives, excipients, and formulations is possible.

Some common disadvantages for all compaction simulators are as follows:

- Impact of bulk flow on the rotary tablet press cannot be simulated.
- Although the available systems do have paddle feeding systems available, they cannot directly simulate the effect of the force feeder on a rotary tablet press.
- Centrifugal force due to the turret rotation cannot be simulated.
- Although it is easy to train users for basic machine operation, it requires expertise in data manipulation and interpretation for maximum utilization of the machine.

Some advantages of using hydraulic compaction simulators are as follows:

- Under constant stress/strain conditions, compaction studies can be performed.
- The machine can be operated under load control.
- Customized profiles/waveforms, including triangle (sawtooth) profiles, can be created and followed precisely.
- Formulations can be stressed beyond rates that current manufacturing presses are capable of achieving.
- It is possible to simulate both single-ended and double-ended compaction events.

Some disadvantages of hydraulic compaction simulators are as follows:

- Accurate simulation of a given rotary tablet press's profile requires considerable effort and time.
- There are general problems associated with hydraulic systems such as space, cooling requirements, and potential leakage (newer hydraulic systems use vacuum seal cartridges to isolate any potential leaks).
- Proper tooling alignment can be time consuming (which may or may not be a concern with the newer hydraulic systems).
- The seismic effect resulting from rapid changes in velocity of the mass in front of the load cells (punch holder plus punch itself), especially at the beginning and end of a cycle, is more pronounced at higher operating speeds (which may or may not be an issue with the newer hydraulic systems).
- There is a tendency for overcompensation by the lower punch during simulation of a single-ended compaction event (16). Newer machines can be fitted with lower punch locks for high-speed single-ended tests that can minimize/eliminate this problem.

Some advantages of mechanical linear compaction simulators are as follows:

- No programming or waveform is required for precise simulation of rotary tablet press punch displacement during the compaction event.
- Displacement profiles of any compression roller would be identical to that of the rotary press since the same roll diameter is used.

Some disadvantages of mechanical linear compaction simulators are as follows:

- Compression rollers for the individual rotary presses to be simulated must be purchased individually and the compression rollers must be physically changed to simulate different machines (although many may be used interchangeably between different machines).
- Distance between precompression and compression rolls (and therefore compaction event) is fixed and may not be representative of the actual machine (25).

- Initial powder loss (≤2%) due to inertial effects of first carriage movement stages may occur (25).
- Some inertial effects on the punches and LVDTs can cause tilting of the punches and LVDTs and distortion of the data.
- Multilayer tablet manufacture is not easily achievable.
- It cannot be used to simulate a single-ended compaction event.

Some advantages of mechanical rotary compaction simulators are as follows:

- Ease of operation and size and space requirements are similar to an R&D rotary tablet press.
- Change of tooling, alignment, calibration, and deformation is easy to perform.
- It is possible to simulate both single-ended and double-ended compaction events.
- Containment options are available (for compression zone and the entire machine).
- The punch displacement profile is pre-programmed and selected through the software making it easy to change the machines that will be simulated.
- The punch displacement profile is independent of the tooling used.
- The manufacture of clinical supplies is possible with options for sorting tablets, etc.
- Batchwise manufacturing is possible and changes to operating conditions can be done quickly.

Some disadvantages of mechanical rotary compaction simulators are as follows:

- Multilayer tablet manufacture is not programmable (although manually it is possible, but achieved with difficulty).
- The sawtooth profiles are not as precise as the sawtooth profiles achieved on a hydraulic simulator.

CRITICAL ISSUES: CALIBRATION OF THE TRANSDUCERS AND CORRECTION FOR SYSTEM DEFORMATION

Calibration of the Transducers

The integrity of the data collected during a compaction event depends on, among other factors, the accuracy of the transducers, the quality and the frequency of the transducer calibration, the distance between the transducers and the punch tips, and the method used to correct for the system deformation. Although chapter 5 addresses the topic of instrumentation used in tabletting research in more depth, some of the critical issues with respect to compaction simulators will be highlighted here.

When compared to the mechanically driven simulators, the issue with respect to the calibration and location of the LVDTs is more critical for the hydraulic simulators because the LVDTs are used for both controlling and monitoring the punch movements when the hydraulic machines are operated under position control. The process of relating the output of a measuring system (such as LVDTs) to the magnitude of the input parameter (i.e., punch displacement) requires that the input magnitude be known (known standard, known accuracy, known input).

$$\text{Known input}(x) \rightarrow \text{LVDT} \rightarrow \text{output}[y = y(x)]$$

The calibration relationship is then inverted to measure an unknown input magnitude based on the system output magnitude.

$$\text{Unknown input}(x) \rightarrow \text{LVDT calibration} \rightarrow \text{output}[x_{\text{cal}} = x(y)]$$

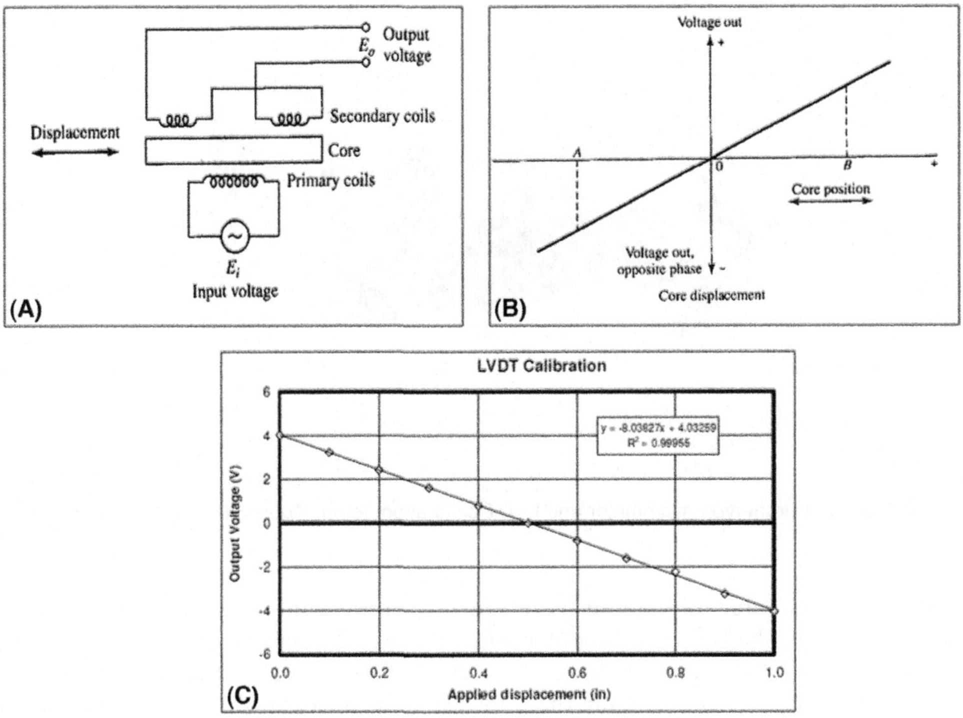

Figure 4 Overall process of LVDT calibration. (**A**) Schematic layout; (**B**) inputs and outputs; and (**C**) example set of data for calibration. *Abbreviation*: LVDT, linear variable differential transducer.

It is important to know that the above relationship is not generally linear toward the upper and lower limits of the LVDTs. Therefore, it is important to locate the LVDTs on the simulators in a way that the actual compaction zone of the materials will be corresponding to the linear region of the calibrated curves as practically as possible. The accuracy of the data analysis depends on the reliability of the data collected, which in turn is directly dependent on not only the integrity of the calibration and positioning of the LVDT, but also the quality of the actual LVDT that may differ from one brand and/or model to another. Problems related to the nonlinearity of the LVDTs can be eliminated by the use of digital or laser displacement transducers; however, such transducers have not yet been utilized in the field of compaction simulators.

The overall process of LVDT calibration is schematically shown in Figure 4A, B, and C. The known inputs could be provided by either using a set of slip gauges (in compliance with the NIST standards in the United States) or drum-type mechanical or digital micrometers (again, in compliance with the standards established by the national institutions). An example of a drum-type micrometer and LVDT calibration set up is shown in Figure 5. The use of the drum-type micrometer is practically impossible when an in-situ calibration of the LVDTs on the simulator is desired.

System Deformation

The precision and accuracy of measurement of both punch force and displacement, including the calibration of the transducers, are important considerations since a small error in measurement of either parameter can result in a much larger error in the calculation of the derived measurements.

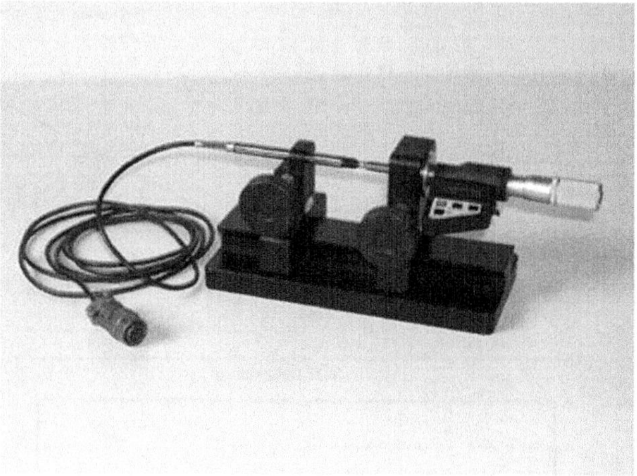

Figure 5 Example of a drum-type micrometer and LVDT calibration setup. *Abbreviation*: LVDT, linear variable differential transducer.

Since it is not possible to place an LVDT directly on the tip of the punch, it is typically attached to the punch shaft or punch holder. Displacement measurements further along the punch (or on the holder) should be comparable to the movement experienced at the punch tip. However, the system deformation along the punch may be different than that experienced at the punch tip and must be taken into account and corrected for to make sure that reported punch displacement values are accurate. For this reason, the system deformation for a compaction simulator must be calculated to take into account any elastic deformation that may be experienced by the tablet punches, punch holders or any other parts of the tablet press connected in series with the punches and involved in the compaction event. The observed system deformation is dependent on the magnitude of the applied force, the punch velocity as well as the specific tooling being used although, in general, the contribution of the tablet tooling alone to the overall system deformation is negligible.

Not much information is available in the literature with regard to the calculation of or the correction for the deformation that occurs during the compaction event. Most authors neglect to mention deformation in their experimental accounts and those who do consider deformation, usually only mention that it has been evaluated or corrected for with no specifics given about the methods involved in this correction.

Çelik and Marshall (16) used a punch-on-punch method to account for deformation with a hydraulically driven compaction simulator. They interchanged the upper and lower punches to measure the bedding in effects of the punches into the punch holders. When the bedding in effect was taken into account, they found that the total distortion of the upper system was larger than that of the lower system. They also observed that removal and replacement of the punches resulted in significantly different deformation values, therefore, once the calculations were made, the punches were not removed from their holders throughout their study.

Lloyd et al. (26) investigated the effects that changing the punch types and loading rates would have on displacement measurement errors on a hydraulically driven compaction simulator. These workers used a fourth-order polynomial equation to compensate for the deformation and load data and found that at loads above 4 kN, the major source of error for displacement measurements was deformation of the punches and holders. The authors suggested that a specific correction equation would be needed for each different type of tooling being used.

Holman and Marshall (27) discussed the calibration requirement for LVDTs on a hydraulically driven compaction simulator. It was observed that a nonlinear relationship existed between the LVDTs and the voltage output over the range specified by the manufacturer. When a plot of displacement versus voltage was examined visually, the relationship appeared to be linear. However, a plot of the residuals indicated that there was a trend being followed, therefore, subsequent equations were examined until a randomness in the residuals resulted. This occurred using a cubic equation. Upon inspection of the cubic equation, they observed that the *y*-intercept was a number close to zero, which should be expected since there would be no change in voltage if there was no change in displacement. The punch deformation that occurred on compaction was also found to be nonlinear, but was adequately described using a second-degree polynomial. The authors felt that the nonlinear elastic behavior of the system may be due to several factors, including (*i*) variability of the actual area of contact between the punch faces and between the punch heads and the end plated of the punch holders with applied force, and (*ii*) the nonlinear elastic deformation of the screw threads that hold the end plate in place on the punch holders.

Cocolas (28) used a punch-on-punch method under both upper and lower punch load control to measure the system deformation of a hydraulically driven compaction simulator (Fig. 6). For measurement of the upper punch, the lower punch position was adjusted to zero (based on the LVDT reading) and the upper punch would then be brought into contact with the lower punch using load control. Once the punches were in contact, the force was increased from approximately 0.025 to 40 kN and back down to near zero. The same procedure was followed for the lower punch using lower load control, with the upper punch position brought to zero. Cocolas pointed out that the deformation measured using this method was the deformation of the entire system, not of each punch separately. He measured the total deformation as the difference between the upper and lower LVDT readings when the force was being applied. By plotting deformation versus average applied force (which was the

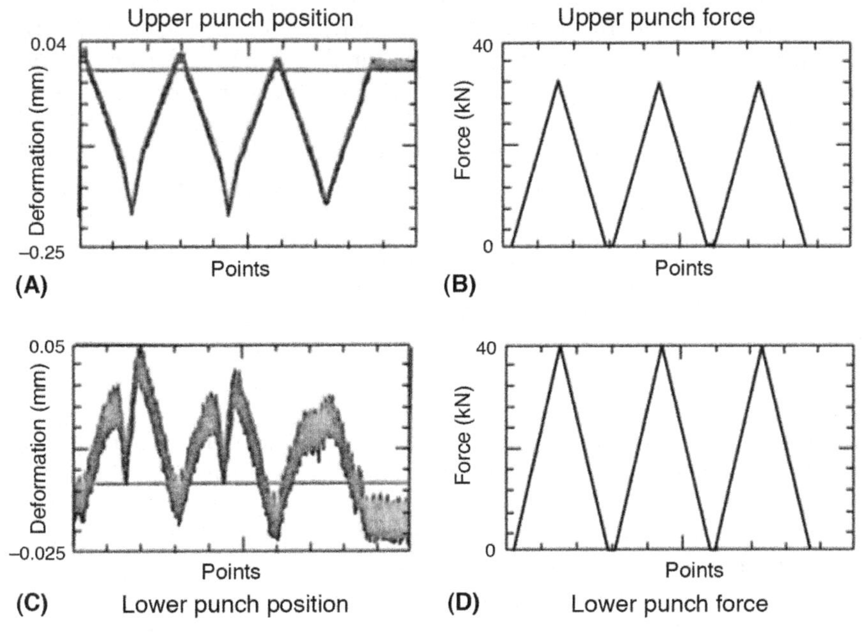

Figure 6 System deformation under upper load control. *Source*: From Ref. 28.

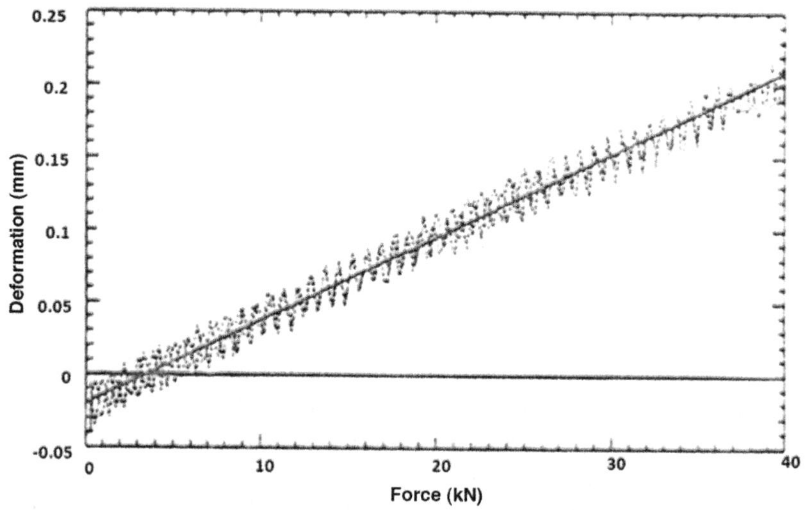

Figure 7 Calibration under upper load control; solid line is the best-fit linear regression. *Source*: From Ref. 28.

average between the upper and lower measured applied forces), the slope of the line would define their relationship once analyzed by linear regression (Fig. 7). The author ignored the intercept, stating that it was an artifact of the LVDT position reading not being exactly zero and inconsequential since there was no deformation at zero force.

Levin et al. (29) reported that in a mechanical linear compaction simulator a system deformation can be performed by using metal tablets of various thicknesses at different forces and speeds. The observed punch separation measurements were compared to the reference metal tablet thicknesses and corrected using a multivariable regression analysis. These authors also performed a dynamic calibration using two metal tablets with different thicknesses at two speeds and two force levels. The resulting punch separation data were again compared to the reference metal tablet thicknesses and corrected using multivariate regression analysis.

In a mechanical rotary compaction simulator the general procedure for system deformation involves setting up the machine with flat-faced tooling without the die in place. The punches are compressed onto themselves (punch to punch) at increasing compression pressure. Since the actual thickness between the punches is zero during punch to punch compression any deviations in the measured thickness are due to the deformation of the tooling and the machine (to the point where the displacement transducers are located). A calibration curve is established for the upper and lower displacement system that can be loaded into the profile to determine the corrected compression thickness of the tablets measured in the die. Since the expansion of steel will vary with temperature it is important to perform the machine deformation measurement at the same temperature as the tablet characterization.

Measuring the in-die tablet height with the two original transducers, not only the punch deformation, but also machine deformation and the small movements of the machine parts, must be taken into account. Fretter (23) found that with increasing force, the distance measured by the transducers deviated from a straight line, which the authors suggested was most likely caused by the play of the machine parts. Over the entire force range, the machine deformation could be approximated using a fourth-degree polynomial where the residuals of the curve were ~10 μm. This allowed in-die tablet thickness measurements between 5 and 45 kN, with an error of 10 μm.

TABLET PRESS SIMULATION

Depending on the type, compaction simulators compress materials by following punch position (displacement) profiles or load profiles, that is, waveforms. The latter can only be applied by the hydraulically driven compaction simulators, which also offer the most flexibility in terms of the types of profiles that can be followed by the punches. Figure 8 shows examples of some possible punch profiles. Note that it is also possible to combine different punch profiles for the upper and lower punches to follow when using a hydraulic compaction simulator.

The mechanical linear compaction simulator does not follow an input waveform, instead the compaction event occurs using the compression rollers from the specific tablet presses to be simulated.

The mechanical rotary cam compaction simulator requires the input of the theoretical waveform for the specific tablet press to be simulated. Later in this section, theoretical waveforms are discussed in more detail.

As already mentioned, hydraulically driven compaction simulators require the input of a waveform for the punches to follow. The most commonly used punch position or punch load control waveforms are either a sawtooth profile with a constant velocity or a sine wave profile, which has the shape of a sine wave from 0° to 180°. By providing the system with the coordinates of punch position with respect to time, the punches will be moved accordingly. To actually simulate the displacement profile of a tablet press, two methods of determining punch movement are generally used: (*i*) actual waveforms, which are collected through instrumentation of the tablet press via LVDTs, slip rings, radiotelemetry, infrared telemetry, etc., and (*ii*) theoretical waveforms, which are equations derived based on a knowledge of the geometry of the press and punches and the operating conditions.

Problems involved with using these methods include the fact that most theoretical equations have been derived when no material was present in the die. This would be expected to lead to discrepancies when an actual compact is being formed (30). In addition, when actual waveforms are being employed, consideration must be made for the deformation that is occurring during the compaction event. If the deformation is not taken into account when using the waveform on a system such as a hydraulic simulator, the so-called actual waveform will not really be simulated.

Theoretical waveforms describing the punch displacement for single station tablet presses (31,32) and rotary tablet presses (33–35) have been described in the literature.

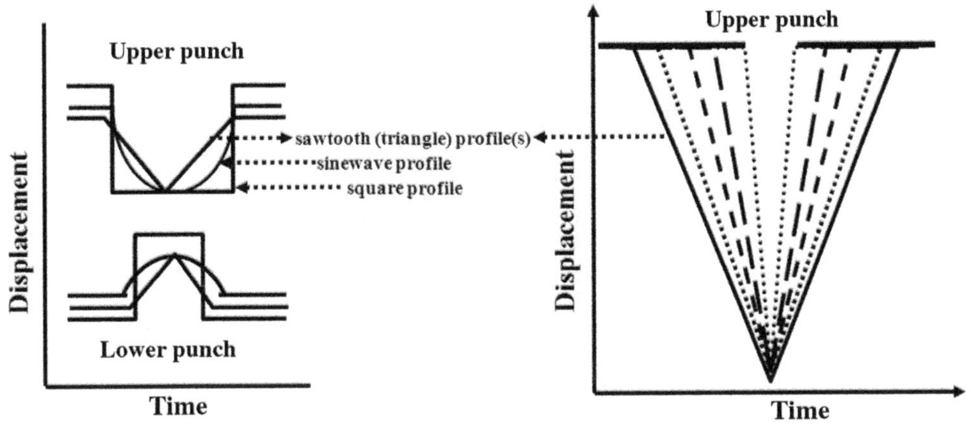

Figure 8 Examples of punch profiles that can be followed by a hydraulic compaction simulator.

Figures 9 and 10 outline the critical variables that are considered in deriving the machine punch displacement.

In Figures 9 and 10 where r_1 is the radius of the compression roller, r_2 is the vertical curvature of the punch head, r_3 is the radial distance between the turret and punch axes, and x and x_1 are the horizontal distance between the vertical center lines of the compression roller

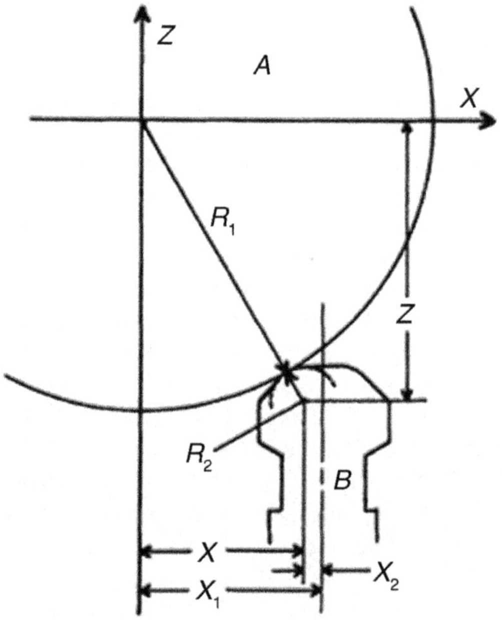

Figure 9 Diagram of punch head contacting pressure roller from Rippie and Danielson. *Source*: From Ref. 33.

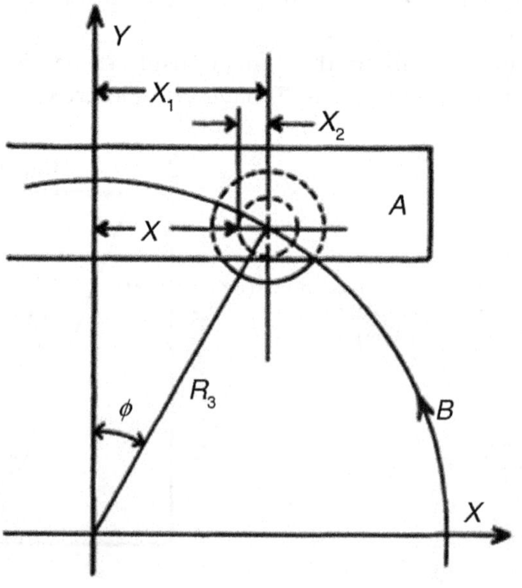

Figure 10 Diagram of rotary press radius of punch travel from Rippie and Danielson. *Source*: From Ref. 33.

and upper punch, respectively, and x_2 the center of the vertical curvature of the punch head. ϕ is the horizontal angle measured from the punch axis to the vertical center line of the roller and is a function of the turret angular velocity.

Armstrong and Palfrey (36) did a comparative study between theoretically predicted and actual punch movements using an Apex eccentric press. Deviation from the theoretical displacement pattern was found to increase with increased force and decrease with increased machine speed. The magnitude of deviation depended on the force, speed, material being compacted and the power of the motor driving the press.

Çelik et al. (37) conducted a simulation study in an attempt to mimic a single-ended compaction process. In this study, they performed the compaction tests using three model powders first on an F3 single station tablet press followed by using a hydraulic compaction simulator. Using the F3 single station press, tablets were made under varying pressures while operating at rates of 42 and 89 tpm. The F-press was fitted with 8-mm diameter, flat-faced, round F tooling. The compaction data, which consisted of the forces on the upper and lower punches, and the displacement of the upper punch, were simultaneously collected using a digital storage oscilloscope for both speeds. The blank displacement profiles of the upper punch, that is, when there is no powder on the F3 press, at these operating rates were also obtained to be used to program a compaction simulator. The authors then compacted the same model powders using the compaction simulator, which was fitted with exactly the same tooling, at the two different pressure levels utilizing the two single-ended compaction profiles obtained from the F3 press. An identical digital storage oscilloscope was employed to monitor the compaction data consisting of the forces on and the displacements of the upper and lower punches. Subsequently, the acquired data from both presses were analyzed using specialized software. Whenever the simulator was used, the authors corrected the displacement measurements of the upper and lower punches for the distortion of the punches as described in section "System Deformation." The postcompaction tests included the measurements of weight, thickness, diameter and breaking force of the resulting compacts. The same set of equipment (balance, micrometer, and hardness tester) was used to test the compacts produced from both the compaction simulator and the F3 press.

The blank displacement profiles of the upper punches of the F3 and the simulator operating at 42 tpm are presented in Figure 11. The superimposed curves suggest that the

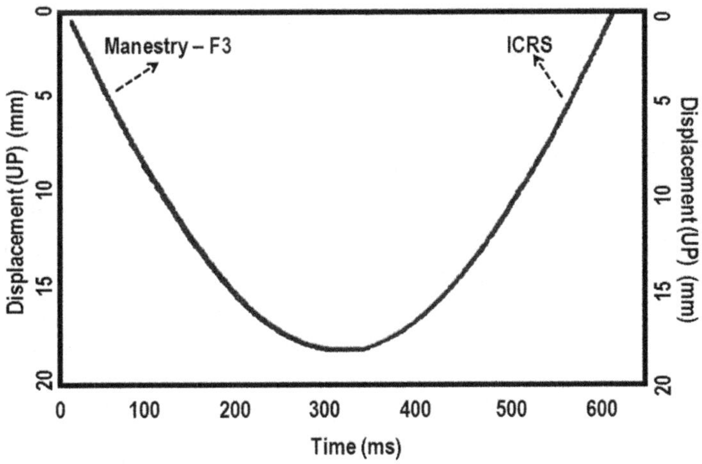

Figure 11 Upper punch displacement profiles of the Manesty F3 press and compaction simulator at 42 tpm. *Source*: From Ref. 37.

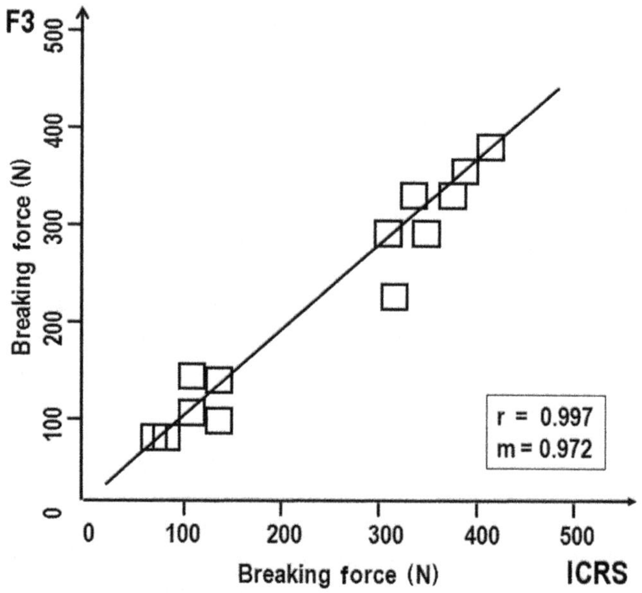

Figure 12 Correlation of tablet breaking force results obtained from the simulation of the Manesty F3 press using the compaction simulator *Source*: From Ref. 37.

simulator was able to follow the given profile precisely. The lower punch of the compaction simulator was programmed to stay at a fixed position, since the lower punch of the F3 press does not move during the loading stage as this type of press is controlled mechanically. However, as reported earlier, the lower punch of a hydraulically driven simulator (especially, older generation simulators) can move slightly downward and then upward during the compaction of a material.

Figure 12 shows the linear correlation between the compact breaking force data obtained from the F3 press and the simulator in this study. Although the breaking strength data were then calculated using Fell and Newton's equation (38). The slope of the line is very close to the ideal value, which is equal to 1. Similarly, a linear relationship was observed between the energy parameters (such as total work of compaction, net work of compaction, etc.) obtained from the two machines (Fig. 13).

From the findings of this work, the authors concluded that simulators can be employed to mimic the compaction event of single station presses.

Pedersen et al. (39) used high-speed video to record displacement profiles on a Manesty Betapress. These authors compared the actual videotaped displacement profiles with the displacement profiles predicted by the Rippie/Danielson equation. They concluded that this equation may be applicable under certain machine conditions, but, overall, should be used with caution.

Ruegger (40) used a hydraulic compaction simulator to simulate the theoretical punch displacement profiles as calculated from Rippie and Danielson's equation as modified by Hoblitzell and Rhodes (34). The author also obtained actual punch displacement profiles from a Manesty Betapress using an instrumented punch system equipped with an infrared telemetry device.

Ruegger investigated several aspects of tablet press simulation: (*i*) how closely the theoretical equations could predict the punch displacement profiles, (*ii*) how closely the

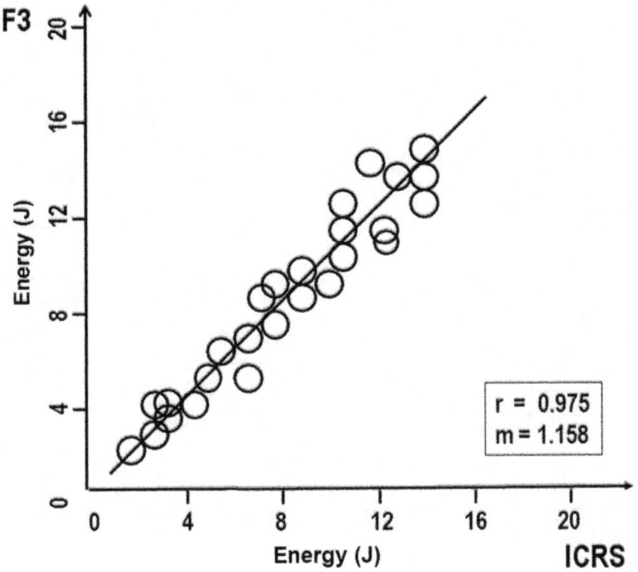

Figure 13 Correlation of energy results obtained from the simulation of Manesty F3 press using a hydraulic compaction simulator. *Source*: From Ref. 37.

hydraulic compaction simulator truly simulates the Manesty rotary tablet press punch displacement profiles, and (*iii*) whether the data collected using the hydraulic compaction simulator with the simulated Manesty and theoretical waveforms could be correlated to the actual Manesty data.

The following equation was used to determine the theoretical punch displacement of the Manesty Betapress.

$$z = [(r_1 + r_2)^2 - (r_3 \sin \omega t - r_p)^2]^{0.5}$$

where z = the displacement of the punch,

r_1 = the radius of the compression roller,
r_2 = the radius of the vertical curvature of the punch head,
r_3 = the radius of the pitch circle,
r_p = the radius of the punch head flat, and
ω = the angular velocity of the turret.

Figure 14 shows the calculated theoretical punch displacements compared to the simulated theoretical punch displacements of the Manesty Betapress operating at 70 rpm. These authors found differences between the simulated and theoretical displacements, which increased as the force was increased. The fact that the theoretical punch displacement is calculated without taking any deformation effects into account, that is, it is derived for an empty die, was suggested to be a factor in the differences between the displacements observed with and without material in the die. When the punches actually come into contact with the powder, the punches and any machine parts connected in series to the punches are deformed to a certain extent. Additionally, during decompression, the elastic expansion of the tablet may affect the punch displacement, that is, during unloading, as the punch moves out of the die, the tablet expands to a certain extent when the force is removed and the tablet is actually exerting

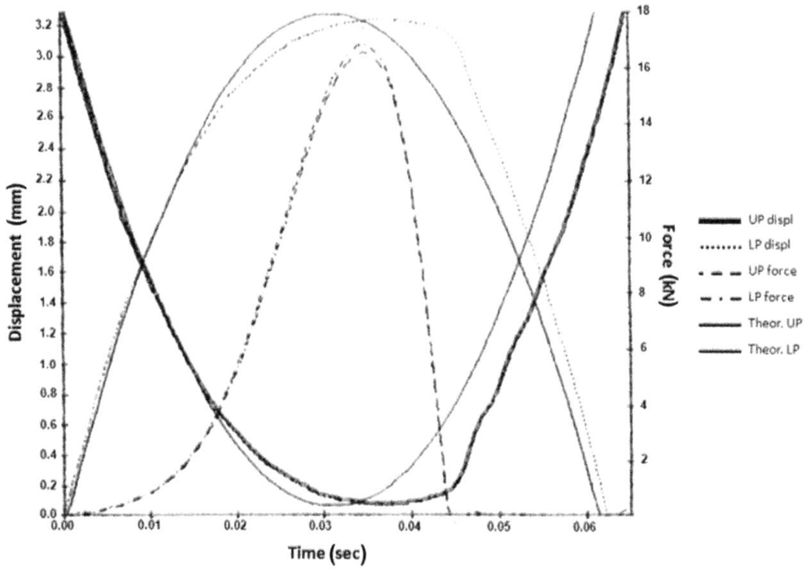

Figure 14 Punch displacement comparison. *Source*: From Ref. 40.

a force on the punch for a short time during decompression. This is not accounted for in Rippie and Danielson's equation.

Figure 15 shows the actual Manesty waveform when the press was operating at 70 rpm compared to a blank run on the hydraulic compaction simulator and a run with the model powder mixture for the simulation of this waveform. The hydraulic compaction simulator can very closely duplicate the actual Manesty profile as shown; however, when the powder is present in the compaction simulator die, there is a deviation in the displacement, especially during decompression. This problem was solved by altering the actual input profile into the hydraulic compaction simulator until the displacement of the simulator punches was equivalent to the actual displacement of the Manesty punch.

Figure 16 shows the sequential changes in the waveform being produced by the hydraulic compaction simulator at 70 rpm, that is, from what was wanted to what was actually produced on the hydraulic compaction simulator, based on the data points that were entered. With each subsequent input change, the simulator was able to simulate the punch displacement of the Manesty machine more precisely.

When the breaking strength values for the tablets produced on the two machines were compared, they were found to not be equivalent. The authors suggested that the feedframe on the Betapress was creating additional mixing action with the lubricant (magnesium stearate), which could not be directly simulated on the hydraulic compaction simulator. To confirm this theory a mixing study was conducted and the tablet breaking strength values were compared for the original material, the material removed from the feedframe and the material subjected to additional mixing times up to 30 minutes.

The authors showed that when comparing the tensile strength values for the different samples, after simulation it could be seen that the feedframe was causing some overlubrication of the formulation, but not to the same extent as the material when mixed for an additional 30 minutes (Fig. 17).

Although the pressure–tensile strength curves for the hydraulic compaction simulator and the Manesty tablet press did not overlap and the slope of the lines was not close to the

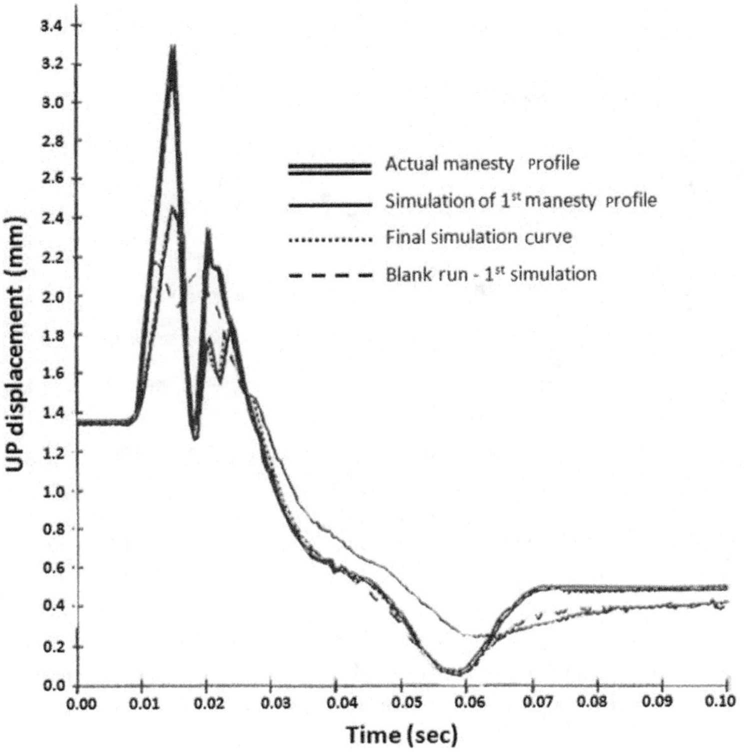

Figure 15 Manesty displacement compared with blank run on the simulator. *Source*: From Ref. 40.

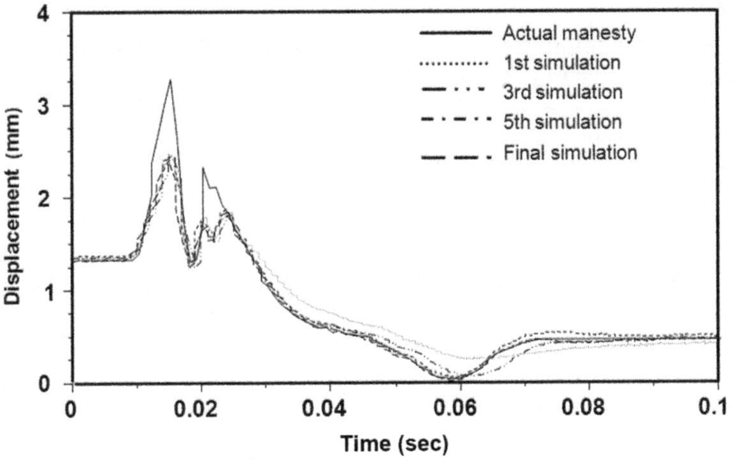

Figure 16 Sequential changes to input waveform. *Source*: From Ref. 40.

ideal value of 1, nevertheless, a very good correlation (r^2 correlation coefficient above 0.95) was achieved. This suggests that the results obtained from the compaction simulator can be used successfully to predict the material behavior on the rotary tablet press.

Analogous to the findings for the simulation of the actual Manesty, the r^2 correlation coefficients were all good (>0.989) for the simulation of the theoretical profiles. Suggesting that

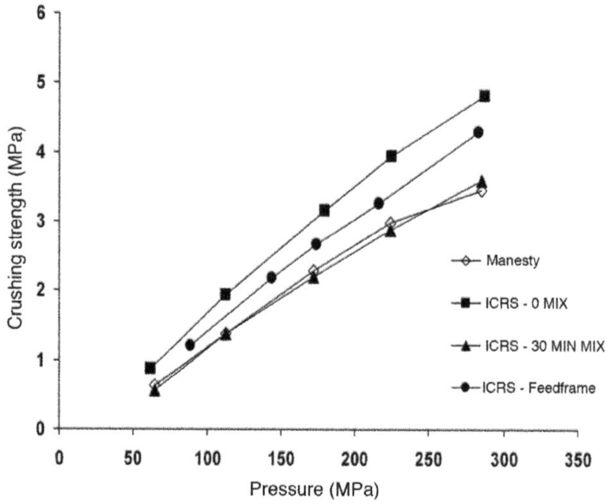

Figure 17 Crushing strength comparison of original blend to feedframe material and after 30 minutes of mixing. *Source*: From Ref. 40.

despite the differences in the actual and theoretical profiles, the theoretical profiles can still be used to gain information with regard to how the formulation will behave on the actual tablet press.

PRACTICAL APPLICATIONS OF COMPACTION SIMULATORS

One of the frequently asked questions regarding compaction simulators is how they should be utilized in compaction studies to get the maximum information and/or benefit. There are several approaches that one might take when using a compaction simulator (depending on the type of information that is desired), some of which are listed below:

1. Fundamental material characterization with respect to compaction properties
2. Robust tablet formulation development
3. Simulation of the tablet press that will be used to manufacture the formulation

In fact, using any type of compaction simulator, one can mimic the compaction event of almost any given tablet press, running under any conditions (within the restrictions of the systems) as shown earlier in this chapter. As realized in those examples, if one takes the approach to develop a formulation, he/she should bear in mind that it will be highly likely that the compaction behavior of that formulation will be affected by the speed and the amount of pressure applied. As a consequence of this, ideally, that formulation should be run at many different speeds and pressure levels for a given model of a given press. This will require even more effort if the same formulation were to be manufactured on different types of tablet presses. It is obvious that this time-consuming approach goes against the philosophy of commissioning a compaction simulator for rapid formulation development.

If the ultimate goal when using a compaction simulator is to mimic a specific tablet press, then, regardless of the time and effort required, these machines should be used to simulate a given press. If the purpose is to develop robust formulations and compare the compaction behavior of different materials (excipients, APIs, formulations), then simulating a press would not be the right approach. As described in chapter 7, if the compaction behavior of three different materials (with different bulk densities) would be compared simulating a rotary tablet press running at a given rpm, in reality, although the rpm speed is the same, the compaction test conditions would not be identical since the punch-tip velocities, at the time the upper punch initially contacts the powder sample, would differ for sample to sample. The use of sawtooth (triangle, constant velocity) profiles would be the most preferred approach for the comparative compaction studies. This approach is also the best way of using the compaction simulators to rapidly develop a robust tablet formulation that would run on any given tablet machine.

If the goal of using the compaction simulator is to study stress/strain relationships, waveforms with extended dwell times would be needed. Such waveforms are best applied by the hydraulically driven compaction simulators since they can follow profiles under extended constant stress (or constant strain) conditions.

When it is desired to study the strain rate sensitivity of materials, then the use of sawtooth profiles would be best for moving punches at a constant speed throughout the compaction event. It should be noted that even sawtooth profiles have their limitations since the velocity is constant except at the maximum penetration point where the punch stops momentarily and the direction of the movement is reversed with the same speed of punch displacement.

As mentioned earlier, simulators have been used for a variety of purposes such as establishing whether a particular formulation has scale-up potential, studying the effects of different methods for massing, drying, and blending, assessing changes in compactibility that may be encountered when transferring a material from one machine to another, and establishing alternative ingredients that may be used in a formulation (Table 3).

The following section gives some examples of how compaction simulators have been utilized in different phases of product development or for the characterization of material properties and compaction modeling.

Preformulation/Material Characterization

Yang et al. (46) used a hydraulic compaction simulator following a theoretical displacement profile mimicking a Manesty Betapress to characterize the compaction properties of several polyethylene oxide (PEO) polymers prior to formulation development for a modified release tablet formulation. Regardless of particle size fractions or molecular weight, the PEO materials were found to have overlapping Heckel profiles. The authors suggested that the shape of the profiles was characteristic of plastically deforming materials, which they also confirmed by calculating the strain rate sensitivity of the bulk materials. It was also observed that the PEO materials underwent significant axial recovery upon decompression and ejection of the compacts, which was directly proportional to the compaction pressure that was applied. This information was used for the selection of excipients during the formulation development phase.

Larhrib and Wells (65) used a hydraulic compaction simulator with a sawtooth profile to study the effect of cooling rate on the compaction properties of PEG 10000. They could not find a good correlation between the degree of crystallinity and the tablet strength. They observed that the slow-cooled sample was more crystalline and had higher compactibility, producing stronger tablets than the untreated and quench-cooled samples. Despite the fact that the untreated sample had intermediate crystallinity between the slow-cooled and quench-cooled samples, it had the worst compaction properties. They were able to show a correlation between plasticity and work of failure—the slow-cooled PEG had the highest plasticity and higher values for work of failure while the untreated material showed the least plasticity and lowest values for work of failure.

Table 3 Selected Literature Review of Research Articles Using Different Compaction Simulators

Displacement control				
Sine wave		Sawtooth		
Double ended	Single ended	Double ended	Single ended	Mechanical linear
Hunter, Fischer, Pratt, and Rowe (1976) (15)	PV Marshall, York, and Richardson (1986) (51)	Pitt, Newton, and Stanley (1990) (54)	Roberts and Rowe (1985/1986/1987) (18, 69–73)	Picker (2000) (82)
Mann, Bowen, Hunter, Roberts, Rowe, and Tracy (1981) (41)	Inchley (1991) (52)	Çelik, Mollan, and Chang (1991) (21)	Bateman, Rubinstein, and Wright (1987) (74)	Kuppuswamy, Anderson, Augsburger and Hoag (2001) (83)
Mann, Roberts, Rowe, and Hunter (1982) (42)	Van Veen, van der Voort Maarschalk, Bolhuis, Zuurman, and Frijlink (2000) (53)	Garr and Rubinstein (1991) (55–57)	Bateman, Rubinstein, Rowe, Roberts, Drew, and Ho (1989) (75)	Durig, Salzstein, Skinner, Harcum, Grasso, and Lau (2002) (84)
Mann, Roberts, Rowe, Hunter, and Rees (1983) (43)		Çelik and Maganti (1993, 1994) (58, 59)	Pedersen and Kristensen (1995) (76)	Picker (2003) (85, 86)
Hunter (1984) (17)		Çelik and Okutgen (1993) (60)	Akande, Rubinstein and Ford (1997) (64)	Durig, Harcum, Lusvardi and Skinner (2003) (87)
Bateman, Rubinstein and Thacker (1989) (44)		Cocolas and Lordi (1993) (61)	Cunningham, Sinka, and Zavaliangos (2004) (77)	Betz, Meyer, Puchkov, and Leuenberger (2004) (88)
Çelik and Marshall (1989) (16)		Mollan and Çelik (1994) (62)	Sinka, Cunningham and Zavaliangos (2004) (78)	Tye, Sun and Amidon (2005) (89)
Holman (1991) (96)		Çelik, Ong, Chowhan and Samuel (1996) (63)	Wu, Ruddy, Bentham, Hancock, Best and Elliott (2005) (79)	Kimura, Betz and Leuenberger (2007) (90)
Rubinstein, Bateman, and Thacker (1991) (20)		Akande, Rubinstein and Ford (1997) (64)	Wu, Best, Bentham, Hancock, Bonfield (2005) (80)	Sun and Hou (2008) (91)
PV Marshall, York, and Maclaine (1993) (45)		Larhrib and Wells (1997) (65)	Wu, Best, Bentham, Hancock, Bonfield (2006) (81)	Sun, Hou, Gao, Ma, Medina and Alvarez (2009) (92)
Yang, Venkatesh and Fassihi (1996) (46)		DeCrosta, Schwartz, Wignent and Marshall (2001) (66)		
Yang, Venkatesh and Fassihi (1997) (47)		Wang, Guillot, Bateman and Morris (2004) (67)		
Kim, Venkatesh and Fassihi (1998) (48)		Gabaude, Guillot, Gautier, Saudemon and Chulia (1999) (68)		
Ruegger and Çelik (2000) (49, 50)				
	No mention of whether single or double ended			
	Pitt and Newton (1987) (19)			
	Ho and Jones (1988) (93, 94)			

Source: Mechanical rotary compaction simulator not included—equipment is new, and limited publications are available at the time of this publication.

Van Veen and coworkers (53) compared the tensile strength of tablets containing two plastically deforming materials, sodium chloride and pregelatinized starch when made as single component tablets compared to their binary mixtures. A linear relationship was found between yield pressure and the amount of pregelatinized starch added to the sodium chloride. The nonlinearity observed when evaluating the porosity and tensile strength of the binary tablets was related to the lack of interparticulate bonding that had occurred between the two materials in the binary mixtures.

Formulation Development

Çelik et al. (63) studied the effect of particle size and shape of a drug substance with two different crystal habits and the compaction properties of binary mixtures with microcrystalline cellulose. A hydraulic simulator was used to make the tablets following a double-ended sawtooth constant velocity displacement profile. The total work of compaction (TWC) was used to correlate the energy involved in the compaction process with the mechanical strength of the different mixtures. With drug substance alone, slightly higher TWC values were obtained with the smaller particle size drug substance particles. It was concluded that the smaller particles would have increased interparticulate bonding during compaction and therefore would require more energy during the compaction event, resulting in increased mechanical strength. In the binary mixtures with microcrystalline cellulose, the compact strength increased with increasing amounts of microcrystalline cellulose and this was reflected in an increase in the TWC as well. In fact, a linear correlation was found between the increase in strength and TWC, suggesting a change in the deformation mechanisms of the formulations as increasing amounts of microcrystalline cellulose were added to the drug substance.

Yang and others (47) used a hydraulic compaction simulator to correlate the compaction properties of the individual components of a triple-layer tablet to the final triple combination. They found that there was a mixture of plastic and brittle fracture in the individual layers and that it was important to have a good understanding of the compaction properties of each layer. By having similar deformation properties in the three layers the authors felt that this led to a more robust tabletting process.

Compaction Process Variables

Akande and coworkers (64) studied the effect of a single compaction event versus a double compaction event (pre + main compaction), using a sawtooth profile, on the compaction properties of a 1:1 acetaminophen:microcrystalline cellulose mixture. They found that once the minimum tablet porosity was achieved, additional increases in main compaction pressure did not continue to increase tablet strength. The addition of a precompression pressure increased tablet strength compared to a main compaction alone, which could be correlated to an increase in the total energy of compaction.

Ruegger and Çelik (49) used a hydraulic compaction simulator to evaluate the impact of the ratio and magnitude of precompaction and main compaction pressures on the tensile strengths of tablets made from pregelatinized starch, a binary mixture of dicalcium phosphate and microcrystalline cellulose, direct compression acetaminophen and direct compression ibuprofen. They found that the results were material dependent, that is, DCP/MCC and PGS tablet crushing strength values did not change as long as the total force applied was the same (regardless of whether the PCP > or < the MCP). Conversely, the DC APAP and DC ibuprofen were found to have increased tablet strength and decreased capping tendency when the PCP < MCP. No significant difference was found for any of the materials when the time interval between the precompaction event and the main compaction event was varied from 30 to 500 milliseconds.

Miscellaneous Studies

Several researchers utilized compaction simulators for the development of models to predict the physicomechanical characteristics of pharmaceutical materials (77–81).

DeCrosta and coworkers (66) used a hydraulic compaction simulator with a constant strain (displacement) waveform, several model compounds and a compaction calorimeter to look at the work and heat generated during the decompression and ejection phases of the tabletting event. The authors indicated that the use of the constant strain waveform allowed the separation of the unloading and compaction events. Acetaminophen was the only material with positive heat values most likely due to its elastic nature, which resulted in bonds breaking during unloading. Brittle materials were observed to have the most work during the decompression phase. The work of ejection was also higher for the brittle materials and shown to increase with increasing compaction force.

Wang et al. (67,95) used atomic force microscopy to establish the rank order for work of adhesion for ketoprofen > ibuprofen > flurbiprofen. They then used a hydraulic compaction simulator to confirm if the rank order predicted by AFM would be equivalent to the punch adhesion observed during the compaction event. Takeoff force and visual observation of the punch faces were used to rank order the punch face adhesion for the pure drug compacts. They found that the rank order of the compacted drug substance and formulations of each profen compound followed the rank order predicted by the AFM.

CONCLUSION

Compaction is the final process for uncoated tablets. This seemingly simple process (i.e., squeezing powders in a confined space) is in fact an extremely complex one. If it is achieved properly, many problems inherent to the formula ingredients or inherited from the previous processes can be eliminated or minimized. However, if this process can also be the cause of many other problems, if enough attention is not paid to the understanding of compaction behavior of what is being pressed. Today, several different types of instrumented presses are utilized by the pharmaceutical scientists to develop robust tablet formulations. Among these machines, compaction simulators are the most integrated and versatile tabletting research tools. They allow the scientist to conduct experiments for in-depth analysis of compaction characteristics of the pharmaceutical materials very efficiently in terms of time, cost and the amount/quality of knowledge gathered. The flexibility of employing the simulators from early formulation studies to troubleshooting in manufacturing makes such machines indispensible, especially in the light recent PAT/QbD phenomena.

REFERENCES

1. Brake EF. Development of methods for measuring pressures during tablet manufacture. MS Diss. Purdue University, 1951.
2. Brokedon, William patent number 9977, December 8, 1843, U.K., Shaping pills, lozenges and black lead by pressure dies.
3. Higuchi T, Arnold RD, Tucker SJ, et al. The physics of tablet compaction I. a preliminary report. J Am Pharm Assoc Sci 1952; 41:93–96.
4. Higuchi T, Nelson E, Busse LW. The physics of tablet compaction III. Design and construction of an instrumented tabletting machine. J Am Pharm Assoc Sci Ed 1954; 43:344–348.
5. Gagnon LA. A correlation of physical aspects in tablet production. PhD Thesis. Purdue University, 1957.
6. Markowski SL. The evaluation of tablet lubricants using a conventional-type tablet machine instrumented with strain equipment. PhD Thesis. Purdue University, 1958.
7. Train D. An investigation into the compaction of powders. J Pharm Pharmacol 1956; 8:745–761.
8. Shotton E, Deer JJ, Ganderton D. The instrumentation of a rotary tablet machine. J Pharm Pharmacol 1963; 15:106T–114T.

9. Knoechel EL, Sperry CC, Lintner CJ. Instrumented rotary tablet machines II. Evaluation and typical applications in pharmaceutical research, development, and production studies. J Pharm Sci 1967; 56(1):116–130.
10. Wray PE. The instrumented rotary tablet machine. Drug Cosmet Ind 1969; 105(3):58–68b, 158–160.
11. Shlanta S, Milosovich G. Compression of pharmaceutical powders I. theory and instrumentation. J Pharm Sci 1964; 53(5):562–564.
12. Varsano J, Lachman L. Compressibility of pharmaceutical solids I. instrumentation employed and preliminary results obtained. J Pharm Sci 1966; 55(10):1128–1133.
13. Rees JE, Hersey JA, Cole ET. Simulation device for preliminary tablet compression studies. J Pharm Sci 1972; 61(8):1313–1315.
14. Çelik M, Travers DN, Buttery TC. Computer logged changes of strain movements in compacts of direct compression bases and granules produced by wet massing techniques. 3rd Pharm Tech Int Conf 1983; 80–89.
15. Hunter BM, Fischer DG, Pratt RM, et al. A high speed compression simulator. J Pharm Pharmacol 1976; 28(suppl):65P.
16. Çelik M, Marshall K. Use of a compaction simulator system in tabletting research. Drug Dev Ind Pharm 1989; 15(5):759–800.
17. Hunter BM. Interpretation of data produced by tablet machine simulators. Proceedings of post-graduate course in tabletting. London, UK; 1984:54–63.
18. Roberts RJ, Rowe, RC. The effect of punch velocity on the compaction of a variety of materials. J Pharm Pharmacol 1985; 37:277–384.
19. Pitt KG, Newton JM. The effect of punch velocity on the tensile strength of aspirin tablets. J Pharm Pharmacol 1987; 39:65P.
20. Rubinstein MH, Bateman SD, Thacker HS. Compression to constant thickness or constant force: producing more consistent tablets. Pharm Tech 1991; 15(9):150–158.
21. Çelik M, Mollan MJ, Chang N. The reworkability of microcrystalline formulations. 10th Pharm Tech Conf Bologna, Italy 1991; 1(1):239–262.
22. Neuhaus T. Investigation and optimization of the Presster, a linear compaction simulator for rotary tablet presses. PhD Thesis. University of Bonn, 2007.
23. Fretter B. The Stylcam 200R, a rotary tablet press simulator optimizing the instrumentation: correcting for machine deformation upon measuring in-die tablet height. 6th World Meeting on Pharmaceutics, Biopharmaceutics and Pharmaceutical Technology, Barcelona, Spain. April 7–10, 2008.
24. Bateman SD. High speed compression simulators in tabletting research. Pharm J 1988; 240(6481): 632–633.
25. Gunterman G. Pfizer, TTS Workshop No. 88 from Art to Science. June 2005.
26. Lloyd J, York P, Cook GD. Punch elasticity compensation in the calibration of displacement measurements on a compaction simulator. J Pharm Pharmacol 1991; 43(suppl):80P.
27. Holman LE, Marshall K. Calibration of a compaction simulator for the measurement of tablet thickness during compression. Pharm Res 1993; 10(6):816–822.
28. Cocolas HG. The evaluation of compaction data obtained using the integrated compaction research system. PhD thesis. Rutgers University, 1993.
29. Levin M, Tsygan L, Dukler S, et al. Calibration of Presster for the measurement of tablet thickness during compression. AAPS Annual Meeting, Baltimore, 2004.
30. Armstrong NA, Palfrey LP. The effect of machine speed on the consolidation of four directly compressible tablet diluents. J Pharm Pharmacol 1989; 41(3):149–151.
31. Armstrong NA, Abourida NMAH, Gough AM. A proposed consolidation parameter for powders. J Pharm Pharmacol 1982; 35:320–321.
32. Charlton B, Newton JM. The theoretical estimation of punch velocities and displacements of single-punch and rotary tablet machines. J Pharm Pharmacol 1984; 36:645–651.
33. Rippie EG, Danielson DW. Viscoelastic stress/strain behavior of pharmaceutical tablets: analysis during unloading and postcompression periods. J Pharm Sci 1981; 70(5):476–482.
34. Hoblitzell JR, Rhodes CT. Determination of a relationship between force-displacement and force-time compression curves. Drug Dev Ind Pharm 1990; 16(2):201–229.
35. Muñoz Ruiz A, Jimenez-Castellanos MR, Cunningham JC, et al. Theoretical estimation of dwell and consolidation times in rotary tablet machines. Drug Dev Ind Pharm 1992; 18(9):2011–2028.
36. Armstrong NA, Palfrey LP. Punch velocities during the compaction process. J Pharm Pharmacol 1987; 39:497–501.
37. Çelik M, Goldman D, Okutgen E, et al. Simulation of an F-3 single station press using an integrated compaction research system. Pharm Res 1992; 9(10):S134.

38. Fell JT, Newton JM. The tensile strength of lactose tablets. J Pharm Pharmacol 1968; 20:657–659.
39. Pedersen MA, Rubinstein MH, Bateman SD, et al. Measuring punch time-displacement profiles on a rotary tableting machine using a high speed video system. Pharm Tech 1993; 17(4):44–52.
40. Ruegger C. An investigation of the effect of compaction profiles on the tableting properties of pharmaceutical materials. PhD Thesis. Rutgers, The State University of New Jersey, 1996.
41. Mann SC, Bowen SD, Hunter BM, et al. The influence of punch tolerance on capping. J Pharm Pharmacol 1981; 33:25P.
42. Mann SC, Roberts RJ, Rowe RC, et al. The influence of precompression pressure on capping. J Pharm Pharmacol 1982; 34(suppl):49P.
43. Mann SC, Roberts RJ, Rowe RC, et al. The effect of high speed compression at sub-atmospheric pressures on the capping tendency of pharmaceutical tablets. J Pharm Pharmacol 1983; 35:44P.
44. Bateman SD, Rubinstein MH, Thacker HS. Properties of paracetamol tablets produced using high precompression pressures. J Pharm Pharmacol 1989; 41(suppl):32P.
45. Marshall PV, York P, Maclaine JQ. An investigation of the effect of the punch velocity on the compaction properties of ibuprofen. Powder Tech 1993; 74:171–177.
46. Yang L, Venkatesh G, Fassihi R. Characterization of compressibility and compactibility of poly (ethylene oxide) polymers for modified release application by compaction simulator. J Pharm Sci 1996; 85(10):1085–1090.
47. Yang L, Venkatesh G, Fassihi R. Compaction simulator study of a novel triple-layer tablet matrix for industrial tableting. Int J Pharm 1997; 152:45–52.
48. Kim H, Venkatesh G, Fassihi R. Compaction characterization of granular pectin for tableting operation using a compaction simulator. Int J Pharm 1998; 161:149–159.
49. Ruegger C, Çelik M. The influence of varying precompaction and main compaction profile parameters on the mechanical strength of compacts. Pharm Dev Tech 2000; 5(4):495–505.
50. Ruegger C, Çelik M. The effect of compression and decompression speed on the mechanical strength of compacts. Pharm Dev Tech 2000; 5(4):485–494.
51. Marshall PV, York P, Richardson R. The effect of duration of compression on the axial recovery properties of compacts of a crystalline drug substance. J Pharm Pharmacol 1986; 38:47P.
52. Inchley AJ. Evaluation of an equation to calculate material tensile strength for doubly convex, curved, circular tablets. 10th Pharm Tech Conf. Bologna, Italy 1991; 1(2):522–532.
53. Van Veen B, van der Voort Maarschalk K, Bolhuis GK, et al. Tensile strength of tablets containing two materials with a different compaction behaviour. Int J Pharm 2000; 203:71–79.
54. Pitt KG, Newton JM, Stanley P. Effects of compaction variables on porosity and material tensile strength of convex-faced aspirin tablets. J Pharm Pharmacol 1990; 42:219–225.
55. Garr JSM, Rubinstein MH. Compaction properties of a cellulose-lactose direct excipient. Pharm Tech 1991; 15(4):76–80.
56. Garr JSM, Rubinstein MH. The effect of rate of force application on the properties of microcrystalline cellulose and dibasic calcium phosphate mixtures. Int J Pharm 1991; 73:75–80.
57. Garr JSM, Rubinstein MH. An investigation into the capping of paracetamol at increasing speeds of compression. Int J Pharm 1991; 72:117–122.
58. Çelik M, Maganti L. Compaction studies on pellets I. Uncoated pellets. Int J Pharm 1993; 95:29–42.
59. Çelik M, Maganti L. Compaction studies on pellets II. Coated pellets. Int J Pharm 1994; 103:55–67.
60. Çelik M, Okutgen E. Feasibility study for the development of a prospective compaction functionality test and establishment of a compaction data bank. Drug Dev Ind Pharm 1993; 19(17 and 18): 2304–2334.
61. Cocolas HG, Lordi N. Axial to radial pressure transmission of tablet excipients using a novel instrumented die. Drug Dev Ind Pharm 1993; 19(17-18):2473–2497.
62. Mollan M, Çelik M. Tabletability of maltodextrins and acetaminophen mixtures. Drug Dev Ind Pharm 1994; 20(20):3131–3149.
63. Çelik M, Ong JTH, Chowhan ZT, et al. Compaction simulator studies of a new drug substance: effect of particle size and shape, and its binary mixtures with microcrystalline cellulose. Pharm Dev Tech 1996; 1(2):119–126.
64. Akande OF, Rubinstein MH, Ford JL. Examination of the compaction properties of a 1:1 acetaminophen:microcrystalline cellulose mixture using precompression and main compression. J Pharm Sci 1997; 86(8):900–907.
65. Larhrib H, Wells JI. Compression of thermally treated polyethylene glycol 10000. Int J Pharm 1997; 153:51–58.

66. DeCrosta MT, Schwartz JB, Wignent RJ, et al. Thermodynamic analysis of compact formation; compaction, unloading, and ejection II. Mechanical energy (work) and thermal energy (heat) determinations of compact unloading and ejection. Int J Pharm 2001; 213:45–62.
67. Wang JJ, Guillot MA, Bateman SD, et al. Modelling of adhesion in tablet compression. II. Compaction studies using a compaction simulator and an instrumented tablet press. J Pharm Sci 2004; 93(2): 407–417.
68. Gabaude CM, Guillot M, Gautier J-C, et al. Effects of true density, compacted mass, compression speed, and punch deformation on the mean yield pressure. J Pharm Sci 1999; 88(7):725–730.
69. Roberts RJ, Rowe RC. Brittle fracture propensity measurements on 'tablet-sized' cylindrical compacts. J Pharm Pharmacol 1986; 38:526–528.
70. Roberts RJ, Rowe RC. The effect of the relationship between punch velocity and particle size on the compaction behavior of materials with varying deformation mechanisms. J Pharm Pharmacol 1986; 38:567–571.
71. Roberts RJ, Rowe RC. Brittle/ductile behavior in pharmaceutical materials used in tableting. Int J Pharm 1987; 36:205–209.
72. Roberts RJ, Rowe RC. The compaction of pharmaceutical and other model materials—a pragmatic approach. Chem Engr Sci 1987; 42(4):903–911.
73. Roberts RJ, Rowe RC. Source and batchwise variability in the compressibility of microcrystalline cellulose. J Pharm Pharmacol 1987; 39:70P.
74. Bateman SD, Rubinstein MH, Wright P. The effect of compression speed on the properties of ibuprofen tablets. J Pharm Pharmacol 1987; 39(suppl):66P.
75. Bateman SD, Rubinstein MH, Rowe RC, et al. A comparative investigation of compression simulators. Int J Pharm 1989; 49:209–212.
76. Pedersen S, Kristensen HG. Compaction behavior of 4-hydroxybenzoic acid and 4-esters compared to their mechanical properties. Eur J Pharm Biopharm 1995; 41(5):323–328.
77. Cunningham JC, Sinka IC, Zavaliangos A. Analysis of tablet compaction. I. Characterization of mechanical behavior of powder and powder/tooling friction. J Pharm Sci 2004; 93(8):2022–2039.
78. Sinka IC, Cunningham JC, Zavaliangos A. Analysis of tablet compaction. II. Finite element analysis of density distributions in convex tablets. J Pharm Sci 2004; 93(8):2040–2053.
79. Wu C-Y, Ruddy OM, Benthan AC, et al. Modelling the mechanical behaviour of pharmaceutical powders during compaction. Powder Tech 2005; 152:107–117.
80. Wu C-Y, Best SM, Bentham AC, et al. A simple predictive model for the tensile strength of binary tablets. Eur J Pharm Sci 2005; 25:331–336.
81. Wu C-Y, Best SM, Bentham AC, et al. Predicting the tensile strength of compacted multi-component mixtures of pharmaceutical powders. Pharm Res 2006; 23(8):1898–1905.
82. Picker K. Three-dimensional modeling to determine properties of tableting materials on rotary machines using a rotary tableting machine simulator. Eur J Pharm Biopharm 2000; 50:293–300.
83. Kuppuswamy R, Anderson SR, Augsburger L, et al. Estimation of capping incidence by indentation fracture tests. AAPS PharmSciTech 2001; 3(1):1–12.
84. Durig T, Salzstein RA, Skinner GW, et al. Advanced structure-function properties of ethylcellulose: implications for tablet compactibility. AAPS Annual Meeting, Toronto, Ontario, November, 2002.
85. Picker K. The 3-D model: comparison of parameters obtained from and by simulating different tableting machines. AAPS PharmSciTech 2003; 4(3):55–61.
86. Picker K. The 3-D model: does time plasticity represent the influence of tableting speed? AAPS PharmSciTech 2003; 4(4):article 66.
87. Durig T, Harcum WW, Lusvardi KM, et al. Compaction characteristics of high ethoxyl, low viscosity ethylcellulose. AAPS Annual Meeting, Salt Lake City, Utah, October 2003.
88. Betz G, Meyer A, Puchkov M, et al. Investigation of granulation and comparison of tablet compaction properties of polymorphs of mannitol after wet granulation with direct compressible mannitol using a compaction simulator. Proc Int Pharm Biopharm Pharm Tech, Nuremberg, Germany, March 15–18, 2004.
89. Tye CK, Sun C, Amidon GE. Evaluation of the effects of tableting speed on the relationships between compaction pressure, tablet tensile strength, and tablet solid fraction. J Pharm Sci 2005; 94(3):465–472.
90. Kimura G, Betz G, Leuenberger H. Influence of loading volume of mefenamic acid on granules and tablet characteristics using a compaction simulator. Pharm Dev Tech 2007; 12:627–635.
91. Sun C, Hou H. Improving mechanical properties of caffeine and methyl gallate crystals by cocrystallization. Cryst Growth Des 2008; 8(5):1575–1579.
92. Sun C, Hou H, Gao P, et al. Development of a high drug load tablet formulation based on assessment of powder manufacturability: moving toward quality by design. J Pharm Sci 2009; 98(1):239–247.

93. Ho AYK, Jones TM. Rise time: a new index of tablet compression. J Pharm Pharmacol 1988; 40:74P.
94. Ho AYK, Jones TM. Punch travel beyond peak force during tablet compression. J Pharm Pharmacol 1988; 40:75P.
95. Wang JJ, Li TL, Bateman SD, et al. Modeling of adhesion in tablet compression. I. Atomic force microscopy and molecular simulation. J Pharm Sci 2004; 92:798–814.
96. Holman LE. The compressibility of pharmaceutical particulate systems. An illustration of percolation. Int J Pharm 1991; 71:81–94.

7 Compactibility functionality test

Eşay Okutgen

INTRODUCTION

Probably, the most important aspect of pharmaceutical formulation development is the selection of suitable excipients that would assist in providing the intended use of the final dosage form. The inert ingredients can profoundly influence the properties of the dosage form, especially when the drug concentration is small. In selecting excipients, finding accurate and comparable information on the physicochemical properties, uniformity, purity, and safety of excipients in a reliable reference source is highly critical for the formulation scientists. This type of information and the existence of standard test methods to measure such properties are important to assure consistent quality and functioning of different excipients as well as one excipient from lot-to-lot. When the dosage form is a tablet, as it is the case for a majority of the products on the market today, it is possible to add to the crucial information list above the assessment and comparison of the compaction behavior of the inactive ingredients.

It is a well-known fact that the diluents, binders, antiadherents, glidants, and lubricants are considered as the group of excipients that affect the compaction characteristics of the tablet the most. Hence, the information related to the assessment and comparison of the compaction behavior of excipients belonging to these groups is especially desired among the tablet formulators. Yet even today, the formulators lack both a single source of reference and the existence of a standard test method to assess and compare the compaction behavior of excipients that play a critical role in the tablettability of a powder composition.

BRIEF HISTORY OF EXCIPIENT MONOGRAPHS

Until their merge in 1975, the monographs of excipients appeared separately both in the U.S. Pharmacopeia (USP) and the National Formulary (NF). A decision was then made to include the monographs of the excipients only in the NF with the exception of those substances that function both as an active ingredient and an excipient (such as mannitol and talc). These latter substances were placed in the USP and cross-referenced in the NF.

In 1974, the Katalog Pharmazeutischer Hilfsstoffe (Catalog of Pharmaceutical Excipients), which contained monographs of almost 100 Swiss pharmacopeial and nonpharmacopeial excipients, was published jointly by three Swiss pharmaceutical companies Ciba-Geigy, Hoffman LaRoche, and Sandoz (1). This catalog contained general information, suppliers list, tests, and specifications obtained from the literature or measured in the laboratories of the above Swiss companies.

Later in 1986, the *Handbook of Pharmaceutical Excipients* (2) was published under the direction of the American Pharmaceutical Association and the Pharmaceutical Society of Great Britain with the contributions of about 150 scientists representing the academia and the industry in the United States and the United Kingdom. The committee that worked on the *Handbook of Pharmaceutical Excipients* developed standard test methods to evaluate over 30 physical properties of excipients.

As one of the pioneers of economical globalization, many pharmaceutical companies have long been operating in different parts of the world. These companies take into account the regulatory requirements in different countries when they develop new products or reformulate the existing ones. For a long time, one of the problems encountered by the formulators had been the lack of harmonized standards for excipient specifications and testing worldwide. All national pharmacopeias contain specifications and test methods for excipients, however, many of the standards were not unified among the various pharmacopeias. This issue has been largely resolved since the foundation of the International Pharmaceutical Excipients Council in 1991 to develop excipient harmonization worldwide. In the same year, the USP-NF also formed a special advisory panel to develop internationally applicable physical test methods for basic powder properties (e.g., density, particle size, particle shape) as well as for applied powder properties (e.g., fluidity and compactibility).

Development of internationally applicable physical test methods for compactibility is especially important today because of the ever increasing number of directly compressible excipients introduced into the market in recent decades. In most cases, when working with these highly specialized materials, one excipient product cannot easily be replaced by another, or even with another grade of the same material.

WHERE WE ARE TODAY

Today, among the available reference books on excipients, The *Handbook of Pharmaceutical Excipients* is one of the most commonly used by the formulation scientists. Since its first publication in 1986, the second (in 1994), the third (in 2000), the fourth (in 2003), the fifth (in 2005), and the sixth editions (in 2009) became available (2–7). The book contains essential data on the physical properties of excipients such as boiling point, melting point, solubility, hygroscopicity, moisture content, moisture adsorption isotherms, bulk and tap density, true density, particle size distribution, specific surface area, flowability, rheology, and, in some cases, compaction hardness profiles. The number of monographs has increased in each edition, from 145 monographs in the first edition to 340 excipient monographs in the fifth one. More than 100 laboratories have contributed to the monographs. However, it should be noted here that, despite the early concerns regarding the lack of harmonized standards for excipients and efforts to address them, in many cases, the testing and/or the data presentation methods have continued to be inconsistent between the monographs *within the same edition* of this important reference book. For example, in the second edition where 200 monographs are contained, the moisture content data for the excipients were obtained from 31 different laboratories, and almost none of them used exactly the same test conditions. Later editions did not seem to eliminate this problem. This and other inconsistent testing methods, even for such basic powder properties, would be a concern for any end user of these data.

The *Handbook of Pharmaceutical Excipients* has also been the *only* reference source containing compaction information for *some* of the tabletting excipients. In fact, despite being the only reference source for these types of data, the number of monographs with information on compaction behavior actually decreased as the more recent editions of The *Handbook of Pharmaceutical Excipients* became available over the years (8). Therefore, the lack of internationally applicable physical test methods for compactibility as well as the lack of a reliable reference source containing complete, accurate, and comparable information for the compaction behavior of tabletting excipients, especially the ones used mainly as diluents and binders, perseveres today. In cases where the information on compaction behavior of excipients exists, the inconsistencies between the testing and/or data presentation methods often make comparison between materials difficult.

COMPARISON OF COMPACTION DATA

The formation of tablets from a blend of powder (or granular) material basically involves the combination of two phenomena occurring at the same time; compressibility and compactibility. As described earlier in this book, compressibility is the ability of a material to be reduced in its volume, whereas compactibility is the ability of a material to form a coherent compact under an applied pressure, and, therefore, the latter possibly is the most critical powder property in regards to tabletting.

Early formulators often selected traditionally known excipients they were familiar with for development of solid dosage forms. The success of the selected material was sometimes a mere coincidence and the reasons behind it were not well understood. This was partly due to the lack of information available on the compaction characteristics, and other physicomechanical properties of the tabletting excipients.

Compaction studies gained increasing importance in tablet formulation development process after the introduction of instrumented single- and multistation tablet presses, and universal testing machines (9–11). More systematic investigations have been facilitated since the compaction simulators became available (12–20). The compaction simulators, as described in detail in chapter 6, are presses specifically designed to mimic the exact cycle of any tablet compaction process in real time and to record all important parameters during the cycle (12). These machines can utilize all the standard Industrial Pharmaceutical Technology Standards Manual (IPT) tooling as well as any kind of specialized tooling.

Despite the fact that the mechanisms involved during compaction have been the subject of numerous publications (12,20,21), data obtained from two or more studies usually are not comparable. As highlighted in several articles by Çelik and his coworkers in the late 1980s and early 1990s (12,22,23), this is due to the fact that at least one of the parameters involved in the compaction process, such as the equipment (i.e., type of press and tooling), test conditions (e.g., range of applied force, compaction speed, lubrication, etc.), the parameters monitored, and the methods used to analyze the obtained data (e.g., Heckel equation, work of compaction, etc.) vary widely in these studies.

As mentioned earlier, *The Handbook of Pharmaceutical Excipients* (2–7) has long been the only reference source available containing compaction information for some of the tabletting excipients. The compaction data presented in the then-available first edition had been criticized by Çelik and Okutgen (23) on the basis that inconsistent methods were employed to generate information for this valuable reference source (i.e., seven different compaction techniques were applied to twelve different excipients). Taking mannitol as an example excipient, these authors demonstrated that, using The *Handbook of Pharmaceutical Excipients* (2) as a reference source, it would be impossible to determine whether the significant batch-to-batch and supplier-to-supplier variations described for mannitol (Fig. 1) were due to the inconsistency of the compaction test conditions (listed in Table 1) or the actual properties of the tested materials. Similar inconsistencies in test conditions continued to exist in the later editions of the *Handbook of Pharmaceutical Excipients* (3–7).

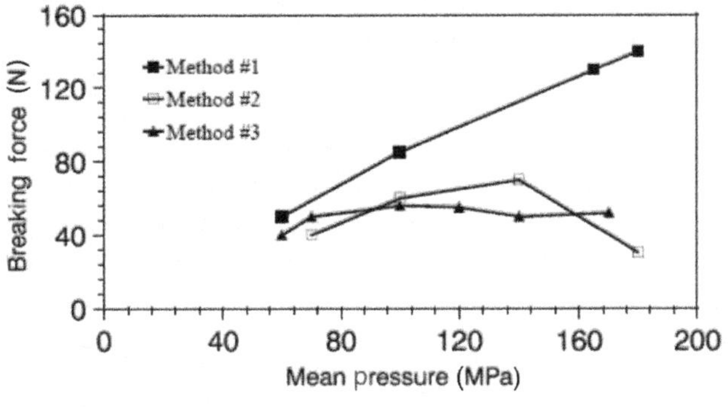

Figure 1 Breaking force versus mean applied pressure plots for different batches/suppliers of mannitol. *Source*: From Ref. 2.

Table 1 Summary of Compaction Test Conditions for Different Batches/Suppliers of Mannitol

Parameter	Method 1	Method 2	Method 3
Supplier company	Triangle import and export	Triangle import and export	Atlas
Lot number	N/A	1233R8060791	2022BO
Tablet press	Manesty E2	Carver press	Stokes B2-16
Tooling diameter	½ in.	½ in.	7/16 in.
Tooling concavity	N/A	flat faced	standard concave
Lubrication	External	External	Internal
Speed	50 tablets/min	N/A	N/A
Tablet weight (mg)	600	500	500
True volume (cm^3)	0.43	0.33	0.33
Data presentation	Breaking force (N) vs. compression pressure (MPa)	Breaking force (N) vs. compression pressure (MPa)	Breaking force (N) vs. compression pressure (MPa)

Source: From Ref. 2.

Despite the efforts of the USP-NF advisory panel to develop internationally applicable test methods for powder properties, no concrete progress has been made by this regulatory body in terms of providing an official standardized compactibility test method. Since the establishment of this panel in 1991, only a draft version of the proposed compactibility test was published (more than a decade ago) in 1999 (24). On the other hand, the FDA established the Scale-Up and Post-Approval Changes (SUPAC) guidelines in mid-1990s (25), followed by the Process Analytical Technology (PAT) initiative in early 2000s (26), both of which resulted from extensive studies involving academia and pharmaceutical industry.

COMPACTIBILITY TEST FACTORS

In early 1990s, the scientists at (then) Pharmaceutical Compaction Research Laboratory & Information Center of Rutgers University conducted independent feasibility studies for the development of a standard "tablettability" testing method to determine the compactibility of pharmaceutical powders (23). This work was carried out by assessing the factors that are believed to play an important role in the process of powder compaction, such as compaction equipment and tooling, tablet weight, range of compaction pressure, type of applied punch-displacement profile, lubrication, compaction parameters to be monitored, types of postcompaction tests, types of data evaluation method(s), and comparison of pass-fail versus fingerprinting criteria. The same study also intended to test the feasibility of establishing a "Compaction Data Bank," which was envisioned to contain performance "fingerprint" information on the compaction behavior of all tabletting excipients, generated under the same predetermined compaction conditions. It was anticipated for such a data bank to serve as an invaluable reference source for tablet formulation scientists.

The 1999 USP Advisory Panel Report on compactibility test (24) discusses various actions to consider when designing a compactibility test, including test procedures for die-wall lubrication, determination of compaction height, and assessment of low-load rate and optional high-load rate tests, as well as various ways to present the test results. But the USP-NF did not move forward with the goals proposed in its draft version, and, therefore, more progress must be made before a standard compaction functionality test is established and widely used.

In 2009, Çelik opined on the proposed recommendations of the USP-NF advisory panel and the work done at Rutgers (27).

In the remaining sections of this chapter, the discussions will be carried out by referring these three publications as the "Rutgers Study" (23), "USP-NF Report" (24), and "Çelik" (27) while comparing the compactibility test factors and their recommended test procedures.

Equipment

This is probably the most critical factor involved in the development of a compactibility test method since it is absolutely necessary that the information obtained from one machine is comparable to information obtained from other machines. The suitable equipment used for performing compaction studies are compared in Table 2.

Their relatively high cost aside, the compaction simulators offer superior advantages compared to the conventional tabletting equipment that can be used for compactibility testing. One other advantage that is not covered in Table 2 is that the compaction simulators provide maximum amount of information with minimum amount of material using only a few grams. Compaction simulators are discussed in more detailed in chapter 6.

The USP-NF Report (24) recommends that the tooling set should comprise a cylindrical die and 10-mm-diameter flat-faced punches, arranged as shown in Figure 2 and the test equipment must be capable of applying loads up to 30 kN. Hence, the USP-NF Report also

Table 2 Comparison of Tabletting Equipment

Feature	Single-station press	Multistation press	Isolated punch and die set	Compaction simulator
Mimics typical production conditions	No	Yes	May be	Yes
Mimics cycles of many presses	No	No	May be	Yes
Requires small amount of material	Yes	No	Yes	Yes
Easy to instrument	Yes	No	Yes	Yes
Easy to set up	Yes	No	May be	May be
Database in literature	Yes	Yes	Some	Some
Used for stress strain studies	No	No	Yes	Yes
High cost	Yes	No	May be	No

Source: From Ref. 12.

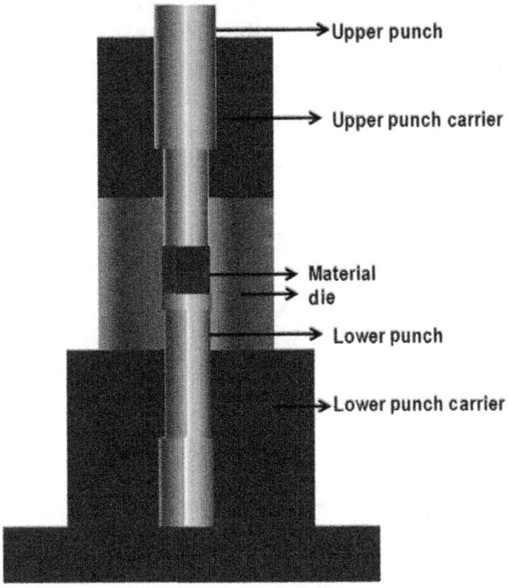

Figure 2 Tooling set for the proposed USP-NF compactibility test. *Source*: Adapted from Ref. 27.

suggests that "the test must be one that can be readily carried out with a minimum of sophisticated equipment and it must be highly reproducible." As a result, the device (Fig. 2) resembles a hydraulic Carver press (which is an inexpensive general purpose compression equipment used) or an Instron-type universal testing instrument (which is commonly used and moderately priced equipment for studying mechanical properties of materials). Both are used in a variety of fields. Çelik argued that, if the proposed device is a Carver-type equipment, then developing a robust formulation based on the test results obtained from such a press would be highly challenging (27). Even if an Instron-type press is intended, still both presses can only apply compression at a much slower rate than encountered in pharmaceutical applications. Therefore, more sophisticated testing equipment could have been proposed.

The equipment used in the Rutgers Study (23) was a compaction simulator, referred to as Integrated Compaction Research System (ICRS), fitted with standard 10.3 mm round, flat-faced BB tooling. Unlike the single- and multistation tablet presses, which operate under displacement (strain) control and produce a sine-wave punch displacement profile, compaction simulators can be programmed to operate under load control as well as under displacement control, and the resulting profiles can be customized. Under displacement control, the punches travel a predetermined distance whereas under load control, the punches try to apply forces following a specified load profile. The applied force versus time profiles obtained when the ICRS was operated under load control and the punch position versus time profiles obtained when the system was programmed to follow a "sawtooth" profile are demonstrated in Figures 3 and 4, respectively. As can be seen from these figures, the system is capable of following a displacement profile more accurately.

At Rutgers Study, a compaction simulator or its simplified version was recommended as the equipment to be used for compactibility testing method (23).

Sample Weight

The USP-NF Report (24), recommends using constant true volume of 0.3 cm^3 instead of the common practice of keeping the tablet weight constant. Using constant true volume is definitely the right approach especially when comparing different materials. This is due to the fact that the distance traveled by the punch into the powder bulk (which is responsible for the amount of compaction force) is interrelated with the volume of the solid present in the die, and not with its weight (28). However, 0.3 cm^3 true volume may not be attainable for materials with very low densities (e.g., less than 0.25 g/cm^3), especially when using a 10-mm-diameter flat-faced round tooling recommended by the USP-NF Report. This is due to the fact that such a

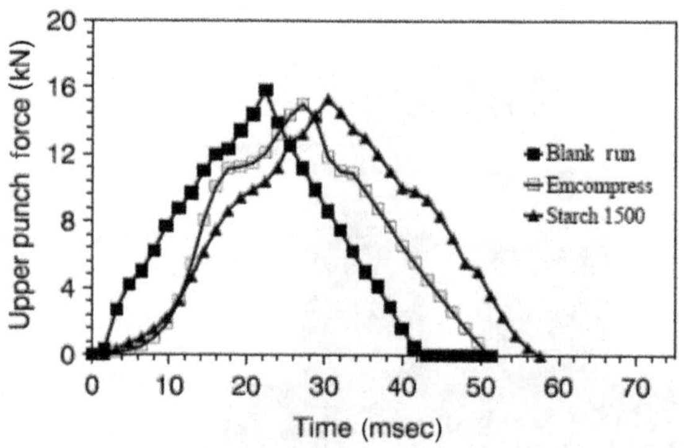

Figure 3 ICRS upper punch force versus time profiles (loading rate 15 kN/sec). *Source*: From Ref. 23.

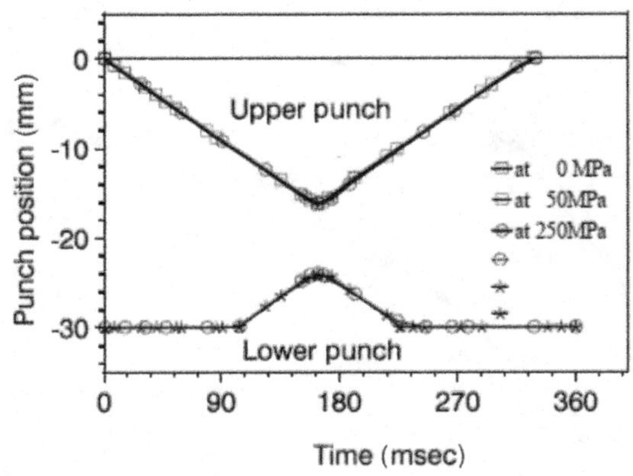

Figure 4 ICRS upper punch position versus time profiles (Emcocel 90M compacts made at 100 mm/sec). *Source*: From Ref. 23.

sample would not be able to fit into the die cavity without tapping, which would obviously influence the result of compaction testing.

In the Rutgers Study (23), after testing a number of materials, using almost the same size of tooling recommended by the USP-NF Report, the use of 0.25 cm^3 constant true volume was determined to be more applicable for materials with a wide range of bulk densities.

Lubrication

Ideally, it would be desirable to study the intrinsic compaction behavior of pharmaceutical powders by performing compaction tests using a thoroughly cleaned punch and die set without any type of lubrication. However, this is often tedious and impractical, especially in cases in which the material exhibits serious sticking and picking behavior. One possible approach is the external lubrication of the surfaces of the die wall and the punch tips, but this practice can cause inconsistencies during application of the lubricant. For example, it would be difficult to achieve a uniform distribution of lubricant on surfaces of the die cavity. The compaction results can also vary from one operator to another, from one test to another, and from one day to the next.

The remaining alternative, internal lubrication has its own drawbacks. It is difficult to determine the optimal type and amount of lubricant, as well as the optimal mixing time. For example, while the use of 0.5% magnesium stearate could be a suitable ratio for samples with rough particle surfaces, it could very well overlubricate other samples having much smoother surfaces. In other words, although two identically lubricated materials may offer distinctly different results, the results do not necessarily provide any insight into the materials' intrinsic compactibility.

The USP-NF Report (24) recommends that the die-wall lubrication conditions be standardized by either external or internal lubrication. The recommended external lubrication requires applying a saturated solution of magnesium stearate in acetone to the previously acetone-cleaned surfaces. The recommended internal lubrication involves mixing the sample with 0.5% magnesium stearate.

Using both an internal and an external lubrication method for making the compacts of two commonly used fillers microcrystalline cellulose and lactose anhydrous, the Rutgers Study (23) showed that internal lubrication method produced more consistent results and it was more reliable (Fig. 5). Hence, despite the above mentioned drawbacks, the internal lubrication method by mixing the test sample with 0.5% magnesium stearate was suggested for use in compaction functionality testing due to the practicality of this method.

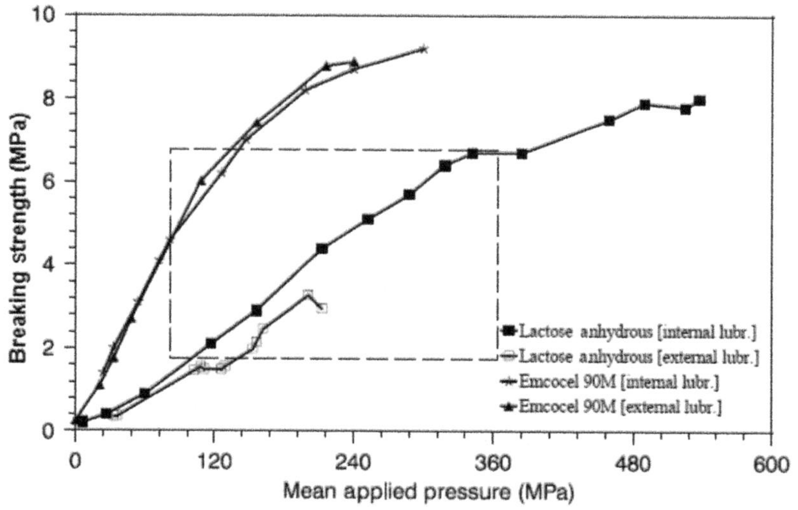

Figure 5 The effects of internal and external lubrication on the breaking strength of the compacts of the excipients used (punch velocity 100 mm/sec). (The "box" denotes a common pharmaceutical range.) *Source*: From Ref. 23.

Çelik (27) recently offered an alternative internal lubrication method in which a predetermined ratio of the (specific) surface area of magnesium stearate to the (specific) surface area of the test sample is employed. This method would eliminate problems related to the differences in the surface areas of the samples. In theory, one complete layer of lubrication would be sufficient. However, the ideal ratio that will be enough to cover the surface of the formulation and yet would not have any significant impact on the mechanical strength or the drug release of the final tablet may very well be between 1 and 1.5. This remains to be determined with further studies. If minimum amount of lubricant (W_L) in a formulation (or single substance) is required, then the following equation, which was originally proposed by Sadek et al (29) for the minimum amount of required glidant in a formulation, can also be utilized.

$$W_L(\%) = 6 \times \left[\frac{d - \rho_L}{D - \rho_P}\right] \times 100$$

where d and D are the diameters of the lubricant particle and the host particle, respectively, and ρ_L and ρ_P are the true densities of the lubricant, and the host particle, respectively.

Compaction Load

The USP-NF Report (24) recommends applying compaction pressures as many as 12 levels (Table 3) with an accuracy of 0.5% for five replicates at each compaction level. However, the practicality of this recommendation was found to be questionable by Çelik (27). It must be remembered that most pharmaceutical tablet presses operate on "displacement" mode, not on "load" mode. This means that the operator sets the distance between the two punches to control the thickness of the tablet, and the amount of applied pressure depends on the material being compacted. For example, on a press with punches set at a given distance apart, the applied load (or pressure) would be much different when compacting microcrystalline cellulose than it would be for dicalcium phosphate dehydrate. To make five replicate compacts of different samples for each of the required pressure levels, the operator would have to

Table 3 Compaction Loading Levels According to the USP-NF Report

Level	Compaction pressure (MPa)	[Log (pressure)]	Corresponding applied load (kN) (approximate)
1	8.0	0.903	0.63
2	11.25	1.051	0.88
3	16.0	1.204	1.26
4	22.5	1.352	1.77
5	32.0	1.505	2.51
6	45.0	1.653	3.53
7	64.0	1.806	5.03
8	90.0	1.954	7.07
9	128.0	2.107	10.05
10	180.0	2.255	14.14
11	256.0	2.408	20.11
12	360.0	2.556	28.27

Source: From Ref. 27.

produce considerable amount of compacts via trial-and-error to determine which setup conditions (i.e., the distance between the two punches) would apply the desired load on a given sample.

The requirement for a predetermined applied pressure would only be an easy task only if a servocontrolled hydraulic machine were recommended. Such machines can run in both pressure and displacement modes, and the operator can easily set a predetermined level of load to be applied on the sample, causing punch displacement to differ from one material to another, or set the punch movement, causing the amount of the applied load to differ.

As an alternative and more operator-friendly approach, repeating the compaction tests over a wide range of pressures instead of applying an exact amount of predetermined pressure has also been recommended by Çelik (27).

The Rutgers Study (24) was more in line with the latter recommendation where the compaction pressures ranged from relatively low values (about 25 MPa) to very high values (about 500 MPa). This was due to the fact that a linear relationship between the tablet strength and the applied pressure would not always exist and, as it was demonstrated in the study, the rank order of breaking strength values for a group of different materials changed when the spectrum of applied pressure was changed (Fig. 6).

Compaction Speed

The USP-NF Report (24) recommends a low-load rate (i.e., slow speed) test and an optional high-load rate test (i.e., high speed). According to this report, for a low-load rate test, the compaction load should be applied in 100 to 300 milliseconds and be removed in 50 to 100 milliseconds. If the optional high-load rate test is performed, then the compaction load should be applied in 10 to 20 milliseconds and be removed in 10 to 20 milliseconds. In another words, the total contact time (during which the sample is in contact with both the upper and the lower punch) should be 150 to 400 milliseconds for the recommended low-load rate test and 20 to 40 milliseconds for the optional high-load rate test. Although it is not specified in the proposal, one can assume that the shorter times correspond to the minimum applied pressures and longer times correspond to the maximum applied pressures, as it will take longer to attain a higher pressure than a lower one at the same speed.

However, the recommended low-load rate values do not represent commonly encountered contact times in standard tablet manufacturing practices whereas, the optional high load rate values are within the practical range.

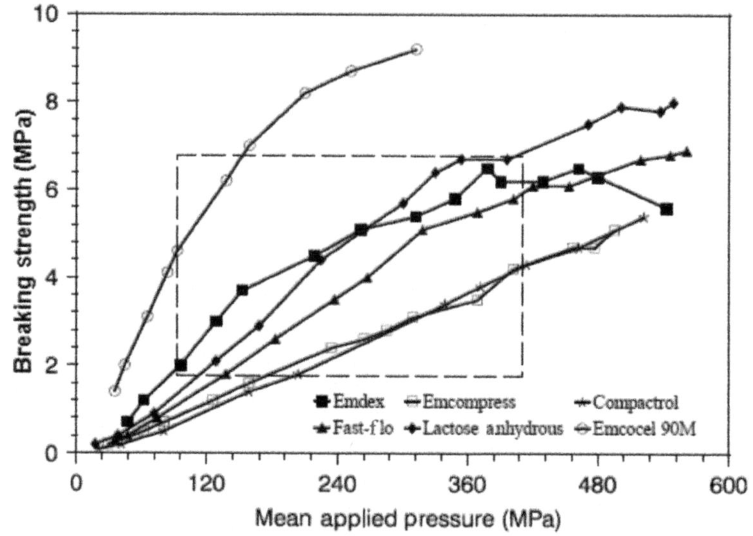

Figure 6 Compaction pressure range studied in the Rutgers Study (punch velocity 100 mm/sec). (The "box" denotes a common pharmaceutical range.) *Source*: From Ref. 23.

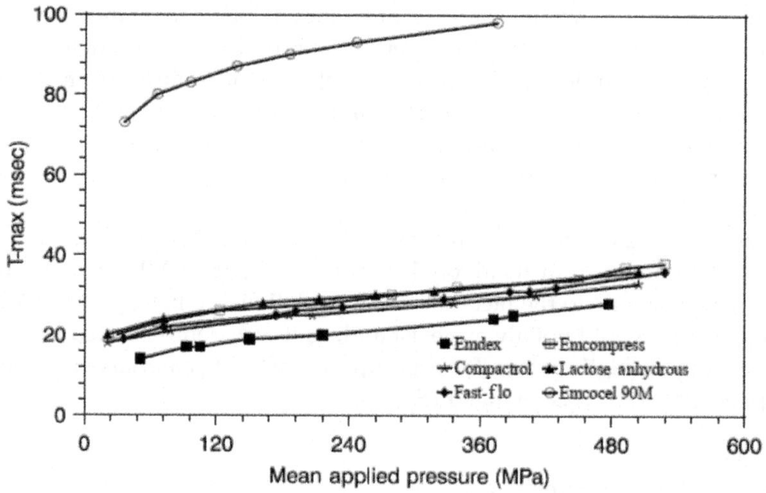

Figure 7 T_{max} (time to attain maximum pressure) versus maximum applied pressure plots (punch velocity 100 mm/sec). *Source*: From Ref. 23.

The Rutgers Study (23) recommends the use of sawtooth profiles, instead of the sine-wave profiles since the former result in a constant velocity regardless of the type or thickness of the material compacted. However, the results obtained from a sawtooth profile need to be comparable with those obtained from a sine-wave profile, at least in terms of a time parameter, such as the contact time or the T_{max} (time to attain maximum pressure). The comparison of T_{max} *versus* mean applied pressure profiles for the compacts of selected excipients made at a punch velocity of 100 mm/sec is presented in Figure 7. When the T_{max} values of the compacts are compared for any given pressure, the variation was ±5 milliseconds (except for Emcocel 90M). When the whole pressure range is examined, it can be seen that the T_{max} values of compacts ranged from 12 to 40 milliseconds, which are well within the range encountered in high speed

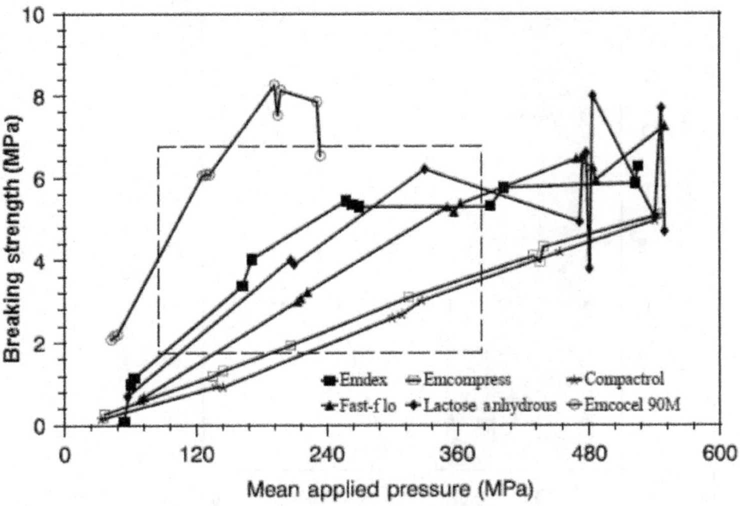

Figure 8 Compaction profiles attained in the Rutgers Study at high punch velocity (300 mm/sec). (The "box" denotes a common pharmaceutical range.) *Source*: From Ref. 23.

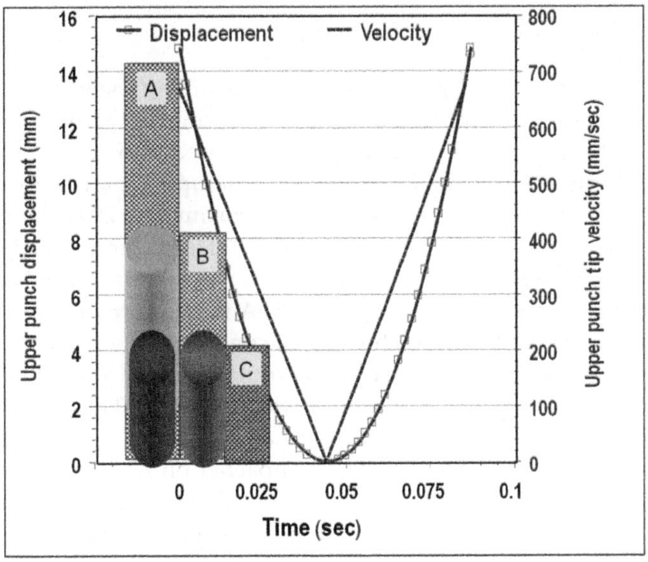

Figure 9 Simulated profile of an upper punch on a Manesty Beta tablet press running at 103 rpm obtained using the Rippie–Danielson method. *Source*: From Refs. 27 and 30.

multistation press operations (again with the exception of microcrystalline cellulose for which the T_{max} values were as high as 100 milliseconds). The Rutgers Study also recommends the application of a sawtooth profile at 300 mm/sec punch velocity to serve as a stress test that can provide information on the robustness of materials in terms of their time dependency.

The compaction profiles of a number of excipients attained in the Rutgers Study at 100 and 300 mm/sec constant punch velocities are shown in Figures 6 and 8, respectively.

Recently, Çelik (27) emphasized the importance of the punch profiles by comparing three hypothetical samples A, B, and C. The initial filling height was given as 14 mm for sample A, 8 mm for sample B, and 4 mm for sample C (Figs. 9 and 10). As the true volumes of these

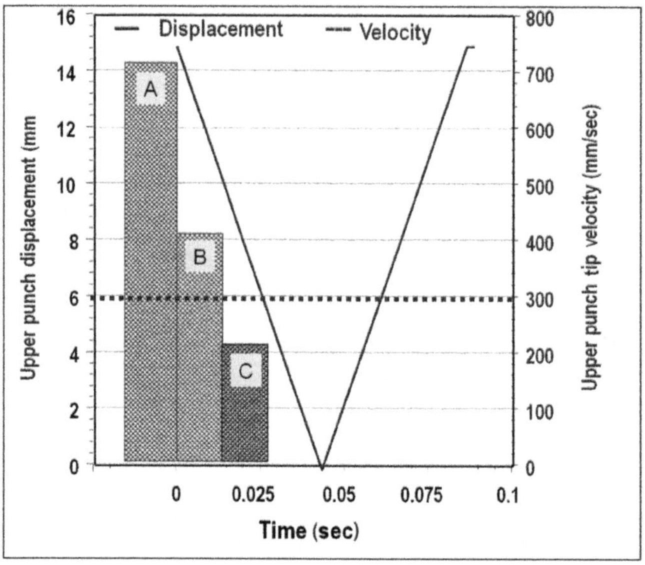

Figure 10 Profile of a sawtooth (*triangle*) upper punch running at a constant speed of 300 mm/sec. *Source*: From Ref. 27.

samples were assumed the same, differences in their filling heights would be due to differences in their bulk densities. If these materials were compacted on a rotary tablet press running at, say, 103 rpm, one could assume that the compaction test conditions would be identical. However, the punch-tip velocity at the time the upper punch initially contacts the powder sample would be different for each sample (approximately 700 mm/sec for sample A, 400 mm/sec for sample B, and 200 mm/sec for sample C). The difference is due to the fact that the punch-tip velocity decreases as the punch gets closer to the maximum penetration point.

Under these circumstances, interestingly, sample A would form a (soft) plug at 400 mm/sec, which matches the velocity of the punch at initial contact with sample B. Sample A would then undergo further compression, and sample B would form a (soft) plug at 200 mm/sec, which matches the punch velocity at initial contact with sample C. This means, in practice, the compactibility test conditions would not be the same, since there would be significant differences in the velocities if the samples were compacted on a rotary press running at the same speed. Any significant differences in the tip velocities would result in improper comparison of compaction behavior of pharmaceutical materials possessing wide range of different bulk densities.

Çelik (27) pointed out that if these samples were run on a press where the punch speed was constant (Fig. 9)—where the upper punch had a sawtooth profile—they would undergo compaction at the same constant speed, regardless of the differences in the initial heights of the three samples. In Figure 10, this speed was calculated to be 300 mm/sec. Therefore, the same test conditions could be applied to all three materials.

Monitored Parameters and Data Evaluation

There are numerous ways of evaluating the data obtained from a compaction study in which the displacement of and forces on the upper and lower punches, and the ejection force are monitored. No universal method among them has emerged due to the challenges involved in the comprehensive analysis of the powder systems compacted.

The USP-NF Report (24) recommends the use of three factors—consolidation, compressibility, and compaction-rate sensitivity—and represents the test results using plots showing average breaking force versus (log) applied pressure and average porosity versus (log) applied pressure. Based on the equipment proposed, the required data analysis and representation methods are adequate, but not necessarily complete.

The Rutgers Study (23) accepts the fact that a simple and universally applicable compactibility test should not require measurements of parameters that are complicated in terms of instrumentation and validation such as the measurement of force transmitted to the die wall. This study recommends the measurements of upper and lower punch force, upper and lower punch displacement, and ejection force to be the basic parameters monitored, and the measurement of dimensions and breaking force of the tablets to be the minimum postcompaction tests to be included. It is also recommended that the postcompaction tests to be performed twice; one immediately after ejection and the other following a predetermined storage period on replicate tablets. The gathered data, the study suggests can be presented in, among many other methods, the breaking strength versus applied pressure profiles, examples of which are given in Figures 6 and 8.

CONCLUSION

Since the compaction characteristics of pharmaceutical materials significantly influence the robustness of the final dosage form, it is vital to use up-to-date equipment and test methods to obtain reliable knowledge on such characteristics. However, despite this obvious need and the initiation of efforts two decades ago to develop internationally applicable standard test methods for powder properties (including a compactibility test) by the USP-NF advisory panel, the fact remains that there is still no established test method required by the pharmacopeias that is capable of comparing the relative tablettability features of different materials and different lots of the same material with sufficient sensitivity. The scientific community today still tries to approach the problem through individual efforts and with in-house remedies.

The compactibility test to be utilized for characterization of compaction behavior of pharmaceutical powders proposed by the USP-NF Report in 1999 does not adequately take into account the current advances in compaction technology while the recommendations emerged from the Rutgers Study published in 1993 may not be fully acceptable by all in the scientific community. However, they are still very relevant today and perhaps the Rutgers recommendations are increasingly more applicable. One controversial issue may be the recommendation of the use of a compaction simulator. Although having many far-reaching advantages over the other available equipment, the compaction simulators are expensive and require a specific know-how to operate. As a result, they are not readily available in every tablet formulation laboratory. Nevertheless, it has been encouraging to see over the years that various types of compaction simulators with improved and simplified designs have become available from manufacturers, and more pharmaceutical companies have been acquiring these valuable tools for their internal tablet formulation research and development efforts. If this trend continues, the emergence of even more simplified compaction simulator designs with more affordable price tags may become a reality in near future, which would be the ultimate test equipment to be proposed for a pharmaceutical compactibility functionality test method.

REFERENCES

1. Ciba-Geigy AG, Hoffmann-La Roche F, Sandoz AG. Katalog Pharmazeutischer Hilfsstoffe. Basel, Switzerland, 1974.
2. Handbook of Pharmaceutical Excipients. American Pharmaceutical Association and the Pharmaceutical Society of Great Britain. Washington D.C., London, 1986.

3. Wade A, Weller PJ, eds. Handbook of Pharmaceutical Excipients. 2nd ed. American Pharmaceutical Association and the Pharmaceutical Press. Washington D.C., London, 1994.
4. Kibbe AH, ed. Handbook of Pharmaceutical Excipients. 3rd ed. American Pharmaceutical Association and the Pharmaceutical Press. Washington D.C., London, 2000.
5. Rowe RC, Sheskey PJ, Weller PJ, eds. Handbook of Pharmaceutical Excipients. 4th ed. American Pharmaceutical Association and the Pharmaceutical Press. Washington D.C., London, 2003.
6. Rowe RC, Sheskey PJ, Owen S, eds. Handbook of Pharmaceutical Excipients. 5th ed. American Pharmaceutical Association and the Pharmaceutical Press. Washington D.C., London, 2005.
7. Rowe RC, Sheskey PJ, Quinn ME, eds. Handbook of Pharmaceutical Excipients. 6th ed. American Pharmaceutical Association and the Pharmaceutical Press. Washington D.C., London, 2009.
8. Çelik M. Eye on excipients. Tablets Capsules 2009; 7(3):37–40.
9. Higuchi T, Nelson E, Busse LW. Physics of Tablet Compression: III. Design and Construction of an Instrumented Tableting Machine. J Pharm Assoc Sci Ed 1954; 43(6):344.
10. de Blaey CJ, Polderman J. Compression of pharmaceuticals II. Registration and determination of force-displacement curves using a small digital computer. Pharm Weekblad 1971; 106:57.
11. Walter JT, Augsburger LL. A computerized force-displacement instrumentation system for a rotary press. Pharm Technol 1986; 10(2):26.
12. Çelik M, Marshall K. Use of a compaction simulator system in tableting research. I. Introduction to and initial experiments with the system. Drug Dev Ind Pharm 1989; 15(5):759–800.
13. Mollan M, Çelik M. Tabletability of maltodextrins and acetaminophen mixtures. Drug Dev Ind Pharm 1994; 20(20):3131–3149.
14. Çelik M, Ong JTH, Chowhan ZT, et al. Compaction simulator studies of a new drug substance: effect of particle size and shape, and its binary mixtures with microcrystalline cellulose. Pharm Dev Tech 1996; 1(2):119–126.
15. Yang L, Venkatesh G, Fashsihi R. Characterization of compressibility and compactibility of poly (ethylene oxide) polymers for modified release application by compaction simulator. J Pharm Sci 1996; 85(10):1085–1090.
16. Ruegger C, Çelik M. The influence of varying precompaction and main compaction profile parameters on the mechanical strength of compacts. Pharm Dev Tech 2000; 5(4):495–505.
17. Ruegger C, Çelik M. The effect of compression and decompression speed on the mechanical strength of compacts. Pharm Dev Tech 2000; 5(4):485–494.
18. Picker K. The 3-D model: comparison of parameters obtained from and by simulating different tableting machines. AAPS PharmSciTech 2003; 4(3):55–61.
19. Cunningham JC, Sinka IC, Zavaliangos A. Analysis of tablet compaction. I. Characterization of mechanical behavior of powder and powder/tooling friction. J Pharm Sci 2004; 93(8):2022–2039.
20. Wu CY, Ruddy OM, Benthan AC, et al. Modelling the mechanical behaviour of pharmaceutical powders during compaction. Powder Tech 2005; 152:107–117.
21. Wray P. The physics of tablet compaction revisited. Drug Dev Ind Pharm 1989; 18:627.
22. Çelik M, Lordi NG. The pharmaceutical compaction research laboratory & information center. Pharm Technol 1991; 15(3):112–116.
23. Çelik M, Okutgen E. A feasibility study for the development of a prospective compaction functionality test and the establishment of a compaction data bank. Drug Dev Ind Pharm 1993; 19(17–18):2309–2334.
24. Marshall K. Report and recommendations of the USP Advisory Panel on physical test methods: Compactability test. Pharm Forum 1999; 25(3):8293–8297.
25. U.S. Department of Health and Human Services, Food and Drug Administration, Center for Drug Evaluation and Research (CDER), "SUPAC-IR: Immediate-Release Solid Oral Dosage Forms: Scale-Up and Post-Approval Changes: Chemistry, Manufacturing and Controls, In Vitro Dissolution," 1995.
26. U.S. Department of Health and Human Services, Food and Drug Administration, Center for Drug Evaluation and Research (CDER), Center for Veterinary Medicine (CVM), Office of Regulatory Affairs (ORA). Guidance for industry: PAT—a framework for innovative pharmaceutical development, manufacturing, and quality assurance. September 2004.
27. Çelik M. Eye on excipients. Tablets Capsules 2009; 7(4):33–37.
28. Marshall K. Monitoring punch forces and punch movements as an aid to developing robust tablet formulations. Drug Dev Ind Pharm 1989; 15:2153–2176.
29. Sadek HM, Olsen JL, Smith HL, et al. A systematic approach to glidant selection. Pharm Technol 1982; 6(2):43–62.
30. Rippie EG, Danielson DW. Viscoelastic stress/strain behavior of pharmaceutical tablets: analysis during unloading and postcompression periods. J Pharm Sci 1981; 70(5):476–482.

8 Compaction properties of directly compressible materials

Gerad K. Bolhuis and Hans de Waard

INTRODUCTION

Compacted tablets are produced from granulations or powder mixtures, made by the following general techniques: wet granulation or wet massing combined with tray drying or fluid bed drying, wet granulation and drying all in one step, dry granulation by roller compaction or slugging, and dry blending (direct compaction). Of these, the oldest is the classic wet granulation and drying technique in two steps using tray drying. In spite of enormous improvements in wet granulation techniques (high-shear granulation, fluid bed granulation, extrusion granulation, continuous granulation, and all-in-one granulation), tablet production by direct compaction has increased steadily over the years because it offers economic advantages through its elimination of the wet granulation and drying steps.

A question of terminology must be addressed first. The phrase usually used to describe tablet production without prior treatment of the particulate solids is "direct compression" and this is extended into the actual names of some of the solids, for example, compressible sugar USP/NF. However, as mentioned earlier in chapter 7, the true meaning of "compression" is the reduction of volume under pressure, which has a much wider application than just to tablet preparation. "Compactibility" is the ability of a substances to yield a compact (or tablet) of adequate strength (1). Thus, the term "excipients for direct compaction" describes these materials in a much more precise manner, and is the phrase chosen, whenever possible, in this chapter.

Although materials for direct compaction can be both excipients and active materials, in this chapter only excipients (filler-binders) will be discussed.

DIRECT COMPACTION

Since the introduction in the early sixties of the last century of spray-dried lactose as the first excipient specially designed for direct compaction (2), other directly compressible excipients, commonly referred to as filler-binders, appeared on the pharmaceutical market (Table 1). The introduction of the very effective filler-binder microcrystalline cellulose (Avicel PH) in 1964 resulted in an increased interest in the production of tablets by direct compaction. An important boost for the increasing interest in direct compaction was the more or less simultaneous introduction of superdisintegrants. Conventionally disintegrants, used in much higher concentrations than superdisintegrants, cannot be used in direct compaction, because of their high lubricant sensitivity (see chap. 9, section "Effect of Host Material Properties on Film Formation").

In a survey conducted in 1992 of 58 pharmaceutical companies in the United States, the respondents were asked about the company policy toward direct compaction (3). 41.4% indicated that direct compaction was the method of choice, and 41.1% indicated that they used both direct compaction and wet granulation. Only 1.7% of the respondents indicated that they never used direct compaction and 15.5% indicated that the process was not recommended. Now, in 2011, direct compaction can be regarded a preferred manufacturing process as the continually modernizing pharmaceutical industry strives to improve its manufacturing cost and productivity (4).

Although the majority of filler-binders for direct compaction is used for the production of common, disintegrating tablets, some filler-binders are used for the production of chewable tablets or lozenges. Examples are polyols, dextrates, and compressible sugar. A number of

Table 1 Year of Introduction of Some Directly Compressible Filler-Binders

Year	Filler-binder
1963	Spray-dried lactose
1964	Microcrystalline cellulose (Avicel PH)
	Dicalcium phosphate dihydrate (Emcompress)
	Direct compression starch (STA-Rx 1500)
1967	Spray-crystallized dextrose/maltose (Emdex)
1983	γ-Sorbitol (Neosorb)
1988	Ludipress
1990	Cellactose
1994	Agglomerated lactose (Tablettose)
1996	Silicified microcrystalline cellulose (Prosolv SMCC)
1998	StarLac
2005	Isomalt (galenIQ 720 and 721)

Table 2 Steps in the Production of Tablets via Wet Granulation and by Direct Compaction, Respectively

Wet granulation	Direct compaction
1. Weighing	1. Weighing
2. Mixing	2. Mixing
3. Moistening	
4. Wet screening	
5. Drying	
6. Dry screening	
7. Admixing disintegrant, lubricant	3. Admixing disintegrant, lubricant
8. Compressing	4. Compressing

filler-binders are now recommended for the production of orally disintegrating tablets (ODTs). In the European Pharmacopoeia they are referred to as orodispersible tablets (5). ODTs disintegrate and dissolve in the mouth either on or beneath the tongue or in the buccal cavity without water within 60 seconds or less (6). Directly compressed ODTs often contain mannitol as a filler-binder, crospovidone as a disintegrant, and silicon dioxide as an anticracking agent (7).

Advantages and Disadvantages of Direct Compaction

Compared to wet granulation processes, direct compaction offers a number of advantages. It requires fewer unit operations in production that means less equipment and space, lower labor costs, less processing time, and lower energy consumption (Table 2).

The elimination of the wet granulation step increases the stability of drugs that can degrade by moisture and/or heat. Another advantage of direct compaction is that the tablets generally disintegrate into primary particles, rather than into granules. The increased surface area for dissolution may result in a fast drug release for some drugs and some drug products. For the majority of tablet formulations, the addition of a disintegrant is necessary to obtain a fast tablet disintegration. It has been demonstrated that the efficiency of the disintegrant is strongly dependent on the nature of the filler-binder (8).

The direct compaction process also has a number of limitations. Tablets containing a high dose of an active ingredient that has poor compactibility, poor flow properties, and/or low bulk density cannot be prepared by direct compaction, because filler-binders have a limited dilution potential (uptake capacity), and tablet size and weight are limited. However, if an active ingredient is more compressible and flowable, a greater proportion can be carried successfully by a filler-binder.

The direct compaction process generally involves mixing a drug substance with excipients prior to compaction. Because of differences in density of the drug substance particles and excipient particles, direct compaction blends are subject to segregation during transfer steps from the mixer to drums, tote bins, hoppers, and so on. The procedure of sampling for analysis must be well defined so that it does not introduce a major error in determining homogeneity of the powder blend. Segregation during the handling of the powder blend before compression and during sampling is a major disadvantage of the direct compaction method. Careful consideration must be given to particle size distribution and density of the drug substance and excipients. One way to limit segregation is to match the particle size distribution and the particle density of the drug substance and excipients. However, many low-dose drug substances are reduced in particle size to obtain uniform dosage or to obtain a large effective surface area in the dosage form for rapid dissolution. The excipients, especially filler-binders, are chosen so that the powder blend exhibits good flow properties. In some cases, premixing of the fine drug substance particles with large particles of a filler-binder (9) leads to mixing by random adhesion and random mixing (often referred to as ordered mixing) (10). This phenomenon has been referred to as interactive mixing because of the interparticulate forces that predominate gravity under the right conditions. The primary interparticulate forces are the long-range forces: van der Waals, electrostatic, and surface adsorption bonding. In other cases, mixing by spray coating (solvent deposition) may be necessary to avoid segregation and to obtain the homogeneity of the low-dose drug substance in the powder mix.

Generally, the physical and physicomechanical properties of drug substances and excipients in a directly compressible powder mixture need better definition and controls than the materials used in wet granulation. This is because the poor physicomechanical properties of the drug substance are not altered as they are in the wet granulation process; with wet granulation, the solution of the binder (usually a polymer) agglomerates the drug substance and the excipient particles, leaving a thin coat of polymer around the particles and the agglomerated mass. Because of variations in the physical and physicomechanical properties, it has been demonstrated that lot-to-lot variations in directly compressible materials can seriously interfere with tableting properties.

Another disadvantage of direct compaction is the high cost of a number of filler-binders (e.g., mannitol and coprocessed products, as compared to the binders and fillers used in wet granulation). However, the added cost of directly compressible excipients is outweighed by the savings realized by eliminating unit operations such as wet granulation and drying.

FILLER-BINDERS FOR DIRECT COMPACTION

Requirements for a Directly Compressible Filler-Binder

The most important requirements for a directly compressible filler-binder are listed below.

- High compactibility to ensure that the compacted mass will remain bonded after the release of the compaction pressure. Only a few excipients can be compressed directly without elastic recovery. Most directly compressible filler-binders have undergone physical modification to improve tableting properties—mainly compactibility, flowability, and apparent density (see section "Improving Properties for Direct Compaction"). Compactibility of a filler-binder or blends of filler-binders with other materials is generally tested by plotting tablet breaking force as a function of the applied load (compaction profiles).
- Good flowability to ensure that the powder blend flows homogeneously and rapidly and leads to uniform die filling. Although there are different test methods for flowability, the best method is the determination of tablet weight variation under production conditions.
- Good blending properties to avoid segregation.
- Low lubricant sensitivity (see chap. 9).

High dilution potential (or uptake capacity), defined as the amount of an active ingredient that can be satisfactorily compressed into tablets with a given filler-binder.
Good stability, that is, the ability to remain unchanged chemically and physically and to remain fully active and effective during storage. Some filler-binders are only stable when stored under certain conditions. Because of hygroscopicity, most sugar-based products and polyols, with the exception of most lactoses cannot be stored at high humidity conditions (11). Moreover, the attraction of moisture by hygroscopic filler-binders is often detrimental to the stability of active ingredients in the tablet. Another example of a filler-binder in which stability plays a role is dicalcium phosphate dihydrate, which easily loses its water of hydration when stored above 40°C (see section "Dicalcium Phosphate").
Inertness, that is, filler-binders should not accelerate the chemical and physical degradation of the active ingredient or excipients caused by compression or storage conditions.
Compatibility with all substances in any formulation in which it is a part.
Noninterference with the biological availability of active ingredients.
Promotive effect on tablet disintegration (if desired).
Promotive effect on drug release (if desired).
Capability of being reworked without loss of flow or compressibility characteristics.
Worldwide continuous availability.
Batch-to-batch reproducibility of physical and physicomechanical properties (constant quality).
Relatively cost effective.

Not a single excipient fulfills all the optimum requirements. Nevertheless, some products, particularly coprocessed substances, have adequate tableting properties for use as a single filler-binder. It is common to use a combination of two (seldom more) filler-binders to obtain a mixture with adequate tableting properties, including good stability and an acceptable cost (12). It should be realized, however, that the properties of the components of a blend can result in synergistic or antagonistic effects with respect to one or more tableting properties (12). Such interactions can be visualized by experimental designs, for example, simplex lattice designs (13,14).

Improving Properties for Direct Compaction

Materials composed of individual unmodified particles are often not suitable as filler-binders because of lack of flowability (e.g., native starches), lack of binding properties (e.g., α-lactose monohydrate 100-mesh), lubricant sensitivity (native starches), etc. For this reason, filler-binders with enhanced physicomechanical properties designed for direct compaction into tablets have been developed in the past decennia. These improvements were accomplished by using different techniques varying from simple screening to crystal engineering.

Although binding properties can be improved by physical modifications such as dehydration, partial pregelatinization and coating, the flow properties of these products are often still insufficient. For this reason, more and more filler-binders are now produced by methods that combine physical modification and agglomeration (see sections "Granulation and Agglomeration" and "Spray Drying").

Grinding and/or Sieving

Most directly compressible materials are prepared by crystallization. The crystal size, and, in part, the crystal shape, are selected by sieving or in some cases, after grinding. The particle size and shape depend on the grinding process (if ground) or on the mesh size and opening shape of the sieve used (if sieved). Although the purpose of sieving and grinding materials for direct compaction is primarily to control flow properties, the compactibility may also alter because of changes in particle properties, such as surface area and possibly surface activation (15).

Crystallized lactose monohydrate is either sieved or it is first ground and then sieved to make different sieve fractions available to the customers. For direct compaction, the unmilled, sieved 100-mesh quality is recommended because of its better flowability. The powdered grades, for example, lactose 200- or 450-mesh, are ground lactoses that have poor flow properties and are intended for use in wet granulation. Dicalcium phosphate dihydrate is commonly milled after crystallization for use in the wet granulation process. For direct compaction, however, only the unmilled larger crystal varieties (having at least 25% by weight of particles larger than 125 μm), can be used, because they have better flow and compaction properties (16).

Granulation and Agglomeration

Granulation and agglomeration represent the transformation of small-sized, cohesive, poorly flowable powders into a flowable and directly compressible form. When primary particles have binding properties of their own, the addition of a binder is not necessary. For example, agglomerated lactose (Tablettose) and directly compressible lactitol (Finlac DC) are prepared without the addition of a binder. When primary particles do not have binding properties, a binder is added to aid in the granule formulation. Examples are Xylitab 100 and Xylitab 200, containing 3% polydextrose and 1.5% carboxymethylcellulose, respectively. Granulation results in nearly sperical-shaped particles. The bulk density depends on the granulation method used: fluid bed granulation results in granules with a lower bulk density as compared with granules prepared in a high-shear mixer (17).

Agglomerated filler-binders have several advantages. In addition to the improvement of the flowability, agglomeration by granulation may improve compactibility of brittle materials because of an increased fragmentation propensity (17). An additional and important advantage of agglomerated filler-binders is the lower lubricant sensitivity as compared with the starting materials (18). As agglomerates are usually brittle, they break at low punch forces. As a consequence, lubricant sensitivity is relatively low because during mixing only the outside of the agglomerates will be covered with lubricant. The primary particles may have brittle or ductile deformation behavior, but this has no further consequences for their lubricant sensitivity. However, agglomeration can also have disadvantages: granulation of powdered cellulose or starch improves the flowability, but increases the lubricant sensitivity to binding (see sections "Powdered Cellulose," "Native Starches," and "Modified Starches," respectively). Agglomeration is commonly a result of physical modification. Examples are special crystallization techniques (section "Special Crystallization Techniques"), and spray drying (section "Spray Drying"). In some cases these techniques are combined with a sieving or grinding step.

Special Crystallization Techniques

The conditions of crystallization determine to a large extent the solid state properties of directly compressible materials. Controlled crystallization would impart free-flowing properties to excipients and to drug substances, but not necessarily self-binding properties.

If polymorphism exists, the compactibility of the polymorphic forms may be quite different because of the internal arrangement of the molecules within the unit cells of crystals. The forces applied to the sample are not transmitted uniformly and the gliding of the molecules may be more or less difficult depending on the crystal structure. Plastic deformation may occur depending on the dislocations and slip planes in the crystals. Crystalline substances are subject to such deformations depending on the symmetry within the crystal lattice. The crystal structure that has a greater degree of symmetry will be more prone to deformation on compression and compaction. The symmetry within the crystal structure diminishes in the following order: cubic, hexagonal, tetragonal, rhombohedral, orthorhombic, monoclinic, and triclinic (19). Thus, cubic crystals such as sodium chloride and potassium chloride can be compacted directly.

For example, α-lactose monohydrate is obtained by crystallization from a solution at temperatures below 93°C. Anhydrous α-lactose is obtained in a stable form by a special thermal dehydration process of sieved α-lactose monohydrate at a carefully controlled temperature and water vapor pressure. Anhydrous β-lactose is obtained by crystallization from a supersaturated solution at temperatures exceeding 93°C. The compactibility of anhydrous α-lactose is much better than that of the α-lactose monohydrate (see section "Anhydrous α-Lactose").

The effect of the structure of polycrystalline aggregates on their compactibility has been demonstrated for alumina trihydrate (20). Mosaic particles with a more disordered structure than radial particles show a much better compactibility. Recrystallization can change the properties for direct compaction, as illustrated for potassium chloride (21) and sodium chloride (22). After recrystallization, the particles are irregular in shape with rounded edges, show a better compactibility, and exhibit reduced friction with punches and dies (23).

Spray Drying

Spray drying involves three steps: (*i*) atomization of an aqueous solution or suspension into a spray, (*ii*) contact between spray and hot air in a drying chamber resulting in moisture evaporation, and (*iii*) recovery of the dried product from the air. Because of the spherical nature of liquid particles after evaporation of water, the resulting spray-dried material consists of porous, spherical agglomerates of solid particles that are fairly uniform in size. The particle size distribution of the spray-dried material is controlled by the atomization process and the type of drying chamber. For crystalline materials rapid cooling of a solution and a high rate of crystallization during spray drying results in porous solids with an imperfect crystal structure that contains amorphous material. Lattice defects imply a high deformability of the crystalline component, and the amorphous material is commonly ductile and, hence, deforms plastically (24). Another advantage of the spray-drying technology is the improved flowability caused by the spherical nature of the agglomerated particles.

Several directly compressible materials are produced by spray drying. Lactose was the first pharmaceutical excipient to successfully exemplify the spray-drying technology. The superior binding ability of spray-dried lactose, when compared with α-lactose monohydrate, has been attributed to the amorphous lactose that exhibits plastic flow on compression (see section "Spray-Dried Lactose"). Examples of other filler-binders prepared by spray drying are Avicel PH (microcrystalline cellulose), Eratab (modified rice starch), Maltrin M510 (maltodextrin), TRI-CAFOS S (tricalcium phosphate), Parteck SI (γ-sorbitol), Emdex (cocrystallized dextrose-maltose), and different types of directly compressible saccharose.

Pregelatination

Pregelatinized starch has been chemically or mechanically processed to rupture all or part of the granules in water. It is then dried to yield an excipient material suitable for direct compaction. Compressible starch is produced by partial hydrolyzation of corn starch. Free amylopectin improves binding properties, whereas free amylose gives a product better disintegration properties than its starting material (see section "Compressible Starch").

Dehydration

Increased binding properties of α-lactose monohydrate occur with increasing thermal or chemical dehydration of the solid (25). Desiccation with methanol gives a much steeper increase

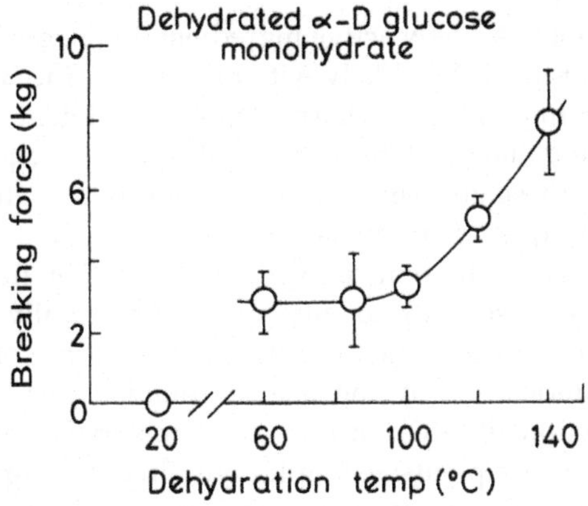

Figure 1 Breaking force of tablets, compressed from fully dehydrated D-glucose monohydrate, treated at different temperatures. *Source*: From Ref. 26.

in binding properties than thermal treatment. During dehydration, a gradual transition within each particle from α-lactose monohydrate into anhydrous α-lactose is observed. The anhydrous product has not only much better compactibility, but also better flow properties than α-lactose monohydrate (see section "Anhydrous α-Lactose"). Change in binding properties of the hydrates by thermal or chemical dehydration has been reported for a number of other materials. These include glucose monohydrate (Fig. 1), citric acid monohydrate, calcium sulfate dihydrate, calcium monohydrogen phosphate dihydrate, and dicalcium phosphate dihydrate (26).

Coprocessing

An increasing number of coprocessed filler-binders have appeared on the pharmaceutical market in recent years. This is because of the benefits of these products, even if compared with the agglomerated filler-binders (18). Because of their excellent overall tableting properties, coprocessed excipients have been even referred to as "multipurpose excipients" (27).

Coprocessed filler-binders are manufactured by coprocessing two (seldom more) materials, usually one that is ductile and the second being brittle. For all products, the coprocessed material has better flow and compaction properties than a physical blend of its constituents. Van der Voort Maarschalk and Bolhuis (18) gave a possible explanation for the improved compaction properties. Most coprocessed products consist of a large amount of brittle material, such as α-lactose monohydrate, and a smaller amount of a plastically deforming material such as cellulose, fixed between or on the particles of the brittle material. Plasticity, present in ductile materials, is important for the creation of a large area of contact. However, ductile materials are sensitive to lubricants and have a high compaction speed sensitivity due to their viscoelasticity. During postcompaction relaxation of tablets compressed from ductile materials, stored elastic energy will break bonds and increase tablet porosity, which results in a lower tablet strength (28). In coprocessed filler-binders, the plastic material will be responsible for the good bonding properties because it creates a continuous matrix with a large surface area for bonding. The presence of a large amount of brittle material prevents the storage of too much elastic energy during compression, which results in a smaller stress relaxation. Moreover, the lubricant sensitivity will be low. Breakage of the coprocessed particle and the presence of a large amount of brittle material will prevent the formation of a coherent lubricant network.

New Filler-Binders

The regulatory aspects of excipients must also be borne in mind. Because excipients are components of medicines, a new excipient would have to be assessed for safety in the same extensive and expensive manner as a new active ingredient. In the case of the latter, such costs can probably be recouped, but this is less likely with excipients, since the economic margin on these is so much lower. The International Pharmaceutical Excipient Council (IPEC) and the Pharmaceutical Quality Group have published guidelines for excipients (29,30). Excipient manufacturers should follow these guidelines and have effective quality systems and process controls to provide consistent, quality products.

Given the costs of safety evaluation, it would appear that the development of new chemical entities exclusively for use as excipients is not economically feasible (31). It is, thus, likely that future development of excipients will take the form of chemical or physical modification of existing materials, or by the use of combinations of existing materials to improve or extend their functionality (32). In rare cases, materials usually regarded as active ingredients can also be used as excipients: examples are directly compressible calcium lactate pentahydrate (see section "Calcium Lactate Pentahydrate") and inulin (see section "Inulin").

Classification of Filler-Binders

There are several possibilities for classifying filler-binders. A straightforward method is based on application. Thus, there are filler-binders for dispersible tablets, for tablets designed to be swallowed intact, for lozenges, chewable tablets, and tablets that incorporate some method of controlled release. A second method is to classify filler-binders by their chemical constitution. A drawback of this is that materials that are chemically similar can have very different tableting properties. Nevertheless this method is widely used, (33,34) and forms the framework of "Materials for Direct Compaction" in the first edition of this book.

An attempt has been made by Van der Voort Maarschalk and Bolhuis (18) to devise a physicomechanical classification of filler-binders, based on the behavior of particles under a compressive load. Van der Voort Maarschalk and Bolhuis (18) identified three classes of behavior under compaction (Fig. 2). Brittle materials with a high fragmentation propensity break during the particle rearrangement phase, which occurs at low punch pressures. This means that fragments of the original particles can be distributed at random within a compact. Materials with a low fragmentation propensity break mainly after the rearrangement process. Thus, though the original particle is broken into fragments, these fragments stay more or less together. The third class of substances—ductile materials—deform rather than fragment and can be regarded as materials with a very low fragmentation propensity.

An advantage of this classification is that it gives an immediate indication for the lubricant and compaction speed sensitivity of excipients. It is well known that the addition of a

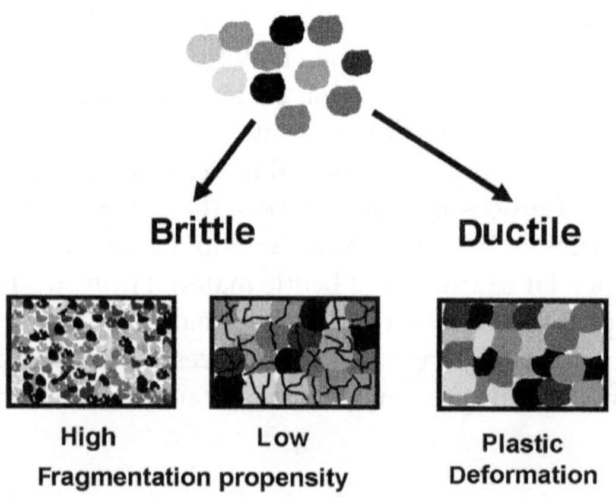

Figure 2 Particle rearrangement during compression of single-particle excipients. *Source*: From Ref. 18.

lubricant to a direct compaction formulation can decrease the physical strength of the tablets, though the extent of that decrease depends on the filler-binder in question. This is termed lubricant sensitivity. Another feature of some filler-binders is that tablet strength decreases as the rate of application of the compressive force is increased, and this is called compaction rate sensitivity. Substances that show high lubricant sensitivity are also likely to be speed sensitive. The early fragmentation of materials with a high fragmentation propensity such as dicalcium phosphates will destroy any lubricant film formed during the mixing stage. Bonding between freshly formed, clean surfaces will be better than that of surfaces contaminated with a lubricant film. Materials with a low fragmentation propensity, such as α-lactose monohydate, break after the rearrangement stage. The lubricant film will be damaged, but since fragments are not dispersed, a more or less coherent lubricant matrix to be retained. This matrix works as a network of weak bonds, resulting in a tablet strength that is significantly lower than that of a compact made from nonlubricated material.

Compaction speed sensitivity is also dependent on particle deformation behavior. As an effect of their viscoelastic behavior, ductile materials commonly have much higher speed sensitivity than brittle materials.

In this chapter, the filler-binders will be classified by their chemical constitution.

CELLULOSE

Cellulose forms the backbone of many excipients used in marketed drug products. Cellulose is built up of repeating glucose units joined by β-1,4 glucosidic bonds. Numerous investigations show that in native cellulose, numerous anhydroglucose units are combined to form a single cellulose molecule. Chain length determination is very difficult, and only average values are given, referred to as the average degree of polymerization (DP).

Pharmaceutical grades of cellulose are obtained by either mechanical or chemical processing, or by both. Attached to carbon atoms #2, #3, and #6 of each anhydroglucose unit is a hydroxyl group capable of undergoing chemical reactions such as esterification or etherification. The reaction of hydroxyl groups varies for stearic reasons. Products can be prepared in which one, two, or three hydroxyl groups per anhydroglucose unit have undergone reaction.

The amount of partial esterification or etherification is given for each substituent separately as the degree of substitution (DS). It ranges from 0 to 3 and is equivalent to the average of converted hydroxyl groups per anhydroglucose unit. Pure cellulose can be ground mechanically with additional treatment by hydrochloric acid. The resulting powder is cellulose powder or microcrystalline cellulose.

In addition, various cellulose-based products with improved tableting properties have been introduced in the past decennium: silicified microcrystalline cellulose (SMCC), low crystalline powdered cellulose (LCPC), powdered cellulose with a cellulose II lattice, and powdered cellulose coated with silica.

Microcrystalline Cellulose

Microcrystalline cellulose is described in the NF as a purified, partially depolymerized cellulose prepared by treating α-cellulose, obtained as a pulp from fibrous plant material with mineral acids. The cellulose fibers in the starting material are composed of millions of microfibers. In the microfibers, two different regions can be distinguished: a paracrystalline region, which is an amorphous and flexible mass of cellulose chains, and a crystalline region, which is composed of tight bundles of cellulose chains in a rigid linear arrangement (35). As a result of controlled hydrolysis, the amorphous fraction has largely been removed, yielding aggregates of the more crystalline portions of cellulose fibers. After purification by filtration and spray drying, dry, porous microcrystals are obtained. By controlling the atomization and

Table 3 Avicel PH Microcrystalline Cellulose Types

Grade	Nominal particle size (μm)	Moisture (%)	Loose bulk density (g/cm^3)
PH-101	50	3.0–5.0	0.26–0.31
PH-102	100	3.0–5.0	0.28–0.33
PH-105	20	≤5.0	0.20–0.30
PH-102 SCG	150	3.0–5.0	0.28–0.34
PH-200	180	2.0–5.0	0.29–0.36
PH-301	50	3.0–5.0	0.34–0.45
PH-302	100	3.0–5.0	0.35–0.46
PH-103	50	≤3	0.26–0.31
PH-113	50	≤2	0.27–0.34
PH-112	100	≤1.5	0.28–0.34

Abbreviation: SCG, special coarse grade.
Source: From FMC Biopolymer.

drying conditions, particle size distribution can be varied (36,37). Microcrystalline cellulose occurs as a white odorless, tasteless, crystalline powder composed of porous particles or as an agglomerated product.

Microcrystalline cellulose was marketed first in 1964 by FMC Biopolymer (Pennsylvania, U.S.) under the name Avicel PH in four different particle size grades, each with different properties. The basic types are Avicel PH-101 and Avicel PH-102 (38). Other brands of microcrystalline cellulose (e.g., Compricel, Emcocel, Pharmacel, and Vivapur) are also available on the pharmaceutical market. More qualities of microcrystalline cellulose are available from the Far East and South America. The equivalence of microcrystalline cellulose obtained from different suppliers has been reported by Doelker (39).

As a solution to the well-known disadvantages of the basic 101 and 102 types of microcrystalline cellulose, several new varieties, having a lower moisture content or better flow properties, have been introduced during the last years (Table 3). An excellent review is written by Doelker et al. (40).

Microcrystalline cellulose is one of the most used filler-binders in direct compaction. Its popularity in direct compaction is because of its extremely good binding properties as a dry binder (Fig. 3). It also works as a disintegrant and a lubricant and has a high dilution potential in direct compaction formulations. In addition to its use in direct compaction formulations, microcrystalline cellulose is used as a diluent in tablets prepared by wet granulation, as a filler for capsules, and for the production of spheres.

Scanning electron micrographs show that Avicel PH-101 has a matchstick-like or rod-like structure, whereas Avicel PH-102 is a mixture of primary particles and agglomerates (Fig. 4A) (38,41). The primary particles are composed of fibrils with a radius of 10 to 15 nm with a hollow axis of about 2 nm. These intraparticular pores account for 90% of the total surface area (38). Measurement of the interparticulate porosity by mercury porosimetry, liquid penetration techniques, and scanning electron microscopy demonstrates that the modal pore radius decreases logarithmically with an increase in compaction pressure. A theoretical zero porosity would be reached at a pressure of 300 MPa (42). The intraparticulate pore size distribution shows no change at 20 MPa, indicating that the internal pores do not collapse (42,43).

Conflicting results of the specific surface area of microcrystalline cellulose have been reported. Using nitrogen adsorption and the BET equation, apparent specific surface area of Avicel PH-101 values between 1.0 and 1.5 m^2/g have been reported (44–46). Marshall and Sixsmith (38) concluded that a specific surface area of 1.3 m^2/g, obtained by mercury porosimetry, represents the external interparticle surface area. Water vapor sorption with application of the BET equation has led to values of 138 and 150 m^2/g (44,45). These values

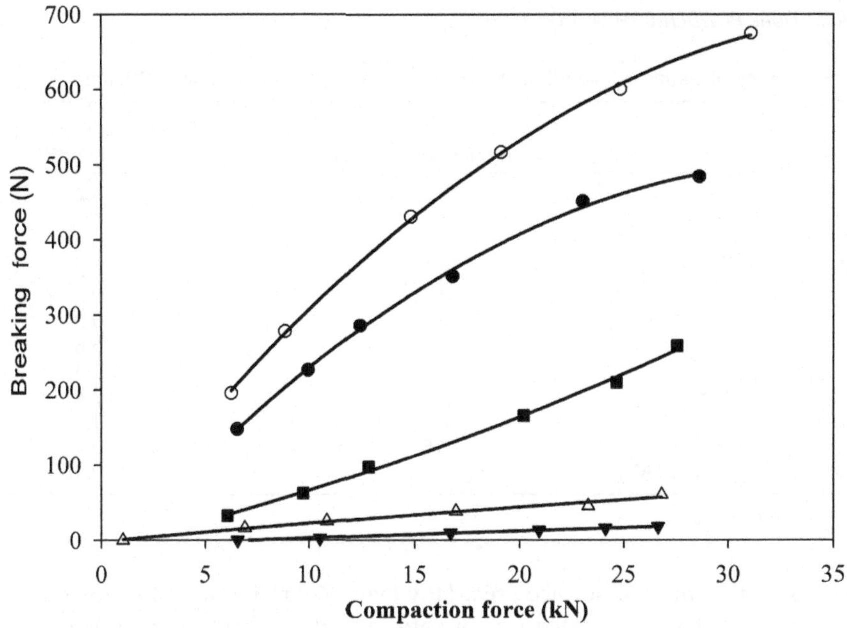

Figure 3 Compaction profiles of various filler-binders, lubricated with 0.5% magnesium stearate. (●) microcrystalline cellulose (Pharmacel 102); (○) silicified microcrystalline cellulose (Prosolv SMCC 90); (▼) modified starch (STARCH 1500); (Δ) dicalcium phosphate dihydrate (DI-CAFOS); (■) calcium lactate pentahydrate (Puracal DC). 500 mg, 13 mm tablets, compaction simulator at 300 mm/sec.

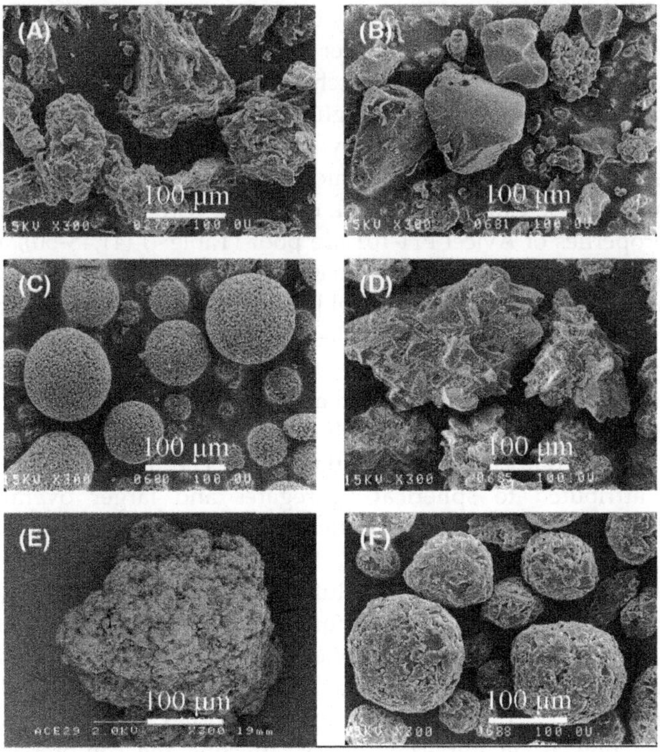

Figure 4 Scanning electron micrographs of directly compressible filler-binders: (**A**) microcrystalline cellulose (Avicel-PH 102); (**B**) compressible starch (STARCH 1500); (**C**) modified rice starch (Eratab); (**D**) dicalcium phosphate dihydrate (Emcompress); (**E**) calcium lactate pentahydrate (Puracal DC); (**F**) dextrates (Emdex).

Table 4 Flow Properties of Series of Filler-Binders, Lubricated with 0.5% Magnesium Stearate

Filler-binder	Flow through aperture (mm)	Variation coefficient of tablet weight (%)
Avicel PH-101	>18	3.2
Avicel PH-102	5	1.0
Elcema G250	5	0.35
STARCH 1500	>18	2.2
Emcompress	2.5	0.9
Compactrol	5	0.3
Emdex	5	0.6
Neosorb 20/60	5	0.5
α-Lactose monohydrate 100-mesh	2.5	0.4
Tablettose	2.5	1.2
Spray-dried lactose (SuperTab 11SD)	2.5	0.4
Anhydrous β-lactose (SuperTab 21AN)	2.5	0.45

Source: Adapted from Ref. 11.

were ascribed to be the internal surface area. However, Zografi et al. (47) showed that these high values do not reflect a true surface area but are caused by penetration of water vapor into the amorphous parts of cellulose and interaction with individual anhydroglucose units. Using nitrogen and krypton adsorption following several methods of sample pretreatment, they found a specific surface area of 1.3 m^2/g with no evidence of extensive microporosity. The crystallinity of Avicel PH-101 determined by X-ray diffraction (44,46) and infrared measurement (44), is about 63%. For Emcocel 50M, the crystallinity is somewhat greater (46).

At room temperature and relative humidity between 40% and 50%, Avicel PH-101 has a moisture content of about 5% to 6% (47). Water vapor sorption occurs by the same basic mechanism as in other types of cellulose and starches. This can be described by the sorption of one tightly bound water molecule at an anhydroglucose unit, followed by a second less tightly bound water molecule, with further additional layers of bulk water. In contrast to starch, water is only sorbed in the noncrystalline regions of microcrystalline cellulose (47).

Microcrystalline cellulose exhibits low bulk densities, with values ranging from 0.32 to 0.45 (37). The flow properties of Avicel PH-101 are poor (Table 4) (41,48–50), as explained by the more or less long-drawn, matchstick-like form of the particles, the particle size distribution, and the low bulk density (41,49). Although Avicel PH-102 has better flow properties because it is a mixture of agglomerates and primary particles, its flowability is only moderate (Table 4). The flowability of Avicel PH-103 and Avicel PH-105 is less than that of Avicel PH-101 and has been ascribed to differences in moisture content or particle size distribution (51). FMC Biopolymer has introduced Avicel PH-200 with a mean particle size of about 180 μm to improve its flow in direct compaction formulations (52). The improved flow properties of Avicel PH-200 are attributed to spherical aggregates and larger overall particle size distribution. Some drawbacks of these large particle size microcrystalline cellulose are higher lubricant sensitivity and lower carrier capacity.

During compaction, microcrystalline cellulose is thought to undergo stress relief deformation by several mechanisms. At low compression forces, stress relief is dominated by a slight elastic phase (53). This has been explained by its hollow microfibrillar structure (54). At higher force, there is either further deformation (55) or permanent deformation by nonspecific plastic flow (56). Following stress relaxation studies, David and Augsburger (57) suggested that plastic flow is an important factor affecting the compressibility of microcrystalline cellulose. Plastic deformation is facilitated by the presence of slip planes, dislocations, and the small size (a few thousand by a few hundred Ångström units) of the individual microcrystals. The hydrolysis portion of the production process is responsible for the formation of the slip planes and plastic deformation of the particles on a microscale. The spray-drying operation is responsible for the deformability of microcrystalline cellulose on a macroscale (36). The plastic behavior of microcrystalline cellulose is deduced from force-displacement curves (58), Heckel plots (59,60), and creep analysis (61). Using mercury porosimetry, Vromans et al. (62) show that there is no alteration in the quantity of small pores in tablets compressed from microcrystalline cellulose, which indicates plastic deformation under pressure. The plasticity of microcrystalline cellulose increases with increasing compressing force (51), which is accompanied by a decrease in viscoelasticity (63). Force versus displacement plots on diametral compression to fracture indicates that microcrystalline cellulose shows stress relief by time-dependent yielding (64). The yield pressure increases with punch velocity, because of a reduction in the amount of plastic deformation caused by the time-dependent nature of plastic flow (63,65). This increase in mean yield pressure at increasing compression speeds is reflected by the high strain rate sensitivity (SRS) value for microcrystalline cellulose (65). Armstrong and Palfrey (66) show that the reduction in tablet strength at increasing compression speed is caused by a reduction in porosity of the compacted powder bed.

The strong binding properties are caused by hydrogen bonds between hydrogen groups on the plastically deformed, adjacent cellulose particles (56). The presence of hydrogen bonds is demonstrated by Hüttenrauch (67). The hydrogen bonds on the extremely large surface area are brought into close contact during plastic deformation. This is the reason for the extremely good compactibility of microcrystalline cellulose, better than that of any other directly compressible filler-binder (41,68,69). The exceptionally good binding properties of microcrystalline cellulose are reflected by its extremely high bonding index, whereas its brittle fracture index (BFI) is extremely low (70).

Recently, several authors studied the effect of production parameters on physical and mechanical properties of microcrystalline cellulose powder. A correlation between DP and compactibility was found by Shlieout et al. (71). An increasing DP leads to an increase in the hardness of microcrystalline cellulose tablets. This effect was explained by a decreased elastic deformation behavior. In other work, however, it was demonstrated that not the DP, but the origin of the raw materials and the production method influenced the compaction properties of microcrystalline cellulose (72). Suzuki and Nakagami (73) pulverized microcrystalline cellulose to vary the crystallinity. They showed that the compactibility parameters *B* value and yield pressure, calculated from Heckel plots, were lowered as the degree of crystallinity became smaller.

The compactibility of microcrystalline cellulose depends on its moisture content. It has been suggested that, at its equilibrium moisture content of 5%, most of the water will be within the porous structure of microcrystalline cellulose and a large proportion of this bound moisture is expected to be hydrogen-bonded to small bits of cellulose within the particle. During plastic deformation, the moisture within the pores should act as an internal lubricant and facilitate slippage and flow within the individual microcrystals (45). The presence of an optimum amount of water will prevent elastic recovery by forming bonds though hydrogen bond bridges (74). The compactibility of microcrystalline cellulose decreases with a reduction of its moisture content (74). The strongest compacts are produced when microcrystalline cellulose contains 7.3% moisture. However, at a lower moisture content, the compacts capped when compressed at 163 MPa. The capping tendency of microcrystalline cellulose-based formulation is reduced by increasing the moisture content from 3.2% to 6.1% (75). This is attributed to the strengthening of interparticle binding forces and the reduction of elastic recovery by lowering of the yield point. The effect of water on the compactibility of

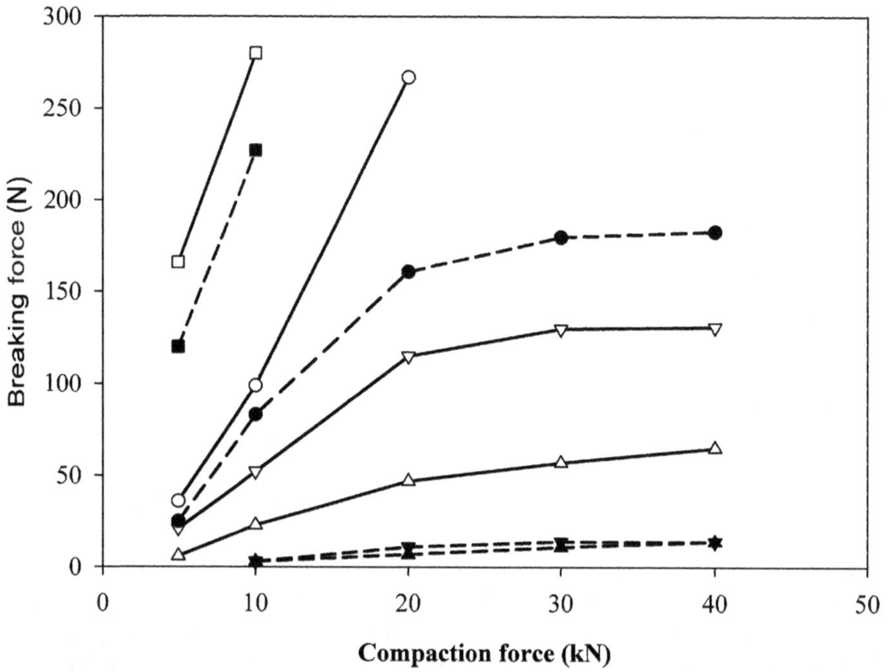

Figure 5 Compaction profiles of different cellulose and starch products, both unlubricated (*open symbols*) and lubricated with 1% magnesium stearate (*closed symbols*). (□, ■) microcrystalline cellulose (Avicel PH-102); (Δ,▲) powdered cellulose (Elcema G250); (○, ●) modified rice starch (Eratab); (▼, ▽) compressible starch (STARCH 1500). 500 mg, 13 mm tablets, hydraulic press, contact time 5.1 seconds.

microcrystalline cellulose is strongly dependent on its thermodynamic state in the excipient, and is similar to the effect of water on the compactibility of starch (76). The particle size of microcrystalline cellulose has only a small effect on its compactibility (59,60). A maximum compactibility has been found for the 80 to 100 μm fraction (77).

As an effect of their plastic behavior, celluloses are sensitive to mixing with lubricants (78). Because of the extremely high compactibility of unlubricated microcrystalline cellulose, softening has practical significance only after relatively long mixing times (Fig. 5). Alternative lubricants such as hydrogenated vegetable oil (Lubritab) or sodium stearyl fumarate (Pruv), have much less effect on the tablet strength following mixing (61). More information of the effect of mixing with lubricants on binding properties of filler-binders is given in chapter 9.

Microcrystalline cellulose has a high dilution potential, defined as the ability of a given quantity of an excipient to bind a specified amount of an active ingredient into an acceptable tablet (36). This high dilution potential is attributed to low bulk density that impacts high covering power, and to broad particle size distribution that allows optimum packing density.

In addition to binding properties, microcrystalline cellulose has lubricating and disintegration properties (36). The lubricating properties are attributed to the very low residual die wall pressure, as an effect of stress relief by elastic or viscoelastic relaxation within the die, as well as to the very low coefficient of friction (70,79,80). From radial movement studies of the outer die wall on compression and recompression of microcrystalline cellulose compacts, Travers and Cox (64) concluded that the self-lubricating properties of microcrystalline cellulose tablets may be due to their elastic properties. With no shear yielding or failure, they regain their original dimensions when the axial force is removed and little or no "locked in" stress remains.

Disintegration of microcrystalline cellulose tablets is attributed to the penetration of water into the hydrophilic tablet matrix by means of capillary action of the pores and by a subsequent disruption of the hydrogen bonds (81). Penetration measurements demonstrate that microcrystalline cellulose tablets exhibit extremely fast aqueous penetration. The widening of the pores by the breaking of the hydrogen bonds result in a ratio between water uptake and original pore volume up to 20 (82). The lack of disintegration in fluids with a low dielectric constant confirms the theory that disintegration is caused by the breaking of hydrogen bonds (68). An increase in compaction load produces an increase in the disintegration time of tablets as an effect of the decreased water penetration into the tablets (49,81,83).

Plain microcrystalline tablets (56,84) or tablets containing microcrystalline cellulose as a filler-binder (85) softened and had a tendency to swell when aged under increased relative humidity. This moisture uptake is probably caused by high intraparticulate porosity (38). The rate of moisture uptake is directly proportional to the relative density of plain microcrystalline tablets (85). The absorbed moisture causes a disruption of the hydrogen bonding and hence a decrease in tablet strength (74). Although it has been reported that the softening of tablets was reversible on removal of the high humidity (56), other results do not confirm this phenomenon (85).

Because of the poor flow properties and the low bulk density, microcrystalline cellulose is not used solely as a primary filler-binder in a tablet formulation, but is instead mixed with an inexpensive filler with good flowability such as α-lactose monohydrate 100-mesh or dicalcium phosphate dihydrate (14,86,87).

Silicified Microcrystalline Cellulose

A major development has been the introduction of SMCC. Although it is a coprocessed filler-binder, this product is discussed in this section because there are major differences between SMCC and other coprocessed excipients. The latter usually contain two components, both of which are fillers or filler-binders, whereas SMCC is a combination of a filler-binder and a glidant. SMCC is marketed by JRS Pharma (Germany) as Prosolv SMCC (88). It is produced by coprocessing 98% microcrystalline cellulose with 2% colloidal silicon dioxide. The excipient is available in two particle size grades, Prosolv SMCC 50 and Prosolv SMCC 90 and in a high density quality: Prosolv HD 90. In direct compaction, SMCC is 10% to 40% more compactible than regular MCC (Fig. 3) and has a lower lubricant sensitivity. The flow rate of Prosolv SMCC 90 was found to be equivalent to that of the PH-200 grade of MCC (89). In addition to a better flowability, SMCC has a higher bulk density than regular MCC, which can be explained by its improved flowability and packing properties (90). Studies using helium pycnometry, laser light scattering, particle size analysis, Fourier transform infrared spectroscopy, gas adsorption, X-ray powder diffraction, solid state NMR, calorimetry, water vapor sorption and Raman spectroscopy have all shown that silicification appears to have no discernible effect on the primary chemical and polymorphic characteristics of microcrystalline cellulose (89,91,92).

The specific surface area of SMCC was found to be about five times higher than that of microcrystalline cellulose, and the pore volume size distributions calculated from nitrogen adsorption isotherms showed that the total pore volume was greater for SMCC (90). This effect was explained by the very high specific surface area of colloidal silicon dioxide. The pore size distribution characteristics determined by a mercury porosimeter were very similar for SMCC and MCC with a comparable particle size distribution (92). This suggests that bulk modification of MCC does not occur during silicification, and that the colloidal silicon dioxide, either by providing surface modification or by modifying strengthening interactions, is primarily responsible for the improvements in functionality, in particular tablet strength. Scanning electron microscopy studies together with electron probe microanalysis have demonstrated that silicon dioxide is primarily located in the surface of SMCC, but some silicon dioxide was detected in the internal regions of some particles (93). The presence in the surface is an important observation, since this may alter characteristics such as interfacial strength and interactions with magnesium stearate. In a comparative study of the mechanical properties of unlubricated compacts of MCC and SMCC, it was found that at relatively slow compaction rates, compacts with a similar relative density were found, which suggests that the two materials exhibit a comparable compaction behavior (94). Not only the tensile strength but also the stiffness and energy of failure were greater for compacts prepared from SMCC than for compacts prepared from MCC or blends of MCC and colloidal silicon dioxide. From these results, the authors concluded that the strength enhancement by silicification of MCC may be a consequence of an interfacial interaction rather than modification of the bulk MCC properties. In a more recent study, Van Veen et al. (95) studied the compaction mechanisms of unlubricated and lubricated MCC and SMCC. They found that neither colloidal silicium dioxide nor magnesium stearate facilitates the densification of MCC during compaction. The slightly higher relaxation of SMCC tablets showed that colloidal silicium dioxide has more negative than positive effect on interparticulate bonding. However, for lubricated MCC a larger increase in tablet relaxation at high compression speed was found than for lubricated SMCC tablets, so the decrease in tablet strength was larger for MCC tablets than for SMCC tablets when lubrication was applied. An examination of the tablet strength of tablets compressed from physical mixtures of MCC with increasing concentrations of colloidal silicium dioxide proved the slightly negative influence of silicon dioxide on the tablet strength of unlubricated MCC tablets and the positive effect of colloidal silicon dioxide addition on the strength of lubricated MCC tablets. The authors showed that coprocessing of MCC with colloidal silicon dioxide showed no extra contribution on the tablet strength of lubricated tablets above physical mixtures. The positive effect of colloidal silicium dioxide on the compactibility of MCC was elucidated by an interaction between magnesium stearate and colloidal silicon dioxide. Only the part of colloidal silicon dioxide that is fixed on the surface of the SMCC particles (about 20–30% of the 2% colloidal silicon dioxide in SMCC) is working effectively in relation to the negative effect of magnesium stearate as lubricant on tablet strength (95).

Mužiková and Nováková (96) compared the tableting properties of high-density SMCC (Prosolv HD 90) with those of SMCC (Prosolv SMCC 90). Compared with regular SMCC, the high-density SMCC had better flow properties and a smaller sensitivity to the rate of tableting. A disadvantage of high-density SMCC is the higher lubricant sensitivity, an effect that was attributed to the larger size of Prosolv HD 90.

Powdered Cellulose

Powdered cellulose NF is purified, mechanically disintegrated cellulose, prepared by processing α-cellulose obtained as a pulp from fibrous plant materials. Powdered cellulose is also known as microfine cellulose or cellulose flocs. JRS Pharma markets powdered cellulose under the name Elcema in powder (P050, P100), fibrous (F150), and granular form (G250). A well-known type is Elcema G250, consisting of granules prepared from the P100 quality, without the addition of a binder, and with a mean particle size of about 250 μm. This product has good flow properties (Table 4). Solka Floc from International Fiber Corp. (New York, U.S.) is available in different grades under which a granular form (Solka Floc Fine Granular) with a mean particle size of 239 μm. When it is not in an agglomerated form, powdered cellulose has poor flow properties, so it is an unsuitable direct compaction excipient when used alone (97,98).

Although powdered cellulose has binding properties, the binding is inferior to that of microcrystalline cellulose (Fig. 5) because of differences in the manufacturing process (36). Powdered cellulose deforms plastically, as shown by considerable stress relaxation following compaction (99,100) or by the change in the stress density profile using different dwell times of compression (100,101). Similar to microcrystalline cellulose, hydrogen bonds are formed between adjacent, plastically deformed cellulose particles, although scanning electron micrographs show that mechanical interlocking may also play a role (102). The lower compactibility, when compared with microcrystalline cellulose, is caused by a lack of slip planes and dislocations in the cellulose granules.

The lubricant sensitivity of powdered cellulose is strongly dependent on the size and form of the particles. The binding properties of the granulated form (Elcema G250) show a dramatic decrease after mixing with lubricants such as magnesium stearate (Fig. 5) (41,103). The powdered and fibrous Elcema qualities are less sensitive to magnesium stearate because of poor flow properties that retard or impart a formation of a lubricant film during mixing (98).

Because of the inferior binding properties of powdered cellulose when compared with those of microcrystalline celluloses, modified powdered celluloses with improved compaction properties have been described. One of these is low crystallinity powdered cellulose (LCPC). It is prepared by controlled decrystallization and depolymerization of cellulose with phosphoric acid (104). The powder and mechanical properties of different batches of LCPC, ranging in crystallinity from 15% to 45% were compared with those of different types of microcrystalline cellulose and powdered cellulose by Kothari et al. (105). Just like microcrystalline cellulose, LCPC consists of aggregates of particles. Further LCPC aggregates showed a smoother surface and were more densely packed than the microcrystalline cellulose products. Although no definite relationship was observed between crystallinity and the true density or moisture content of the various materials, LCPC picked up higher moisture content at a given vapor pressure compared with the higher crystallinity products microcrystalline cellulose and powdered cellulose. The yield pressure of LCPC, calculated from Heckel plots, was significantly lower than that of microcrystalline and powdered cellulose products. This suggests that LCPC undergoes plastic deformation at relatively lower compression pressures. Tensile strength values of tablets of LCPC were comparable to those for microcrystalline cellulose tablets. The disintegration times for LCPC tablets were much shorter than for microcrystalline cellulose tablets. This effect was explained by the difference in crystallinity between the two materials as well as the ease of accessibility for water molecules to enter and interact with free hydroxyl groups (105).

Another new cellulose-based tableting excipient, referred to as UICEL, has been developed by the same group by treating cellulose powder with an aqueous solution of sodium hydroxide and subsequently precipitating it with ethyl alcohol (106). In contrast to microcrystalline cellulose, UICEL shows a cellulose II lattice, whereas microcrystalline cellulose belongs to the cellulose I polymorphic form. Both crystallinity and DP were lower than for microcrystalline cellulose. Like microcrystalline cellulose type 102, the new product consists of a mixture of aggregated and nonaggregated fibers. Compared to Avicel PH-102, UICEL is denser and less ductile. Although the compactibility of UICEL is much smaller than that of microcrystalline cellulose type 102, the tablet strength is high enough for pharmaceutical practice. A definite advantage of UICEL over microcrystalline cellulose is the much shorter tablet disintegration time. Even tablets compressed at high forces disintegrate within a few seconds, so that UICEL has the potential to be used in the design of fast disintegrating tablets (106).

STARCH AND STARCH DERIVATIVES

Starches from several natural sources and their common derivatives are both well-known and safe and have been extensively investigated in tablet formulations for various purposes. However, poor flow and high lubricant sensitivity make native starches less favored in direct compaction (107). Chemical, mechanical and physical modifications of natural starches have been made to improve their direct compaction properties, both for immediate release and controlled release tablets.

Modified starches include those in which the grain has been more or less completely split, such as pregelatinized starch and esterified and etherified starches (known in the pharmaceutical field as carboxymethyl starches and hydroxyethyl starches) and granulated or agglomerated starches (referred to as "modified starches" in section "Modified Starches").

Native Starches

Starch from commonly used sources consists of two polysaccharides, amylose and amylopectin, that are based on a glucose monomer. Amylose is a linear polymer and represents approximately 27% by weight, while amylopectin has a branched structure and represents about 73% by weight. The amylose component resembles cellulose, except for the stereochemical configuration of linkage between monomer units. Starch is insoluble in water because the two polymers (amylose and amylopectin) are intermolecularly associated in the crystal lattice.

Starch is an important excipient for tablet formulations. It is widely used as a disintegrant, diluent and as a binder in the form of starch paste that is used in the wet granulation process.

Although unlubricated native starches have good compression characteristics, their poor flow properties and their high lubricant sensitivity make them less suitable for use as excipients in tablets prepared by direct compaction. Although it is generally accepted that starch undergoes plastic deformation under pressure (108), the tendency of starches to total and pure plastic deformation seems to be dependent on particle size and particle shape distributions (109,110). According to the values of yield pressure, corn starch is most prone to plastic flow with only small elastic recovery. Potato starch also flows plastically with ease, whereas barley and wheat starches are more elastic.

The compactibility of starches is dependent on the equilibrium moisture content, which is dependent on the relative humidity of the atmosphere under which the powders are stored. Maximum tablet strength is obtained at 60% to 70% relative humidity, corresponding to equilibrium moisture content of about 10% w/w (107). It is generally accepted that water sorption onto starches (just as on cellulose) occurs in different stages. At first, water becomes tightly bound to anhydroglucose units throughout the starch grain until a 1:1 water: anhydroglucose unit stoichiometry (11.1% w/w water) is obtained. Between 1:1 and 2:1 stoichiometry, the water molecules become less tightly bound to the water molecules already bound to anhydroglucose units. Water absorbed at a stoichiometry larger than 2:1 is even less tightly bound and has the properties of bulk water (76). Water absorbed within the starch particles influences the compaction properties by affecting the degree of viscoelasticity. Water levels below 1:1 stoichiometry reduce the compactibility of starch. Less tightly bound water (stoichiometry between 1:1 and 2:1) is needed to provide plasticity to the starch system, whereas bulk water decreases the ability to form bonds between the starch particles, probably due to the formation of a water film.

In a study about tableting properties of potato, corn, wheat, and barley starch, corn starch was best in compactibility, whereas potato starch was the best with respect to flowability (109). The latter is, however, extremely susceptible to mixing with magnesium stearate (107). This high lubricant sensitivity is an effect of the plastic behavior of starch products under compression (see chap. 9). In another comparative evaluation of several native starches, evidence showed that rice starch had better compaction properties and worse flow properties when compared to corn, potato, and tapioca starches. Moreover, the binding capacity of rice starch was insensitive to mixing with magnesium stearate (107). The lack of lubricant

Table 5 Tablet Strength and Magnesium Stearate Susceptibility of Starches and Starch Granulations After Storage at 19°C and 44% RH

	Strength		
Starch type	Unlubricated[a]	Lubricated[b]	Strength reduction ratio
Starch as such			
Potato	68 ± 10	0	0
Tapioca	70 ± 4	39 ± 4	0.6
Maize	56 ± 8	30 ± 7	0.5
Rice	102 ± 20	104 ± 11	1.0
Granulations			
Potato	47 ± 3	<10	<0.2
Tapioca	54 ± 2	<10	<0.2
Maize	27 ± 4	<10	<0.2
Rice	139 ± 11	50 ± 8	0.4

[a]Breaking force (N), unlubricated, compression force 20 kN (except rice starch as such: 10 kN).
[b]Breaking force (N), lubricated with 0.5% magnesium stearate, compression force 20 kN (except rice starch as such: 10 kN).
Source: From Ref. 111.

sensitivity of rice starch is caused by its poor flowability, which impairs the formation of a lubricant film over the particles during mixing with a lubricant. These physicomechanical properties are attributed to the fine particle size of rice starch as compared to other starches.

To improve tableting properties for direct compaction, starch can be physically modified in different ways. For example, special pregelatination process improves tableting properties (see section "Compressible Starch"). Another possibility for improvement of tableting properties of starch is agglomeration (107). Although granulation with 2.5% starch paste of the same kind of starch improves flowability, tablets from granulated potato, tapioca, and maize starch have lower tablet strength than tablets prepared from starches without wet granulation with starch paste (Table 5). However, wet granulation of rice starch with starch paste improves both flow and compaction properties (107). Because of their better flow properties, starch granulations are more susceptible to mixing with magnesium stearate than starches prepared without wet granulation. After mixing starches for 30 minutes with 0.5% magnesium stearate in a Turbula mixer at 90 rpm, only tablets prepared with rice starch granules show sufficient breaking force (Table 5) (111). It has been demonstrated that the lubricant sensitivity of starch granules is dependent, among other factors, on properties of the granules such as bulk density and flowability, which are determined by the method of wet granulation (98). Commercially available starch granulations are described in section "Modified Starches."

Compressible Starch

Pregelatinized starch USP/NF or pregelatinized starch Ph Eur is starch that has been chemically or mechanically processed to rupture all or part of the granules in the presence of water, and subsequently dried. Although pregelatinized starch is primarily a binder in wet granulation, it can be modified to make it compressible and flowable in character. A special pregelatinized starch for direct compaction, commonly called *directly compressible starch* or *compressible starch*, is partially hydrolyzed corn starch, marketed in 1964 as STA-Rx 1500 by Staley (Illinois, U.S.) and now available as STARCH 1500 from Colorcon (U.K.). More recent, Cerestar (U.K.) introduced partly gelatinized agglomerated starch granules with the brand name C*Pharm DC 93000. Chemically, compressible starch does not differ from starch USP.

STARCH 1500 is prepared by subjecting corn starch to physical compression or shear stress in high moisture conditions causing an increase in temperature and a partial gelatinization of some of the starch. The product consists of both individual starch grains and aggregates of starch grains bonded to the hydrolyzed starch (Fig. 4B) (112). During the manufacturing process, some of the hydrogen bonding between amylose and amylopectin are partially ruptured, so that the product contains 5% free amylose, 15% free amylopectin, and 80% unmodified starch (113). The free amylopectin provides cold water solubility and aids the binding properties, whereas the free amylose and unmodified starch are responsible for disintegration properties. Manudhane et al. (114) and Sakr et al. (115) showed that STARCH 1500 has many advantages over starch USP with respect to binding, disintegration, and dissolution properties.

Although STARCH 1500 has been claimed to be a free-flowing material with better flow properties than corn starch, the flow properties are poor when compared with other filler-binders because of the large specific surface of the powder, resulting in strong cohesion between particles (Table 4) (41). The fluidity can be improved by the addition of 0.25% colloidal silica.

An average particle size for STARCH 1500 cannot be given, because the range of particle sizes is very large and the granule shape varies (116).

Because STARCH 1500 has lubricating properties of its own, it can be compressed without the addition of a lubricant. These self-lubricating properties are caused by the same factors as discussed for microcrystalline cellulose (64). The binding properties of unlubricated STARCH 1500 are rather good (Fig. 5). Density-stress (Heckel plots) and stress-relaxation studies show that STARCH 1500 exhibits extensive, yet slow, plastic deformation during compression. This agrees with microscopic observations after compaction, in which discrete primary particles with approximately the same size as before compaction could be identified, and the reported reduction of the specific surface after compaction (117). Changes in contact time during compaction, therefore, have a marked effect on tablet properties (66,100,118). As compared with other plastically deforming materials, the strength of STARCH 1500 tablets is low. This effect may be because plastic deformation is too slow to produce adequate interparticle binding during rapid compression. In addition, during compaction at high strain rate, a large proportion of the total deformation will be elastic (100). When elastic, and plastic deformation and interparticles binding occur during compression and when elastic recovery occurs on decompression and ejection, interparticle bonds are not formed rapidly enough to prevent brittle fracture (reducing tablet strength and possibly causing capping) (119).

Contradicting results were published on the effect of particle size distribution of STARCH 1500 and binding properties. McKenna and McCafferty (59) found an increase in tablet strength with a decrease in particle size. This effect was explained by an increased amount of plastic flow and packing with decreasing particle size. Alderborn and Nyström (120) found only a decline in tablet strength with an increase in particle size up to 125 μm; above this size, the tablet strength increased with particle size, and leveled off for larger particle size fractions. These changes were attributed to differences in particle shape: particles below 125 μm consisted of fairly smooth primary particles; larger particles exhibited a rougher surface and seemed to consist of very small particles, which were aggregated or fused together. The tableting properties of STARCH 1500 can be improved by combining it with another filler-binder such as microcrystalline cellulose (114) or dicalcium phosphate dihydrate (121).

Because of its plastic behavior under pressure, STARCH 1500 is extremely sensitive to mixing with lubricants (Figs. 3 and 5). After mixing it with 0.5% magnesium stearate in a Turbula mixer at 90 rpm, the tablet breaking force decreased from 18 kg down to zero within 10 minutes (Fig. 2 in chap. 9). For this reason, the use of alkaline stearates should be avoided or kept below 0.25%. When STARCH 1500 is used as a filler-binder in formulations, stearic acid and hydrogenated vegetable oil are acceptable alternatives to magnesium stearate (114). Additional information about the lubricant sensitivity of filler-binders can be found in chapter 9.

STARCH 1500 retains its disintegration characteristics despite partial pregelatinization. Although plain starch acts as a disintegrant primarily through wicking and elastic recovery of the deformed grains, STARCH 1500 also acts by swelling in the presence of moisture (112). This could be why disintegration properties of STARCH 1500 are comparable with or even better than those of plain starch (114).

Another pregelatinized starch, C*Pharm DC 93000, is prepared by the spray drying of partially cooked starch (122). Both its spherical shape and a favorable particle size distribution lead to good flow properties. The compression profile shows that even when lubricated with 0.5% magnesium stearate, tablets with a remarkably high breaking force, three times higher than obtained with other starch products, can be produced (123,124). The compaction speed dependence was found to be relatively low. In addition to the binding properties, C*Pharm DC 93000 acts as a disintegrant. The decreased crystallinity of starch obtained by gelatinization favors particle bonding but at the same time reduces tablet disintegration time. C*Pharm DC 93000 has a special ratio of both a crystalline and an amorphous structure. It is recommended as a filler-binder, disintegrant, and diluent in wet and dry granulation as well as in direct compaction.

Modified Starches

A modified rice starch, also referred to as spray-dried rice starch, was developed as Eratab by Erawan Pharmaceutical Research and Laboratory Co., Ltd. (Bangkok, Thailand) and introduced in 1992. Modified rice starch is produced by spray drying of rice starch and is composed of aggregates of rice starch spherical grains (Fig. 4C). Modified rice starch is a dry, white, odorless, and tasteless powder. The majority of particles have a size range between 75 and 150 μm. Water content is about 11.3% and bulk density is about 0.55 g/cm^3.

The flow properties of modified rice starch, measured as angle of repose, Hausner ratio, Carr's compressibility indices and flow rate through a given orifice, are better than those of native rice starch and STARCH 1500 (125,126).

Modified rice starch has excellent binding properties (Fig. 5) because of the compactibility of native rice starch (which is better than native corn starch). Like native starches, the compactibility of modified rice starch is dependent on its moisture content. The compaction behavior of modified starch was characterized using Heckel analysis by Mitrevej et al. (127). It was found that the modified starch agglomerates undergo deaggregation at low compaction forces followed by plastic deformation under load. The elasticity was small as compared with STARCH 1500.

Mixing modified rice starch with magnesium stearate causes a reduced binding capacity, but the decrease in breaking force is much smaller than that of STARCH 1500 (Fig. 5) (127). The reduced lubricant sensitivity is an effect of deaggregation of the agglomerates during early stages of compression. An already formed lubricant film will be destroyed during the deaggregation process. In addition to the better flow properties, advantages of modified rice starch, as compared with STARCH 1500, are a tablet breaking force remaining after lubrication with magnesium stearate.

In tablet formulations, modified rice starch can be used as a single filler-binder or blended with other excipients such as α-lactose 100-mesh or anhydrous β-lactose. Combining with microcrystalline cellulose should be avoided because of the poor flowability of the blend and slow disintegration of the tablets. Data for stored tablets containing model drugs and modified rice starch at 31°C and 75% RH shows that modified rice starch can be used in production of tablets for countries with a tropical climate (125).

Starch Acetates

Starch acetates are prepared by partially reacting the hydroxyl groups of native barley starch with acetic acid anhydride in an esterification reaction. Starch acetates with four different degrees of substitution (DS 0.34; 1.19; 2.1 and 2.9) have been studied and compared with different commercially available filler-binders (128). Although the particles were relatively small, their flow properties were found to be suitable for direct compaction. The compaction properties increased with the DS: the tensile strength of starch acetate tablets with DS 1.19 or higher were rather similar to tablets prepared from Avicel PH-101. In the surfaces of the low DS value starch acetate tablets, interparticulate borderlines were still visible after compression. In the higher DS value tablets, borderlines and individual starch particles were hardly visible, pointing to a complete plastic deformation. Tablets prepared from starch acetates with DS 0.34 or DS 1.19 completely disintegrated within a few minutes and showed fast release of a test drug. Tablets containing higher DS starch acetate did not disintegrate at all, and demonstrated sustained release dissolution profiles. These effects were explained by an increase in hydrophobicity, a decrease in tablet porosity, and a decrease in swelling properties with increasing DS. The authors concluded that starch acetates can, depending on the DS, be used as filler-binders or as controlled release excipients.

Maltodextrins

Maltodextrins are composed of water-soluble glucose polymers obtained by partial hydrolysis of starch with acid and/or enzymes, whereby the basic polymeric structure is retained. The dextrose equivalent (DE) value, defined as the amount of reducing sugars as percentage of dextrose in the dry substance, is less than 20. The retardant effect was not exhibited with mixtures containing a water soluble drug substance. Mollan and Çelik (129,130) compared a spray-dried maltodextrin (Maltrin M510), three fluidized-bed agglomerated maltodextrins (Maltrin M500, Malta*Gran TG, and Malta*Gran 10) and an experimental roller-compacted maltodextrin. Maltrin and Malta*Gran are brand names for maltodextrins from Grain Processing Corp. (Nebraska, U.S.) and Zumbro IFP (Minnesota, U.S.), respectively. The commercially available maltodextrins underwent plastic deformation and formed strong tablets, but showed high lubricant sensitivity. Tablets compressed from the experimental roller-dried maltodextrin were stronger and less sensitive to lubrication than those of the other maltodextrins. This effect was attributed to the larger surface area, the higher bulk density and more fragmentary failure of the roller-compacted product. It has been shown that maltodextrins easily sorb and desorb moisture from the atmosphere and that the moisture content of the maltodextrin strongly influences both compaction and postcompaction behavior (131). Compaction behavior of the maltodextrins was more fragmentary under conditions of low humidity and became more plastically deforming as the moisture content increased.

Just as all starch derivatives, maltodextrin has a high lubricant sensitivity. Another objection is the retardant effect of hydrophobic lubricants on drug release of tablets containing water-insoluble active ingredients (132). The disintegration time of tablets containing maltodextrins were found in general to be prolonged, an effect that was attributed to the formation of a rate limiting "gel" layer around the tablets (131).

CALCIUM SALTS

Different calcium salts are used as filler-binder for direct compaction. Of the inorganic salts, dicalcium phosphate dihydrate ($CaHPO_4.2H_2O$) is the most common, but also anhydrous dicalcium phosphate, tricalcium phosphate, calcium sulfate dehydrate, and calcium silicate are marketed for direct compaction. More recently, an organic calcium salt, calcium lactate pentahydrate, has been introduced as a filler-binder.

Dicalcium Phosphate

Dicalcium phospate dihydrate is used in large quantities both as a source of calcium and phosphorus and as an excipient in the nutritional and health industry uses dicalcium phosphate dihydrate in large quantities (133). The calcium phosphates are synthesized by a complicated procedure using phosphoric acid and slaked lime (134,135). Dicalcium phosphate dihydrate is available in a free-flowing form, for example, as Emcompress from JRS Pharma, Di-Tab from Innophos (New Jersey, U.S.), and DI-CAFOS from Budenheim (Germany). In addition to its use as a directly compressible filler-binder, dicalcium phosphate dihydrate is used as a filler in capsules and in tablets prepared by wet granulation. An advantage of using dicalcium phosphates in tablets for vitamin or mineral supplement is the high calcium and phosphorous content.

In addition to the hydrate form, anhydrous dicalcium phosphate is also used as a directly compressible filler-binder. It is available as Anhydrous Emcompress from JRS Pharma, DI-CAFOS AN from Budenheim, and A-Tab from Innophos. Recently, Fuji Chemical Industry (Tokyo, Japan) introduced Fujicalin, a dicalcium phosphate anhydrous, consisting of porous spheres.

Di-Tab is a brand of unmilled dicalcium phosphate dihydrate, whereas Emcompress is a unique form of dicalcium phosphate dihydrate in which particle size distribution is controlled to ensure flowability. According to one investigator, these two products have very similar properties and perform equally well on compression (136), although another investigator has observed some differences (137). Dicalcium phosphate dihydrate and anhydrous dicalcium phosphate meet the specifications of (anhydrous) dibasic calcium phosphate (or synonyms) in all main pharmacopoeias.

Emcompress consists of aggregates of small primary particles (crystallites) (Fig. 4D). The particle size distribution lies primarily between 75 and 420 μm, with an average particle size of about 130 μm. The favorable particle size and high density give Emcompress its excellent flow properties (Table 4) (41,49,138).

Dicalcium phosphate dihydrate is slightly alkaline with a pH of 7.0 to 7.4, which precludes its use with active ingredients that are sensitive to even small amounts of alkali. For example, acidic active ingredients such as ascorbic acid cannot be combined with dicalcium phosphate dihydrate.

During compression, dicalcium phosphate dihydrate undergoes considerable fragmentation as evidenced by photographs (117,139). The brittle behavior of dicalcium phosphate dihydrate during compaction was shown by scanning electron micrographs of the surface of fracture of a tablet (78). Heckel plots of Emcompress also exhibit brittle failure during diametrical compression (78,99,101,117). The brittle nature of Emcompress was also confirmed by stress-relaxation experiments and Heckel plots by Rees and Rue (99). As expected, Emcompress showed minimal stress relaxation, and Heckel plots showed that increasing the time of material under compression (dwell time), there was no effect on compressibility. Therefore, because of its brittle nature, the tensile strength of dicalcium phosphate dihydrate tablets is hardly affected by an increase in tableting speed (66,99). Fragmentation is achieved so rapidly that prolonging exposure to force has no further effect (66). Duberg and Nyström (117) introduced the strength-isotropy ratio (between axial and radial tensile strength) as a method for the characterization of fragmentation tendency during compaction. High strength-isotropy ratios, as found for dicalcium phosphate dihydrate, reflect a large amount of fragmentation during compaction. The fragmentation propensity of dicalcium phosphate dihydrate was also demonstrated by the increase in tablet surface area, measured by permeametry (140) or mercury intrusion porosimetry. Using the latter technique, a comparison of pore volumes at different pore sizes of tablets compressed at two compaction loads, showed an increase in the

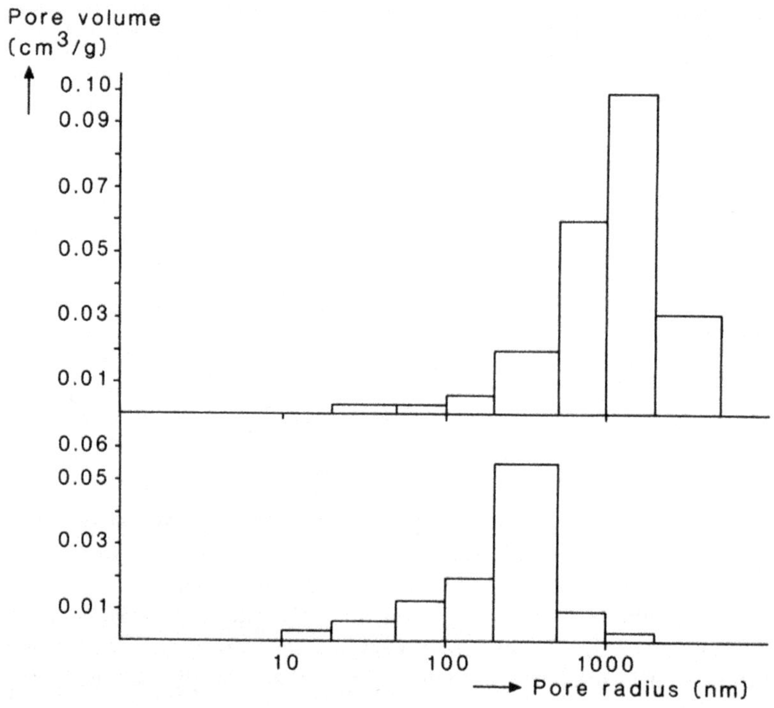

Figure 6 Comparison of pore volumes at different pore sizes of tablets compressed from dicalcium phosphate dihydrate at a compaction load of 37.8 (*upper part*) and 302 MPa (*lower part*), respectively. *Source*: From Ref. 62.

number of small pores during compression (Fig. 6). In a detailed evaluation of the compression cycle of Emcompress, using the Heckel equation for both the compression and decompression phases, Duberg and Nyström (141) found that the powder is extensively fragmented even at relatively low loads. The particles continue to fragment at intermediate loads and subsequently appear to undergo mainly plastic deformation, as indicated by a time difference between maximum load and minimum porosity. Using deagglomeration experiments with a laser diffraction spectrometer, Herzog (135) showed that particles larger than 63 μm show brittle fracture during compression. On the other hand, particles with a mean diameter of 6 μm do not fragment at all.

As compared with other filler-binders, the binding properties of dicalcium phosphate dihydrate are moderate (Fig. 3), which is an effect of its brittle nature. Fracture creates a large number of interparticulate contact points and the compaction load per unit area of interparticulate contact is decreased, reducing the strength of the bonds that are produced (119). The creation of a large number of contact points implies that a comparatively weak type of bonding is involved, for example, molecular van der Waal's forces (142).

Unlubricated dicalcium phosphate dihydrate tablets are difficult to eject from dies. Free axial compaction of large compacts demonstrates that this effect may be caused by shear failure and rebonding along the shear cones (64). Investigations on ejection force of dicalcium phosphate dihydrate, carried out on a rotary press, have shown that with the addition of 0.25% magnesium stearate, the ejection force increases significantly with the number of pressings, indicating insufficient lubrication (143). With the addition of 0.5% or 1.0% magnesium stearate, the ejection force remains constant even after 500 pressings.

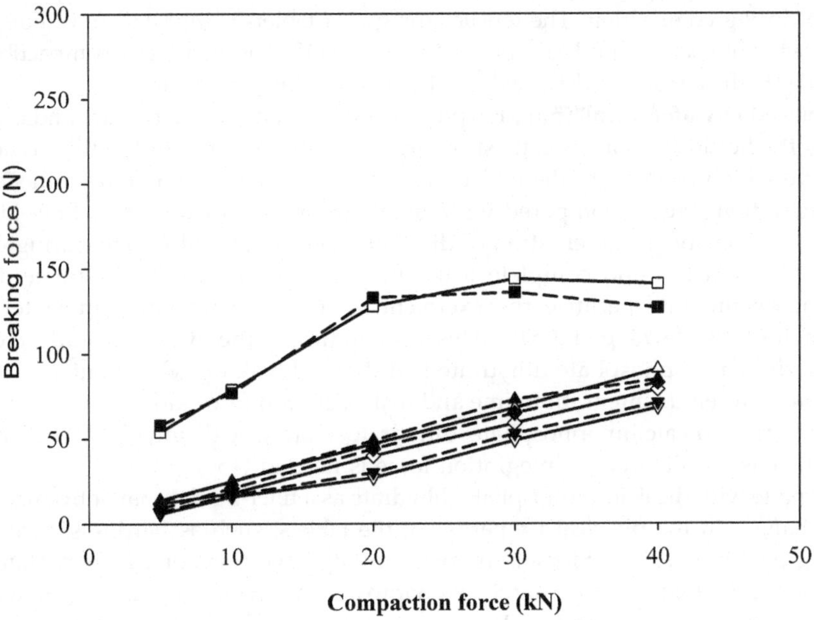

Figure 7 Compaction profiles of different calcium phosphates, both unlubricated (*open symbols*) and lubricated with 1% magnesium stearate (*closed symbols*). (□, ■) tricalcium phosphate (TRI-CAFOS S); (Δ, ▲) dicalcium phosphate dihydrate (Emcompress); (◇, ◆) dicalcium phosphate dihydrate (DI-CAFOS); (▼, ▽) anhydrous dicalcium phosphate (DI-CAFOS AN). 500 mg, 13 mm tablets, hydraulic press, contact time 5.1 seconds.

One advantage of dicalcium phosphate dihydrate as a filler-binder is that lubricants such as magnesium stearate have practically no effect on its binding properties (Fig. 7) (103). This effect is explained by the assumption that clean, lubricant-free surfaces are created by fragmentation of dicalcium phosphate dihydrate during the process of consolidation and compaction (78) (see also chap. 9).

Conflicting results were published regarding the effect of particle size of dicalcium phosphate dihydrate on compressibility of the material. Gillard et al. (144) and Khan and Rhodes (145) found that by using the same compaction pressure, coarse powders rather than fine powders give harder tablets. Large particles are more easily keyed and interlocked than small particles, whereas, because of high lattice energy and high resistance to activation, friction between the particles is less important (15). On the other hand, Alderborn and Nyström (120) reported that for Emcompress, the tablet strength was almost independent of the particle size. Herzog (135) showed a minimum hardness for tablets compressed from particles of about 200 μm. The increase in tablet hardness with an increase in particle size was attributed to the structure of the larger particles, which are agglomerates of small primary particles. Because of the brittle behavior, the shape of Emcompress particles has practically no effect on their compaction properties (146).

Most types of anhydrous dicalcium phosphate have a slightly higher compactibility than types of dicalcium phosphate dihydrate (134,135). The compactibility of DI-CAFOS AN, however, is similar to the compactibility of DI-CAFOS or Emcompress (Fig. 7). Fujicalin, however, is a new product manufactured by restricted crystal growing synthesis to reduce crystal size of the product followed by a spray-drying process yielding porous spheres with a high specific surface area (147). Schlack et al. (148) compared Fujicalin with dicalcium phosphate dihydrate (DI-CAFOS). Using Heckel analysis, both products had high yield pressures, which points to fragmentation during compaction. The tensile strength of tablets compacted from Fujicalin was higher by a factor of more than 4 compared to DI-CAFOS. The increased compactibility was attributed to the high porosity and the extremely high specific surface area.

When placed in water, dicalcium phosphate dihydrate tablets are rapidly and completely penetrated by the liquid (8). This is caused by the hydrophilic nature of the filler-binder (149) and the relatively high porosity of the tablets. The small effect of 0.5% magnesium stearate on the water penetration rate, as compared with tablets compressed from other filler-binders, is attributed to the extensive fragmentation of dicalcium phosphate dihydrate during compression (82). In spite of a fast and complete water penetration, dicalcium phosphate dihydrate tablets do not disintegrate because the excipient is relatively insoluble in water and no disintegration force is developed (150). This also indicates the type of bonds created on compaction of dicalcium phosphate dihydrate that do not break up when contact with water. The correlation between disintegration time and disintegration force kinetics indicates that, for the disintegration of dicalcium phosphate dihydrate tablets, a disintegrant with an active mechanism such as swelling or disintegration force is needed (150).

Tablets made with dicalcium phosphate dihydrate as a filler-binder may introduce on aging unfavorable changes in the physical properties of the tablets, such as hardness, disintegration time and drug dissolution time. Storage in low-humidity conditions of tablets containing 84% dicalcium phosphate dihydrate, 10% starch, 5% naproxen and 1% magnesium stearate results in an increase in tablet hardness, and a decrease in disintegration and dissolution rates (151). However, on aging under high humidity, hardness, disintegration, and dissolution rates decreased at all initial moistures at which tablets were compacted. The decrease in drug dissolution rate after aging under low humidity is because of the limited dissolution and recrystallization of calcium phosphate in the available water in the tablet, whereas the decrease in the dissolution rate on aging under high humidity is ascribed to the expansion and contraction and general opening of the structure of the starch grains and their bonding, via water molecules, to calcium phosphate dihydrate (151). The idea that the changes in tablet hardness are caused by the disintegrant cannot be excluded, because during storage of disintegrant-free tablets, compressed from dicalcium phosphate dihydrate lubricated with 0.5% magnesium stearate for eight weeks at 20°C and 50% or 85% RH, no change in tablet strength was observed (11).

Delattre and Jaminet (152) saw a decrease in tablet strength of about 40% after storage for one month at 45°C, whereas Horhota et al. (153) found no change in strength and thickness of tablets and a decrease in dissolution rate under similar conditions. During storage at 23°C and 44% RH of tablets containing dicalcium phosphate dihydrate with amaranth as a tracer and sodium alginate as a disintegrant, the disintegration and dissolution rates decreased without changes in hardness and tablet weight; during storage at 45°C and 75% RH, however, the tablets showed blotching, substantial weight loss, and complex changes in disintegration and dissolution (154). The changes at elevated temperatures were attributed to loss of water of hydration; changes at 25°C were thought to be caused by other factors, such as case hardening.

The course of dehydration of dicalcium phosphate dihydrate at elevated temperatures is an extremely complicated process, dependent on water-vapor pressure. It was demonstrated that dicalcium phosphate dihydrate easily loses its hydrated water when stored above 40°C, and that this process was irreversible (155,156). In general, this may decrease chemical stability of drugs in tablet formulations containing dicalcium phosphate dihydrate (157,158), possibly after solubilizing the drug in accumulated free water because of dehydration within the drug/excipient system (156).

The disadvantage of dehydration of dicalcium phosphate dihydrate can be avoided by using anhydrous dicalcium phosphate, because this product contains no water that can be lost. Moreover, anhydrous dicalcium phosphate does not absorb water to form the dihydrate.

Emcompress offers the best potential for direct compaction when used in combination with microcrystalline cellulose (152) or starch (86). The strength of tablets prepared from blends of Emcompress with Avicel PH-102 depends on the compaction force and the percentage of Avicel (12).

Tricalcium Phosphate

Another calcium phosphate for direct compaction is tricalcium phosphate (tribasic calcium phosphate USP), available as Tri-Tab (Innophos) and TRI-CAFOS S (Budenheim). Actually, the commercially available product is hydroxyapatite—$Ca_5(OH)(PO_4)_3$—also commonly and erroneously referred to as tricalcium phosphate (134). It is a variable mixture of calcium and phosphates.

Tricalcium phosphate can be used as a filler-binder in direct compaction and as a filler in tablets prepared by wet granulation. Its higher calcium load when compared with dicalcium phosphate dihydrate may be advantageous when used in vitamin tablets.

Carstensen and Ertell (134) reviewed the synthesis, physical properties, and chemical properties of tricalcium phosphate. TRI-CAFOS S is probably prepared by spray drying, whereas Tri-Tab is prepared by granulation or a compaction technique (135). Both products are built up of very fine and porous primary particles with a mean size of 2.2 and 1.5 μm, respectively (135). Compared with Tri-Tab, TRI-CAFOS S has a much lower bulk density. Extremely good flow rates were found for Tri-Tab and these results were related to the high density of the excipient (159).

TRI-CAFOS S has better compactibility than the different types of DI-CAFOS (133,135) (see also Fig. 7). The compressibility of Tri-Tab is lower, because of the hardness of the agglomerates, which imparts fragmentation (135). Tricalcium phosphate obeys the Heckel equation only if the particle density of the agglomerated tricalcium phosphate is 1.92 g/cm^3 (rather than the true density 3.1 g/cm^3 of the individual crystals), and is used to determine the relative density of the compacts (160,161). Comparing mercury intrusion porosimetry data with nitrogen adsorption data shows that a large proportion of the solid has pores smaller than 60 Å. These micropores, which are attributed to the pore space in the individual crystals, are not compressible in conventional pressure ranges and are therefore not considered as part of the Heckel pore space. The pore space that is lost during compression is between 0.5 and 2 μm in diameter. It was demonstrated that bonding occurs not because of fracture or deformation of the individual crystals, but rather because of the deformation of agglomerates under pressure (162). This view is supported by the work of Patel et al. (159), which demonstrates that the compactibility of tricalcium phosphate is time-dependent. A slow eccentric press yields harder tablets at a given compression force than a much faster rotary press, which is an indication that the mechanism of consolidation is plastic deformation. On the contrary, Herzog (135) shows by Heckel analysis that TRI-CAFOS S and Tri-Tab fragment during compression.

As the chemical nature of tricalcium phosphate is not precisely defined, Pontier et al. (163) investigated the compaction properties of a series of apathic calcium phosphates differing in their Ca/P molar ratio from 1.500 (i.e., tricalcium phosphate) to 1.667 (i.e., hydroxyapatite). The results were compared to those of Tri-Tab. The best compactability was found for the product with a Ca/P molar ratio of 1.500. Uncalcinated hydroxyapatite had similar compaction properties than Tri-Tab but is easier to use due to lack of sticking on die and punches. The cohesion properties of calcinated hydroxyapatite and β-tricalcium phosphate are insufficient for use in direct compaction.

The water content of tricalcium phosphate has a profound effect on its compactibility; its optimum lies between 4% and 5%. Higher or lower values lead to capping and a decreased binding capacity (135).

It has been reported that, in contrast to the dicalcium phosphates, the addition of magnesium stearate has a profound effect on the compactibility of tricalcium phosphate. At 2% magnesium stearate concentration, for instance, tablets cannot be compacted at all (134). This lubricant sensitivity was assumed to be an indication for the previously noted plastic behavior under pressure. Own experiments show, however, that TRI-CAFOS S has a low lubricant sensitivity (Fig. 7).

A serious drawback of the tricalcium phosphates is their high sticking tendency with dies and punches. This sticking tendency, which results in high ejection forces, is attributed to the fine structure of the primary particles (135). Another drawback of using tricalcium phosphate as an excipient in tablets is the effect of aging under both low and high humidities on tablet hardness, disintegration time, and dissolution rate. The moisture content of tribasic calcium phosphate-based tablets at the time of compression, and the moisture that was gained or lost during aging, has a significant effect on the hardness, disintegration time, and dissolution rate of the drug. Some differences and some similarities between the responses of tricalcium phosphate tablets and dibasic calcium phosphate dihydrate tablets are reported (151,164).

Calcium Sulfate Dihydrate

Another inorganic salt used for direct compaction but not commonly used in pharmaceuticals is a specially processed grade of calcium sulfate dihydrate NF marketed (among others) as Compactrol by JRS Pharma. Calcium sulfate dihydrate is a commodity item and is produced in commercial quantities for the construction industry. The pharmaceutical grade is an inexpensive excipient used as a filler, also known as terra alba. Compactrol has good flow properties (see Table 4), but has poor compactibility (11). Physical stability studies of tablets prepared with Compactrol, show that tablet hardness decreases at accelerated storage conditions, but that disintegration time and drug dissolution rates are hardly affected (165).

Calcium Lactate Pentahydrate

Recently a directly compressible form of calcium lactate has been introduced as a filler-binder for direct compaction (166). Calcium lactate is a widely used source of calcium in pharmaceutical preparations and has, in comparison with other organic calcium salts, a good water solubility and bioavailability. Two different varieties—calcium lactate trihydrate and calcium lactate pentahydrate—are described in pharmacopoeias. A special free-flowing form of calcium lactate pentahydrate, marketed as Puracal DC by Purac (The Netherlands), has been designed especially for direct compaction. It is a granular material prepared by a special spray-drying technique (Fig. 4E).

Directly compressible calcium lactate pentahydrate has good flow properties. It flows freely through an orifice with a diameter of 2.5 mm and has a Hausner ratio of 1.20. The compaction properties are good, and much better than those of calcium lactate trihydrate or dicalcium phosphate dihydrate (166). In spite of the spray-drying process, directly compressible calcium lactate pentahydrate is a brittle crystalline material with no measurable traces of amorphous material. In spite of this brittle nature, directly compressible calcium lactate pentahydrate has a high lubricant sensitivity, which is a consequence of its lower fragmentation propensity. Even when lubricated with 0.5% magnesium stearate, the compactibility of directly compressible calcium lactate pentahydrate is better than that of the calcium phosphate dihydrate (Fig. 3). As a consequence of its crystalline structure, directly compressible calcium lactate pentahydrate has a low compaction speed sensitivity. The disintegration time of tablets containing calcium lactate pentahydrate is determined by dissolution of the excipient, but can be shortened by the addition of a suitable disintegrant such as croscarmellose sodium. In a number of formulation examples, it was shown that tablets containing directly compressible calcium lactate pentahydrate as the main or as an additional filler-binder have a short disintegration time and a fast drug release (166).

Calcium Silicate

A special calcium silicate for direct compaction is RxCIPIENTS FM1000 from Huber Engineered Materials (Georgia, U.S.). It complies with the monographs of calcium silicate in the USP/NF and JPE. It is specially designed for ODTs. It is commonly used in combination with about 70% compressible sugar, a polyol (e.g., mannitol), or microcrystalline cellulose and with crospovidone as a disintegrant (167). Tablets prepared from these blends disintegrate in the mouth in prime particles in 5 to 30 seconds (168). In contrast to many other ODTs, the friability of tablets containing RxCIPIENTS FM1000 is low.

POLYOLS

Nowadays there is an increasing interest in sugar substitutes such as mono- and disaccharide alcohols, also called polyols, in pharmaceutical formulations. One major reason for this interest is the chemical stability of most of the polyols. As they are not containing a carbonyl group, they are not subject to the Maillard reaction, and, hence are chemically more stable than related saccharides. Furthermore, the majority of these polyols can be consumed by diabetics without any significant increase in body glucose, insulin or lactic acid concentration unlike the conventional saccharides such as sucrose, glucose and lactose. Additionally, their sweetness and noncariogenic characteristics provide advantages in a number of tablet formulations. Because of the excellent mouthfeel and pleasant taste, polyols are commonly used in lozenges and chewable tablets. There is, however, an increasing use in common disintegrating tablets and in ODTs.

Polyols are polyhydric alcohols that are prepared from sugars by reducing aldehyde or keton sites to a primary or secondary hydroxyl group through catalytic hydrogenation, enzymatic conversion or fermentation, or through a combination of these. Sorbitol and mannitol were the first polyols available in special directly compressible forms; nowadays also special forms of isomalt, lactitol, and xylitol are marketed as filler-binders for direct compaction (169).

Sorbitol

Sorbitol was the first sugar alcohol, used in direct compaction. Sorbitol is produced by high-pressure catalytic hydrogenation of glucose in the presence of Raney nickel. Chemically, it is an isomer of mannitol. The most significant differences between sorbitol and mannitol are their hygroscopicity and aqueous solubility; sorbitol is hygroscopic at relative humidities above 65%, whereas mannitol is nonhygroscopic, and the aqueous solubility of sorbitol is higher than that of mannitol.

Different forms of sorbitol are manufactured depending on crystallization and purification conditions, and these forms are available commercially. Pharmaceutical grade sorbitol is available in several different physical forms from various suppliers. Shangraw et al. (112) showed that there were considerable differences in shape and structure of different sorbitol products resulting in different compaction behavior. Sorbitol exists in five different anhydrous crystalline polymorphic forms (A-, B-, Δ-, E-, and Γ-sorbitol), two hydrate forms (sorbitol hydrate I and II) and one amorphous form (170). Different sorbitol samples including commercial products were investigated and it was found that the most stable form, Γ-sorbitol, previously referred to as γ-sorbitol, had the best compaction properties, but showed a longer time for disintegration and dissolution (171,172). The compactibility of γ-sorbitol was dependent on the particle structure produced by the manufacturing process (171).

Γ-Sorbitol is prepared by spray drying or crystallization and marketed by different companies. The first products on the pharmaceutical market were Sorbitol Instant (Karion

Figure 8 Scanning electron micrographs of directly compressible filler-binders: (**A**) and (**B**) γ-sorbitol (Parteck SI 150); (**C**) mannitol (Pearlitol SD 200); (**D**) isomalt (galenIQ 721); (**E**) Sugartab; (**F**) StarLac.

Instant, now with the brand name Parteck SI) from Merck (Germany) and Neosorb P in different particle size fractions from Roquette Frères (France). Parteck SI is produced by a special spray-drying process that causes the sorbitol to crystallize in an interwoven, filamentary microstructure (Fig. 8A, B). It is available in as a fine powder (SI 150) and as a coarser grade (SI 400). Neosorb P is a crystallized γ-sorbitol, prepared by cooling a melt and seeding it with sorbitol crystals (173). Two spray-dried sorbitols with different particle size distributions, Sorbogem SD 250 and SD 500, are marketed by SPI Pharma (Delaware, U.S.).

As an effect of the regular form (Fig. 8A) and favorable particle size distribution, Parteck SI has excellent flow properties (Table 6).

Comparative evaluation of two types of γ-sorbitol (173) showed that the tableting properties were strongly dependent on particle structure, particle size distribution, and bulk density of the products. Spray-dried γ-sorbitol (Parteck SI) could be compressed into much harder tablets than the γ-sorbitol prepared from a melt (Neosorb P). Schmidt and Vortisch (174) also studied the influence of the manufacturing method on compactibility of eight different types of commercially available sorbitol. All products showed different tableting properties and these differences could not be explained because of the differences in physical properties such as particle size, bulk and tapped density, and polymorphism. The differences in compaction behavior were caused by the method of manufacturing resulting in differences in particle shape and surface properties. Spray drying results in irregular and porous particles that show high compactibility. Sorbitol products prepared from a melt or by crystallization from a solution have lower compactibility.

Table 6 Physical Properties of Polyols for Direct Compaction

	Isomalt[a]	Lactitol[b]	Mannitol[c]	Sorbitol[d]	Xylitol[e]
Mean particle size (μm)	220	125	152	360	209
Specific surface area (m^2/g)	0.32	0.95	0.69	1.22	0.29
Bulk density (g/cm^3)	0.46	0.58	0.52	0.45	0.63
Tapped density (g/cm^3)	0.52	0.71	0.56	0.51	0.74
Hausner ratio	1.14	1.22	1.09	1.13	1.17
Flow through aperture (mm)	2.5	>18	2.5	5	>18
Water solubility at 20°C (g/100 g)[f]	40	36	15	67	38

[a]galenIQ 721.
[b]Finlac DC.
[c]Pearlitol SD 200.
[d]Parteck SI 150.
[e]Xylitab 300.
[f]Handbook of Pharmaceutical Excipients (190).
Source: From Ref. 169.

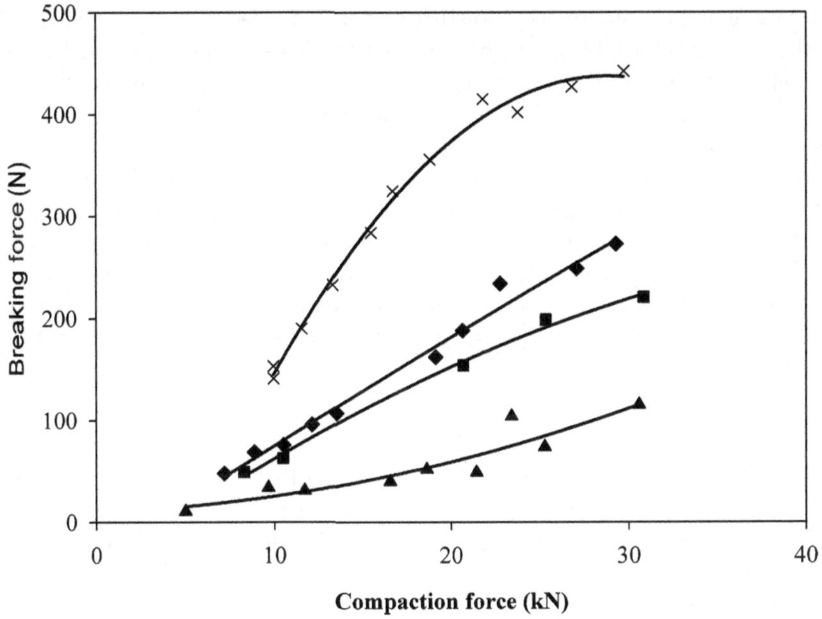

Figure 9 Compaction profiles of different polyols for direct compaction, lubricated with 0.5% magnesium stearate. (x) γ-sorbitol (Parteck SI 150); (◆) isomalt (galenIQ 720); (■) mannitol (Pearlitol SD 200); (▲) lactitol (Finlac DC). 500 mg, 13 mm tablets, compaction simulator at 300 mm/sec. *Source*: Adapted from Ref. 169.

In a comparative evaluation on polyols as filler-binders for disintegrating tablets for direct compaction (169), it was found that Parteck SI 150 has extremely good binding properties (Fig. 9), better than those of the other polyols examined. This is caused by both the high plasticity and the open crystalline matrix of small interwoven filamentary crystals (Fig. 8B). However, at higher forces the tablets show capping, just as was reported previously by Deurloo et al. (175).

Addition of 0.5% or even more magnesium stearate has only a small effect on the compactibility of γ-sorbitol for direct compaction (169,173). This is caused by the granular structure of the products: a lubricant film, formed during the mixing process, will be destroyed by fragmentation of the brittle agglomerates during the early stages of compaction (18).

Both types of commercially available γ-sorbitol form ordered mixes (powder mixtures where fine drug substance particles randomly adhere to the large particles of the excipient) with different B-vitamins (176). A large amount of the drug substance particles can adhere to the irregular particles of Parteck SI as opposed to fewer particles adhering to the more regular particles of Neosorb P 20/60. For six different antibiotics, Parteck SI had greater adhesion capacity and binding strength than crystalline sorbitol (177).

The hygroscopicity of sorbitol limits its use in tablet formulations. A high percentage of sorbitol cannot be used in formulations containing hygroscopic drugs or drugs that degrade readily in the presence of moisture. During storage at 50% relative humidity (RH) and room temperature, the strength of tablets prepared with Neosorb P 20/60 did not change. However, during storage at 85% RH, the tablets liquified (11). Parteck SI tablets increased their weight after aging up to 65% RH, and the weight gain was higher than for tablets compressed from anhydrous β-lactose or compressible sugar, but less than the tablets compressed from crystalline sorbitol (178). The lower hygroscopic behavior of Parteck SI tablets compared with tablets made with crystalline sorbitol was attributed to the smooth tablet surface, which may seal the tablet core against atmospheric moisture to some degree.

During storage, sorbitol tablets can become harder. This effect is caused by dissolution and recrystallization of sorbitol during aging, forming a reinforcing network throughout the tablet core (179). The inclusion of pregelatinized starch in sorbitol tablets can prevent recrystallization and an increase in tablet strength (179).

Direct compaction grades of γ-sorbitol can be used for the production of lozenges, chewable tablets, and disintegrating tablets; examples of the formulations (including a number of the so-called problem drugs) can be found in the literature, for example, Basedow et al. (178), Schmidt (180), and Darrouzet (181). For disintegrating tablets, a superdisintegrant with a high swelling power, such as croscarmellose sodium, must be used. The use of crospovidone appeared to change tablet properties during aging, especially under accelerated relative humidity conditions (178).

Mannitol

Mannitol is a polyol isomer of sorbitol and the commercial products are prepared by catalytic hydrogenation of carbohydrate solutions of glucose and/or fructose. It is often obtained with its isomer, sorbitol, from which it is isolated. Mannitol and sorbitol have a high negative heat of solution and sweet taste that make them the most useful excipients for lozenges and chewable tablets. Because of its poor flowability and binding properties, unmodified mannitol cannot be used for tablet production by direct compaction. For this reason, the product has been modified to improve its tableting characteristics.

For mannitol three anhydrous polymorphic crystal forms (α-, β-, and δ-mannitol) and one hemihydrate have been defined (182). Debord et al. (183) studied the compactibility of different polymorphic forms of mannitol. They concluded that the α-form had the best compactibility. The particle shape had a great influence on the compactibility. For the same particle size, a granulated powder had better compaction properties than native crystals.

Nowadays different directly compressible spray-dried or granulated forms of mannitol differing in particle size and density are available. Most known are Pearlitol DC/SD from Roquette Frères, Mannogem EZ from SPI Pharma, and Parteck M from Merck.

Scanning electron micrographs show that the granules of Pearlitol SD 200 are built up of very small crystalline needles (Fig. 8C). As an effect of the particle form and favorable particle size distribution Pearlitol SD 200 has excellent flow properties (Table 6).

Although the compactibility is lower than for γ-sorbitol, the compaction profile (Fig. 9) shows that Pearlitol SD 200 has good compaction properties (169). The agglomerates fragment during compaction whereas the primary particles exhibit plastic deformation and will form a large surface for bonding (184). As an effect of the granular structure, the lubricant sensitivity of Pearlitol 200 SD is low (169).

An important advantage of mannitol over sorbitol it its lower hygroscopicity. Sangekar et al. (86) reported that tablets compressed from granular mannitol picked up less moisture after exposure to high humidity than tablets compressed from anhydrous Celutab, granular sorbitol, or direct compaction sugar. A problem for mannitol, however, is its sticking tendency, resulting in pollution of die and punches (169). As compared with other polyols, tablets compressed from mannitol have shorter disintegration times. This is caused by the lower solubility of mannitol, as compared with other polyols (Table 6), enabling a certain amount of water penetration into the tablets (169). Directly compressible mannitol products are commonly used in formulations for effervescent and chewable tablets. There is an increased interest, however, for application in common, disintegrating tablets and ODTs.

Lactitol

Lactitol is produced by the catalytic hydrogenation of lactose. It is approximately 0.4 times as sweet as sucrose and is widely used as a replacement for sucrose in foodstuffs. Granulated lactitol is a form of lactitol, designed for use as a direct compaction tablet diluent, and it is marketed as Finlac DC by Xyrofin (New York, U.S.). Granulated lactitol is prepared by a water granulation process. The product is composed of microcrystalline agglomerates with a mean particle size of 125 μm.

The flow properties, measured as tablet weight variation coefficient and Hausner ratio are good (Table 6) (169,185). The first authors found, however, that the flowability through small orifices is poor (Table 6), which may be caused by the hygroscopicity of lactitol.

Different results were found for the compactibility of granulated lactitol. Armstrong (185) found that the compressibility of granulated lactitol is higher than for many other direct compaction excipients of carbohydrate origin. On the other hand, in a recent publication it was reported that directly compressible forms of sorbitol, isomalt and mannitol had a better compactibility than granulated lactitol (Fig. 9) (169). During compaction granulated lactitol undergoes extensive fragmentation (185).

Granulated lactitol needs only a low concentration of lubricant (about 0.2%) and has a low lubricant sensitivity. The latter may be attributed to the granular structure of the product, since during compression, any lubricant film will be destroyed as a consequence of particle fragmentation. Being freely soluble in water, tablets compressed from granulated lactitol erode from the outside rather than disintegrate. Both the addition of a disintegrant and differences in concentration of magnesium stearate had an insignificant effect on disintegration time. Reworking had little effect on tablet strength (169,185).

Xylitol

Because of the sweet taste—xylitol is the sweetest sugar alcohol—and its high negative heat of solution, xylitol is a good candidate as an excipient for chewable tablets. Garr and Rubinstein (186) investigated fundamental direct compaction properties of a milled xylitol powder with a mean particle size of 170 μm. They found that the mean yield pressure, radial tensile strength and elastic recovery of xylitol tablets were independent of the compression speed, indicating that the dominant mechanism of consolidation is fragmentation. For 12.5 mm diameter xylitol tablets, the tensile strength increased almost linearly with the compaction force up to a force of 22 kN. At higher compaction forces, tensile strength decreased as a result of elastic relaxation.

Agglomerated xylitol is now available as Xylitab Danisco (UK). Xylitab 100 is granulated with 3% polydextrose, Xylitab 200 and the coarser Xylitab 200 CG are granulated with 1.5%

sodium carboxymethylcellulose, and Xylitab 300 is granulated with a xylitol solution and is intended for use in blends with other filler-binders.

In an investigation of Xylitab 100 and Xylitab 200, it was shown that the flow properties, measured as tablet weight uniformity were excellent (187). In recent work it was shown that the flow properties of Xylitab 300 are dependent on the method used (Table 6). In spite of the low Hausner ratio, there was no free flow through an orifice of 18 mm. The latter was attributed to the slight hygroscopicity of xylitol (169).

Tablets with a reasonable hardness could be compressed from Xylitab 300 lubricated with 0.5% magnesium stearate at a low compaction speed. As an effect of the granular structure, the lubricant sensitivity was absent. The presence of the lubricant had even a significantly positive effect on the tablet strength. This effect was attributed to the antiadhesive properties, reducing tablet damage during ejection (169). At high compaction speed, however, Xylitab 300, lubricated with 0.5% magnesium stearate could not be compressed because of capping problems.

Isomalt

Isomalt is a mixture of two dissacharide alcohols, derived from the hydrogenation of isomaltulose, a reducing disaccharide obtained through an enzymatic rearrangement of the sucrose molecules. The principal components are the disaccharide alcohols 1-*O*-α-D-glucopyranosyl-D-mannitol dihydrate (GPM) and 6-*O*-α-D-glucopyranosyl-D-sorbitol (GPS). Isomalt is the only polyol produced from sucrose. It meets the specification of Isomalt in both Ph Eur and USP/NF. Ndindayino et al. (188) evaluated four grades of sieved isomalt for their physical characteristics. Only one fraction of sieved isomalt (Palatinit C) exhibited potential characteristics for direct compaction. Evaluation by Heckel plot analysis showed that isomalt exhibits plastic behavior and undergoes elastic recovery primarily in the die. The binding properties were acceptable after addition of 1% magnesium stearate, but there was lack in uptake capacity of active ingredients. Moreover, good flow could only be achieved after addition of colloidal silica. To improve its tableting properties, isomalt was melt-extruded prior to compression (189). After fusion, crystalline isomalt was transformed into an amorphous form. Although the tableting properties of amorphous isomalt were dramatically improved, the extruded isomalt powder showed agglomeration problems due to recrystallization of the amorphous phase into a stable crystalline form in the presence of atmospheric moisture.

Recently, two good flowing, agglomerated forms of isomalt containing small primary particles, especially designed for direct compaction have been introduced by BENEO-Palatinit (Germany): galenIQ 720 and galenIQ 721 (Fig. 8D). The agglomerates are prepared by milling the starting product sieved isomalt and a subsequent agglomeration process. The difference between galenIQ 720 and galenIQ 721 is the ratio GPS: GPM. In type 720 the ratio GPS:GPM is 1:1 and in type 721 the ratio is 3:1. As the water solubility of GPS is higher than for GPM, the products have different solubilities (190). Both galenIQ 720 and 721 have excellent flow properties, expressed as Hausner ratio and flow through orifices (191). As an effect of the favorable particle form (Fig. 8D) and size distribution (Table 6), the flow through small orifices was better than for directly compressible forms of lactitol and xylitol (169).

In recent work, the compaction properties of sieved, milled and agglomerated isomalt were compared (191). Milling of sieved isomalt results in a product with an increased surface area, which leads to an enormous increase in compactibility. The flow properties of milled isomalt are too bad for use as filler-binder in direct compaction. Agglomeration of milled isomalt, resulting in the commercial products galenIQ 720 and 721 improves both flowability and lubricant sensitivity whereas the good compaction properties of milled isomalt are maintained.

In a comparative evaluation between directly compressible forms of different polyols, Bolhuis et al. (169) found that both isomalt (galenIQ 720 or 721) and mannitol (Pearlitol 200 SD) were the most convenient materials for direct compaction. Their compaction properties were good and no capping was observed (Fig. 9). However, mannitol gave more adhesion problems than isomalt. Just as found for other polyols, with the exception of mannitol, disintegration of tablets containing agglomerated isomalt was strongly dependent on the compaction force used, that is, tablet porosity (169). When the porosity is too small, water will not penetrate into the tablet and the tablet erodes by dissolution of isomalt from the outside of the tablet. This erosion process is independent of the compression force used and so fast that it guarantees a "disintegration" (measured without disks) within a couple of minutes. Fast disintegrating, immediate release tablets can be produced using, for example, 3% crospovidone as a superdisintegrant (192).

LACTOSE

Lactose is a natural disaccharide, produced from cow's milk that contains about 4.6% lactose, corresponding to about 38% of its dry solids. Chemically, lactose consists of one galactose unit and one glucose unit. It exists in two isomeric forms, α-lactose and β-lactose, and can be either crystalline or amorphous. Crystalline α-lactose occurs in the monohydrate and the anhydrous forms. The crystalline β-lactose exists in the anhydrous form only. Pure amorphous lactose is not available commercially and is generally present in modified forms of lactose in varying amounts. The modified forms of lactose may contain varying amounts of water depending on the amorphous content.

Crude α-lactose monohydrate is produced from casein or cheese whey by evaporation and crystallization. For the production of pharmaceutical grade lactose, the crude product is subsequently purified by recrystallization. Depending on the temperature of crystallization, different types of lactose can be obtained (Fig. 10); below 93.5°C, lactose precipitates as α-lactose monohydrate, and above 93.5°C as anhydrous β-lactose. α-Lactose monohydrate can be processed further either by dehydration into the anhydrous α-lactose form, or by spray

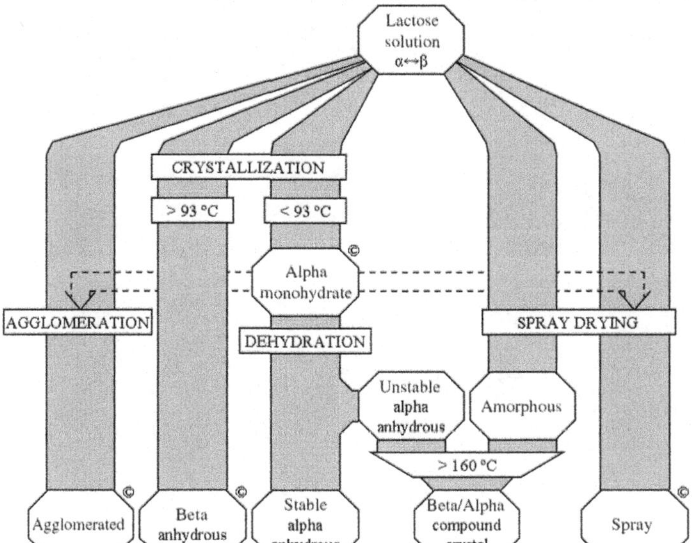

Figure 10 Schematic representation of the production of different types of lactose.

drying into the spray-dried form, or by granulation into the agglomerated form. All different types of lactose have different compaction properties.

α-Lactose Monohydrate

Lactose is probably the most widely used diluent in tablet formulations, and the most common form used is α-lactose monohydrate, often referred to as *lactose*, or *hydrous lactose*, or *regular lactose*. It is generally used in powdered (ground) form as a filler for tablets, prepared by means of a wet granulation technique. In direct compaction, coarse, regular grade, and sieved crystalline fractions of α-lactose monohydrate (particularly the 100-mesh quality) are used because of their good flowability.

α-Lactose monohydrate contains one mole of water per mole of lactose, that is, it contains 5% water of crystallization. The total moisture content is about 5.2%; about 0.2% is free moisture. Sieved lactose is available in mesh sizes ranging from 45/50 to 325. The 100-mesh lactose has a particle size distribution that is almost similar to that of a product previously described as lactose EFK (extra fine crystalline) and microcrystalline lactose in French publications. Because of their high packing density and regular shape of the particles, both products have excellent flowability (Fig. 11A and Table 4) (11,41,138).

When compared with other filler-binders, α-lactose monohydrate exhibits relatively poor binding properties (Fig. 12). Experiments based on Heckel and other stress-density equations (193–195), stress relaxation studies, and particle size measurements after compaction (196) showed that α-lactose monohydrate consolidates mainly by fragmentation. In other studies,

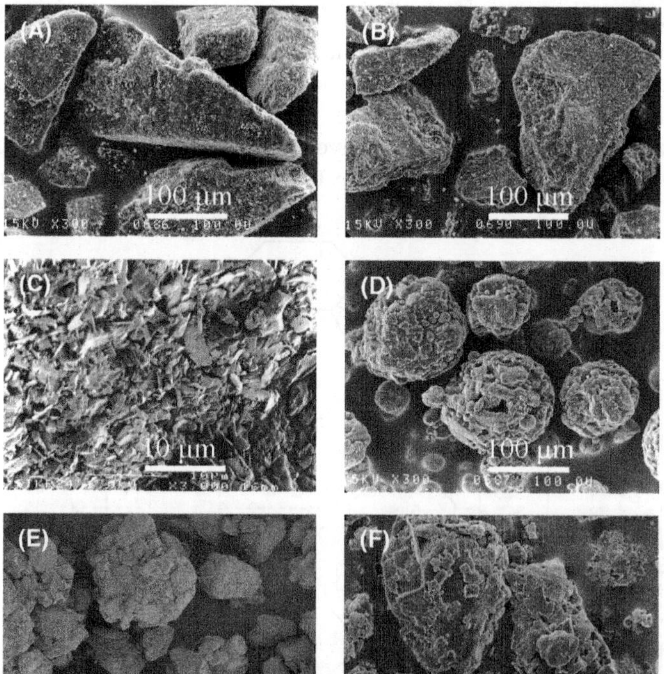

Figure 11 Scanning electron micrographs of directly compressible filler-binders: (**A**) α-lactose monohydrate 100 Mesh; (**B**) and (**C**) anhydrous β-lactose (SuperTab 21AN); (**D**) spray-dried lactose (SuperTab 11SD); (**E**) agglomerated lactose (Tablettose 100); (**F**) Cellactose 80.

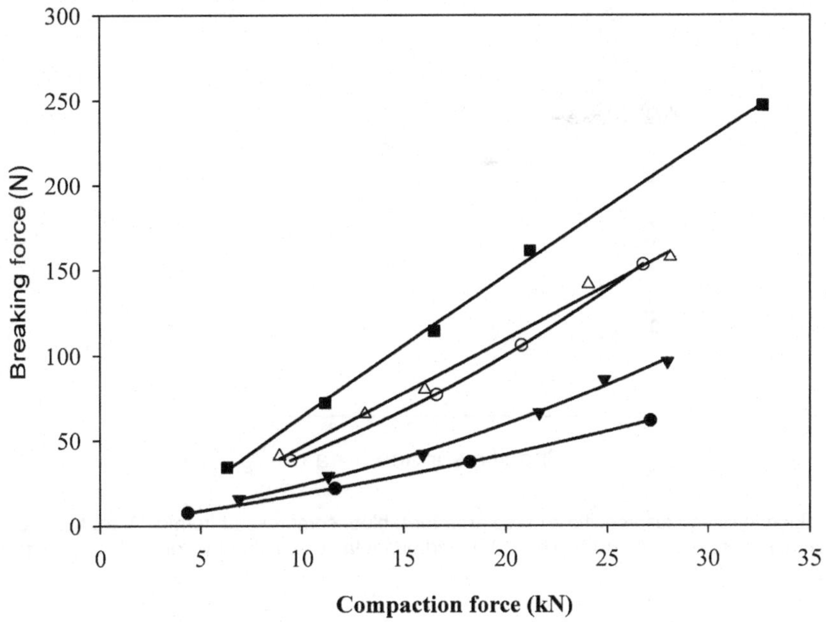

Figure 12 Compaction profiles of different lactoses for direct compaction, lubricated with 0.5% magnesium stearate. (●) α-lactose monohydrate 100 Mesh; (○) anhydrous β-lactose (SuperTab 21AN); (▼) agglomerated lactose (Tablettose 100); (△) spray-dried lactose (SuperTab 11SD); (■) spray-dried lactose (SuperTab 14SD). 500 mg, 13 mm tablets, compaction simulator at 300 mm/sec.

consolidation by fragmentation and plastic deformation were described (57,65,197,198). The fragmentation behavior was confirmed by the fast attainment of an ultimate porosity after multiple compressions (79) and by the relative low strain rate sensitivity (65). Because of its higher yield pressure Roberts and Rowe (65) concluded that crystalline α-lactose monohydrate was higher in brittleness than spray-dried lactose and anhydrous β-lactose, respectively, and this was attributed to the more angular nature of α-lactose monohydrate as compared with the aggregated collection of smaller crystals of the directly compressible lactose. Roberts and Rowe (60) demonstrated that for powdered grades of α-lactose monohydrate, an increase in particle size resulted in an increase in relative density for a given applied force. This may be caused by the occupation of the interparticulate voids between larger crystals by small particles. During compaction, fragmentation of the larger crystals and filling of the remaining interstices led to a further increase of the relative density. Furthermore, for smaller particle size material, the increased frictional forces associated with particle sliding can restrict densification. The more uniform size crystalline α-lactose monohydrate showed a higher rate of densification than the powdered grades, which may be caused by the high stress concentrations at contact points, as a consequence of the angularity of the crystals, from which cracks can easily develop. The fractured particles filled the interparticle voids as the fracturing process proceeded during the application of compression force (60).

Duberg and Nyström (117) evaluated different techniques for the evaluation of fragmentation tendency during compaction of α-lactose monohydrate and other materials. From the results of scanning electron microscopy, surface area measurements of starting powder and tablets (krypton adsorption), isotropic ratio, and Heckel plot analysis, it was concluded that the particles of α-lactose monohydrate compacted by plastic deformation and fragmentation, although the increase in tablet surface area and the isotropic ratio showed that

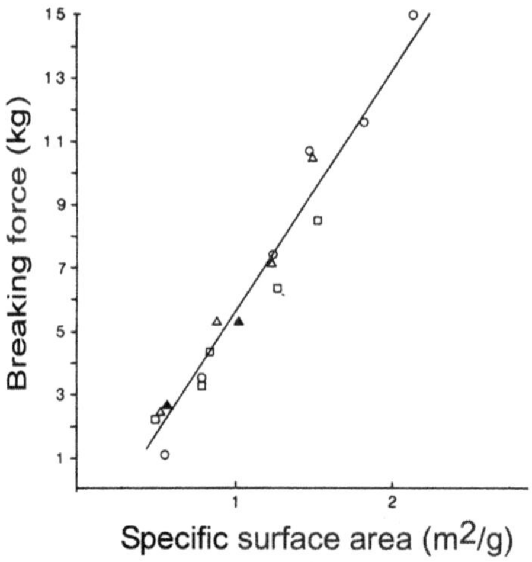

Figure 13 Breaking force versus specific surface area for tablets compressed from different types of crystalline lactose: (□) α-lactose monohydrate 100 Mesh; (○) anhydrous α-lactose; (Δ) anhydrous β-lactose; (▲) crystalline β-lactose. *Source*: From Ref. 62.

the particles deformed to a large extent by fragmentation (117,199). Further work at Uppsala University showed that the fragmentation propensity of α-lactose monohydrate could be rank ordered as intermediate using a permeametry technique (140). A similar linear increase in tablet pore surface area with an increase in compaction force was found using mercury intrusion porosimetry for all types of crystalline lactose (α-lactose monohydrate, anhydrous β-lactose, and anhydrous α-lactose) (62). Assuming a proportionality between the change in pore surface area and the actual binding surface area, the authors found a unique linear relationship between strength and the pore surface area of tablets, compressed from different types of crystalline lactose at different forces (Fig. 13). The unique relationship was valid for the whole range of particle sizes (32–400 μm) of α-lactose monohydrate (200). The proportionality between strength of brittle lactose tablets and the internal specific surface area was elucidated by a theoretical model (201). Assuming that a tablet is made up of spherical isometric particles and that the strength of all types of crystalline lactose tablets is caused by van der Waals dispersion forces acting at the coordination points of the particles, a proportionality between tensile strength and internal specific surface of the tablet is obtained.

From the proportionality between strength and pore surface area of tablets, it was concluded that fragmentation was the predominant mechanism of consolidation for all types of crystalline lactose and that the presence of water of crystallization, or the α/β-ratio, or the degree of crystallinity had no influence on the binding mechanism (62). A relation between particle texture and compactibility was also found. Using a gas adsorption apparatus to determine the specific surface areas of crystalline lactose powders, Vromans et al. (202) demonstrated that the binding capacity was directly related to the powder surface area. Thus, the compactibility of crystalline lactose appears to depend on particle texture, which is determined by the manufacturing conditions. Slow crystallization of lactose produces single crystals with low powder surface area and poor compaction properties (e.g., α-lactose monohydrate). Rapid crystallization by dehydration (see section "Anhydrous α-Lactose") or roller drying (see section "Anhydrous β-Lactose") results in aggregates of microcrystals (anhydrous lactose). Because of their higher powder surface area, these irregular crystals have better binding capacity compared with the regular crystals of α-lactose monohydrate.

The strength of tablets compressed from α-lactose monohydrate increases with a decrease in particle size of the excipient (200,203,204). Fell and Newton (205) prepared tablets from three particle size fractions of crystalline lactose at two compaction speeds. Calculation of the densification because of particle rearrangement by the Heckel and Cooper & Eaton methods, showed that the smaller fraction underwent the largest rearrangement. Vromans et al. (204) showed that compression of small particles resulted in tablets with increased thickness, compared with compacts obtained from large particles. The observed differences were caused by the higher porosity of tablets compressed from the small particles. For a number of sieve fractions of unmilled crystalline α-lactose monohydrate, fragmentation diminished with decreasing initial particle size, which may increase the role of plastic and elastic deformation (200,206). However, initial differences in particle surface area did not decrease the fragmentation, so that even larger surface areas were available for bonding from small particles, resulting in stronger compacts (200).

With a decrease in particle size, yield pressure increased and the strain reduction ratio sensitivity index (SRS index) decreased (195), which pointed to a reduction in the amount of fragmentation (60). The transition from a brittle crystal to a purely ductile material was thought to occur at a medium particle size of approximately 20 μm (60). The change in the mechanism of consolidation at a decreasing particle size of α-lactose monohydrate was confirmed by Vromans et al. (24) by demonstrating that particle fractions of 1 to 8 μm and 8 to 16 μm did not meet the unique proportionality between breaking force and tablet pore surface area, which exists for larger particles of α-lactose monohydrate.

Riepma et al. (207) demonstrated that tablets compacted from α-lactose monohydrate showed a small time-dependent moisture uptake when exposed to an ambient humid atmosphere (45% RH). Moisture sorption was found to reach a plateau within 10 minutes after compaction, which was accompanied by a decrease in both strength and specific BET surface area of the tablets. Subsequent storage of the tablets in a dry atmosphere resulted in an increase in tablet breaking force but not in a change in specific surface area. The tablets showed no moisture sorption and no change in breaking force or specific surface area when transferred immediately after ejection from the die into a dry atmosphere. These results indicate that contact points between lactose particles in a tablet dissolved when exposed to a humid atmosphere and recrystallized on exposure to lower relative humidity conditions. It was suggested that the cause of irreversible decrease in specific surface area of the tablets on exposure to high relative humidity conditions was blockage of the very narrow pores in the tablets by sorbed moisture.

Bolhuis et al. (11) showed that the strength of α-lactose monohydrate tablets, initially measured 30 minutes after ejection from the die, did not change during storage for eight weeks at 20°C/50% RH or at 20°C/85% RH.

Tablets compressed from α-lactose monohydrate without a lubricant, disintegrate very quickly in water as a result of rapid liquid uptake and fast dissolution of the bonds (208). The presence of a hydrophobic lubricant has a strong inhibiting effect on water penetration and hence on disintegration time. This effect can easily be counteracted by the addition of microcrystalline cellulose (12) or a high-swelling disintegrant such as sodium starch glycolate or croscarmellose sodium (8,208).

In practice, α-lactose monohydrate 100-mesh is often combined with microcrystalline cellulose (12,87). This combination results in a strong synergistic effect on disintegration time, whereas the breaking force increases proportionally to the percentage of microcrystalline cellulose (12). Therefore, the combination of lactose 100-mesh and microcrystalline cellulose is one of the more popular blends in direct compaction. Using a systematic optimization method according to a simplex lattice design, optimum tablet formulations can be obtained from α-lactose monohydrate 100-mesh in combination with microcrystalline cellulose, anhydrous β-lactose, or spray-dried lactose (14,209).

Anhydrous α-Lactose

The binding capacity of α-lactose monohydrate is increased dramatically by thermal or chemical dehydration of the crystals (25). During treatment, a gradual transition within each particle from the hydrous to the anhydrous form was observed. The thermally dehydrated product showed strongly increased binding properties with decreasing water content, whereas desiccation by methanol gave a steeper increase in the tablet strength (25).

During dehydration, α-lactose monohydrate changes from single crystals into aggregates of anhydrous α-lactose particles. In contrast to the large differences in compactibility between α-lactose monohydrate and anhydrous α-lactose, compaction of the two types of lactose resulted in tablets with almost equal porosities at the same compaction load, resulting in superimposed Heckel plots (25,62). Large differences were found between the pore volume distributions of the two types of tablets. Mercury porosity measurements showed that tablets from anhydrous α-lactose exhibited a greater number of pores, much smaller in size, than α-lactose monohydrate. The steeper increase in breaking force with compaction load and a corresponding increase in pore surface area of tablets lead to the increased fragmentation during compression of the anhydrous product. This effect explains the large increase in binding capacity with increasing dehydration (25,62). Wong et al. (210) compared the deformation characteristics of single crystals of anhydrous α-lactose and α-lactose monohydrate. Indentation hardness testing showed that the anhydrous crystals were much softer, less elastic, and had a lower resilience than the monohydrate crystals. The anhydrous crystals were less anisotropic than the monohydrate crystals. Stress-strain data of the single crystals and photographs of single crystal deformations showed that the anhydrous crystals withstood a lower maximum load, exhibited lower displacement prior to destructive cracking, and thus required less work to break than the corresponding monohydrate crystals. The monohydrate crystals underwent much more pronounced splitting and fragmentation (spalling) than the anhydrous crystals, which tended to crush by gradual localized cracking at the point of contact. It was concluded that α-lactose monohydrate crystals are hard, elastic, brittle, and strong; compression initiates the progression of large cracks that result in the breaking of both small and large fragments from the crystals. The anhydrous crystals, which are softer, weaker, and less elastic, undergo brittle fracture much more readily and at lower stresses than the monohydrate.

A disadvantage of using anhydrous α-lactose in direct compaction formulations is the relatively slow disintegration of the tablets. When in contact with water, the tablets dissolve during the disintegration process. This effect, which is caused by poor water penetration into the tablets, is caused by a combination of small pore diameters and precipitation of dissolved anhydrous α-lactose in the course of the water penetration process into α-lactose monohydrate (211).

Anhydrous α-lactose was manufactured and marketed by DMV (The Netherlands) as Pharmatose DCL 30. In spite of the excellent flowability and binding properties (11), the product was withdrawn from the market because relatively slow disintegration turned out to be a major disadvantage.

Anhydrous β-Lactose

Another form of lactose that was especially designed for direct compaction is anhydrous β-lactose, also referred to as anhydrous lactose. It is described in all main pharmacopoeias. The commercial products consist of agglomerated extremely fine crystals (Fig. 11B, C), produced by roller drying of a solution of pharmaceutical grade α-lactose monohydrate followed by subsequent comminution and sieving (212). Commercial products contain about 80% anhydrous β-lactose; the remaining material is anhydrous α-lactose. The water content is less than 0.5%. Roller-dried anhydrous β-lactose is marketed as Lactose Anhydrous DT by Sheffield Pharma Ingredients (Illinois, U.S.), part of the Kerry Group, SuperTab 21AN, and SuperTab 22AN (former brand names Pharmatose DCL 21 and DCL 22) by DMV-Fonterra (Germany) Excipients and Lactopress Anhydrous by FrieslandCampina/Domo-Pharma (The Netherlands).

With respect to flow properties, contradicting results have been reported. The flow properties were judged as good (11,213,214), fair (215), and less than optimum (41,48), respectively. The differences may be caused by the differences in particle size distribution of different materials. In Table 4 both flow through small orifices and tablet weight variation coefficient point to a good flowability for SuperTab 21AN. SuperTab 22AN contains a smaller amount of fines and hence even a better flowability than SuperTab 21AN (216).

The compaction profiles in Figure 12 show that anhydrous β-lactose (SuperTab 21AN) has improved compaction properties when compared to α-lactose monohydrate 100-mesh. The good compactibility was attributed to the particle structure, resulting in a large surface for bonding (41). On the basis of the small deformation before failure during diametrical compression testing of anhydrous β-lactose, Rees and Rue (99) concluded that anhydrous β-lactose is a brittle material. This was confirmed by the relatively small stress relaxation and the absence of a change in nonrecoverable deformation (NRD) with a change in platten rate (100). The lower yield pressure and the higher strain-rate sensitivity of anhydrous β-lactose, as compared with α-lactose monohydrate, showed that anhydrous β-lactose was less brittle than α-lactose monohydrate (65).

Using mercury porosimetry, Vromans et al. (62) showed that the pore size distribution of tablets compressed from roller-dried β-lactose was strongly dependent on compaction load. The shift from larger to smaller pores with increasing compaction force was attributed to the fragmentation of the particles during compaction. In addition, the breaking force was related to the specific pore surface area (Fig. 13) and the β-lactose exhibited the same relationship as other crystalline lactoses. The fragmentation propensity of roller-dried β-lactose is related to the morphology of the particles, which in turn is affected by the method of manufacturing. Nowadays it is assumed that the agglomerates fragment during compaction, but that very small primary particles (Fig. 11C) deform by plastic deformation, because they are smaller than the brittle/ductile transition particle size of about 20 μm (60). Anhydrous β-lactose has a lower lubricant sensitivity than α-lactose monohydrate. This is caused by the granular structure of anhydrous β-lactose: a lubricant film, formed during mixing with magnesium stearate will be destroyed during the fragmentation of the granules in early stages of the compaction process (18).

The presence of about 20% anhydrous α-lactose causes a prolonged disintegration time, when tablets from anhydrous β-lactose are compressed at moderate to high compression forces. This phenomenon is caused by a combination of small pore diameters and precipitation of dissolved anhydrous α-lactose in the course of the water penetration process into α-lactose monohydrate (211). The disintegration time can be enhanced by the incorporation of a disintegrant such as crospovidone, which enables a rapid progress of the water penetration front in the tablets (208).

Anhydrous β-lactose is not hygroscopic. Immediately after ejection, tablets compressed from anhydrous β-lactose show a similar moisture uptake as tablets containing α-lactose monohydrate (see section "α-Lactose Monohydrate"), resulting in a decrease in breaking force and specific tablet surface area within 10 minutes after ejection (207). Storage for eight weeks at 20°C/85% RH resulted in a small increase in hardness (11). Because of the low moisture content, anhydrous β-lactose is an ideal excipient for moisture-sensitive drugs.

Spray-Dried Lactose

Spray-dried lactose was the first product made in a special form for direct compaction. In addition to microcrystalline cellulose, it played a paramount role in the early acceptance of direct compaction as a valuable tableting technique. In spite of the improved binding and flow properties as compared with α-lactose monohydrate (2), the early quality of spray-dried lactose had a number of problems, particularly discoloration, caused by the presence of contaminants in the mother liquor [mainly 5-(hydroxymethyl)-2-furaldehyde] (217), and the reaction with primary amines (218). Spray-dried lactose quality was improved to overcome the problem of discoloration. A better understanding of the relationship between particle structure, physical properties, and mechanisms of consolidation and compaction enabled the development of improved spray-dried lactoses for direct compaction.

Spray-dried lactose is produced by spray drying a suspension of α-lactose monohydrate (primary particles) in a saturated aqueous solution of lactose (219). It can be considered as a coprocessed product, because the suspension consists of two phases: particles of α-lactose monohydrate and a solution of lactose in a mutarotated form. In solution lactose exists as an equilibrium of approximately 60% β-lactose and 40% α-lactose, the two stereo isomers being differentiated by the orientation of the hydroxyl group at the carbon atom 1 in the glucose ring of lactose. The ratio of α-lactose and β-lactose is substantially unaltered by spray drying, especially at higher temperatures (220). The resulting spheres of spray-dried lactose are built up of two different components: about 80% to 85% microcrystals of α-lactose monohydrate, and 15% to 20% amorphous lactose. The amorphous lactose is the result of quick dehydration of the dissolved lactose in the suspension during spray drying. Spray-dried lactose typically contains 9% to 12% β-lactose, most of which is present in the amorphous matrix (221). In contrast to the earlier products that contained a fairly high proportion of large crystals of α-lactose monohydrate because of partial crystallization before spray drying (41,222–224), the later materials contain only a few small individual crystals of α-lactose monohydrate (Fig. 11D). Spray-dried lactose has a moisture content of about 5%; a small part of this is free moisture; the remainder is water of crystallization of α-lactose monohydrate. The spray-dried lactose available commercially include Lactose-316 Fast Flo (Sheffield Pharma Ingredients), SuperTab 11SD and 14SD (former brand names Pharmatose DCL 11 and DCL 14) by DMV-Fonterra Excipients, FlowLac 90 and 100 (Meggle, Germany), and Lactopress Spray-dried from FrieslandCampina/Domo-Pharma.

Because spherical droplets of lactose suspension are spray-dried, the dry material produced by the process is spherical in nature (Fig. 11D), which accounts for excellent flow properties (Table 4).

The binding properties of spray-dried lactose are good (Fig. 12 for SuperTab 11SD), although different spray-dried products from different suppliers have been reported to behave differently (215). In a comparative evaluation of several directly compressible filler-binders on a high speed rotary press, Deurloo et al. (175) concluded that the tableting properties of Lactose-316 Fast Flo and SuperTab 11SD were better than those of the other products. Spray-dried lactose seems to be less influenced by the rate of compression than other types of lactose (175,215).

The compressibility of spray-dried lactose is a function of the primary particle size of α-lactose monohydrate and the amount of amorphous lactose. α-Lactose monohydrate is a brittle material that fragments on compaction with relatively poor binding properties (196) (see section "α-Lactose Monohydrate"); the compression behavior changes from brittle to ductile when the particle size is reduced to 45 μm or less (225). The amorphous lactose is considered to form a binding layer on the particle surface. Because the primary particles in spray-dried lactose are generally smaller than 45 μm and amorphous lactose is a material with plastic deformation behavior, it may be expected that, after initial fragmentation of the spherical

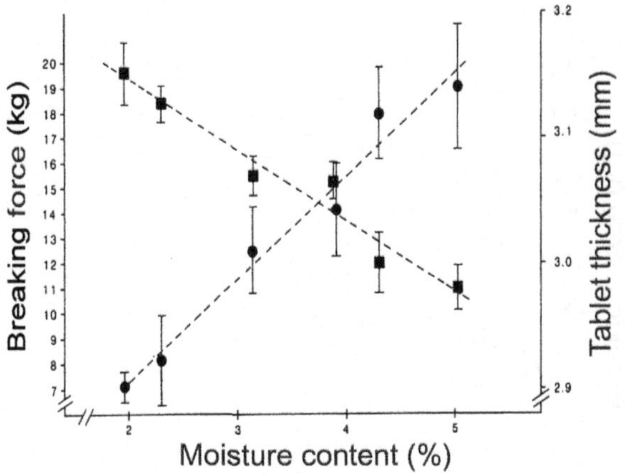

Figure 14 Compact strength (●) and thickness (■) of tablets of amorphous lactose, compressed at a compaction load of 75 Mpa, as a function of the moisture content. *Source*: From Ref. 226.

particles, spray-dried lactose deforms mainly by plastic deformation. This is supported by the low yield pressure (203) and by different tableting indices (70).

Because α-lactose monohydrate has only poor binding properties, the amorphous part of the excipient must be responsible for the increase in compactibility (203,214). For this reason, the tableting properties of pure amorphous lactose were investigated (226). Fully amorphous lactose prepared by spray drying exhibited an increasing compactibility after water sorption, and the compact thickness decreased simultaneously (Fig. 14). Using mercury porosimetry data, it was calculated that, in contrast to crystalline lactose products, the tablet pore surface area did not change with increasing compaction forces. From these observations, and from the high lubricant sensitivity, it was concluded that fully amorphous lactose deforms mainly by plastic flow. In contrast to fully amorphous lactose, the compactibility of the spray-dried lactose (containing 15-20% amorphous lactose) was not affected by moisture uptake, and the lubricant sensitivity was lower (226). The pore surface area increased with increasing compaction load. From these results, it was concluded that the consolidation of spray-dried lactose is determined by the fragmentation of the crystalline α-lactose monohydrate, whereas the binding is largely determined by the amorphous lactose (226). Further investigations by Vromans et al. (24) with a number of experimental spray-dried lactose products containing different amounts of amorphous lactose demonstrated that the compactibility of the samples was a function of the primary particle size of α-lactose monohydrate and the amount of amorphous lactose (Fig. 15). Only about 15% of amorphous lactose was enough to increase the compactibility of fine particles of α-lactose monohydrate. The amorphous lactose was considered to form a binding layer on the particle surface.

On the basis of the study of Vromans et al. (24) the compaction properties have been improved by decreasing the primary particle size from 34 to 20 μm (221). Spray-dried lactose prepared from primary particles with an average of 20 μm is marketed as SuperTab 14SD by DMV-Fonterra Excipients. Figure 12 shows that the improved product has better compaction properties than regular spray-dried lactose. The increase of the compactibility of α-lactose monohydrate at decreasing primary particle size was explained by both the larger surface for bonding and the change from brittle to ductile when the particle size is reduced (221,227). The addition of a lubricant caused a significant decrease in breaking force for both types of spray-dried lactose. The lubricant sensitivity is limited, however, by the initial fragmentation

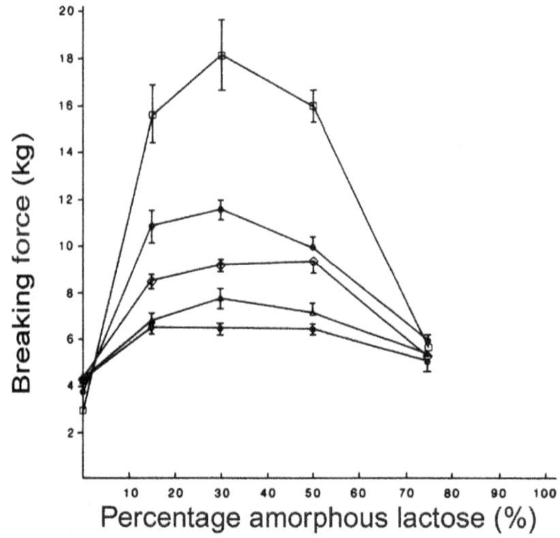

Figure 15 Breaking force of tablets of experimental spray-dried lactose samples, compressed at a compaction pressure of 75 Mpa. The water content of the amorphous lactose is about 2%. Starting materials: (□) 1–8 μm; (●) 8–16 μm; (◇) 16–24 μm; (Δ) 24–32 μm; (○) 32–45 μm. *Source*: From Ref. 24.

of spray-dried lactose spheres during consolidation and compression, in particular for the improved product SuperTab 14SD.

The presence of amorphous lactose in spray-dried lactose may be problematic if it is stored in open containers at relative humidities over 50%. In that case, the amorphous part (which contains just as the solution, an equilibrium of 60% β-lactose and 40% α-lactose) slowly crystallizes to α-lactose monohydrate. This process decreases the compaction properties. It was recently discovered, however, that under certain conditions of temperature and humidity, a rapid crystallization of amorphous lactose can occur (228). In this case, the amorphous part crystallizes not to α-lactose monohydrate but to a stable mixed α/β crystal form of lactose with the same ratio of β- to α-lactose as in the amorphous form. The compactibility of this stabilized spray-dried lactose was comparable to that of the original spray-dried lactose containing amorphous lactose (221).

Agglomerated Lactose

The moderate binding properties of α-lactose monohydrate can be improved by conversion into a granulated form (212). The first agglomerated lactose on the market was Tablettose, produced in a special granulation process by Meggle (229). It is made up almost entirely of aggregated crystals of α-lactose monohydrate (Fig. 11E); it contains no amorphous lactose. Nowadays Tablettose is available in three types with different particle size fractions: Tablettose 70, 80, and 100. Because of the smaller primary particles, Tablettose 100 exhibits a greater tablet hardness compared to Tablettose 70 or 80. Another commercially available agglomerated lactose is SuperTab 30GR (former brand name Pharmatose DCL 15) from DMV-Fonterra Excipients. Freund, Japan, is marketing two agglomerated lactose products, Dilactose S and R. These two products do comply with the JP, but not with the USP/NF and Ph Eur because their total water content is less than 4.5%. Although both Tablettose 100 and SuperTab 30GR are agglomerated lactose, meeting the specifications of lactose monohydrate USP/NF and Ph Eur, the compaction behavior of SuperTab 30GR is better than that of Tablettose 100 (Fig. 16). This improvement, which could neither be explained from the powder surface area nor from the powder bulk density, is caused by differences in the two production processes (230). During the production of SuperTab 30GR, part of the dissolved lactose is precipitated as β-lactose. For

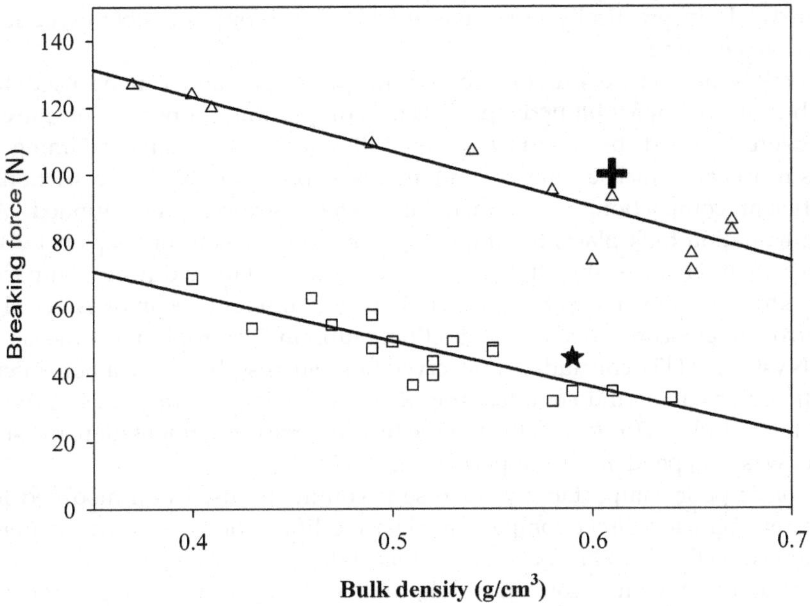

Figure 16 Breaking force of tablets compressed from different lactose granule fractions (125–250 μm; 212–425 μm) versus bulk density of the granulations before compression. The granules were prepared from different α-lactose monohydrate (□) or anhydrous β-lactose (Δ) powders by various wet granulation techniques with only water as a binder and by dry granulation, respectively. The figure includes the relationship between the bulk density of Tablettose and the breaking force of Tablettose tablets (★) and between SuperTab 30GR tablets (✚).

this reason, SuperTab 30GR contains approximately 12% β-lactose, whereas Tablettose contains only about 3%. Because β-lactose undergoes more fragmentation and deformation under load than does α-lactose monohydrate, a larger intergranular bond surface area will be formed. The effect is strengthened because the β-lactose is located at the surface of the SuperTab 30GR granules. The effect of β-lactose on the binding properties of SuperTab 30GR is confirmed by examining the relationship between granule bulk density and tablet breaking force (17). SuperTab 30GR fits the relationship for β-lactose, whereas Tablettose fits the relationship for α-lactose monohydrate (Fig. 16) (230).

OTHER SUGARS AND OLIGOSACCHARIDES

Sugar-based excipients have enjoyed widespread acceptance because of their safety and pleasant taste. Compressible sugar is commonly used for lozenges and chewable tablets, because of the sweet taste and good mouthfeel. Nowadays they are increasingly used as one of the components in ODTs. There are many types of compressible sugars today and most of them are composed of sucrose granulated with small amounts of modified dextrins to make the sucrose more compressible. In fact these sugars are coprocessed materials, but because of their properties, they are discussed in this section.

Compressible Sugar

Sucrose is a nonreducing disaccharide composed of D-glucosyl and D-fructosyl moities and occurs almost universally in all components of practically every existing phanerogam. However, major production sources are sugar beets, sugar cane and to a lesser extent sorghum and sugar maple. The latter two sources normally provide sucrose containing syrups. As an item of commerce, crystalline sucrose represents the highest volume organic compound produced worldwide in practically pure state (>99.5%). The only possible exception to this statement is ethylene oxide.

Large crystals of sucrose provide good flow properties and can be used for direct compaction, but the compactibility is poor. The compaction properties of pure sucrose (saccharose) were studied by Hardman and Lilley (231,232) using scanning electron microscopy, surface area measurements, and mercury porosimetry. Large frictional forces prevented efficient compaction, even when large shear stresses were applied. Although extensive fragmentation took place, the fragments tended to remain in heaps between larger particles rather than to move into the pore space. Fragmentation during compaction was confirmed by stress-density measurements (118) using different techniques (Heckel plots, scanning electron microscopy, strength reduction ratio, and isotropic ratio measurements), Duberg and Nyström (117) concluded that crystalline sucrose behaved as an intermediate between plastic deformation and complete fragmentation, which corresponds to the intermediate brittle fracture index (70). As an effect of the high degree of fragmentation, the strength of sucrose tablets was independent of the particle size (120).

Because of its poor compactability, sucrose is commonly used in a modified form that makes it more efficient for direct compaction. The modified form is known as *Compressible Sugar* and is described the USP/NF. Several products fall into this category; they are composed of 95% to 98% sucrose and may contain starch, maltodextrin, or invert sugar and a suitable lubricant.

Compressible sugar is available from different sources. NuTab from Chr. Hansen Ingredient Technology (New Jersey, U.S.) contains sucrose, about 4% invert sugar, and small percentages of corn starch and magnesium stearate. It is prepared by a roller compaction process and then broken up to the desired particle range. NuTab is available in medium and coarse grades. Di-Pac from Domino Speciality Ingredients (Maryland, U.S.) consists of 97% sucrose and 3% modified dextrins and is prepared by a cocrystallization process. Each granule is composed of hundreds of sucrose microcrystals, glued together by the dextrins. Sugartab (Fig. 8E) from JRS Pharma is a nonpharmacopeial product because the invert sugar content exceeds 3%. The same counts for directly compressible sugar from SPI Polyols (Delaware, U.S.): it is a coprocessed excipient consisting of 95% sucrose and 5% sorbitol.

Mendes et al. (233) evaluated NuTab as a chewable filler-binder for direct compaction in combination with several active ingredients and 1.0% magnesium stearate. Generally, good tablets could be prepared with NuTab. Shah at al. (136) found for NuTab and Di-Pac almost similar compaction profiles. The compaction profile of Sugartab in Figure 17 shows that the product has reasonable binding properties. David and Augsberger (57) investigated the effect of time of compression and the presence of a lubricant on the tablet strength of various materials including Di-Pac. It was concluded that particle fracture played a more dominant role during compaction of Di-Pac or lactose than during compaction of microcrystalline cellulose or compressible starch. However, other authors mentioned that, in contrast to plain sucrose (which consolidates mainly by fragmentation), plastic deformation in combination with brittle fracture played a large role during compaction of compressible sugar (234). Because of the porous structure and large surface area of cocrystallized sucrose (e.g., Di-Pac), its compactibility is strongly dependent on the moisture content. The equilibrated free moisture content of 0.4% appears optimum, while higher moisture content may produce hard candy or troche tablets (212).

The coprocessed directly compressible sugar from SPI Polyols, containing 95% sucrose and 5% sorbitol consists of agglomerates with an open crystalline matrix. The compaction properties are better than those of Di-Pac, which was attributed to the presence of sorbitol (235).

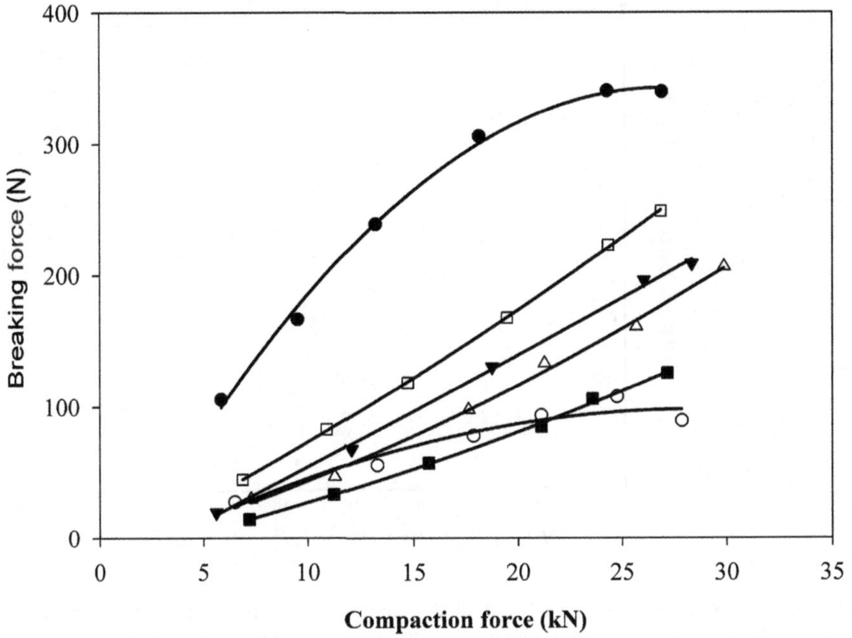

Figure 17 Compaction profiles of dextrates (Emdex), compressible sugar (Sugartab) and different coprocessed filler-binders, lubricated with 0.5% magnesium stearate. (●) dextrates (Emdex); (○) compressible sugar (Sugartab); (▼) Cellactose 80; (□) Microcelac 100; (Δ) Ludipress; (■) StarLac. 500 mg, 13 mm tablets, compaction simulator at 300 mm/sec.

Tablets containing compressible sugar will soften during exposure to high humidities because of the hygroscopic nature of sucrose (86). Compressible sugar containing invert sugar (e.g., NuTab) is more hygroscopic than when it contains malto-dextrins (e.g., Di-Pac) because of the absence of reducing sugars. Placebo Di-Pac tablets pick up less than 0.1% moisture during storage for 30 days at 25°C/75% RH (234). Compressible sugar has few incompatibilities, but is incompatible with primary and with many secondary amines.

Because of the high solubility of sucrose, the tablets do not disintegrate but dissolve during disintegration or dissolution testing.

Dextrose

Dextrose, or D-glucose, is a monosaccharide commonly produced by acid hydrolysis of starch. It is available in the anhydrous and monohydrate forms. Bolhuis and Lerk (41) showed that coarse fractions of dextrose monohydrate can be compressed directly when mixed with 1.0% magnesium stearate. However, the tablets capped at moderate compaction loads. The effect of moisture on the compaction properties of anhydrous dextrose and dextrose monohydrate was examined by Armstrong et al. (236). An increase in moisture content of anhydrous dextrose produced a corresponding increase in tensile strength and tablet toughness up to a moisture level of 8.9%. At a higher water content level, tablet strength fell sharply (Fig. 18). The increase in tensile strength was believed to be caused by an increased lubrication effect and recrystallization of dissolved anhydrous dextrose during compaction. The same moisture content resulted in lower tensile strength values when dextrose monohydrate was used compared to the anhydrous dextrose. Any increase in moisture content obtained by exposure to elevated humidities led to a reduction in both tensile strength and toughness. The differences in the compressional properties of the two types of dextrose were attributed to the different

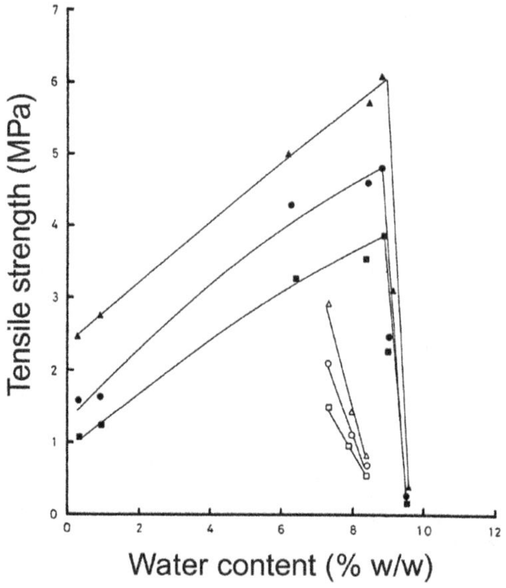

Figure 18 Relationship between tensile strength of tablets compressed at 14 (□,■); 19 (○,●); and 25 (Δ, ▲) kN and their water content. Closed symbols: anhydrous dextrose; open symbols: dextrose monohydrate. *Source*: From Ref. 236.

mechanism by which water is held, and to differences in the mobility of water. In dextrose monohydrate, the water of crystallization was thought to be locked into the crystal structure, and therefore not available to encourage bond formation, whereas in anhydrous dextrose, the moisture should exist in a relatively mobile form. Any increase in moisture content after exposure to a high humidity resulted in a reduction in both tensile strength and toughness.

Maltose

Maltose is a nonhygroscopic sugar that is widely used in food products. Crystalline maltose is available in both spray-dried and crystalline forms. Spray-dried crystalline maltose is used as a directly compactible tablet excipient and marketed by SPI Polyols as Advantose 100. As a result of the spray-drying process, this crystalline maltose consists almost entirely of almost spherical particles with very good flow properties. The material is highly compactible and shows no capping when compacted at high forces (237). If blended with other filler-binders such as microcrystalline cellulose or granular mannitol, crystalline maltose improves compaction and flow properties and decreases capping tendencies when subjected to high loads (237). Another advantage of crystalline maltose is its low lubricant sensitivity. Spray-dried crystalline maltose can be used for both chewable and disintegrating tablets.

Dextrates

The USP/NF describes dextrates as a purified mixture of saccharides resulting from the controlled enzymatic hydrolysis of starch. It may be hydrous or anhydrous. It has a dextrose equivalent between 93.0% and 99.0%. A spray-crystallized dextrose in the hydrous form is marketed by JRS Pharma as Emdex (formerly available as Celutab in hydrous and anhydrous forms). It consists of 90% to 92% dextrose, 3% to 5% maltose and a remaining mixture of higher glucose polymers. The anhydrous form is currently not commercially available. Emdex is recommended for chewable tablets because of its sweet taste and high negative heat of solution, which creates a cooling mouthfeel.

Emdex is composed of porous spheres (Fig. 4F) that have excellent flow properties in spite of sticking together of some particles (41). Scanning electron micrographs indicate that each spherical granule consists of randomly arranged flat microcrystals bound together by minute amounts of higher saccharides and interspersed with void spaces of various shapes and sizes.

Using an instrumented rotary press, Ridgway et al. (238) reported good flow and compaction properties without any sign of capping or difficulty in ejection for both hydrous and anhydrous Celutab, lubricated with 1.5% magnesium stearate. Using an instrumented eccentric press, Bolhuis and Lerk (41) reported that in comparison with a series of other filler-binders, hydrous Celutab showed poor compression characteristics when tableted with 0.5% magnesium stearate. However, excellent overall performance as an excipient for direct compaction was exhibited when the lubricant concentration was increased to 1.0%. The compaction profile (Fig. 17) shows that Emdex has very good binding properties. Consolidation was reported to occur by plastic deformation of the porous particles that deformed along many planes (49,212). This was confirmed by force-displacement curves for axial and radial displacements (64) and by the linearity of Heckel plots (118).

Because of the large particle size of Emdex (Fig. 4F), problems with blending were reported. Shangraw (224) advised a preblending of micronized active substances with Emdex before the addition of other excipients. Micronized drug particles lodged in the pores at the surface of Emdex spheres are apparently held in place with sufficient attractive forces to prevent dislodging during the subsequent blending operations.

As an effect of the high solubility of dextrose and the other sugars, tablets prepared from Emdex dissolve, rather than disintegrate, when placed in water (41,238). As the dissolution takes place mainly from the outer surface, which is initially constant, the disintegration time is not greatly affected by the compaction load (41,49).

The large surface area and the presence of polysaccharides make Emdex quite hygroscopic at relative humidities above 75%. Even after storage for eight weeks at room temperature and 50% RH, the breaking force of tablets decreases dramatically, whereas during storage at 85% RH, the tablets liquefy (11).

Inulin

Inulin is a fructan-type carbohydrate that consists of linear chains of fructose units linked via a β-2-1-bond and with a glucose unit at the end. It is widespread used as a diagnostic agent for renal function and is described in the USP and BP. Inulin is an inert, stable nonreducing carbohydrate that shows only moderate hygroscopicity (239). In addition to its use as a diagnostic agent, it is a food ingredient. After oral administration, inulin is not absorbed but it is partially metabolized by bacteria in the colon flora. Recently inulin has been evaluated as a filler-binder for tablets prepared by direct compaction (240).

Inulin as such has a high bonding capacity, which depends on the degree of crystallinity, percentage of included air, particle size and DP. Amorphous inulin with a DP of about 10, prepared by spray drying, has good binding properties, but its plastic deformation behavior leads to a rather high lubricant sensitivity. Using particles with entrapped air can decrease this lubricant sensitivity, since because of their high porosity, these particles fragment before plastic deformation commences. A lubricant film formed during the mixing process will thus be destroyed during particle fragmentation (240). Because of the low DP, spray-dried amorphous inulin is easily soluble in water, which promotes tablet disintegration. Crystalline inulin can only be prepared with a high DP, which makes it poorly soluble in water. Although crystalline inulin has good binding properties, tablet disintegration is slow.

Bolhuis et al. (241) introduced hollow inulin particles as excipient for direct compaction. The hollow particles collapse, followed by a plastic deformation of the fragments. The fragmented particles not only have an increased compactibility, but also a much reduced lubricant sensitivity as the lubricant film is destroyed during fragmentation of the hollow particles.

COPROCESSED PRODUCTS

In recent years, a number of excipient mixtures were introduced utilizing particle engineering in the design of combination products for direct compaction. It was demonstrated that the use of specially prepared excipient mixtures, defined in this chapter as coprocessed products, can combine the advantages of the starting materials while overcoming their respective disadvantages (242). More information is given in section "Coprocessing." Examples are Ludipress from BASF (Germany), Cellactose, Microcelac, and StarLac from Meggle.

Although spray-crystallized dextrose-maltose (Emdex) and compressible sugar are coprocessed products in reality, they are commonly considered as single components and are official in the USP/NF. For this reason, they are described in the sections on single components (see sections "Compressible Sugar" and "Dextrates"). In fact, spray-dried lactose (see section "Spray-Dried Lactose") is also a coprocessed excipient, because it is nowadays produced from two different materials: α-lactose monohydrate is suspended in a lactose solution, which contains lactose in a mutarotated form.

Acceptance of new coprocessed products within the pharmaceutical industry is highly influenced by the balance between performance and price, because the production costs of some products is high. This is particularly the case where expensive unit operations such as jet-milling (see section "Advantages and disadvantages of Direct Compaction") are necessary.

Coprocessed Lactose and Polyvinylpyrrolidone

Ludipress

Ludipress (BASF) is a coprocessed product consisting of 93.4% α-lactose monohydrate (filler), 3.2% polyvinylpyrrolidone (Kollidon 30; binder) and 3.4% crospovidone (Kollidon CL; disintegrant). It is a free-flowing powder, produced by coating of lactose particles with polyvinylpyrrolidone and crospovidone. The spherical particles are made up of a large number of small crystals with smooth surfaces. The particle size ranges from 10 to 600 μm; no more than 25% of the particles smaller than 100 μm. As an effect of the favorable form and particle size distribution, the flow properties, measured as Hausner ratio and angle of repose were excellent (243).

The binding properties of Ludipress, lubricated with 0.5% magnesium stearate are good (Fig. 17) and were found to be much better than those of the physical mix (243). The compactibility was attributed to the plastic deformation behavior of the amorphous components povidone, crospovidone, and amorphous lactose, which is generated during the production process. After milling, Ludipress can be recompressed with only a minor loss of binding properties. The dilution potential of Ludipress, with respect to paracetamol, is lower than that of Avicel PH-101, Elcema G250 or Elcema P050 (244). As an effect of the presence of crospovidone, the disintegration of Ludipress tablets is fast when compressed at low forces. An increase in compression force causes a strong decrease of tablet porosity and a consequent impeded water uptake and an increase in disintegration time (243).

Ludipress LCE

Ludipress LCE (BASF) is composed of 96.5% α-lactose monohydrate as filler and 3.5% povidone (Kollidon 30) as binder, and unlike Ludipress does not contain the disintegrant

Kollidon CL. It is completely soluble in water, and is particularly suitable for the production of lozenges, chewable tablets, effervescent tablets and modified release formulations. The compaction profile of placebo tablets containing 0.5% magnesium stearate exhibits an almost linear increase up to 300 MPa. The compaction properties have been explained by the furrowed surface structure, which causes strong interlocking of the compressed granules. Additionally the binder povidone contributes to the hardness, as it binds the fine lactose particles together (245). In effervescent tablets, it compensates for the poor compactibility of sodium hydrogen carbonate, sodium carbonate and tartaric acid (246).

Coprocessed Lactose and Cellulose

Cellactose 80

Cellactose 80 (Meggle) is a coprocessed product produced by spray drying 75% α-lactose monohydrate with 25% cellulose. In this material, the advantageous properties of the two starting materials are combined in a single form to reduce the negative characteristics of the two components. Because of regular particle shape (Fig. 11F) and favorable particle size distribution, the product has excellent flowability (243) and superior to those of physical blends of agglomerated lactose with either cellulose powder or microcrystalline cellulose (247).

The compaction profile in Figure 17 shows that Cellactose 80 has good compaction properties. Different authors showed that the compaction properties were better than of blends of α-lactose monohydrate and cellulose (247,248). The good compactibility of Cellactose 80 can be attributed to a synergistic effect of consolidation by fragmentation of the lactose, coupled with the concomitant plastic deformation of cellulose (249,250). In a comparative evaluation between Cellactose, α-lactose monohydrate, and a mixture of 25% microcrystalline cellulose with 75% dicalcium phosphate dihydrate, Garr and Rubinstein (249) demonstrated that at compression speeds below 330 mm/sec, the mean yield pressure increased with the compression speed for both Cellactose and lactose, indicating the dominance of plastic deformation over fragmentation. At higher compression speeds, the yield pressure attained a constant level, indicating that consolidation primarily takes place by fragmentation. In contrast, tablets prepared from the microcrystalline cellulose/dicalcium phosphate dihydrate blend showed no stabilization of the mean yield pressure, suggesting that the presence of microcrystalline cellulose makes plastic deformation the dominating consolidation mechanism. It was also demonstrated that the tablet strength of Cellactose did not diminish with an increasing compression speed. In a comparative evaluation of several filler-binders, compressed on a high-speed rotary press, the tableting properties of Cellactose were found to be excellent. In contrast to a physical mix of microcrystalline cellulose and lactose monohydrate, Cellactose has a small lubricant sensitivity (251).

The disintegration of tablets compressed from Cellactose is fast when the tablets are compressed at a load below 100 MPa, but at an increase in compaction load the disintegration time was drastically elevated (243,252).

Microcelac 100

Microcelac 100, marketed by Meggle, is a spray-dried compound containing 75% α-lactose monohydrate and 25% microcrystalline cellulose dry matter, unlike Cellactose 80, which contains powdered cellulose. An advantage of Microcelac 100 over Cellactose 80 is better flowability. In Cellactose 80, some long powdered cellulose fibers are not incorporated in the Cellactose particles. In Microcelac 100 particles, the microcrystalline cellulose fibers are shorter and more entrapped, which promotes a more spherical form. In a comparative evaluation of Microcelac 100 with three different lactoses, mixed with microcrystalline cellulose, Michoel et al. (253) showed that Microcelac 100 had superior flow and binding properties. Figure 17 shows that the binding properties are better than those of Cellactose 80. The good flow properties were attributed to the spherical form of the agglomerates and the improved binding properties to the increased plasticity as an effect of spray drying. Good adhesion to the porous surface of Microcelac 100 decreased segregation of a low-dose drug (folic acid).

Coprocessed Lactose and Starch

StarLac is a spray-dried compound consisting of 85% α-lactose monohydrate and 15% maize starch. It was jointly developed by Roquette and Meggle and is marketed by the latter. With respect to regulatory affairs, the excipient can be considered as a pure physical mixture of α-lactose monohydrate Ph. Eur., USP/NF plus corn starch Ph Eur, USP/NF. SEM photographs show that StarLac powder consists of lactose crystals and starch particles, embedded in amorphous lactose (Fig. 8F). StarLac has excellent flow properties because of the coarse sphere-shaped particles and the narrow particle size distribution, this in contrast to a physical mix of 85% spray-dried lactose and 15% native corn starch (254).

Figure 17 shows that the compaction properties are adequate, but less than those of Cellactose 80. On compaction, the starch creates a fine network, whereas the lactose particles show particle fragmentation. The volume-pressure deformation properties of StarLac were dependent on the lactose properties. Only at high maximum relative density the influence of starch causes a change in these properties. An advantage compared to spray-dried lactose is the higher plastic deformability (255). On the basis of Heckel analysis, Wagner and Dressler (254) found that starch that is bound within the StarLac compound improved the deformation properties to a greater extent than starch that was just physically mixed with large spray-dried lactose particles. Small starch particles may travel into pores of the spray-dried lactose lattice, whereas starch particles within StarLac remain in place. Although the compaction properties of StarLac were somewhat worse than those of a physical blend of spray-dried lactose and corn starch, StarLac was superior with respect to flowability, tablet disintegration time and drug dissolution rate (254,255). The rapid disintegration is caused by the presence of starch within the StarLac granules.

Coprocessed Starch and Cellulose

Coprocessed products containing rice starch and different percentages of jet milled or sieved microcrystalline cellulose were prepared by spray drying (256). Increasing the microcrystalline cellulose proportion imparted greater compressibility but decreased the flow properties as an effect of a less spherical shape and a rougher surface of the particles. The initial particle size of microcrystalline cellulose had no effect on the compactibility of the coprocessed products, but jet milled microcrystalline cellulose, having a particle size range close to the size of rice starch, provided more spherical particles than did sieved microcrystalline cellulose. Compressibility of a product with 70% rice starch and 30% jet-milled microcrystalline cellulose was better than that of spray-dried rice starch (Eratab), Cellactose, and agglomerated lactose (Tablettose) but lower than that of microcrystalline cellulose type 101. Together with the excellent flow properties, the low friability and the good self-disintegrating properties, the authors consider this new coprocessed product as a suitable filler-binder for direct compaction (256).

Other Coprocessed Materials

A number of other coprocessed materials have been described, for example, coprocessed magnesium carbonate/powdered cellulose (257) and calcium carbonate/microcrystalline cellulose (258). Although these products have better compactibilities than physical mixtures, containing the same materials, they are not commercially available.

Recently a number of coprocessed filler-binders with a special functionality have been introduced on the pharmaceutical market. They are intended for chewable tablets or for ODTs.

Compressol SM Coprocessed Polyol (SPI Pharma) is a directly compressible excipient containing sorbitol and mannitol. It combines the low hygroscopicity of mannitol with the high compactibility of sorbitol. Moreover, it requires less lubrication than mannitol. Compressol SM is free flowing and has a compactibility in between those of sorbitol and mannitol. It was especially designed for chewable tablets, prepared by direct compaction.

Advantose FS 95 direct compaction fructose is a coprocessed system of 95% fructose and 5% starch. It is marketed by SPI Pharma. As an effect of coprocessing, the material is free flowing. Advantose FM 95 is recommended for chewable tablets. Although the material can be compressed into firm tablets, they can easily be chewed.

Barcroft CS 90 (SPI Pharma) is a coprocessed filler-binder containing 90% calcium carbonate and 10% starch. It is produced by spray drying, which improves both flowability and compaction properties. It can be used in both chewable and conventional, disintegrating tablets.

Ludiflash (BASF) is coprocessed filler-binder containing 90% mannitol, 5% crospovidone (Kollidon CL-SF), and 5% polyvinyl acetate (Kollicoat SR 30D). It is designed for ODTs. Mannitol acts as fast-dissolving filler with a sweet taste, crospovidone is the disintegrant and polyvinyl acetate acts as a hydrophobic binder for enhanced disintegration. The Ludiflash components comply with leading pharmacopoeia monographs. A combination of a controlled particle size distribution, the particle structure and a high bulk density provide good flow properties. To achieve very short disintegration times, it is important to have tablets with a high porosity, which allows water to penetrate very fast. The careful control of the compression load is very important and should between 50 and 90 MPa (259). Sodium stearyl fumarate is an appropriate lubricant for fast disintegrating formulations based on Ludiflash because it offers a broader compression force range and therefore harder tablets than magnesium stearate (260).

F-MELT is a coprocessed material for use in the preparation of ODTs by direct compaction. It contains carbohydrates, disintegrants, and inorganic ingredients, marketed by Fuji Chemical Industry in Japan, and is prepared by spray drying. The inorganic components are anhydrous dicalcium phosphate (Fujicalin; see also section "Dicalcium Phosphate") and magnesium aluminometasilicate (Neusilin). The components comply with all international pharmacopoeia standards. There are two types: Type M has a higher flowability and type C has a faster tablet disintegration. F-MELT consists of spherical granules with good flow properties. The compactibility is good, but to retain disintegration times below 30 seconds, tablet hardness should be limited.

REFERENCES

1. Alderborn G, Nyström C. Pharmaceutical Powder Compaction Technology. New York: Marcel Dekker, Inc, 1996:vii.
2. Gunsel WC, Lachman L. Comparative evaluation of tablet formulations prepared from conventionally-processed and spray-dried lactose. J Pharm Sci 1963; 52(2):178–182.
3. Shangraw RF, Demarest DA. A survey of current industrial practice in the formulation and manufacture of tablets and capsules. Pharm Technol 1993; 17(1):32–44.
4. Miinea L, Farina J, Kallam M, et al. A new world of excipients for oral solid dosage formulation. Pharm Processing Magazine, March 2009.
5. European Pharmacopoeia. Vol. 1, 6th ed. Strasbourg: Council of Europe, 2007:748.
6. Pfister WR, Ghosh TK. Orally disintegrating tablets. Products, technologies and development issues. Pharm Technol 2005; 29(10):136–150.
7. Strickley RG, Iwata Q, Wu S, et al. Pediatric drugs—a review of commercially available oral formulations. J Pharm Sci 2008; 97(5):1731–1774.
8. Van Kamp HV, Bolhuis GK, De Boer AH, et al. The role of water uptake on tablet disintegration. Design of an improved method for penetration measurements. Pharm Acta Helv 1986; 61(1):22–29.

9. Egermann H. Zur Dosierungsgenauigkeit einzeldosierter, fester Arzneiformen mit niedrigem Wirkstoffgehalt. 8. Mitteilung: Agglomeratzerstorung durch Siebung. Sci Pharm 1979; 47:25–31.
10. Westerberg M, Jonsson B, Nyström C. Physicochemical aspects of drug release. IV. The effect of carrier particle properties on the dissolution rate from ordered mixtures. Int J Pharm 1986; 28(1): 23–31.
11. Bolhuis GK, Reichman G, Lerk CF, et al. Evaluation of anhydrous α-lactose, a new excipient in direct compression. Drug Dev Ind Pharm 1985; 11(8):1657–1681.
12. Lerk CF, Bolhuis GK, De Boer AH. Comparative evaluation of excipients for direct compression II. Pharm Weekbl 1974; 109:945–955.
13. Huisman R, Van Kamp HV, Weyland JW, et al. Development and optimization of pharmaceutical formulations using a simplex lattice design. Pharm Weekbl Sci 1984; 6:185–194.
14. Van Kamp HV, Bolhuis GK, Lerk CF. Optimization of a formulation for direct compression using a simplex lattice design. Pharm Weekbl Sci 1987; 9:265–273.
15. Hüttenrauch R. Modification of starting materials to improve tableting properties. Pharm Ind 1983; 45(4):435–440.
16. Sheth PR, Wiley JH. Calcium phosphates in tablet compressing. Patent number US3134719, 1964.
17. Zuurman K, Riepma KA, Bolhuis GK, et al. The relationship between bulk density and compactibility of lactose granulations. Int J Pharm 1994; 102(1–3):1–9.
18. Van der Voort Maarschalk K, Bolhuis GK. Improving properties of materials for direct compaction. Pharm Technol 1999; 23(5):34–42.
19. Humbert-Droz P, Gurny R, Mordier D, et al. Densification behaviour of drugs presenting availability problems. Int J Pharm Tech Prod Manuf 1983; 4(2):29–35.
20. Veesler S, Boistelle R, Delacourte A, et al. Influence of structure and size of crystalline aggregates on their compression ability. Drug Dev Ind Pharm 1992; 18(5):539–560.
21. Lazarus J, Lachman L. Experiences in development of directly compressible tablets containing potassium chloride. J Pharm Sci 1966; 55(10):1121–1127.
22. Hammouda Y, El-Banna HM, Eshra AH. The use of sodium chloride as a directly compressible filler in therapeutic tablets. Pharm Ind 1975; 37(5):361–363.
23. Shotton E, Obiorah BA. The effect of particle shape and crystal habit on properties of sodium chloride. J Pharm Pharmacol 1973; 55(suppl):37P–43P.
24. Vromans H, Bolhuis GK, Lerk CF, et al. Studies on tableting properties of lactose. VII. The effect of variations in primary particle size and percentage of amorphous lactose in spray dried lactose products. Int J Pharm 1987; 35(1–2):29–37.
25. Lerk CF, Andreae AC, De Boer AH, et al. Increased binding capacity and flowability of α-lactose monohydrate after dehydration. J Pharm Pharmacol 1983; 35(11):747–748.
26. Lerk CF, Zuurman K, Kussendrager K. Effect of dehydration on the binding capacity of particulate hydrates. J Pharm Pharmacol 1984; 36(6):399–399.
27. Goto K, Sunada H, Danjo K, et al. Pharmaceutical evaluation of multipurpose excipients for direct compressed tablet manufacture: comparisons of the capabilities of multipurpose excipients with those in general use. Drug Dev Ind Pharm 1999; 25(8):869–878.
28. Van der Voort Maarschalk K, Vromans H, Bolhuis GK, et al. The effect of viscoelasticity and tabletting speed on consolidation and relaxation of a viscoelastic material. Eur J Pharm Biopharm 1996; 42(1):49–55.
29. The Joint IPEC-PQG Good Manufacturing Practices Guide for Pharmaceutical Excipients, The International Pharmaceutical Excipients Council and The Pharmaceutical Quality Group, 2006.
30. Qualification of Excipients for Pharmaceutical Use, The International Pharmaceutical Excipients Council, 2008.
31. Moreton RC. Tablet excipients to the year 2001, a look into the crystal ball. Drug Dev Ind Pharm 1996; 22(1):11–23.
32. Lesney MS. More than just the sugar in the pill. Today's Chemist at Work 2001; 10(1):38–43.
33. Sheth BB, Bandelin FJ, Shangraw RF. Compressed tablets. In: Lieberman HA, Lachman L, eds. Pharmaceutical Dosage Forms: Tablets. Vol 1, 2nd ed. New York: Marcel Dekker, Inc, 1989.
34. Shangraw RF. Direct compression tableting. In: Swarbrick J, Boylan JC, eds. Encyclopedia of Pharmaceutical Technology. Vol 4. New York: Marcel Dekker, Inc, 1991:85–106.
35. Wallace JT. Cellulose derivatives and natural products utilized in pharmaceutics. In: Swarbrick J, Boylan JC, eds. Encyclopedia of Pharmaceutical Technology. Vol 2. New York: Marcel Dekker, Inc., 1990:321.
36. Lamberson RL, Raynor GE. Tableting properties of microcrystalline cellulose. Man Chem Aerosol News 1976; 47(6):55–61.

37. Sixsmith D. Microcrystalline cellulose as a tablet excipient. Man Chem Aerosol News 1976; 47(8): 27–28.
38. Marshall K, Sixsmith D. Some physical characteristics of microcrystalline cellulose. 1. Powders for pharmaceutical use. Drug Dev Comm 1974; 1(1):51–71.
39. Doelker E. Comparative compaction properties of various microcrystalline cellulose types and generic products. Drug Dev Ind Pharm 1993; 19(17–18):2399–2471.
40. Doelker E, Maswelle D, Veuillez F, et al. Morphological, packing, flow and tableting properties of new Avicel types. Drug Dev Ind Pharm 1995; 21(6):643–661.
41. Bolhuis GK, Lerk CF. Comparative evaluation of excipients for direct compression. I. Pharm Weekbl 1973; 108:469–481.
42. Sixsmith D. The effect of compression on some physical properties of microcrystalline cellulose powders. J Pharm Pharmacol 1977; 29(1):33–36.
43. Marshall K, Sixsmith D. The effect of compressional force on some properties of tablets containing microcrystalline cellulose. J Pharm Pharmacol 1975; 27(suppl):53P–53P.
44. Nakai Y, Fukuoka E, Nakajima S, et al. Crystallinity and physical characteristics of microcrystalline cellulose. Chem Pharm Bull 1977; 25(1):96–101.
45. Hollenbeck RG, Peck GE, Kildsig DO. Application of immersional calorimetry to investigation of solid-liquid interactions: microcrystalline cellulose-water system. J Pharm Sci 1978; 67(11):1599–1606.
46. Pesonen T, Paronen P. Evaluation of a new cellulose material as binding agent for direct compression of tablets. Drug Dev Ind Pharm 1986; 12(11–13):2091–2111.
47. Zografi G, Kontny MJ, Yang AYS, et al. Surface area and water vapor sorption of macrocrystalline cellulose. Int J Pharm 1984; 18(1-2):99–116.
48. Mendell EJ. Direct compression method of producing solid dosage forms. Man Chem Aerosol News 1972; 43(5):47–49.
49. Khan KA, Rhodes CT. Comparative evaluation of some direct compression diluents. Pharm Acta Helv 1976; 51(1):23–26.
50. Rudnic EH, Chilamkarti R, Rhodes CT. Further studies of the potential of recording powder flow meters. Drug Dev Ind Pharm 1980; 6(3):279–289.
51. Marshall K, Sixsmith D. The flow properties of microcrystalline cellulose powders. J Pharm Pharmacol 1976; 28(10):770–771.
52. Erkoboni DF, Patel CI, Mehra DK, et al. Compressional and Tabletting Performance of an Experimental Grade Large Particle Size Microcrystalline Cellulose. Princeton NJ: FMC Corp., 1990.
53. Aulton ME, Tebby HG, White PJ. Indentation hardness testing of tablets. J Pharm Pharmacol 1974; 26(suppl.):59P–60P.
54. Marshall K, Sixsmith D, Stanley-Wood NG. Surface geometry of some microcrystalline celluloses. J Pharm Pharmacol 1972; 24(suppl):138P–138P.
55. Hüttenrauch R, Jacob J. Bedeutung des Pressdrucks für die Verarbeitung mikrokristalliner Cellulose. Pharmazie 1970; 25(10):630–631.
56. Reier GE, Shangraw RF. Microcrystalline cellulose in tableting. J Pharm Sci 1966; 55:(5):510–514.
57. David ST, Augsburger LL. Plastic flow during compression of directly compressible fillers and its effect on tablet strength. J Pharm Sci 1977; 66(2):155–159.
58. Stamm A, Mathis C. Étude de la comprimabilité de différents excipients pour compression directe. Sci Technol Pharm 1976; 5(5):245–253.
59. McKenna A, McCafferty DF. Effect on particle size on the compaction mechanism and tensile strength of tablets. J Pharm Pharmacol 1982; 34(6):347–351.
60. Roberts RJ, Rowe RC. The effect of the relationship between punch velocity and particle size on the compaction behaviour of materials with varying deformation mechanisms. J Pharm Pharmacol 1986; 38(8):567–571.
61. Staniforth JN, Baichwal AR, Ahmed H, et al. The excipient-process equipment interface. Man Chemist 1987; 58(6):33–36.
62. Vromans H, De Boer AH, Bolhuis GK, et al. Studies on tableting properties of lactose. Part 2. Consolidation and compaction of different types of crystalline lactose. Pharm Weekbl Sci 1985; 7:186–193.
63. Travers DN, Çelik M, Buttery TC. A computer aided investigation on strain movements in compacts under constant stress within the die. Drug Dev Ind Pharm 1983; 9(1–2):139–157.
64. Travers DN, Cox M. Studies of the effect of compaction force on displacement of large compacts formed from direct compression matrices. Drug Dev Ind Pharm 1978; 4(2):157–174.
65. Roberts RJ, Rowe RC. The effect of punch velocity on the compaction of a variety of materials. J Pharm Pharmacol 1985; 37(6):377–384.

66. Armstrong NA, Palfrey LP. The effect of machine speed on the consolidation of four directly compressible tablet diluents. J Pharm Pharmacol 1989; 41(3):149–151.
67. Hüttenrauch R. Nachweis von H-Bindungen in Arzneiformen durch Deuteriumaustausch. Beweis der Bindungskräfte in Cellulose Komprimaten. Pharmazie 1971; 26:645–646.
68. Enezian GM. La compression directe des comprimés a l'aide de la Cellulose Microcristalline. Prod et Prob Pharm 1968; 23(4):185–205.
69. Stamm A, Mathis C. Les excipients pour compression directe. Prévision du comportement à la compression. Étude des propriétés des comprimés. Sci Technol Pharm 1977; 6:65.
70. Hiestand HEN, Smith DP. Indices of tableting performance. Powder Technol 1984; 38(2):145–159.
71. Shlieout G, Arnold K, Müller G. Powder and mechanical properties of microcrystalline cellulose with different degrees of polymerization. AAPS PharmSciTech 2002; 3(2):article 11.
72. Dybowski U. Does polymerisation degree matter? Man Chemist 1997; 68(12):19–21.
73. Suzuki T, Nakagami H. Effect of crystallinity of microcrystalline cellulose on the compactability and dissolution of tablets. Eur J Pharm Biopharm 1999; 47(3):225–230.
74. Khan KA, Musikabhumma P, Warr JP. The effect of moisture content of microcrystalline cellulose on the compressional properties of some formulations. Drug Dev Ind Pharm 1981; 7(5):525–538.
75. Ritter A, Sucker HN. Studies of variables that affect capping. Pharm Technol 1980; 3:24.
76. Zografi G, Kontny MJ. The interactions of water with cellulose- and starch-derived pharmaceutical excipients. Pharm Res 1986; 3(4):187–194.
77. Hüttenrauch R, Jacob J, Zöbisch B. Einfluss der Teilchengrosse auf die Tabletteneigenschaften bei der Verarbeitung mikrokristalliner Cellulose. Pharmazie 1972; 27:416–417.
78. De Boer AH, Bolhuis GK, Lerk CF. Bonding characteristics by scanning electron microscopy of powders mixed with magnesium stearate. Powder Technol 1978; 20(1):75–82.
79. Armstrong NA, Abourida NM, Krijgsman L. Multiple compression of powders in a tablet press. J Pharm Pharmacol 1982; 34(1):9–13.
80. Rees JE, Rue PJ. Elastic and visco elastic properties of compressed tablets. J Pharm Pharmacol 1977; 29(suppl):37P–37P.
81. Fox CD, Richman MD, Reier GE, et al. Microcrystalline cellulose in tableting. Drug Cosmet Ind 1963; 92(2):258–261.
82. Lerk CF, Bolhuis GK, De Boer AH. Effect of microcrystalline cellulose on liquid penetration in and disintegration of directly compressed tablets. J Pharm Sci 1979; 68(2):205–211.
83. Ganderton D, Selkirk AB. The effect of granule properties on the pore structure of tablets of sucrose and lactose. J Pharm Pharmacol 1970; 22(5):344–353.
84. Shah MA, Wilson RG. Some effects of humidity and heat on the tableting properties of microcrystalline cellulose formulations. I. J Pharm Sci 1968; 57(1):181–182.
85. Nyqvist H, Nicklasson M. The effect of water sorption on physical properties of tablets containing microcrystalline cellulose. Int J Pharm Tech Prod Manuf 1983; 4(3):67–73.
86. Sangekar SA, Sarli M, Sheth PR. Effect of moisture on physical characteristics of tablets prepared from direct compression excipients. J Pharm Sci 1972; 61(6):939–944.
87. Delattre L, Jaminet F. La formulation de comprimés à faible dosage médicamenteux par compression directe. Labo-Pharma-Probl Techn 1975; 23(248):1021–1026.
88. Allen JD. Improving DC with SMCC. Man Chemist 1996; 67(12):19–23.
89. Sherwood BE, Becker JW. A new class of high-functionality excipients: silicified microcrystalline cellulose. Pharm Technol 1998; 22(10):78–88.
90. Luukkonen P, Schæfer T, Hellén L, et al. Rheological characterization of microcrystalline cellulose and silicified microcrystalline cellulose wet masses using a mixer torque rheometer. Int J Pharm 1999; 188(2):181–192.
91. Buckton G, Darcy P, Sherwood B. Caloric and gravimetric water sorption profiles of silicified microcrystalline cellulose (SMCC). Pharm Res 1996; 13:S198–S198.
92. Tobyn MJ, McCarthy GP, Staniforth JN, et al. Physicochemical comparison between microcrystalline cellulose and silicified microcrystalline cellulose. Int J Pharm 1998; 169(2):183–194.
93. Edge S, Potter UJ, Steele DF, et al. The location of silicium dioxide in silicified microcrystalline cellulose. Pharm Pharmacol Commun 1999; 5(6):371–376.
94. Edge S, Steele DF, Chen A, et al. The mechanical properties of compacts of microcrystalline cellulose and silicified microcrystalline cellulose. Int J Pharm 2000; 200(1):67–72.
95. Van Veen B, Bolhuis GK, Wu YS, et al. Compaction mechanism and tablet strength of unlubricated and lubricated (silicified) microcrystalline cellulose. Eur J Pharm Biopharm 2005; 59(1):133–138.
96. Mužiková J, Nováková P. A study of the properties of compacts from silicified microcrystalline celluloses. Drug Dev Ind Pharm 2007; 33(7):775–781.

97. Shukla JH, Mauih SZ, Mendes RW. Evaluation of powdered cellulose as a direct compression carrier. Drug Dev Ind Pharm 1980; 6(2):161–176.
98. Bos CE, Vromans H, Lerk CF. Lubricant sensitivity in relation to bulk density for granulations based on starch or cellulose. Int J Pharm 1991; 67(1):39–49.
99. Rees JE, Rue PJ. Work required to cause failure of tablets in diametral compression. Drug Dev Ind Pharm 1978; 4(2):131–156.
100. Rees JE, Rue PJ. Time-dependent deformation of some direct compression excipients. J Pharm Pharmacol 1978; 30(10):601–607.
101. Rue PJ, Rees JE. Limitations of the Heckel relation for predicting powder compaction mechanisms. J Pharm Pharmacol 1978; 30(10):642–643.
102. Paronen P, Juslin M. On the flowability of ascorbic acid diluent binary mixtures and the physical properties of corresponding tablets. Acta Pharm Fenn 1983; 92:187–196.
103. Bolhuis GK, Lerk CF, Zijlstra HT, et al. Film formation by magnesium stearate during mixing and its effect on tableting. Pharm Weekbl 1975; 110:317–325.
104. Wei S, Kumar V, Banker GS. Phosphoric acid mediated depolymerization and decrystallization of cellulose: preparation of low crystallinity cellulose—a new pharmaceutical excipient. Int J Pharm 1996; 142(2):175–181.
105. Kothari SH, Kumar V, Banker GS. Comparative evaluations of powder and mechanical properties of low crystallinity celluloses, microcrystalline celluloses, and powdered celluloses. Int J Pharm 2002; 232(1–2):69–80.
106. Kumar V, De la Luz Reus-Medina M, Yang D. Preparation, characterization, and tabletting properties of a new cellulose-based pharmaceutical aid. Int J Pharm 2002; 235(1–2):129–140.
107. Bos CE, Bolhuis GK, van Doorne H, et al. Native starch in tablet formulations: properties on compaction. Pharm World Sci 1987; 9(5):274–282.
108. Führer C, Nickel E, Thiel F. Elektronenoptische Gefügeuntersuchungen von Tabletten. Acta Pharm Technol 1975; 21:149.
109. Juslin M, Kahela P, Paronen P, et al. Comparative evaluation of starches as tablet adjuvants. Acta Pharm Fenn 1981; 90:83–90.
110. Paronen P, Juslin M. Compressional characteristics of four starches. J Pharm Pharmacol 1984; 35(10):627–635.
111. Bos CE. The development of tablet formulations for use in tropical countries. Meppel, the Netherlands: Krips Repro, 1990:42.
112. Shangraw RF, Wallace JW, Bowers FM. Morphology and functionality in tablet excipients for direct compression: part II. Pharm Technol 1981; 5(10):44–60.
113. Technical Bulletin ST/C/001. Colorcon, Inc., Westpoint, PA, USA.
114. Manudhane KS, Contractor AM, Kim HY, et al. Tableting properties of a directly compressible starch. J Pharm Sci 1969; 58(5):616–620.
115. Sakr AM, Elsabbagh HM, Emara KM. STA-Rx^{R} 1500 Starch: a new vehicle for the direct compression of tablets. Arch Pharm Chem Sci 1974; 2:14–24.
116. Wurster DE, Peck GE, Kildsig DO. A comparison of the moisture adsorption-desorption properties of corn starch, USP, and directly compressible starch. Drug Dev Ind Pharm 1982; 8(3):343–354.
117. Duberg M, Nyström C. Studies on direct compression of tablets. VI. Evaluation of methods for the estimation of particle fragmentation during compaction. Acta Pharm Suec 1982; 19(6):421–436.
118. Humbert-Droz P, Mordier D, Doelker E. Méthode rapide de détermination du comportement à la compression pour des études de préformulation. Pharm Acta Helv 1982; 57(5-6):136–143.
119. Rees JE. Deformation and fracture of pharmaceutical materials during the compaction process. Acta Pharm Suec 1981; 18:68–70.
120. Alderborn G, Nyström C. Studies on direct compression of tablets. IV. The effect of particle size on the mechanical strength of tablets. Acta Pharm Suec 1982; 19(5):381–390.
121. Panaggio A, Rhodes CT, Schwartz JB. Properties of mixtures of two tablet matrices. Pharm Acta Helv 1984; 59(2):37–39.
122. Michaud JLM, Provoost DR, Van Bogaert E. Free-flowable, directly compressible starch as binder, disintegrant and filler for compression tablets and hard gelatin capsules. Patent number EP0933079, 1999.
123. Van Bogaert E. Direct compressible starch. A high quality multi-functional excipient. Information Sheet Cerestar, Vilvoorde (Belgium), 2000.
124. Michaud J. Starch-based materials for direct compression. Pharma Chem 2002; 1(6):42–47.
125. Bos CE, Bolhuis GK, Lerk CF, et al. Evaluation of modified rice starch, a new excipient for direct compression. Drug Dev Ind Pharm 1992; 18(1):93–106.

126. Mitrevej A, Sinchaipanid N, Faroongsarng D. Spray-dried rice starch: comparative evaluation of direct compression fillers. Drug Dev Ind Pharm 1996; 22(7):587–594.
127. Mitrevej A, Faroongsarng D, Sinchaipanid N. Compression behavior of spray dried rice starch. Int J Pharm 1996; 140(1):61–68.
128. Korhonen O, Raatikainen P, Harjunen P, et al. Starch acetates—multifunctional direct compression excipients. Pharm Res 2000; 17(9):1138–1143.
129. Mollan MJ, Çelik M. Characterization of directly compressible maltodextrins manufactured by three different processes. Drug Dev Ind Pharm 1993; 19(17–18):2335–2358.
130. Mollan MJ, Çelik M. The effects of lubrication on the compaction and post-compaction properties of directly compressible maltodextrins. Int J Pharm 1996; 144(1):1–9.
131. Mollan MJ, Çelik M. The effects of humidity and storage time on the behavior of maltodextrins for direct compression. Int J Pharm 1995; 114(1):23–32.
132. Papadomitriou E, Efentakis E, Choulis NH. Evaluation of maltodextrins as excipients for direct compression tablets and their influence on the rate of dissolution. Int J Pharm 1992; 86(2–3):131–136.
133. Fischer E. Calcium phosphate excipients in the pharmaceutical industry. Proceedings of the CPhI conference and exhibition, Milan, 1991.
134. Carstensen JT, Ertell C. Physical and chemical properties of calcium phosphates for solid state pharmaceutical formulations. Drug Dev Ind Pharm 1990; 16(7):1121–1133.
135. Herzog R. Calcium phosphate in der Tablettierung, 1991.
136. Shah MN, Carrol MA, Miller LG. An evaluation of alternate sources of directly compressible tablet excipients and acetaminophen. Pharm Technol 1983; 7(2):45–60.
137. Chilamkurti RN, Rhodes CT, Schwartz JB. Some studies on compression properties of tablet matrices using a computerized instrumented press. Drug Dev Ind Pharm 1982; 8(1):63–86.
138. Gillard J, Delattre L, Jaminet F, et al. Étude des paramètres influençant l'écoulement des agents diluants utilisés en compression directe. J Pharm Belg 1972; 27(6):713–742.
139. Khan KA, Rhodes CT. Effect of compaction pressure on the dissolution efficiency of some direct compression systems. Pharm Acta Helv 1972; 47(10):594–607.
140. Alderborn G, Pasanen K, Nyström C. Studies on direct compression of tablets. XI Characterization of particle fragmentation during compaction by permeametry measurements of tablets. Int J Pharm 1985; 23(1):79–86.
141. Duberg M, Nyström C. Studies on direct compression of tablets XVII. Porosity–pressure curves for the characterization of volume reduction mechanisms in powder compression. Powder Technol 1986; 46(1):67–75.
142. Duberg M, Nyström C. Studies on direct compression of tablets. XII. The consolidation and bonding properties of some pharmaceutical compounds and their mixtures with Avicel PH 105. Int J Pharm Tech Prod Manuf 1985; 6:17–25.
143. Schmidt PC, Tenter U. Presskraft und Weg Zeit Charakteristik von Rundlauftablettenpressen. 5 Mitt.: Messung und Auswertung von Ausstosskraften. Pharm Ind 1989; 51(2):183–187.
144. Gillard J, François A, Hermans Y, et al. Influence d'agents désintégrants sur la texture de comprimés a base de monohydrogénophosphates calciques de granulometries homogènes. J Pharm Belg 1973; 28(4):395–409.
145. Khan KA, Rhodes CT. Further studies of the effect of compaction pressure on the dissolution efficiency of direct compression systems. Pharm Acta Helv 1974; 49(7–8):258–261.
146. Wong LW, Pilpel N, Ingham S. Effect of particle shape on the compression of powders. J Pharm Pharmacol 1988; 40(suppl):69P–69P.
147. Takami K, Machimura H, Takado K, et al. Novel preparation of free flowing spherically granulated dibasic calcium phosphate anhydrous for direct tabletting. Chem Pharm Bull 1996; 44(4):868–870.
148. Schlack H, Bauer-Brandl A, Schubert R, et al. Properties of Fujicalin, a new modified anhydrous dibasic calcium phosphate for direct compression: comparison with dicalcium phosphate dihydrate. Drug Dev Ind Pharm 2001; 27(8):789–801.
149. Lerk CF, Schoonen AJM, Fell JT. Contact angles and wetting of pharmaceutical powders. J Pharm Sci 1976; 65(6):843–847.
150. Caramella C, Colombo P, Conte U, et al. Water uptake and disintegrating force measurements: towards a general understanding of disintegration mechanisms. Drug Dev Ind Pharm 1986; 12(11–11):1749–1766.
151. Chowhan ZT. The effect of low- and high-humidity ageing on the hardness, disintegration time and dissolution rate of dibasic calcium phosphate-based tablets. J Pharm Pharmacol 1980; 32(1):10–14.
152. Delattre L, Jaminet F. Étude de quelques facteurs influençant le pouvoir liant d'excipients pour compression directe. Pharm Acta Helv 1974; 49(3–4):108–116.

153. Horhota ST, Burgio J, Lonski L, et al. Effect of storage at specified temperature and humidity on properties of three directly compressible tablet formulations. J Pharm Sci 1976; 65(12):1746–1749.
154. Lausier JM, Chiang C-W, Zompa HA, et al. Aging of tablets made with dibasic calcium phosphate dihydrate as matrix. J Pharm Sci 1977; 66(11):1636–1637.
155. Carstensen JT. Effect of moisture on the stability of solid dosage forms. Drug Dev Ind Pharm 1988; 14(6):1927–1969.
156. De Haan P, Kroon C, Sam AP. Decomposition and stabilization of the tablet excipient calcium hydrogenphosphate dihydrate. Drug Dev Ind Pharm 1990; 16(13):2031–2055.
157. Shah DH, Arambulo AS. A study of the chemical and physical stability of ascorbic acid, folic acid, and thiamine hydrochloride tablets formulated with Emcompress Standard®. Drug Dev Comm 1975; 1(6):495–505.
158. Mielck JB, Rabach H. Einfluss von Tablettier Hilfsstoffen auf die Hydrolyse von Natrium Acetylsalizylat bei erhöhten Temperaturen und geringen Feuchten. Pharm Technol 1984; 30(1): 33–43.
159. Patel NK, Patel BR, Plakogiannis FM, et al. An evaluation of tricalcium phosphate excipients particularly using instrumented rotary and single station tablet presses. Drug Dev Ind Pharm 1987; 13(15):2693–2718.
160. Hou XP, Carstensen JT. Compressibility of anhydrous tricalcium phosphate. J Pharm Sci 1985; 74(4):466–468.
161. Hou XP, Carstensen JT. Compression characteristics of basic tricalcium phosphate ($Ca_3(PO_4)_2$, $Ca(OH)_2$). Int J Pharm 1985; 25(2):207–215.
162. Carstensen JT, Hou X-P. The Athy-Heckel equation applied to granular agglomerates of basic tricalcium phosphate [$3Ca_3PO_4)_2 \cdot Ca(OH)_2$]. Powder Technol 1985; 42(2):153–157.
163. Pontier C, Viana M, Champion E, et al. Apatitic calcium phosphates used in compression: rationalization of the end-use properties. Powder Technol 2003; 130(3):436–441.
164. Chowhan ZT, Amaro AA. The effect of low- and high-humidity aging on the hardness, disintegration time and dissolution rate of tribasic calcium phosphate-based tablets. Drug Dev Ind Pharm 1979; 5(6):545–562.
165. Elsabbagh HM, Elshaboury MH, Abdel-Aleem HM. Physical properties and stability of diazepam and phenobarbitone sodium tablets prepared with compactrol. Drug Dev Ind Pharm 1985; 11(11):1947–1955.
166. Bolhuis GK, Eissens AC, Zoestbergen E. DC Calcium lactate, a new filler-binder for direct compaction of tablets. Int J Pharm 2001; 221(1–1):77–86.
167. Augsburger L, Madieh S, Mehra DK, et al. The role of calcium silicate in developing directly compressible "Fast-Melt" tablets. In: AAPS Annual Meeting; 2003; Salt Lake City, 2003.
168. RxCIPIENTS FM1000 Calcium silicate Application bulletins 1, 2 and 3, Huber Engineered Materials, 2003.
169. Bolhuis GK, Rexwinkel EG, Zuurman K. Polyols as filler-binders for disintegrating tablets prepared by direct compaction. Drug Dev Ind Pharm 2009; 35(6):671–677.
170. Rukiah M, Lefebvre J, Hernandez O, et al. Ab initio structure determination of the C form of D-sorbitol (D-glucitol) by powder synchrotron X-ray diffraction. J Appl Crystallogr 2004; 37(5):766–772.
171. Du Ross JW. Modification of the crystalline structure of sorbitol and its effect on tableting characteristics. Pharm Technol 1984; 8(9):42–53.
172. Guyot-Hermann AM, Draguet-Brughmans M. Gamma sorbitol as a diluent in tablets. Drug Dev Ind Pharm 1985; 11(2–3):551–564.
173. Schmidt PC. Tableting characteristics of sorbitol. Pharm Technol 1983; 7(11):65–74.
174. Schmidt PC, Vortisch W. Einfluss der Herstellungsart von Füll und Bindemitteln auf ihre Tablettierfähigkeit. Pharm Ind 1987; 49(5):495–503.
175. Deurloo MJM, Peeters JPJM, Bolhuis GK, et al. Evaluation of direct compression excipients on a production scale tabletting press. Pharm Weekbl Sci 1992; 14(suppl):F37–F37.
176. Schmidt PC, Benke K. "Übersättigte" geordnete Mischungen auf der Basis von Sorbit. Pharm Ind 1984; 46(2):193–198.
177. Nikolakakis I, Newton JM. Solid state adsorption of antibiotics onto sorbitol. J Pharm Phamacol 1989; 41(3):145–148.
178. Basedow AM, Möschl GA, Schmidt PC. Sorbitol instant an excipient with unique tableting properties. Drug Dev Ind Pharm 1986; 12(11–13):2061–2089.
179. Davis JD. Alteration of hardness gain in sorbitol based tablets. Drug Cosmet Ind 1981; 128(1):38–45.
180. Schmidt PC. Zur Tablettierung von Ascorbinsäure/Sorbit-Mischungen. Pharm Technol 1984; 30(4):302–311.

181. Darrouzet H. Mannitol granular (MG, GG2) and Neosorb powder DC: pharmaceutical excipients for direct compression and solid dosage forms. Proc of the Pharma Tagung, ZDS-Fachschule der Deutschen Susswarenindustrie, Solingen, 1985.
182. Burger A, Henck JO, Hetz S, et al. Energy/temperature diagram and compression behavior of the polymorphs of D-mannitol. J Pharm Sci 2000; 89(4):457–468.
183. Debord B, Lefebvre C, Guyot-Hermann AM, et al. Study of different crystalline forms of mannitol: comparative behaviour under compression. Drug Dev Ind Pharm 1987; 13(9):1533–1546.
184. Tabandeh H, Zolfaghari ME, Bagheri D. The effect of compression pressure on hardness value in Avicel, lactose and mannitol tablets. J Pharm Pharmacol 1998; 50(suppl):192.
185. Armstrong NA. Direct compression characteristics of granulated lactitol. Pharm Technol Eur 1998; 9(2):42–46.
186. Garr JSM, Rubinstein MH. Direct compression characteristics of xylitol. Int J Pharm 1990; 64(2–3):223–226.
187. Morris LE, Moore JC, Schwartz JB. Characterization and performance of a new direct excipient for chewable tablets: xylitab. Drug Dev Ind Pharm 1996; 22(9–10):925–932.
188. Ndindayino F, Henrist D, Kiekens F, et al. Characterization and evaluation of isomalt performance in direct compression. Int J Pharm 1999; 189(1):113–124.
189. Ndindayino F, Henrist D, Kiekens F, et al. Direct compression properties of melt-extruded isomalt. Int J Pharm 2002; 235(1–2):149–157.
190. Rowe RC, Sheskey PJ, Owen SC. Handbook of Pharmaceutical Excipients. 5th ed. London: Pharmaceutical Press, 2006.
191. Bolhuis GK, Engelhart JJP, Eissens AC. Compaction properties of isomalt. Eur J Pharm Biopharm, 2009; 72(3):621–625.
192. Babbel MB, Fritzsching B. Fast dissolving disintegrating tablets with isomalt. Pharm Technol 2009; 21(2):25–28.
193. Hardman JS, Lilley BA. Deformation of particles during briquetting. Nature 1970; 228(5269):353–355.
194. Hersey JA, Cole ET, Rees JE. Powder Consolidation During Compaction. In: Proceedings of the first International Conference on the compaction and consolidation of particulate matter. Powder Technology Publication Series. Vol. 4. London: The Powder Advisory Centre, 1973:165.
195. York P. Particle slippage and rearrangement during compression of pharmaceutical powders. J Pharm Pharmacol 1978; 30(1):6–10.
196. Cole ET, Rees JE, Hersey JA. Relations between compaction data for some crystalline pharmaceutical materials. Pharm Acta Helv 1975; 50(1–2):28–32.
197. Hersey JA, Rees JE. Deformation of particles during briquetting. Nature Phys Sci 1971; 230(12):96–97.
198. Chowhan ZT, Chow YP. Compression behavior of pharmaceutical powders. Int J Pharm 1980; 5(2):139–148.
199. Nyström C, Mazur J, Sjögren J. Studies on direct compression of tablets II. The influence of the particle size of a dry binder on the mechanical strength of tablets. Int J Pharm 1982; 10(3):209–218.
200. De Boer AH, Vromans H, Lerk CF, et al. Studies on tableting properties of lactose. Part III. The consolidation behaviour of sieve fractions of crystalline α-lactose monohydrate. Pharm Weekbl Sci 1986; 8:145–150.
201. Leuenberger H, Bonny JD, Lerk CF, et al. Relation between crushing strength and internal specific surface area of lactose compacts. Int J Pharm 1989; 52(2):91–100.
202. Vromans H, Bolhuis GK, Lerk CF, et al. Studies of tableting properties of lactose. IX. The relationship between particle structure and compactibility of crystalline lactose. Int J Pharm 1987; 39(3):207–212.
203. Fell JT, Newton JM. The tensile strength of lactose tablets. J Pharm Phamacol 1968; 20:657–659.
204. Vromans H, De Boer AH, Bolhuis GK, et al. Studies on tabletting properties of lactose. Part I: the effect of initial particle size on binding properties and dehydration characteristics of lactose. Acta Pharm Suec 1985; 22:163–172.
205. Fell JT, Newton JM. Effect of particle size and speed of compaction on density changes in tablets of crystalline and spray-dried lactose. J Pharm Sci 1971; 60(12):1866–1869.
206. Alderborn G, Nyström C. Studies on direct compression of tablets XIV. The effect of powder fineness on the relation between tablet permeametry surface area and compaction pressure. Powder Technol 1985; 44(1):37–42.
207. Riepma KA, Dekker BG, Lerk CF. The effect of moisture sorption on the strength and internal surface area of lactose tablets. Int J Pharm 1992; 87(1–3):149–159.
208. Van Kamp HV, Bolhuis GK, Lerk CF. Studies on tableting properties of lactose. V. Effects of both lubrication and addition of disintegrants on properties of tablets prepared from different types of crystalline lactose. Acta Pharm Suec 1986; 23(4):217–230.

209. Van Kamp HV, Bolhuis GK, Lerk CF. Optimization of a formulation based on lactoses for direct compression. Acta Pharm Technol 1988; 34(1):11–16.
210. Wong DYT, Wright P, Aulton ME. The deformation of alpha-lactose monohydrate and anhydrous alpha-lactose monocrystals. Drug Dev Ind Pharm 1988; 14(15–17):2109–2126.
211. Van Kamp HV, Bolhuis GK, Kussendrager KD, et al. Studies on tableting properties of lactose. IV. Dissolution and disintegration properties of different types of crystalline lactose. Int J Pharm 1986; 28 (2–3):229–238.
212. Shangraw RF, Wallace JT, Bowers FM. Morphology and functionality in tablet excipients for direct compression: part I. Pharm Technol 1981; 5(9):69–78.
213. Batuyios NH. Anhydrous lactose in direct tablet compression. J Pharm Sci 1966; 55(7):727–730.
214. Nicklasson M, Nyqvist H. Studies on lactose qualities for direct compression. Int J Pharm Tech Prod Manuf 1982; 3(4):115.
215. Whiteman M, Yarwood RJ. The evaluation of six lactose-based materials as direct compression tablet excipients. Drug Dev Ind Pharm 1988; 14(8):1023–1040.
216. Technical Bulletin DMV, Veghel, the Netherlands.
217. Brownley CA, Lachman L. Browning of spray-processed lactose. J Pharm Sci 1964; 53(4):452–454.
218. Castello RA, Mattocks AM. Discoloration of tablets containing amines and lactose. J Pharm Sci 1962; 51(2):106–108.
219. Kussendrager K, De Hoog P, van Leverink J. Some physical properties of spray-dried lactose with respect to stability and compression. Acta Pharm Suec 1981; 18(2):94–95.
220. Buckton G, Chidavaenzi O, Koosha F. The effect of spray-drying on the physical form of lactose. AAPS PharmSciTech 2002; 3(4):1–1.
221. Bolhuis GK, Kussendrager KD, Langridge J. New developments in spray-dried lactose. Pharm Technol 2004, Suppl. Excipients and Solid Dosage Forms, 26–31.
222. Alpar O, Hersey JA, Shoton E. The compression properties of lactose. J Pharm Pharmacol 1970; 22(suppl):1S–7S.
223. Fell JT. The flow and compaction properties of lactose. Pharm Weekbl 1976; 111:681–685.
224. Shangraw RF. Raw materials of natural origin in solid dosage forms. Drug Cosmet Ind 1978; 122(6):68.
225. Roberts RJ, Rowe RC. Brittle/ductile behaviour in pharmaceutical materials used in tabletting. Int J Pharm 1987; 36(2–3):205–209.
226. Vromans H, Bolhuis GK, Lerk CF, et al. Studies on tabletting properties of lactose. VI. Consolidation and compaction of spray dried amorphous lactose. Acta Pharm Suec 1986; 23:231–240.
227. Rassu G, Eissens AC, Bolhuis GK. Tableting properties of improved spray-dried lactose. J Drug Del Sci Technol 2006; 16(6):455–459.
228. Kussendrager KD, Van den Biggelaar HA. Method for producing a crystalline tableting additive, additive thus obtained and use thereof. Patent number WO0208470, 2002.
229. Nürnberg E, Ritsert S. Möglichkeiten der Verpressung von Pulvergemischen und Granulaten fur niedrigdosierte Tabletten auf der PT 3090 IC. Technische Nachrichten Fette 1992, 92.1(1):1–12.
230. Bolhuis GK, Zuurman K. Tabletting properties of experimental and commercially available lactose granulations for direct compression. Drug Dev Ind Pharm 1995; 21(18):2057–2071.
231. Hardman JS. Deformation of particles during briquetting. Nature 1970; 228(October 24):353–355.
232. Hardman JS, Lilley BA. The influence of shear strain on the compaction of powders. In: Proceedings of the first international conference on the compaction and consolidation of particulate matter. Powder Technology Publication Series. Vol. 4. London: The Powder Advisory Centre, 1973:115.
233. Mendes RW, Gupta MR, Katz IA, et al. Nu-Tab as a chewable direct compression carrier. Drug Cosmet Ind 1974; 115(6):42–46.
234. Rizzuto AB, Chen AC, Veiga ME. Crystal structure to enhance pharmaceutical properties of excipient and drug substances. Pharm Technol 1984; 8(9):32–39.
235. Bowe KE. Recent advantages in sugar-based excipients. Pharm Sci Technol Today 1998; 1(4):166–173.
236. Armstrong NA, Patel A, Jones TM. The compressional properties of dextrose monohydrate and anhydrous dextrose of varying water contents. Drug Dev Ind Pharm 1986; 12(11–13):1885–1901.
237. Bowe KE, Billig JL, Schwartz JB. Crystalline maltose. A direct compression pharmaceutical excipient. Pharm Technol Eur 1998; 10(5):34–40.
238. Ridgway K, Lazarou C, Thorpe EE. The properties of tablets manufactured on an automatically-controlled rotary machine. J Pharm Pharmacol 1972; 24(4):265–271.
239. Hinrichs WLJ, Prinsen MG, Frijlink HW. Inulin glasses for the stabilization of therapeutic proteins. Int J Pharm 2001; 215(1–2):163–196.

240. Eissens AC, Bolhuis GK, Hinrichs WLJ, et al. Inulin as filler-binder for tablets, prepared by direct compaction. Eur J Pharm Sci 2002; 15(3):31–38.
241. Bolhuis GK, Eissens AC, Adrichem TP, et al. Hollow filler-binders as excipients for direct compaction. Pharm Res 2003; 20(3):515–518.
242. York P. Crystal engineering and particle design for the powder compaction process. Drug Dev Ind Pharm 1992; 18(6–7):677–721.
243. Schmidt PC, Rubensdörfer CJW. Evaluation of Ludipress as a 'multipurpose excipient' for direct compression. Part I: powder characteristics and tableting properties. Drug Dev Ind Pharm 1994; 20(18):2899–2925.
244. Baykara T, Duman G, Özsener KS, et al. Comparing the compressibility of ludipress with the other direct tableting agents by using acetaminophen as an active ingredient. Drug Dev Ind Pharm 1991; 17(17):2359–2371.
245. Kolter K, Heinz R, Fussnegger B. Compaction properties of Ludipress LCE—a new direct compression excipient. Ludwigshafen (Germany): BASF, 1999.
246. Kolter K, Fussnegger B. Development of tablet formulations using Ludipress LCE as direct compression excipient. Ludwigshafen (Germany): BASF, 1999.
247. Reimerdes D, Aufmuth KP. Tabletting with coprocessed lactose-cellulose excipients. Manuf Chem 1992; 63(12):21–24.
248. Belda PM, Mielck JB. The tabletting behaviour of Cellactose compared with mixtures of cellulose with lactoses. Eur J Pharm Biopharm 1996; 42(5):325–330.
249. Garr JSM, Rubinstein MH. Compaction properties of a cellulose-lactose direct compression excipient. Pharm Technol Int 1991; 3(1):24–27.
250. Armstrong NA, Roscheisen G, Al-Aghbar MR. Cellactose as a tablet diluent. Man Chem 1996; 67(10):25–26.
251. Arida AI, Al-Tabakha MM. Cellactose a coprocessed excipient: a comparison study. Pharm Dev Technol 2008; 13(2):165–175.
252. Casalderry M, Souto C, Concheiro A, et al. Comparison of cellactose with two ad hoc processed lactose-cellulose blends as direct compression excipients. Chem Pharm Bull 2000; 48(4):458–463.
253. Michoel A, Rambout P, Verhoye A. Comparative evaluation of coprocessed lactose and microcrystalline cellulose with their physical mixtures in the formulation of folic acid tablets. Pharm Dev Technol 2002; 7(1):79–87.
254. Wagner KG, Dressler JA. A corn starch/α-lactose monohydrate compound as a directly compressible excipient. Pharm Technol Eur 2003; 15(3):33–40.
255. Hauschild K, Picker-Freyer KM. Evaluation of a new coprocessed compound based on lactose and maize starch for tablet formulation. AAPS PharmSciTech 2004; 6(2):16–16.
256. Limwong V, Sutanthavibul N, Kulvanich P. Spherical composite particles of rice starch and microcrystalline cellulose: a new coprocessed excipient for direct compression. AAPS PharmSciTech 2004; 5(2):30–30.
257. Freitag F, Runge J, Kleinebudde P. Coprocessing of powdered cellulose and magnesium carbonate: direct tableting versus tableting after roll compaction/dry granulation. Pharm Dev Technol 2005; 10(3):353–362.
258. De Lourdes Garzón Serra M, Villafuerte Robles L. Compactibility of agglomerated mixtures of calcium carbonate and microcrystalline cellulose. Int J Pharm 2003; 258(1–2):153–163.
259. Technical Information Ludiflash: BASF, 2007.
260. Kruse S, Gebert S, Meyer-Böhm K, et al. Compression characteristics and lubricant sensitivity of orally disintegrating tablets based on Ludiflash. In 6th World Meeting on Pharmaceutics, Biopharmaceutics and Pharmaceutical Technology. Barcelona, Spain, 2008.

9 Lubrication issues in direct compaction

Gerad K. Bolhuis and Arne W. Hölzer

INTRODUCTION

Lubricants are commonly included in tablet formulations to reduce die wall friction during both compaction and ejection of the tablet. Their presence, however, may cause undesirable changes in tablet properties. Shotton and Lewis (1) investigated the effect of magnesium stearate on the diametrical breaking strength of tablets of crystalline materials and of two simple granulations without binders. They found that the lubricant decreased the strength of all tablets, but at most for the crystalline materials. The softening of tablets by lubricants has been previously reported by Strickland et al. (2), who observed that magnesium stearate and other lubricants, added as dry powder to granules, appeared to adhere to and form a coat around individual granules.

EFFECT OF LUBRICANTS ON TABLET STRENGTH

Bolhuis et al. (3) showed that magnesium stearate forms an adsorbed lubricant film around host particles during the mixing process (Fig. 1). This lubricant film interferes with the bonding properties of the host particles by acting as a physical barrier. This can be seen when the tablet breaking force or tensile strength is plotted as a function of the logarithm of the mixing time with the lubricant. Figure 2 shows that the effect depends strongly on the material used (4). For microcrystalline cellulose (Avicel PH-102), powdered cellulose (Elcema G250), compressible starch (STARCH 1500) and calcium sulfate dihydrate (Compactrol) a strong decrease in breaking force can be seen with an increase in mixing time with the lubricant. On the other hand, the bonding properties of the two kinds of dicalcium phosphate dihydrate are not affected by mixing with magnesium stearate.

In addition to the decreased bonding properties, the addition of hydrophobic lubricants causes increased disintegration times and decreased dissolution rates (1,5–11).

The magnitude of the effect of lubricants on tablet strength is dependent on a large number of factors, for example, the nature and properties of the lubricant, the nature and properties of the other tablet ingredients, and the processing conditions. These factors affect the formation of a lubricant film during the mixing process and/or the adventures of the film during compaction, and these are discussed in the following sections.

FILM FORMATION OF LUBRICANTS DURING MIXING

When a powdered lubricant is added to a tabletting blend and submitted to a mixing action it is distributed either as a free fraction or, when the lubricant is prone to deagglomeration and a following delamination, as a surface film on the base/carrier material. Prolonged mixing time will transfer more lubricant from the free fraction to the surface film. The phenomenon of decreased tablet strength with an increase in mixing time of tablet ingredients with lubricants is caused by the formation of this lubricant film, which interferes with the binding of the particles (3). The decrease in tablet strength has been attributed to weaker bonds after compression between lubricant-lubricant molecules rather than strong excipient-excipient bonds (12).

It has been demonstrated by means of X-ray examination that magnesium stearate can be extended by means of a slight pressure to a continuous film, composed of parallel-oriented molecules (13). Apparently, during a common mixing process, the shear forces are sufficient to

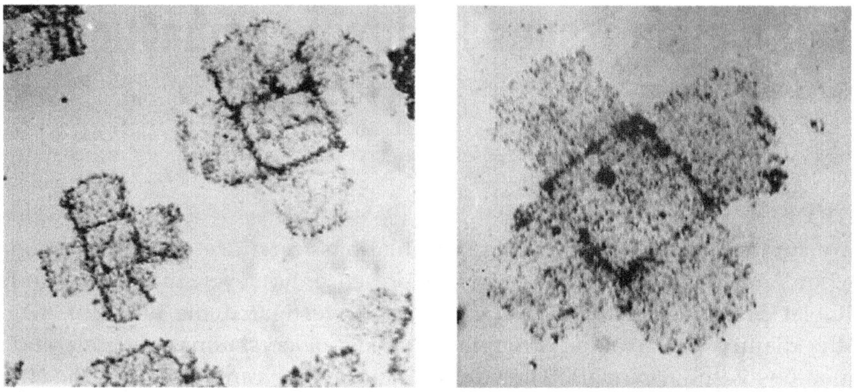

Figure 1 Unfolded magnesium stearate envelopes on water, isolated from sodium chloride crystals, after mixing with magnesium stearate (approximately 30x and 60x). *Source*: From Ref. 3.

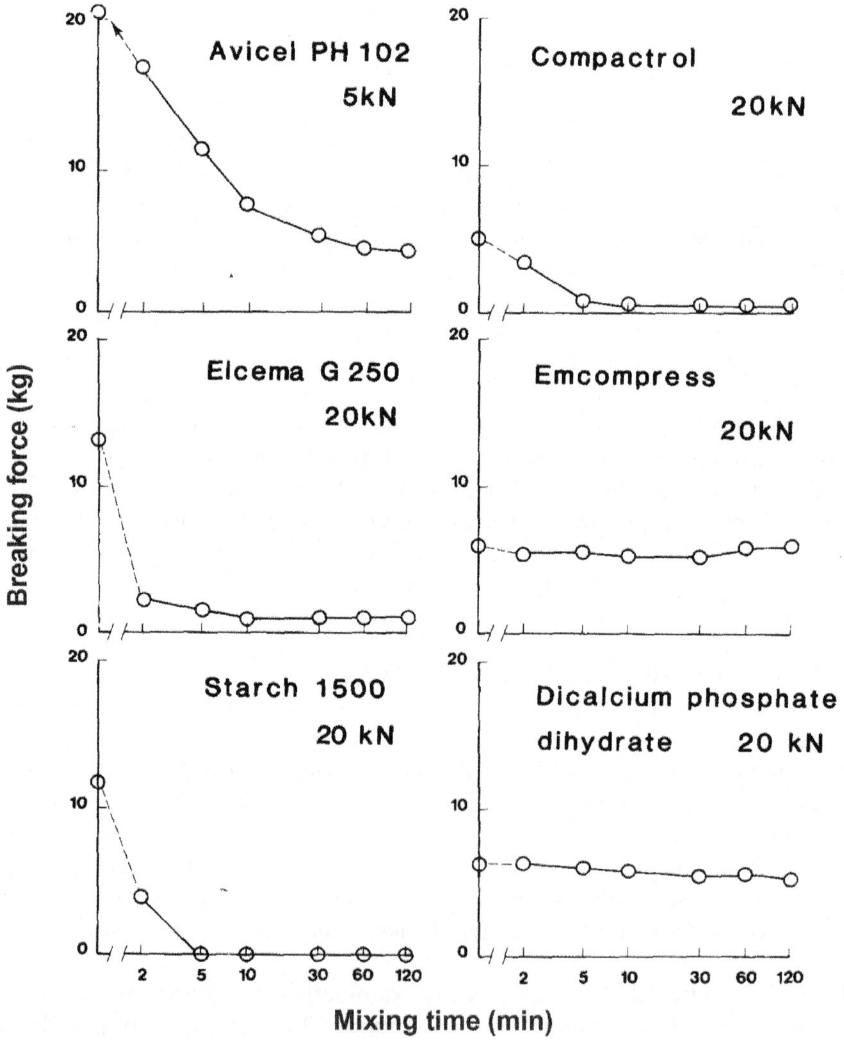

Figure 2 Breaking Force versus lubricant mixing time (log scale) for tablets compressed from filler-binders with 0.5% magnesium stearate. *Source*: From Ref. 4.

Table 1 Percentage of Magnesium Stearate (Specific Surface Area by Weight 15.6 m^2/g), Theoretically Required to Cover a Particulate Solid with a Monoparticulate and with a Monomolecular Layer, Respectively

		Percentage of magnesium stearate		
		Monoparticulate layer		Monomolecular layer
Particulate solid	Specific surface A_w (m^2/g)	W_{MP}(spheric)	W_{MP}(flat)	W_{MM}
Amylose V	0.097	1.9	0.9	0.025
Crystalline sodium chloride	0.010	0.2	0.1	0.003
Powdered sodium chloride	0.115	2.2	1.1	0.03

Source: From Ref. 3.

shear off molecular layers from magnesium stearate particles and form a film on the host particles. It has been postulated that film formation is developed in accordance with the Langmuir pattern and can be regarded as an ordered mixing process, (6,14–16).

In an attempt to assess film formation and understand the causes of deleterious effects of magnesium stearate, a number of investigations have concentrated on studying the lubricant distribution on host particles and conflicting statements are reported regarding the nature of the lubricant film.

Bolhuis et al. (3) calculated the quantity of magnesium stearate needed to cover different host particles with a monoparticulate and a monomolecular layer, respectively, of the lubricant. For amylose, a monoparticulate layer of magnesium stearate will be formed when about 1% to 2% magnesium stearate is present. For a monomolecular layer, only about 0.05% magnesium stearate was calculated to be sufficient. Tablets of amylose admixed with magnesium stearate with a concentration lower than needed for a monoparticulate layer but larger than for a monomolecular layer gave zero tablet strength (Table 1). It was concluded that magnesium stearate forms a monomolecular lubricant film around the amylose particles.

This hypothesis was proved by mixing amylose with a magnesium stearate *tablet*. The strength of tablets compressed from the blends, where samples were taken at different mixing times, showed that zero tablet strength was attained after a weight loss of the magnesium stearate tablet, corresponding to a concentration in the amylose tablets of 0.04%. This concentration was found to be of the same order of magnitude as that required to coat the host particles with a monomolecular layer.

In a theoretical consideration, Shah and Mlodozeniec (6) suggested that the degree and extent of surface coverage of a substrate particle by lubricants can be described by at least three different mechanisms: (*i*) adsorption or surface contact adhesions; (*ii*) diffusion or solid penetration, which includes mechanical interlocking; and (*iii*) delamination or deagglomeration of the lubricating agent to form a film coating (usually discontinuous) on the substrate particles. According to the mechanism of boundary lubrication (2), solid lubricants such as magnesium stearate are adsorbed on the surface of substrate particles and form a uniform surface-adsorbed film in a manner similar to a Langmuir-type adsorption (6). The authors assumed that magnesium stearate particles first adsorb on the surface of host particles, followed by a uniform distribution over the substrate surface as an effect of delamination or deagglomeration of magnesium stearate particles resulting from shear effects upon continued mixing. Diffusion or solids penetration appeared to play a minor role in lubricant spreading during mixing. A continuous coating was thought not to occur during mixing, but only during the compression process (6).

Until now, it was not possible to see lubricant films on host particles by means of scanning electron microscopy (SEM), but isolated films could be detected by transmission electron microscopy (3). Several attempts have been made to detect lubricant films by chemical or physical methods for a better understanding of the nature of these films; most of these investigations have relied, however, on indirect techniques.

Lerk and Bolhuis (17) measured contact angles of sodium chloride/magnesium stearate blends. For binary mixtures of small particle sized powders a linear relationship exists between the cosine of the contact angle of the mixed system and the proportion of the components. On this basis, the addition of small amounts of magnesium stearate with a contact angle of 121° to crystalline sodium chloride with a contact angle of 28° would be expected to increase the contact angle of the blend by at most a few degrees. The observed increase of the contact angle of sodium chloride up to 121°C after mixing with 1% magnesium stearate indicates the existence of a lubricant film (17).

Another indirect technique to determine surface coverage with lubricant is the measurement of the dissolution rate of host particles. Based on the idea that a hydrophobic lubricant decreases the effective drug/solvent interfacial area and thereby decreases the release of the drug, Nicklasson and Brodin (18) calculated the effective surface area coated by lubricants by comparing the initial rates of release of drugs from nondisintegrating drug/ lubricant disks. The observed data were explained by comparing the results with SEM pictures. The areas were found to be dependent on the type and amount of lubricant used and the size of the host particles. The influence of mixing time on the dissolution properties was found to be small: one minute seemed to be enough to distribute the lubricant uniformly. Using 5% magnesium stearate or sodium stearyl fumarate (Pruv), up to 60% of the disk area was coated; with 5% of a 1:1 mixture of talc and stearic acid, the coated area was about 20%. It must be emphasized, however, that film formation of magnesium stearate on the surface of a tablet is not directly related to the film formation of the lubricant mixing stage since the tablet surface film can be influenced by the spreading out of the lubricant at the ejection of the tablet, the affinity of the lubricant to the tablet surface and the die wall, respectively, and the migration of magnesium stearate in the tablet mass during compaction (19). The film formation of magnesium stearate on the surface of acetylsalicylic acid particles has been investigated by applying a flow-through dissolution technique (15). The film formation on the surface of the host particles was increased by increasing mixing time. Dependent on the specific surface area of the magnesium stearate quality used, a maximum coverage of about 70% was obtained after mixing between 10 and 30 minutes.

Rowe (20) determined the extent of the surface coverage of films of magnesium stearate and calcium stearate, respectively, on tablet surfaces with a film-coating/tablet adhesion technique. The results were very similar to those obtained by the rotating disk technique (18), although the substrate materials were widely different.

While most information has accumulated on the physical characteristics of lubricant films, only a few studies have focused on the chemical structure of the in situ film, primarily using elemental microanalysis techniques.

A direct technique to examine visually the distribution of magnesium stearate on host particles is possible by SEM in combination with energy-dispersive X-ray microanalysis (EDAX). Using this technique, Lerk and Bolhuis could not make X-ray maps for magnesium because the low concentration of magnesium produced a Mg-K peak that was drowned in the background radiation (17). More successful were attempts with higher magnesium stearate concentrations where the formation of a magnesium stearate film on a substrate by EDAX analysis has been confirmed visually (21). The thickness of the film was found to be variable and even after 60-minute mixing time magnesium stearate crystals were found in the blend.

More recently, however, Roblot-Treupel and Puisieux (22) confirmed the preferential location of magnesium stearate in the cavities of host particles and the regularization of the surface provided by the lubricant. Lerk and Sucker (23,24) found likewise that part of magnesium stearate was trapped in cavities of the granules.

Hussain et al. (16) estimated the percentage surface coverage of sodium chloride by different magnesium stearate concentrations from the EDAX data. Film formation by lubricants from different manufacturers was examined, and results suggested similarity in mechanism but different degree of host surface coverage for equivalent mixing conditions. Depending on the magnesium stearate used, the infinite surface coverage varied between 4.5% and 15%. From the fact that the EDAX data fitted the Langmuir type of adsorption isotherm, it was suggested that a molecular film was first formed, which, on further blending, was followed up by a gradual build up of a particulate film. This particulate layer may have been initiated at gross defect points on the host particle surface. A simple and direct correlation between lubricant surface area and surface coverage of the host particles was found. Noteworthy is the difference between the estimated infinite surface coverage determined by EDAX, being at maximum 15% (16), and determined by changes in dissolution rates, 60% (15,18). It was suggested that in the dissolution technique the free fraction and loose agglomerates of the lubricant may also contribute to the estimate of percentage surface coverage of the host (16).

Another direct method for measuring magnesium stearate film formation is secondary ion mass spectrometry (SIMS). From static SIMS spectra (sampling depth 20 Å) of sodium chloride tablets, containing magnesium stearate, it has been concluded that the lubricant film tends to be patch wise in nature rather than a continuous mono- or multimolecular film (25). In another study (26), it has been observed that during mixing sodium chloride with magnesium stearate, lubricant film formation appeared to begin instantly, and the maximum rate of film formation occurred during the first few minutes. During prolonged mixing, film formation continued at a much lower rate.

The effect of magnesium stearate on bonding properties depends strongly on the completeness of the film during the bonding stage of the compaction process. The completeness of the film depends on possibilities and velocity of film formation during mixing and the adventures of the film during compaction and consolidation.

The possibility and velocity of film formation is influenced by

- nature and properties of the lubricant,
- nature and properties of the host particles,
- specific surface area of the host particles in relation to the concentration of the lubricant,
- presence of other additives in the blend,
- mixing time and intensity, and
- type, size, and content of the mixer.

The adventures of a formed lubricant film depend on (*i*) the consolidation and compaction behavior of the host particles and perhaps on (*ii*) storage conditions of lubricated products.

EFFECT OF LUBRICANT PROPERTIES ON FILM FORMATION

Nature of the Lubricant

The effect of tablet lubricants on bonding properties of host particles depends on the nature of the lubricant used. Lerk et al. (27) compared stearic acid, different metal soaps of stearic acid and polytetrafluoroethylene (PTFE). Stearic acid and the stearates decreased the breaking force

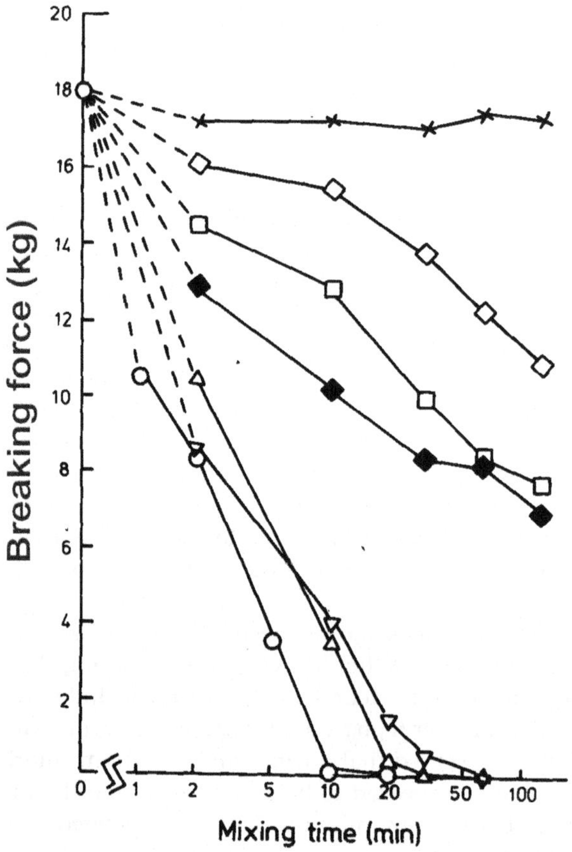

Figure 3 Effect of mixing time on breaking force values of tablets compressed from blends of STARCH 1500 with 0.5% of various lubricants. (x) PTFE; (◇) coarse sodium stearate; (◆) powdered sodium stearate; (□) stearic acid; (▽) aluminium stearate; (△) calcium stearate; (○) magnesium stearate. *Source*: Adapted from Ref. 27.

of STARCH 1500 tablets, but the effect of magnesium, calcium, or aluminium stearate was larger than that of sodium stearate and stearic acid (Fig. 3). It was shown, however, that the results may be influenced by the specific area of the lubricants used. Tablets with the coarse sodium stearate were stronger than tablets with the fine lubricant grade. PTFE did not affect the bonding properties between STARCH 1500 particles, because it has, in contrast to stearic acid and the stearates no laminar structure and hence cannot form a film around the excipient particle. The absence of a lubricant film on sodium chloride particles after intensive mixing with PTFE was confirmed by contact angle measurements (17).

Theoretical calculations of interactions of lubricants and host particles were presented by Rowe (28). Based on a model involving the integrating of the Lennard–Jones pair potential function to predict the properties of a single and a two component system, the intensity of the molecular interactions were predicted for the excipients microcrystalline cellulose and anhydrous lactose and the lubricants magnesium stearate, stearic acid and PTFE, using literature values of their partial solubility parameters. Table 2 shows data for the interaction parameter and strengths of both the adhesive and cohesive interactions within the lubricant for the excipient/lubricant systems studied. For both excipients, the values for the adhesive interactions between excipient and lubricant decrease in the order magnesium stearate > stearic acid > PTFE. Table 2 shows that in the case of magnesium stearate the lubricant-excipient interactions are higher than the lubricant-lubricant interactions. The strong adhesive

Table 2 Interaction Parameters (φ) and Strengths of the Adhesive (A-B) and Cohesive (B-B) Interactions for Microcrystalline Cellulose and Anhydrous Lactose Lubricated with Magnesium Stearate, Stearic Acid, and PTFE

Excipient A	Lubricant B	φ	Strength [adhesive (A-B)]	Interaction (MPa) [cohesive (B-B)]
Microcrystalline cellulose	Magnesium stearate	0.48	85.8	82.8
Microcrystalline cellulose	Stearic acid	0.40	69.2	77.4
Microcrystalline cellulose	PTFE	0.16	22.0	49.0
Anhydrous lactose	Magnesium stearate	0.48	87.1	82.8
Anhydrous lactose	Stearic acid	0.38	65.7	77.4
Anhydrous lactose	PTFE	0.14	19.6	49.0

Source: From Ref. 28.

interactions explain the formation of a monomolecular magnesium stearate film over the excipient. This will decrease the number of the strong cohesive interactions between the excipient particles causing a decrease in tablet strength. For PTFE the adhesive interactions are less than half the cohesive interactions (Table 2), which indicates that this lubricant will tend to concentrate in the interstitial pores between the excipient particles. As an effect, the excipient-excipient interactions and hence the tablet strength will not significantly decrease. Stearic acid will have properties between the two extremes but closer to magnesium stearate than PTFE. This is consistent with studies of the effect of stearic acid and PTFE on tablet strength (27,29–31).

The effect of a decrease in tablet strength during mixing with a lubricant is not limited to stearates and stearic acid. A significant reduction in tablet strengths has been found during mixing of excipients with hydrogenated oils (29,31–33), glycerides (29,32,33), sodium and magnesium lauryl sulfate (32), sodium stearyl fumarate (32,34) and polyethylene glycol 4000 (31,33). On the other hand, just like PTFE, talc and graphite did not affect tablet strength (29). In other work, however, it has been reported that talc may decrease tablet strength (31).

The effect of mixing time with lubricant on both tablet properties and lubricating properties was studied by Hölzer and coworkers, using a large number of lubricants and different test materials (32,35,36). Some results from these papers, completed with unpublished data are shown in the Figure 4. The figure shows the effect of mixing time with 0.1% lubricant on percentage reduction in tensile strength of sodium chloride tablets as compared to the values of unlubricated sodium chloride tablets. Mixing with lubricant was performed in a 2-L Turbula mixer at 42 rpm (20, 200, and 2000 revolutions, respectively); the tablets were compacted at 150 MPa in an eccentric press. Table 3 lists the physical and chemical properties of the lubricants used, ordered by specific surface area. Figure 4 shows that a prolonged mixing time generally increases the reduction in tensile strength but that the magnitude of the effect depends on the lubricant used. Lubricants with a large surface area, for example, the magnesium stearates give a high reduction in tensile strength even when admixing the lubricant as short as 20 revolutions. The triglycerides seem to have a smaller negative effect on tensile strength than the stearates. For two PTFE qualities a different behavior can be seen: Fluon particles are spread out during mixing, which reduces the tensile strength; Teflon PFA has hardly any effect on tablet strength even after prolonged mixing.

Concentration and Specific Surface of the Lubricant

In addition to the nature of origin, the concentration and the specific surface area of the lubricant determine the effect on tablet strength. Using low magnesium stearate concentrations without changing the mixing conditions, film formation will be slower and hence the decrease of the strength of the tablet will be smaller for a given mixing time, as compared to higher lubricant concentrations (3,34). For amylose particles it was shown that the effect of mixing with

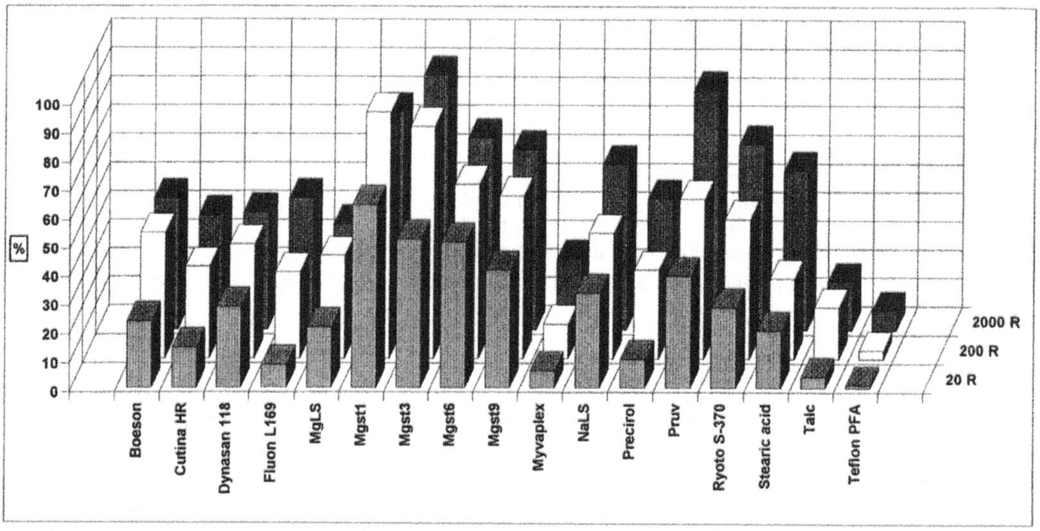

Figure 4 Effect of mixing time on boundary lubrication. Percentage reduction in tensile strength for sodium chloride tablets containing 0.1% lubricant.

Table 3 Physical and Chemical Properties of Lubricants Used in Trials with Sodium Chloride, Ordered by Specific Surface

Lubricant	Description	Area (m^2/g)	Density (g/cm^3)	Melting point (°C)
Mg stearate 1	Salt of fatty acid	6.29	1.13	143
Mg stearate 3	Salt of fatty acid	4.91	1.12	–
Mg stearate 6	Salt of fatty acid	4.52	1.08	134–136
Dynasan 118	Triglyceride	4.46	1.08	71
Talc	Silicate	3.93	3.05	–
Na stearyl fumarate	Salt of fatty acid	2.72	1.14	225
Mg stearate 9	Salt of fatty acid	2.45	1.14	140–154
Fluon L169	Polymer, PTFE	2.06	2.43	>300
Mg lauryl sulfate	Alkyl sulfate	0.88	1.26	241
Na lauryl sulfate	Alkyl sulfate	0.62	1.21	189
Cutina HR	Triglyceride	0.60	1.04	86
Ryoto S-370	Sugar ester	0.54	1.14	71
Boeson VP	Triglyceride	0.34	1.04	>48
Teflon PFA powder	Polymer, PTFE	0.29	2.48	>300
Précirol	Triglyceride	0.26	1.06	63
Myvaplex 600	Triglyceride	0.11	1.02	50–60
Stearic acid	Fatty acid	0.09	1.00	68

magnesium stearate on tablet strength was controlled by the mixing time rather then by the lubricant concentration, providing that the concentration is high enough for the formation of a monomolecular film (3). A similar effect may be expected when the mean particle size of the lubricant particles increases: Mixing excipients with large magnesium stearate particles will slow down the film formation process, as compared to mixing with small lubricant particles. This may be the reason that for a fixed mixing time, the effect of granulated magnesium stearate on tablet strength was found to be lower than that of powdered magnesium stearate (1). Sufficient fine powdered lubricant was transferred to the surface of the base granulate particles to reduce the strength of the bond. This idea was used by Johansson (37,38) who compared magnesium stearate in a granular form with powdered magnesium stearate as a lubricant, with respect to both effect on tablet strength and lubricating properties. Under ideal circumstances, an agglomerate of fine primary lubricant particles should be strong enough to withstand the forces at the lubricant mixing, but during compaction the granules and the released primary particles should be spread out due to the shearing action at the die and punch surfaces. For a fixed mixing time, the granular magnesium stearate showed lubricating properties that are comparable to powdered lubricant without negatively affecting tablet properties when present in concentrations above 1% (37). This effect is supported by the observation that, compared with powdered magnesium stearate, the granular form gives a lower final surface coverage of host particles (15). Increasing the particle size of the magnesium stearate granulation increased the amount of lubricant required to obtain lubrication similar to powdered magnesium stearate (39). The negative effect of the granular lubricant on tablet strength increased however, by a longer mixing time or when the batch size was increased (40). These effects were explained, respectively, by attrition of the lubricant granules after long mixing and by higher shear forces, due to the larger masses involved.

It is often considered that the surface area of the lubricant is the most critical parameter of the material with respect to the deleterious effects on tablet properties. Hölzer and Sjögren (35) found a quantitative relationship between the tensile strength of sodium chloride or lactose tablets and the surface area of sodium stearyl fumarate. A correlation between the lubricant surface area and coverage on the base material after a mixing period of as short as one minute has been reported too (15). For three batches of magnesium stearate differing in morphology, particle size, bulk density, and specific surface area, used in such amounts to develop equivalent lubricating areas, the effect on tablet strength was almost identically (41). From experiments with two different types of magnesium stearate, blended with hydrochlorothiazide as a drug model, Colombo and Carli (42) concluded that on the basis of intrinsic dissolution rate, contact angle, and ejection work data, the crystalline structure of the lubricant has no marked influence on its film-forming properties and on the characteristics of the resulting mixtures. On the other hand, they showed that the fraction of hydrochlorothiazide coated with lubricant was linearly dependent on their specific surface area.

In a study on batch to batch variations of seven commercial magnesium stearates, the differences in effect on tablet properties could be explained to a great extent by differences in surface areas and spreading out the lubricant particles during the short mixing process, both for sodium chloride and anhydrous lactose tablets (36). From these studies (36,41,42), it can be concluded that lubricant surface area seems to be a critical parameter. In another study, however, using different kinds of magnesium stearate with a different specific surface area no direct correlation between the lubricant surface area and surface coverage of sodium chloride particles could be found (16). Although surface area may play a role during the initial mixing stages, on further mixing the delamination propensity for film formation and the inherent material properties of magnesium stearate as well as host powder surface will influence the total percentage surface coverage (16).

Morphology and Crystal Modification of the Lubricant

For magnesium stearate, it is well documented that large batch variations exist with respect to both chemical and physical properties (36). In addition to chemical differences, structural and crystalline characteristics are likely to be important criteria in determining the rate and extent of surface coverage (43–45).

SIMS analysis indicated a profound difference between commercial magnesium stearates and a high purity magnesium stearate with respect to film generation; the commercial samples covered the host particles to a much greater extent than the pure product (26). The poor film formation of the pure product was attributed to its crystal structure and shearing strength; the more crystalline and high purity lubricant seems to be more resistant to shearing during the mixing process than the less crystalline commercial materials (46).

Magnesium stearate consisting of needles shows, in contrast to magnesium stearate consisting of platelets, a poor distribution over host particles (47,48).

EFFECT OF HOST MATERIAL PROPERTIES ON FILM FORMATION

A prerequisite for the formation of a lubricant film on host particles is a distribution of lubricant particles among the host particles. This means that particle size and flow properties of host particles influence the rate of lubricant film formation. It has been found that the effect of mixing time on tablet strength decreased with increasing particle size of microcrystalline cellulose particles (49). This effect was attributed to the larger shear forces in a mixer, created by the larger particles. Moreover, the flow properties of host particles may have a large effect on the rate of lubricant film formation. When the flow properties of the host particles are extremely poor, the distribution of the lubricant particles and the consequent formation of a lubricant film during mixing will be a very slow process. Bos et al. (50) compared rice starch with extremely bad flow properties with potato starch which has reasonable flow properties. It can be seen from Table 4 that the breaking force values of rice starch tablets were, in contrast to potato starch tablets, not affected by mixing for 30 minutes with 0.5% magnesium stearate. In this study strength reduction, defined as the lubricant sensitivity ratio (LSR) was used as a quantitative measure to express the sensitivity to mixing with a lubricant of tabletting materials. The LSR is the ratio between the decrease in breaking force values of tablets, due to mixing with a lubricant and the breaking force values of unlubricated tablets.

$$\text{LSR} = \frac{\text{BF}_\text{u} - \text{BF}_\text{l}}{\text{BF}_\text{u}}$$

where BF_u and BF_l are the breaking force values of tablets prepared without and with a lubricant, respectively. Prolonged mixing of rice starch for several hours with the lubricant caused a marked decrease of the lubricant sensitivity ratio. When the rice starch was granulated (Table 4), the improved flow properties resulted in an increased lubricant sensitivity.

In a later study, Vromans et al. (51) showed that for different types of lactose, the sensitivity to lubrication was related to the bulk density of the powder. Although bulk density is a secondary parameter, depending on fundamental properties such as true density, particle

Table 4 Flow Properties, Tablet Strength, and Magnesium Stearate Susceptibility of Different Starches and Starch Granulations

Starch (granulation)	Hausner ratio[a]	BF_u (N)	BF_l (N)	LSR
Potato starch	1.20	68 ± 10	0	1.0
Tapioca starch	1.33	70 ± 4	39 ± 4	0.4
Maize starch	1.38	56 ± 8	30 ± 7	0.5
Rice starch	1.38	102 ± 20	104 ± 11	0.0
Potato starch granulation	1.18	47 ± 3	<10	>0.8
Tapioca starch granulation	1.25	54 ± 2	<10	>0.8
Maize starch granulation	1.21	27 ± 4	<10	>0.6
Rice starch granulation	1.21	139 ± 11	50 ± 8	0.6

[a]Unlubricated

Abbreviations: BF_u, breaking force values of the unlubricated tablets; BF_l, breaking force values of the lubricated tablets; LSR, lubricant sensitivity ratio.

Source: Adapted from Ref. 50.

size, shape, texture, and surface roughness, the authors proposed some theoretical considerations. First, a low bulk density is an indication for poor flowability of a powder, which might delay or even prevent the formation of a lubricant film during the mixing process. Second, a lower bulk density will result in a larger contribution to particle rearrangement and consequently higher friction during consolidation. This could disturb an already formed lubricant film and enhance bond formation. A similar relationship between bulk density and lubricant sensitivity ratio was found for granulations based on native starches or on modified celluloses, but the relationships were different for the different investigated materials (52). For the granulations, the flowability of the particulate system was thought to be the predominant mechanism in the sensitivity to lubrication with magnesium stearate.

The consolidation and compaction characteristics of host particles are known to have considerable influence on their susceptibility to lubrication (12,53). De Boer et al. (12) illustrated that the sensitivity of tablet excipients to magnesium stearate depends on the compression behavior and the bonding mechanism of the material. The bonding properties of brittle materials such as dicalcium phosphate dihydrate (Fig. 2) and anhydrous β-lactose (Fig. 5) were hardly influenced by lubrication. The phenomenon was explained by the assumption that clean, lubricant-free surfaces are created by fragmentation of the particles during consolidation of the particle system. On the other hand, a maximum effect of magnesium stearate was found for excipients that undergo complete plastic deformation without any fragmentation under compression and are bonded by cohesion, such as starch and some starch derivatives (for instance STARCH 1500, see Fig. 2).

Most tablet excipients behave, however, in a manner intermediate between complete plastic deformation and complete brittle fracture, so that the lubricant sensitivity depends on the extent of fracture of the particles during compression. This should be the reason that for many materials the tablet strength decreases not to zero but to a minimal level after mixing with magnesium stearate (Fig. 5).

De Boer et al. (12) illustrated the effect of magnesium stearate on excipients with a different consolidation behavior by means of scanning electron micrographs. The upper surface of an unlubricated amylose tablet (Fig. 6A) shows that the particles are plastically deformed, but keep their individuality. When compaction was performed after one-hour mixing with 0.05% magnesium stearate, the particles are plastically deformed too, but the magnesium stearate film on the amylose particles has prevented bonding (Fig. 6B). Similar results were found for crystalline sodium chloride, a plastically deforming material. Comparison of the cross section of lubricated and unlubricated sodium chloride tablets (Fig. 6C, D) shows that the failure occurs mainly across the particles for plain sodium chloride tablets, but around the particles for tablets compressed from blends containing magnesium stearate. No differences could be seen between micrographs of surfaces of dicalcium phosphate dihydrate tablets without and with magnesium stearate, respectively, because of the extensive fragmentation behavior of the excipient.

In more recent work, Riepma et al. (54) showed for different materials that the lubricant sensitivity was not always related to the degree of fragmentation during compaction. In contrast to dicalcium phosphate dihydrate, both α-lactose monohydrate and sodium citrate exhibited a considerable reduction in tablet strength upon lubrication, whereas it is known that intensive fragmentation occurs during consolidation (54,55). Photomicrographs of fractures of lubricated tablets of α-lactose monohydrate and of sodium citrate showed that the tensile failure occurred around the interfaces between the original crystals (54). Realizing that these surfaces are coated with magnesium stearate, it was suggested that a three-dimensional matrix of magnesium stearate is sustained during compression of the particulate system. On the other hand, on photomicrographs of fractures of lubricated tablets of dicalcium phosphate

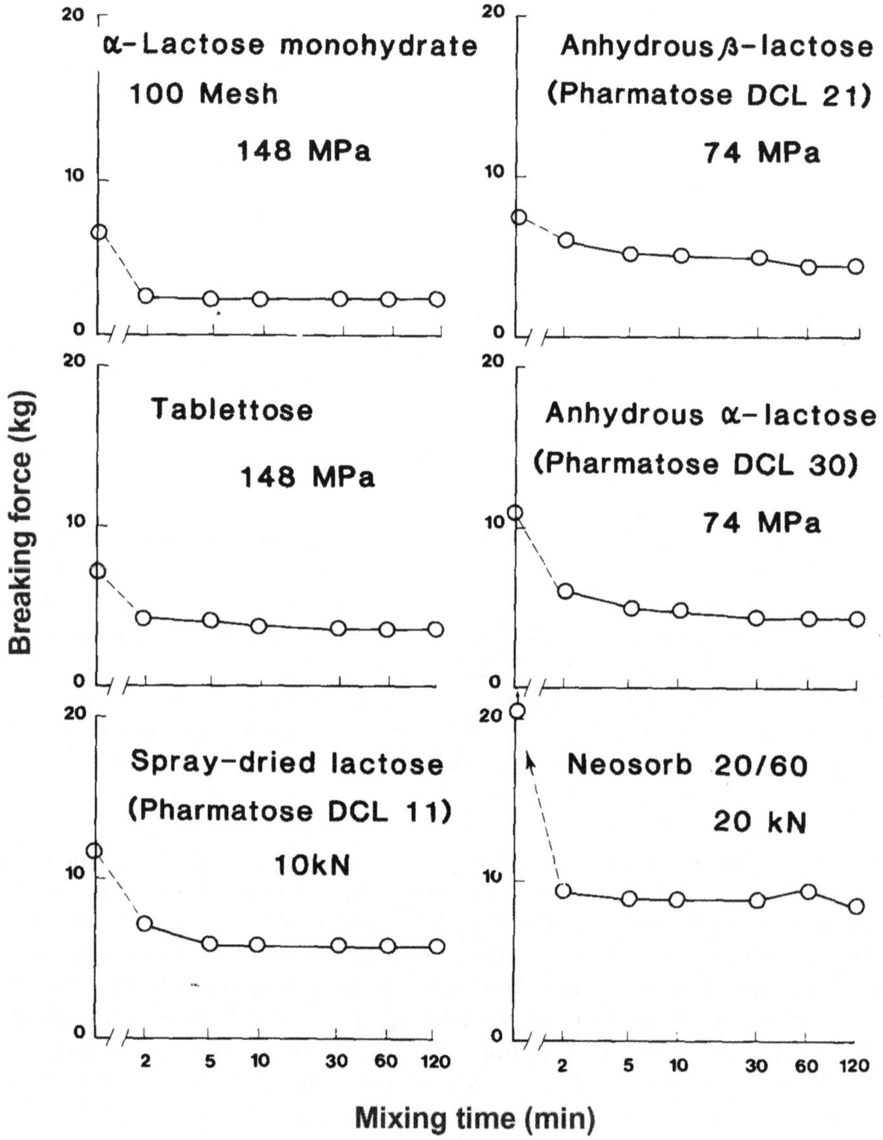

Figure 5 Breaking Force versus lubricant mixing time (log scale) for tablets compressed from filler-binders with 0.5% magnesium stearate. *Source*: From Ref. 4.

dihydrate, no original crystals could be distinguished. This means that addition of the lubricant did not lead to weak bonds between the original particles. Based on these results, the authors presented two different models for the elucidation of the lubricant sensitivity of brittle materials (54). Figure 7 shows the noncoherent matrix model. A coherent network of magnesium stearate (Fig. 7A), created by dry mixing the excipient with the lubricant, is interrupted by fragmentation and consolidation of the particulate system (Fig. 7B). The fat lines in the figure represent the magnesium stearate film on the surface of the original particles. As a consequence, the effect of a lubricant on the compactability of an excipient will be limited by the created lubricant-free surfaces. Figure 8 shows the coherent matrix model. A coherent network of magnesium stearate (Fig. 8A), created by dry mixing the excipient with the lubricant, is sustained during the process of consolidation (Fig. 8B). Fragmentation occurs

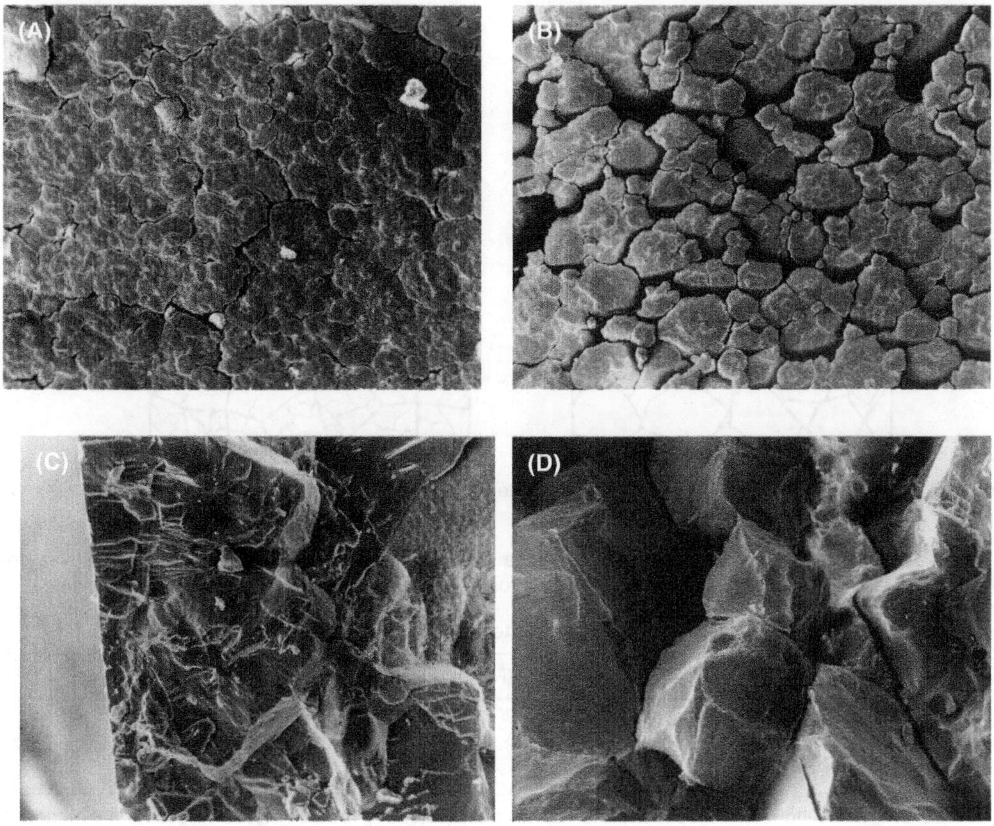

Figure 6 (**A**) Micrograph of the upper surface of a tablet compressed from amylose V (x150 approximately); (**B**) Micrograph of the upper surface of a tablet compressed from a mixture of Amylose V with 0.05% magnesium stearate (x150 approximately); (**C**) Micrograph of the surface of fracture of a tablet compressed from crystalline sodium chloride (x150 approximately); (**D**) Micrograph of the surface of fracture of a tablet compressed from a mixture of crystalline sodium chloride with 0.1% magnesium stearate (x150 approximately). *Source*: From Ref. 12.

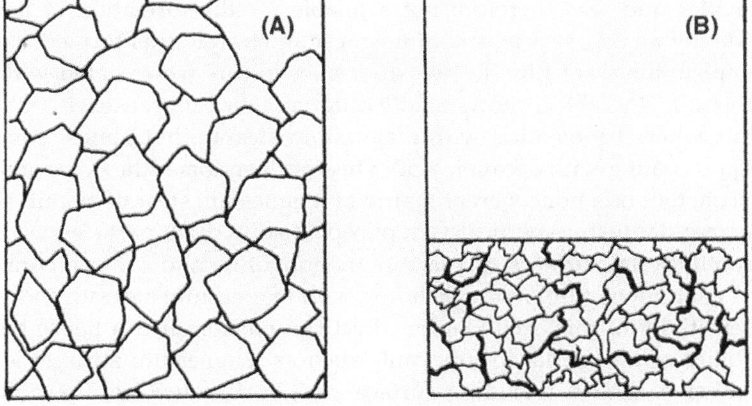

Figure 7 Noncoherent matrix model (**A**) and (**B**) as described in the text. *Source*: From Ref. 54.

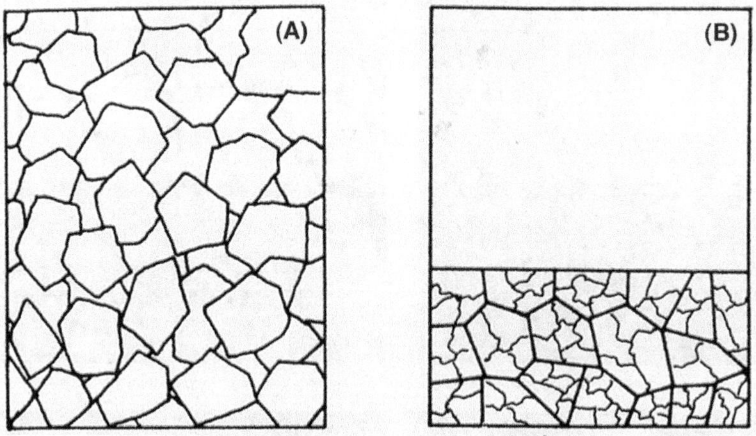

Figure 8 Coherent matrix model (**A**) and (**B**) as described in the text. *Source*: From Ref. 54.

Table 5 Effect of Lubrication with Magnesium Stearate (0.5%) and Decanoic Acid (0.5%), Respectively, on Breaking Force Values of Tablets Compacted at 20 kN from a Fraction (250–300 μm) of Several Materials

	Magnesium stearate			Decanoic acid		
Material	BF_u (N)	BF_l (N)	LSR	BF_u (N)	BF_l (N)	LSR
α-Lactose monohydrate	42	24	0.43	42	21	0.50
Sodium citrate	23	5	0.78	23	11	0.52
Dicalcium phosphate dihydrate	32	29	0.09	32	16	0.50

Abbreviations: BF_u, breaking force values for the unlubricated tablets; BF_l, breaking force values for the lubricated tablets; LSR, lubricant sensitivity ratio.
Source: From Ref. 54.

within the areas surrounded by the lubricant. Consequently, the strength of a compact is principally determined by the structure of a magnesium stearate matrix, created during the process of mixing the excipient with the lubricant.

The formation of a coherent lubricant matrix is strongly dependent on the surface texture of the host particles (4,24). Lerk and Sucker (24) showed that during the mixing process of irregular-shaped granule particles magnesium stearate formed a discontinuous layer around the particles of the granular excipient. Part of the magnesium stearate was trapped into the asperities and cavities and was therefore not available for the formation of a magnesium stearate film. Evidently no coherent matrix of magnesium stearate was formed. To assure the presence of a complete lubricant film, in two studies excipients were coated with carboxylic acid in diethyl ether (51,54). Table 5 shows totally different lubricant sensitivity ratios for three different excipients when dry-blended with magnesium stearate but almost equal lubricant sensitivities on liquid coating with decanoic acid. This result endorses the determining effect of the presence of a coherent or a noncoherent matrix of magnesium stearate within a particulate system. The very irregular texture of dicalcium phosphate dihydrate particles, as compared to the relatively smooth crystal surfaces of α-lactose monohydrate and sodium citrate, prevents the formation of a continuous film on the blending with magnesium stearate and explains the small lubricant sensitivity of this excipient on direct compaction. It can hence be concluded that the susceptibility of a material to lubricants such as magnesium stearate is a complex function of a number of factors including surface area, surface texture, flowability, mixing properties, and consolidation behavior.

EFFECT OF PROCESS CONDITIONS ON FILM FORMATION

The extent of film formation and hence the negative effect of lubricants on the strength of the tablets depends strongly on mixing time (Fig. 2) and the mixing procedure (3,6). The influence of magnesium stearate on tablet strength has been investigated using different types of lab-scale mixers (27,56,57). It was concluded that not only mixing time but also mixing intensity plays a large role (27). When production-scale volumes are used, mixing and shearing intensity increase. As the rate at which an ordered mix is formed increases significantly when the batch size is increased (58), it may be expected that in large industrial mixers, the shear forces, affecting the migration of magnesium stearate from magnesium stearate particles to excipient particles, will be much greater than in lab-scale mixers. Therefore, it may be expected that the formation of a film of magnesium stearate while the tablet ingredients are being mixed with the lubricant proceeds faster in production-scale mixers and will depend on the type of the mixer used and its rotation speed. This means that the tablet properties also will depend on the type, size, load, and rotation speed of the mixer. Few studies have been made, however, of the effect on tablet properties of scaling up of the lubricant mixing process. Johansson (40) studied the effect of mixing with magnesium stearate on tablet strength using three different double cone mixers containing 5-, 25-, and 80-kg material, respectively. The negative effect of the lubricant increased with batch size, which was attributed to the higher shear forces in the tumbling mixer with the larger masses. As in tumbling mixers the energy, required for the formation of an ordered mixture is mainly provided by shear forces within the mixture, larger masses promote the deagglomeration of lubricant particles and the formation of an ordered mixture.

Bolhuis et al. (59) mixed a lactose/microcrystalline cellulose test formulation with 0.5% magnesium stearate in seven different mixers, operating at different mixing velocities (Table 6). The decrease in breaking force values of tablets, compressed from the blends occurred much more quickly in production-scale mixers than in the lab-scale mixers when they operated at the same rotation speed. The critical mixing time, that is, the time required for a decrease of the breaking force values from 180 N (unlubricated tablets) down to an arbitrarily minimal level of 60 N varied between 1.5 and 30 minutes. For the production-scale mixers, the decrease in tablet strength as an effect of lubricant admixing depended more on the rotation speed than on the type or size of the mixer or the size of the load. This can be seen by comparing different production scale mixers operating with speeds between 20 and 26 rpm (Fig. 9). It was found that a Turbula mixer, operating at high rotation speeds can be used to predict the effect of lubricant admixing on tablet strength in production scale mixers. This is

Table 6 Sensitivity of the Test Formulation, When Mixed in Various Mixers at Various Speeds, to the Addition of the Lubricant[a]

Mixer	Capacity (L)	Rotation speed (rpm)	Breaking force half-life (min)	Critical mixing time (min)
Turbula	2	45	1.0	10.5
		90	<1.0	1.5
Cubic	13	20	8.0	30.0
		60	2.7	17.0
Drum	45	10	1.0	9.0
Planetary	90	25	1.0	4.0
		42	<1.0	2.0
Planetary	200	26	2.0	3.8
Planetary	900	10	3.3	7.9
V-shaped	1000	22	1.5	3.6

[a]The sensitivity is reflected both in the breaking force half-life and in the critical mixing time.
Source: From Ref. 59.

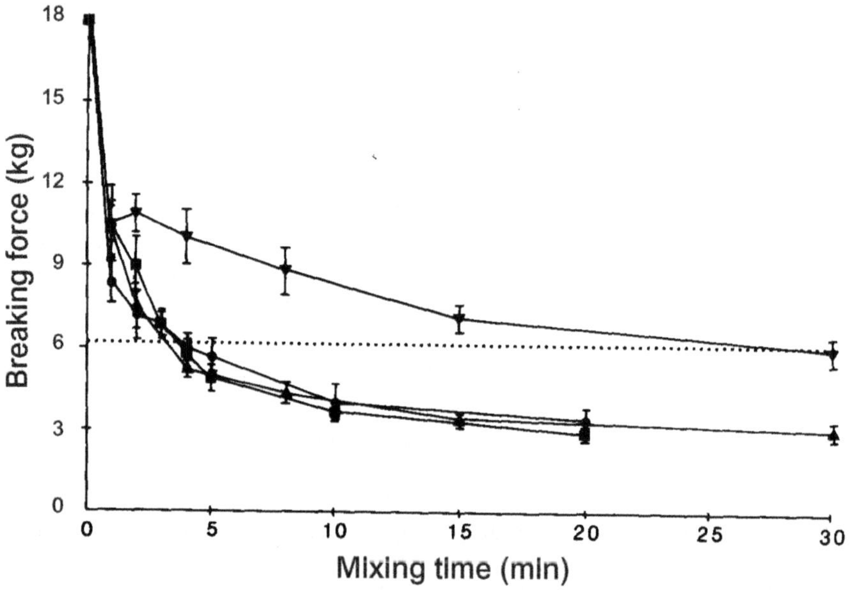

Figure 9 Effect of lubricant mixing time on the breaking force values of tablets compressed from the test formulation. (▼) formulation mixed in a 13-L cubic mixer at 20 rpm; (•) formulation mixed in a 90-L planetary mixer at 25 rpm; (■) formulation mixed in a 200-L planetary mixer at 26 rpm; (▲) formulation mixed in a 1000-L V-shaped mixer at 22 rpm. *Source*: From Ref. 59.

important in preformulation work, where lab-scale mixers are commonly used to determine the lubricant sensitivity of tabletting blends.

EFFECT OF LUBRICANT FILM FORMATION ON LUBRICATION PROPERTIES

Boundary Lubrication

Effective lubricants such as magnesium stearate and sodium stearyl fumarate are examples of boundary lubricants. The theory of Bowden and Tabor shows that effective lubrication is provided only if the lubricant can react with the surface of the die wall to form a layer with strong lateral adhesion (2,60,61). Solid boundary lubricants such as the metal salts of stearic acid, for example, magnesium stearate, form a layered film at the die wall and on punch surfaces on which the tablet slides during ejection. The hydrophobic carbon chains are directed outward from the die surface. Tablet lubricants act as a friction lowering material at the die wall building up the boundary shearing and gliding protection layer and/or as an antiadherent to prevent adhesion, sticking and binding of the tablet material to the tooling surfaces as well as other machine parts. Moreover, lubricants can act as a glidant that promotes the flow of the granulation within the hopper and die cavity. The next sections will show that variation in mixing time with lubricants will not only affect tablet properties such as strength, disintegration time and drug dissolution, as shown in first part of this chapter, but will also have an effect on the lubricating properties of the lubricant.

Effect of Lubricant Mixing Time on Tablet/Die Wall Friction

In pharmaceutical tabletting the desired effect of a lubricant is to have a lubricating layer on those metal surfaces that come into contact with the particles of the tabletting blend. Generally, only small amounts of tablet lubricants are needed to lower the friction forces effectively (34). The lubrication and friction properties have been measured with the aid of force transmission, by the determination of friction forces or friction coefficients, by acoustic emission and by electric power consumption measurements to optimize the amount of lubricant and to minimize the negative effects of lubricants (2,32,60,62–65). The ejection force, calculated per unit contact area between the tablet and the die wall is recommended as a good measure of friction during tabletting, but has the disadvantage that it depends on the compaction load (47,66).

To obtain a measure, which for most materials is independent of load and/or dimension, the friction coefficients at the maximum compaction [μ_1 = ratio of axial force difference (FD) and die wall force at compression maximum (DWF)] and at ejection [μ_2 = ratio of ejection force (EJF)] and radial force at ejection (DWFE)] have been calculated (32,45,63,67,68). It was shown that the ejection force EJF is due to variation in friction coefficient and not due to variation in radial forces at ejection DWFE (32).

Strength of the Lubricant Boundary Layer

In a classical series of papers Strickland et al. showed the mechanism of action and evaluation of 70 materials as tablet lubricants (60). After coating the die wall with boundary type lubricants, the ejection force was measured during tabletting of eight successive tablets of a nonlubricated sodium bicarbonate test granulation. The ejection force increased from 6 up to 8, 10, 20, 43, 65, 88, and 103 kg when the die was prelubricated with solid magnesium stearate. The strength of the lubricant boundary layer formed by magnesium stearate, conditioning the die wall was also investigated by Hölzer and Sjögren (63). First unlubricated sodium chloride tablets were compacted at 115 Mpa upper-punch pressure. The friction coefficient μ_1 measured was 1.4 (Fig. 10). Then material lubricated with 1.0% magnesium stearate was tabletted. The polar parts of the boundary lubricant molecules adhered to the metal surface, forming a resistant layer (a boundary lubricant film) on the die wall. The friction coefficient was reduced and reached a constant value of 0.3. When changing to unlubricated sodium chloride once more, the friction coefficient increased again, but more than 30 tablets were needed before the starting value was reached. This result shows that the boundary lubricant film on the die wall is very resistant to abrasion by sodium chloride.

Effect of Mixing Time on Ejection Force

Müller et al. (62) have shown the influence of mixing time on the remaining force and ejection force of tablets. Two brittle tablet masses, granulatum simplex and crystalline lactose, were

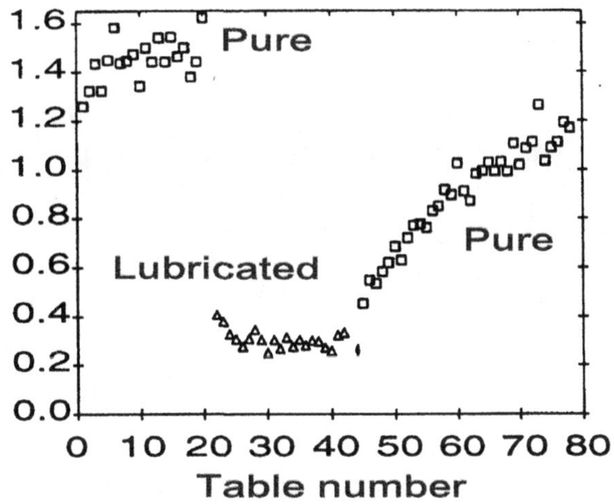

Figure 10 Strength of boundary lubricant film. Friction coefficient (μ_1) at compaction for consecutive sodium chloride tablets.

mixed with different magnesium stearates, used in concentrations between 0.2% and 1%. Minimum values for both remaining and ejection forces were obtained after two to five minutes mixing in a Turbula mixer. Prolonged mixing times had a negligible effect on the lubricating properties.

Ragnarsson et al. (34) studied the influence of mixing on the lubricating properties of magnesium stearate. The ejection force of tablets compressed from blends of sodium chloride, anhydrous β-lactose, or calcium citrate granulate with 0.1% to 2.5% magnesium stearate was measured in an eccentric press. Magnesium stearate reduced the ejection force and this effect was related to the concentration but not to the mixing time except at the lowest concentration of 0.1%. Generally, it seems possible to get a good lubrication even when the lubricant is poorly spread out in the mixture. This means that for the lubricating effect a short mixing time is sufficient when admixing the common used standard concentrations 0.25 to 0.50 w/w percent magnesium stearate to tabletting mixtures or granulations.

The effect of mixing time with lubricant on both tablet properties and lubricating properties was studied by Hölzer and coworkers, using a large number of lubricants and different test materials (32,35,36,63,66,69,70). Some results from these papers, completed with unpublished data are shown in the Figure 11. The figure shows the effect of mixing time (20, 200, or 2000 revolutions) with 0.1% lubricant on the ejection force per unit contact area (EJF/A) of sodium chloride tablets, compacted at 150 MPa in an eccentric press. Mixing with lubricant was performed in a 2-L Turbula mixer at 42 rpm. Table 3 lists the physical and chemical properties of the lubricants used, ordered by specific surface area. Figure 11 shows that a prolonged mixing time generally increases the lubrication efficiency of 0.1% lubricant, expressed as reduction in EJF/A, but that the magnitude of the effect depends on the lubricant used. Lubricants with a large surface area, for example, the magnesium stearates give a high reduction in EJF/A even when admixing the lubricant as short as 20 revolutions. The EJF/A for tablets with lubricants containing agglomerates; for example, the triglycerides Boeson VP, Cutina HR, Dynasan 118, Myvaplex, and Précirol show a great influence of the mixing time. The lubricant particles are spread out efficiently and the sodium chloride crystals are covered much more after a longer mixing time. A similar effect was found for stearic acid; the EJF/A values were reduced 40% to 60% due to the increased surface coverage. Fluon particles are also

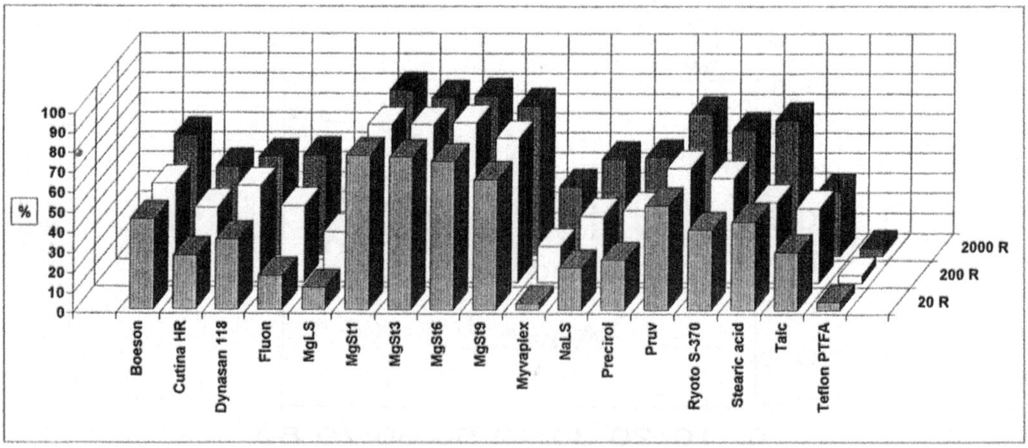

Figure 11 Effect of mixing time on boundary lubrication. Percentage reduction in ejection force per unit contact area for sodium chloride tablets containing 0.1% lubricant.

spread out efficiently and the ejection forces are reduced 20% up to 50%. For another PTFE quality, Teflon PFA, no lubricant effect could be seen, even not at prolonged mixing.

Johansson studied the influence of the lubricant mixing time on the lubrication and tablet properties of four different tablet masses, using magnesium stearate in either powdered or granular form as lubricant (37–40). The EJF/A was used for estimation of the lubrication effect. The adhesion to the punch faces was measured by visual inspection. Tablets with granular magnesium stearate showed values of EJF/A comparable with those of tablets with powdered lubricants at high concentrations (2–5%). Powdered lubricant was better than granular in preventing adhesion to the punch faces at low concentration. The lubricating effect of the powdered lubricant was found to be almost unaffected by the mixing time while the granular lubricant required somewhat longer mixing times to obtain optimal lubrication properties. Increasing the particle size of the magnesium stearate granulation increased the amount of lubricant required to obtain lubrication similar to powdered magnesium stearate (39). Variation in the specific surface area of the starting material could be masked by using them in granular form (39). The results show that film formation is not a prerequisite for good lubrication. The granular magnesium stearate mixtures formed not a film as strong and not as fast as the powdered magnesium stearate lubricant but gave efficient lubrication on the die wall due to the boundary lubrication effect.

Effect of Mixing Time on Friction Coefficients

Hölzer and Sjögren (63) measured DWFE simultaneously with EJF. The ratio of EJF/DWFE was expressed as the dynamic friction coefficient during ejection μ_2. In further work, the authors measured the effect of mixing time with 0.1% of different lubricants (Table 3) on the reduction of the friction coefficient μ_2 of sodium chloride tablets, compacted at 150 MPa in an eccentric press (32). Mixing with lubricant was performed in a 2-L Turbula mixer at 42 rpm. Figure 12 shows that the highest reduction in friction coefficient μ_2 was obtained by the magnesium stearates, followed by sodium stearyl fumarate (Pruv), stearic palmitic sugar ester (Ryoto S-370), stearic acid, and glyceryl palmitic stearate (BoesonVP). PTFE in a coarse grade (Teflon PFA) had very little effect as a lubricant. The effect of mixing time on friction coefficients at ejection (μ_2) in Figure 12 shows similar results as obtained for the EJF/A values

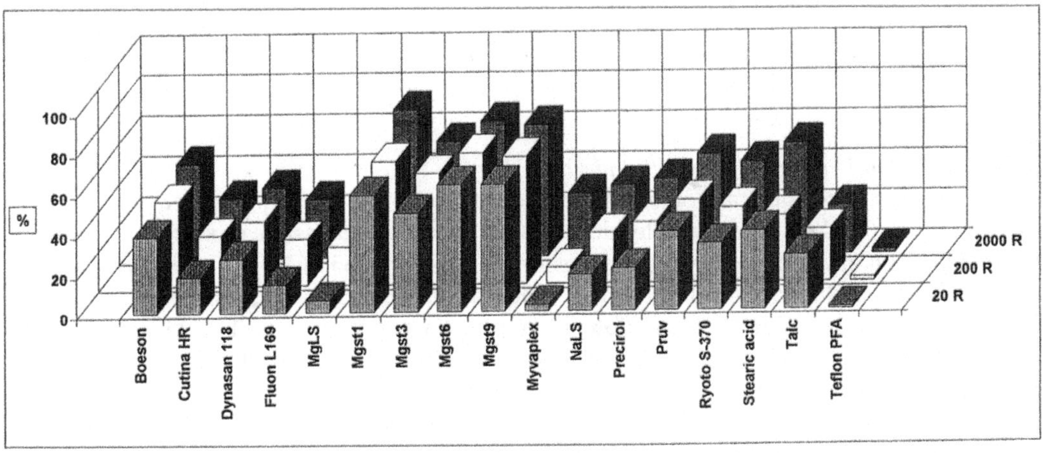

Figure 12 Effect of mixing time on boundary lubrication. Percentage reduction in friction coefficient at ejection (μ_2) for sodium chloride tablets containing 0.1% lubricant.

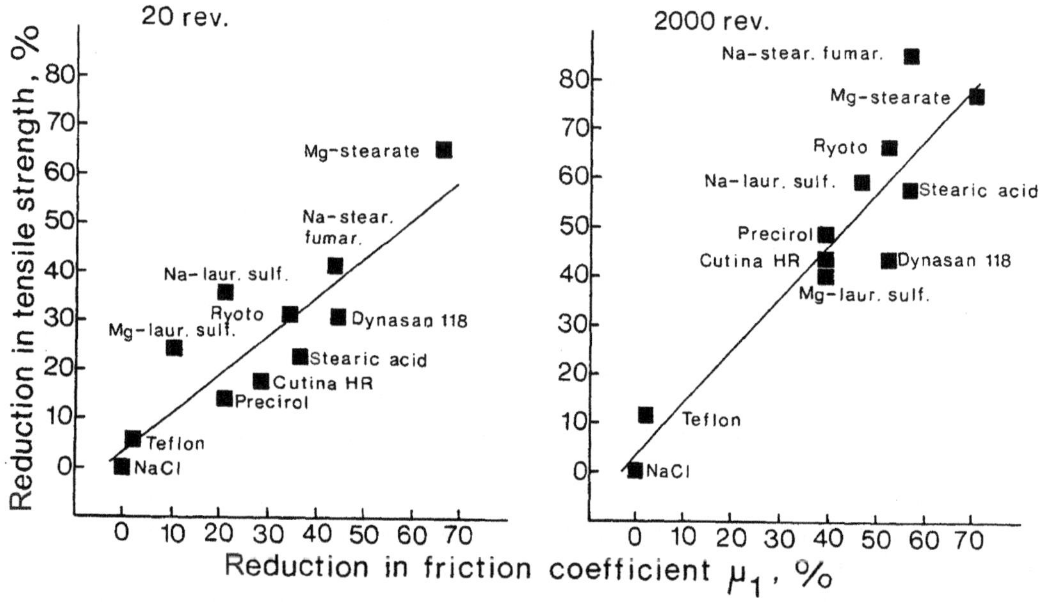

Figure 13 Reduction in tensile strength vs. reduction in friction coefficient (μ_1) for tablets of sodium chloride admixed with 0.1% of different lubricants for 20 and 2000 revolutions. *Source*: From Ref. 32.

in Figure 11. The μ_2 values are reduced even more when the mixing is increased because the ejection forces are reduced more than the radial die wall force during ejection. By measuring both die wall forces and ejection forces during ejection one obtains the possibility to distinguish if the friction is lowered by the lubricant.

Using another technique for the measurement of the dynamic friction coefficient at ejection, μ_2, Kikuta and Kitamori (67) found likewise that an increase in mixing time with the lubricant gave lower friction coefficients. Moreover, adhesion to the punches, estimated as the intercept in the correlation to Coulombs law, decreased on prolonged mixing time (67).

Comparing the Figures 4, 11, and 12 shows that the negative effect of lubricants on tablet strength should be correlated with their positive friction lowering effects. Hölzer and Sjögren (32) showed for mixtures of sodium chloride with 0.1% of different lubricants that the percentage reduction in tensile strength was related to the reduction in friction coefficient μ_1 (Fig. 13).

Power Consumption During Mixing with Lubricant

The influence of a solid lubricant on electric power consumption during mixing and the effect of addition of colloidal silica on particle friction of a directly compressible mixture was studied by Schrank-Junghäni and coworkers in a series of papers (65,71,72). The electric power consumption was reduced from 80 to 40 W when admixing 0.2% magnesium stearate to a lactose granulation for about one minute in a planetary mixer (Fig. 14). Measuring the interparticulate friction allows comparison of the wall and internal friction and provides a classification of different lubricants (72). For stearic acid higher concentrations and longer mixing times are required to reach the same effect on power consumption as for magnesium stearate mixing, because magnesium stearate forms the boundary layer on particulate solids much faster than stearic acid. This can, to some extent, be explained by differences in particle surface area. The results show that power consumption is a valuable tool for the determination

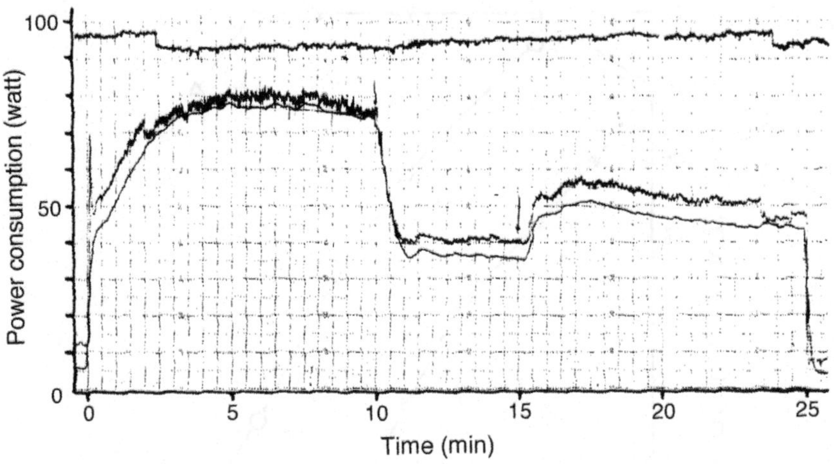

Figure 14 Plot of power consumption versus time for a directly compressible mixture with 0.2% magnesium stearate and 0.2% colloidal silica added five minutes later. *Source*: From Ref. 71.

of the minimum concentration and mixing times required with lubricants, so that the negative effect on tablet properties can be minimized.

Effect of Lubricant Mixing Time on Adhesion

A more severe problem than friction is the adhesion of tablet material on machine parts and tooling, causing sticking and picking on punch and die surfaces. Often a much higher concentration of the lubricant is necessary to reduce the adhesion then to reduce the friction to a minimum level. Methods for studying the antiadhesive properties are (*i*) measuring the adhesion force at the die wall, (*ii*) measuring the push off force from the lower punch surface and (*iii*) by visual inspection of the punch surfaces (64). Only few studies describe the effect of mixing time with lubricant on the antiadhesive properties.

Hölzer and Sjögren (35) found that the adhesion-reducing effect of lubricants can not only be increased by an increasing concentration of the lubricant, but also by prolonged mixing. Arbitrary chosen scores for picking and sticking decreased for anhydrous lactose tablets produced with different concentrations magnesium stearate or sodium stearyl fumarate when the mixing time was increased from 20 to 2000 revolutions in a 2-L cubic mixer at 42 rpm. This effect was attributed to an increased surface coverage and film forming of the boundary lubricant on prolonged mixing (35).

EFFECT OF THIRD COMPONENTS ON THE FILM FORMATION OF LUBRICANTS

The film formation of lubricants during mixing can be influenced by third components. Simultaneous mixing of excipient particles with magnesium stearate and colloidal silica (Aerosil 200) can significantly suppress the negative effect of the lubricant on the bonding properties (Fig. 15). A larger effect was obtained when the host particles were blended with colloidal silica prior to the addition of magnesium stearate (27,73). The addition of colloidal silica after previous mixing of the excipient with magnesium stearate (Fig. 16) may even restore the bonding properties, when the ratio between colloidal silica and magnesium stearate is 4 to 1 (27). Even low, commonly used concentrations of colloidal silica (0.2%) can suppress the deleterious effect of 0.5% magnesium stearate on tablet bonding to a certain extent when the excipient particles were first mixed with the glidant and consequently for a short time with the lubricant.

The interaction between magnesium stearate and colloidal silica has been elucidated by the measurement of contact angles and electron microprobe analysis (EDAX) (17) and by dissolution measurements (17,19). It was shown that a magnesium stearate film on sodium

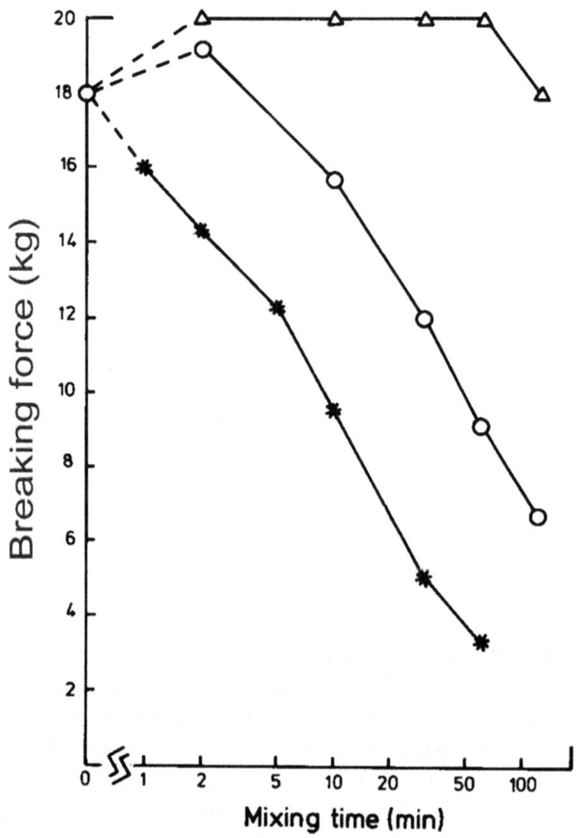

Figure 15 Effect of mixing time on the breaking force values of tablets compressed from blends of STARCH 1500 with magnesium stearate or with magnesium stearate and Aerosil 200. (★) 0.1% magnesium stearate; (○) 0.1% magnesium stearate + 0.1% Aerosil 200; (△) 0.1% magnesium stearate + 0.4% Aerosil 200. *Source*: From Ref. 27.

chloride particles can be stripped from the substitute during mixing with fourfold the amount of colloidal silica under the formation of separate colloidal silica/magnesium stearate spheres. If colloidal silica is admixed together with the lubricant, the surface coverage with lubricant is reduced (19). The effect of premixing of host particles with low concentrations of colloidal silica has been elucidated by competitive inhibition of magnesium stearate molecular layers at the adhesion sites that are occupied by colloidal silica particles (74).

The intensity of the interactions between the components of the ternary powder system microcrystalline cellulose, magnesium stearate, and colloidal silica have been predicted using literature and calculated values for their partial solubility parameters (75). These show that there is a greater interaction between magnesium stearate and colloidal silica than between magnesium stearate and microcrystalline cellulose although both interactions are greater than the cohesive interaction within the lubricant itself. Moreover, the interaction between colloidal silica and microcrystalline cellulose was found to be high, although less than the cohesive interactions within colloidal silica itself. From these results it can be predicted that in a ternary system the microcrystalline cellulose will be preferentially coated by colloidal silica and that the majority of the magnesium stearate will be enrobed by the colloidal silica. On compaction at high shear forces, shearing of the enrobed magnesium stearate will occur first, followed by shearing at the microcrystalline cellulose-colloidal silica interface. Although the model is something crude and oversimplified, the theoretical considerations are consistent with the previously found "protecting" effect of colloidal silica with respect to lubricant sensitivity (75).

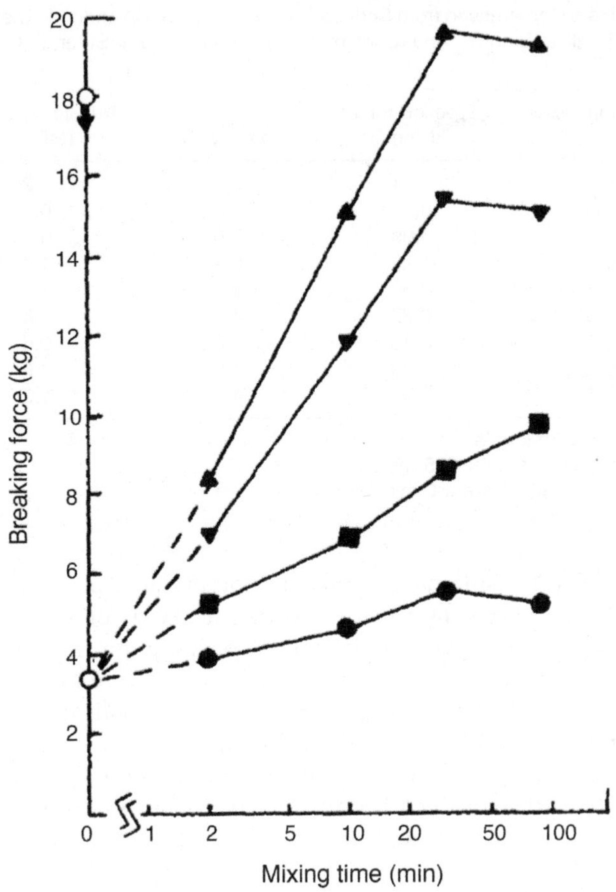

Figure 16 Effect of mixing with Aerosil 200 on the breaking force values of tablets compressed from previously mixed (one hour) blends of STARCH 1500 and 0.1% magnesium stearate. (•) 0.1%; (■) 0.2%; (▼) 0.3%; (▲) 0.4% Aerosil 200. *Source*: From Ref. 27.

The interaction between magnesium stearate and the glidant colloidal silica not only affects the bonding properties of an excipient but also affects disintegration time (34), drug dissolution rate (17), and lubricating properties (15,19,34,71,76). Lerk et al. (27) showed that the ejection force of tablets, compressed from a compressible starch/lactose blend 1:1, lubricated with 0.5% magnesium stearate was not affected by the addition of 0.2% colloidal silica. However, when the concentration of colloidal silica is equal to that of magnesium stearate or even higher, the glidant will have a significant effect on the lubrication properties. Ragnarsson et al. (34) studied the effect of admixing colloidal silica on the tabletting properties of sodium chloride, lubricated with magnesium stearate. Mixing time, sequence of mixing, and the concentrations were varied (Table 7). From the blends, tablets were compacted at 200 MPa maximum upper-punch pressure. Mixing with 0.1% magnesium stearate and 0.5% colloidal silica simultaneously or mixing with 0.1% magnesium stearate after previous mixing with 0.5% colloidal silica restored the bonding properties, but increased at the same the ejection force as compared with tablets containing 0.1% magnesium stearate only. Mixing with 0.5% colloidal silica after previous mixing with 0.1% magnesium stearate or mixing with 0.5% magnesium stearate and 0.5% colloidal silica simultaneously gave rather low ejection forces but could not restore the bonding properties. This means that the positive effect on strength may be counteracted by a negative effect on friction lowering properties. Similar results were reported

Table 7 Properties of Tablets Compressed from Sodium Chloride, Sodium Chloride/Magnesium Stearate Blends, Sodium Chloride/Colloidal Silica Blends, and Sodium Chloride/Magnesium Stearate/Colloidal Silica Blends, Respectively

Magnesium stearate (%)	Colloidal silica (%)	Ejection force (kN/cm^2)	Porosity (%)	Tensile strength (MPa)	Disintegration time (min)
–	–	1.31	5.0	1.3	4.2
0.1[a]	–	0.18	3.6	0.5	8.8
0.5[a]	–	0.09	3.5	<0.1	19.0
–	0.5[a]	1.28	6.1	1.7	5.7
0.1[a]	0.5[a]	0.88	4.9	1.2	8.1
0.5[a]	0.5[a]	0.27	4.2	0.6	19.5
0.1[a]	0.5[b]	0.43	4.8	0.6	9.0
0.1[a]	0.5[c]	0.43	4.8	0.6	8.6
0.1[b]	0.5[a]	0.71	4.8	1.2	7.7

[a]Mixed with sodium chloride, 25 minutes.
[b]Admixed to mixture in footnote a, another 25 minutes.
[c]Admixed to mixture in footnote a, another 50 minutes.
Source: From Ref. 34.

by Shrank-Junghäni et al. (71) using power consumption measurements (Fig. 14). Addition of 0.2% colloidal silica to a direct compression blend, premixed for five minutes with 0.2% magnesium stearate, results in an increase in power consumption from 40 up to 55 W. This indicates increased interparticle friction and the friction toward the mixer vessel walls.

The addition of colloidal silica on both surface coverage with magnesium stearate and tablet ejection force of an acetylsalicylic acid granulation, lubricated with magnesium stearate was investigated by Johansson and Nicklasson (15,19). A flow-through technique was used for calculating the surface coverage of the lubricant. When 1% colloidal silica was admixed after previous mixing with 1% magnesium stearate the surface coverage was not affected but the ejection force/area increased twice. If 1% colloidal silica and 1% magnesium stearate were mixed simultaneously with the granulation, both surface coverage and ejection force/area were affected. The authors concluded that colloidal silica imparts primarily with the free fraction of magnesium stearate partly withdrawn from further coverage of the base material as well as from the lubrication of the die wall. Moreover, it was concluded that the lubricating effect of magnesium stearate is not directly related to the coverage on the surface of the base material.

In contrast to colloidal silica, little has been reported on the interaction between talc and magnesium stearate. In a series of papers Staniforth and Ahmed (73,76,77) showed that, in contrast to colloidal silica, 2% talc cannot restore the bonding properties of microcrystalline cellulose, mixed with 0.5% magnesium stearate (Fig. 17). A preblend of talc and magnesium stearate caused even a larger decrease of the work of failure than magnesium stearate alone. It was concluded that talc becomes coated with magnesium stearate particles, which promotes dispersion of the lubricant, whereas colloidal silica itself coats magnesium stearate particles, preventing the formation of a lubricant film on the base material (73). The ejection force of the tablets, which fell from 700 to 275 N after the addition of 0.5% magnesium stearate, increased to values between 350 and 420 N when 2% talc or 2% colloidal silica were present as a third component. These results indicate that components such as talc and colloidal silica can modify the influence of magnesium stearate on lubrication and compactability indepentely (76).

Lerk and Sucker (23,24) showed that the addition of small amounts of talc forces magnesium stearate out of cavities of a material during the mixing process and hence promotes film formation of magnesium stearate. As the amount of magnesium stearate consumed for film formation is no longer available in the asperities to reduce the friction during the ejection of a compact, a decreased lubrication efficiency was found. Consequently,

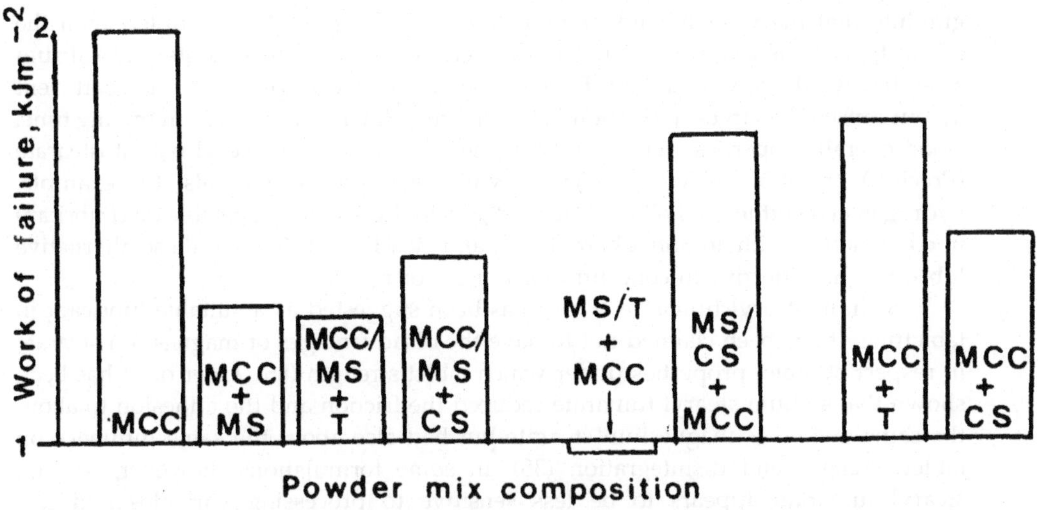

Figure 17 Relationship between powder mixing conditions and tablet work of failure. *Abbreviations*: MCC, microcrystalline cellulose; MS, magnesium stearate; T, talc; CS, colloidal silica. *Source*: From Ref. 73.

the interaction between talc and magnesium stearate is deleterious for both bonding properties and lubrication efficiency.

LIMITING LUBRICANT SENSITIVITY

There are different possibilities to limit the deteriorating effect of magnesium stearate on tablet properties, without affecting the lubrication properties to a large extent.

1. Undoubtedly the best method is *omitting a lubricant* in a tablet formulation and applying alternative lubrication methods, mostly involving modifications to tablet machines. Reported techniques are the addition of exact amount of a suitable lubricant directly onto the punch and die surfaces by electrostatic deposition (77) or by the so-called PKB method (Presskammerbeschichtung) immediately after tablet ejection (78). Other techniques are using bonded PTFE die linings or dies made from steel with lubricant inclusions (79) or cycling compressions of a carrier formulation first containing a lubricant, to create a lubricant film on the die wall, followed by compression of the unlubricated formulation (80). All these methods have been received with limited success or are extremely expensive.
2. The lubricant concentration should not be higher than is strictly necessary. The *minimum concentration* required with lubricants can be determined by means of power consumption measurement (65,71,72). Magnesium stearate reduces the electrical power consumption of a mixer, even when used in very small concentrations. The optimum concentration can also be calculated by means of external friction measurements (81) or based on surface area of both lubricant and host particles (82). Hölzer found for anhydrous β-lactose that the surface ratio between the lubricant and the host particles should be greater than 10% to 15% (83).
3. The choice of *alternative lubricants* seems an attractive solution of the problem of the deleterious effect of magnesium stearate on the bonding properties of tablet excipients. Several comparative evaluations have been carried out between such materials and magnesium stearate, considering both lubrication efficiency and effect on tablet properties (29,32,33,35,84,85). It has been demonstrated for different stearates, hydrogenated vegetable oils, glycerides, sugar esters, PTFE, talc, and

graphite that the time-dependent reduction in tablet strength is coupled with the lubrication efficiency (Fig. 13) (29,32). Generally, magnesium stearate and calcium stearate gave the lowest ejection force or the highest reduction in friction coefficient μ_1, but caused the largest reduction in tablet strength with an increase in mixing time. Good overall properties were found for glycerides such as glyceryl palmitostearate (Précirol) and Boeson VP (29,33,84), hydrogenated vegetable oils, for example, hydrogenated cotton seed oil (Sterotex) (29,84), hydrogenated vegetable oil (Lubritab) in combination with an antiadherent (85), and stearic acid (33). All these alternative lubricants are effective in concentrations of 1% or more.

Sodium stearyl fumarate (Pruv) has been suggested as a suitable lubricant in tabletting. It has been claimed not to have the disadvantages of magnesium stearate in respect of tablet properties under which tablet strength (86). Later on it has been shown that sodium stearyl fumarate reduced the friction and the adhesion to about the same degree as magnesium stearate but had also about the same influence on tablet strength and disintegration (35). In some formulations, however, sodium stearyl fumarate appears to be less sensitive to processing variables and less hydrophobic than magnesium stearate (70). In formulations where magnesium stearate causes problems, sodium stearyl fumarate may be a good alternative (83).

The results from the experiments with sodium chloride tablets (Table 3, Figs. 4, 11, and 12), completed with effect on disintegration time (Fig. 18) can serve as a directive for the choice of alternative lubricants. The negative effects of lubricant on tablet tensile strength and disintegration time are well correlated to the positive effect

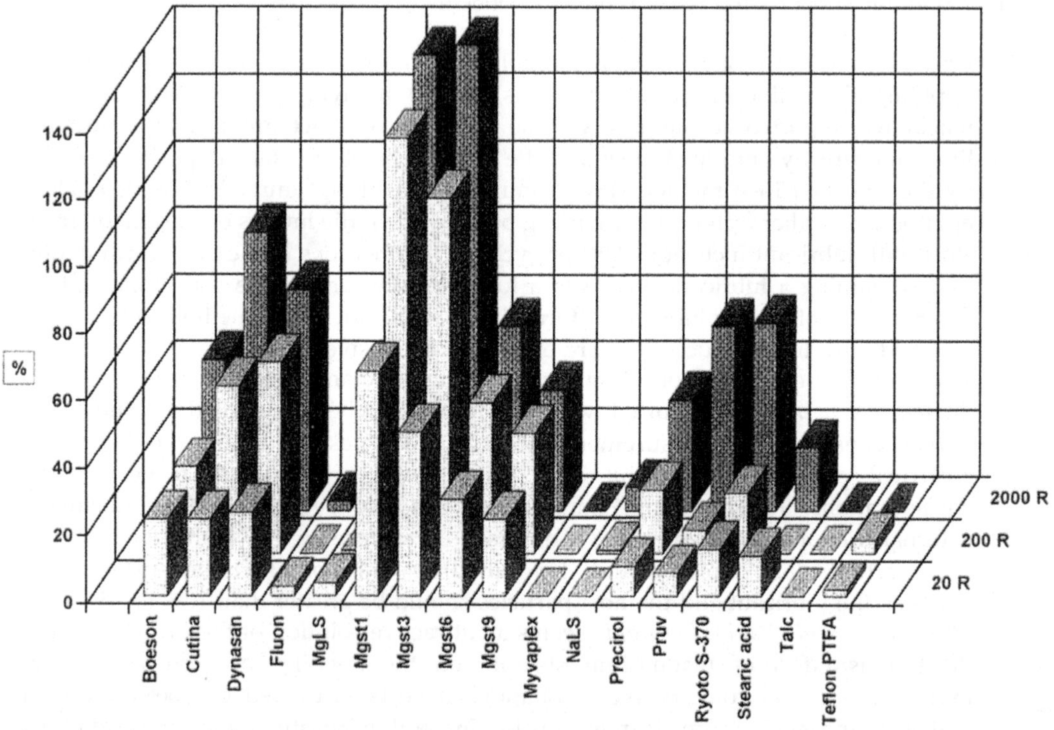

Figure 18 Effect of mixing time on boundary lubrication. Percentage increase in disintegration time for sodium chloride tablets containing 0.1% lubricant.

on lubrication. However, the negative properties are more pronounced when the mixing time is increased. It can be seen that the lauryl sulfates had more effect on tablet strength than on disintegration time; on the other hand, the triglycerides increased disintegration time more than they reduced tablet strength. For adhesive excipients such as anhydrous β-lactose, the antiadhesive properties of lubricants should be taken into account too. Good antiadherents are magnesium stearate, magnesium lauryl sulfate, sodium stearyl fumarate (Pruv) and talc in combination with magnesium stearate.

4. Vezin et al. (87) showed for microcrystalline cellulose that loss of tablet strength arising from *lubricant overmixing* can be substantially reduced by careful independent adjustment of main and precompression forces. This effect was attributed to the greater separation in time of the two distinct compaction events, providing a longer interval in which time-dependent effects such as stress relaxation and escape of air may occur.
5. Another possibility is *changing the mixing procedure*. Drugs and excipients should be mixed without a lubricant first. After the addition of a lubricant, the mixing should be continued for a short period. It has been demonstrated that a short mixing time, resulting in a poor distribution of magnesium stearate, did not impair its lubricating efficiency (34). A similar effect as a short mixing time can be reached by using a coarser magnesium stearate or magnesium stearate granules, as was proposed by Johansson (37–40).
6. The lubricant sensitivity can also be decreased by a proper *choice of tablet excipients*. As mentioned before, the effect of magnesium stearate depends on the nature of the excipient used. The largest reduction in tablet breaking force values can be expected for tablets containing starch or cellulose products (Fig. 2), the smallest effects will be found for tablets containing brittle materials such as dicalcium phosphate dihydrate or anhydrous β-lactose (Fig. 5). It should be realized, however, that most tablet formulations are mixtures of materials with very different compaction properties and in practice it is difficult to anticipate the compressional characteristics of a tabletting mixture. For example, dicalcium phosphate dihydrate, which is regarded as a brittle material, can behave as a plastic material when it is mixed with only 10% of microcrystalline cellulose and 1% magnesium stearate (84).
7. Another possibility to limit the deleterious effect of magnesium stearate on bonding properties is *premixing with colloidal silica* (Aerosil 200). The glidant colloidal silica can delay the film formation of magnesium stearate when tablet ingredients are mixed with colloidal silica first and then with magnesium stearate. When the concentration of colloidal silica is low as compared with the concentration of magnesium stearate, the effect on the lubrication efficiency of magnesium stearate is limited (27).

REFERENCES

1. Shotton E, Lewis CJ. Some observations on the effect of lubrication on the crushing strength of tablets. J Pharm Pharmacol 1964; 16(suppl):111T–120T.
2. Strickland WA, Nelson E, Busse LW, et al. The physics of tablet compression IX: fundamental aspects of tablet lubrication. J Am Pharm Assoc Sci Ed 1956; 45(1):51–55.
3. Bolhuis GK, Lerk CF, Zijlstra HT, et al. Film formation by magnesium stearate during mixing and its effect on tabletting. Pharm Weekbl 1975; 110(16):317–325.
4. Bolhuis GK, Reichman G, Lerk CF, et al. Evaluation of anhydrous α-lactose, a new excipient in direct compression. Drug Dev Ind Pharm 1985; 11(8):1657–1681.
5. Levy G, Gumtow RH. Effect of certain tablet formulation factors on dissolution rate of active ingredient III. Tablet lubricants. J Pharm Sci 1963; 52(11):1139.

6. Shah AC, Mlodozeniec AR. Mechanism of surface lubrication: influence of duration of lubricant-excipient mixing on processing characteristics of powders and properties of compressed tablets. J Pharm Sci 1977; 66(10):1377–1382.
7. Murthy KS, Samyn JC. Effect of shear mixing on in vitro drug release of capsule formulations containing lubricants. J Pharm Sci 1977; 66(9):1215–1219.
8. Bolhuis GK, Smallenbroek AJ, Lerk CF. Interaction of tablet disintegrants and magnesium stearate during mixing I: effect on tablet disintegration. J Pharm Sci 1981; 70(12):1328–1330.
9. Lerk CF, Bolhuis GK, Smallenbroek AJ, et al. Interaction of tablet disintegrants and magnesium stearate during mixing II. Effect on dissolution rate. Pharm Acta Helv 1982; 57(10/11):282–286.
10. Iranloye TA, Parrott EL. Effects of compression force, particle size, and lubricants on dissolution rate. J Pharm Sci 1978; 67(4):535–539.
11. Soininen A, Kuusivuori P. Influence of the length of mixing time on bioavailability of capsule formulations containing magnesium stearate. Acta Pharm Fenn 1980; 89:215–222.
12. De Boer AH, Bolhuis GK, Lerk CF. Bonding characteristics by scanning electron microscopy of powders mixed with magnesium stearate. Powder Technol 1978; 20(1):75–82.
13. Müller BW. Tribologische Gesetzmässigkeiten und Erkenntnisse in der Tablettentechnologie. 2. Mitt.: Untersuchungen über den Aufbau von Schmierstoffschichten. Pharm Ind 1976; 38(4):394–398.
14. Staniforth JN. Order out of chaos. J Pharm Pharmacol 1987; 39(5):329–334.
15. Johansson ME, Nicklasson M. Investigation of the film formation of magnesium stearate by applying a flow-through dissolution technique. J Pharm Pharmacol 1986; 38(1):51–54.
16. Hussain MSH, York P, Timmins P. A study of the formation of magnesium stearate film on sodium chloride using energy-dispersive X-ray analysis. Int J Pharm 1988; 42(1–3):89–95.
17. Lerk CF, Bolhuis GK. Interaction of lubricants and colloidal silica during mixing with excipients. II. Its effect on wettability and dissolution velocity. Pharm Acta Helv 1977; 52(3):39–44.
18. Nicklasson M, Brodin A. The coating of disk surfaces by tablet lubricants, determined by an intrinsic rate of dissolution method. Acta Pharm Suec 1982; 19(2):99–108.
19. Johansson ME, Nicklasson M. Influence of mixing time, particle size and colloidal silica on the surface coverage and lubrication of magnesium stearate. In: Rubinstein MH, ed. Pharmaceutical Technology: Tableting Technology. Vol. 1. Chichester: Ellis Horwood Ltd, 1987:43.
20. Rowe RC. The coating of tablet surfaces by lubricants as determined by a film/tablet adhesion measurement. Acta Pharm Suec 1983; 20(1):77.
21. Pintye-Hódi K, Tóth I, Kata M. Investigation of the formation of magnesium stearate film by energy dispersive X-ray microanalysis. Pharm Acta Helv 1981; 56(11):320–324.
22. Roblot-Treupel L, Puisieux F. Distribution of magnesium stearate on the surfaces of lubricated particles. Int J Pharm 1986; 31(1–2):131–136.
23. Lerk PC, Sucker H. Interaction of magnesium stearate and talc upon tabletting mixtures, I. Effect on ejection force after compaction. Acta Pharm Technol 1988; 34(2):68–72.
24. Lerk PC, Sucker, H. Interaction of magnesium stearate and talc upon tabletting mixtures, II. Effect on wettability of powder blends. Acta Pharm Technol 1988; 34(2):72–76.
25. Davies MC, Brown A, Newton JM. Chemical characterisation of lubricant films. J Pharm Pharmacol 1987; 39(suppl):122P.
26. Hussain MSH, York P, Timmins P, et al. Secondary Ion Mass Spectrometry (SIMS) evaluation of magnesium stearate distribution and its effects on the physico-technical properties of sodium chloride tablets. Powder Technol 1990; 60(1):39–45.
27. Lerk CF, Bolhuis GK, Smedema SS. Interaction of lubricants and colloidal silica during mixing with excipients. I. Its effect on tabletting. Pharm Acta Helv 1977; 52(3):33–38.
28. Rowe RC. Interaction of lubricants with micro crystalline cellulose and anhydrous lactose—a solubility parameter approach. Int J Pharm 1988; 41(3):223–226.
29. Bolhuis GK, Lerk CF, Broersma P. Mixing action and evaluation of tablet lubricants in direct compression. Drug Dev Ind Pharm 1980; 6(1):15–33.
30. Alpar O, Deer JJ, Hersey JA, et al. The possible use of polytetrafluoroethylene (FLUON) as a tablet lubricant. J Pharm Pharmacol 1960; 21(suppl):6S–8S.
31. Jarosz PJ, Parrott EL. Effect of lubricants on tensile strength of tablets. Drug Dev Ind Pharm 1984; 10(2):259–273.
32. Hölzer AW, Sjögren J. Evaluation of some lubricants by the comparison of friction coefficients and tablet properties. Acta Pharm Suec 1981; 18(3):139–148.
33. Delattre L, Gillard J, Jaminet F, et al. Étude comparative d'agents lubrifiants dans des excipients pour compression directe. J Pharm Belg 1976; 31(5):497–508.

34. Ragnarsson G, Hölzer AW, Sjögren J. The influence of mixing time and colloidal silica on the lubrication properties of magnesium stearate. Int J Pharm 1979; 3(2–3):127–131.
35. Hölzer AW, Sjögren J. Evaluation of sodium stearyl fumarate as a tablet lubricant. Int J Pharm 1979; 2(3–4):145–153.
36. Hölzer AW. Batch to batch variations of commercial magnesium stearates: chemical, physical and lubricant properties. Labo-Pharma Probl Technol 1984; 32(338):28–36.
37. Johansson ME. Granular magnesium stearate as a lubricant in tablet formulations. Int J Pharm 1984; 21(3):307–315.
38. Johansson ME. Investigations of the mixing time dependency of the lubricating properties of granular and powdered magnesium stearate. Acta Pharm Suec 1985; 22(6):343–350.
39. Johansson ME. Influence of the granulation technique and starting material properties on the lubricating effect of granular magnesium stearate. J Pharm Pharmacol 1985; 37(10):681–685.
40. Johansson ME. The effect of scaling-up of the mixing process on the lubricating effect of powdered and granular magnesium stearate. Acta Pharm Technol 1986; 3(1):39–42.
41. Frattini C, Simioni L. Should magnesium stearate be assessed in the formulation of solid dosage forms by weight or by surface area. Drug Dev Ind Pharm 1984; 10(7):1117–1130.
42. Colombo I, Carli F. Comparative evaluation of structure and micromeritics properties of magnesium stearate. Il Farmaco Ed Pr 1984; 39(10):329–341.
43. Müller BW. The pseudo-polymorphism of magnesium stearate. Proceedings of the 1st Int Conf Pharm Techn, Volume IV, Paris, 1977:134–141.
44. Miller TA, York P. Physical and chemical characteristics of some high purity magnesium stearate and palmitate powders. Int J Pharm 1985; 23(1):55–67.
45. Miller TA, York P. Frictional assessment of magnesium stearate and palmitate lubricant powders. Powder Technol 1985; 44(3):219–226.
46. Miller TA, York P, Jones TM. Frictional variation of magnesium stearate lubricants. J Pharm Pharmacol 1983; 35(suppl):42P.
47. Steffens KJ. Die physikalischen Eigenschaften von Magnesiumstearat und ihr Einfluss auf das tribologische Verhalten bei der Tablettierung. Thesis, Marburg, 1978.
48. Bolhuis GK. Enkele aspecten van de formulering en bereiding van tabletten met direct comprimeerbare vulbindmiddelen. Krips Repro, Meppel, 1979:76.
49. Van der Watt JG. The effect of the particle size of micro crystalline cellulose on tablet properties in mixtures with magnesium stearate. Int J Pharm 1987; 36(1):51–54.
50. Bos CE, Bolhuis GK, Van Doorne H, et al. Native starch in tablet formulations: properties on compaction. Pharm Weekbl Sci Ed 1987; 9(4):274–282.
51. Vromans H, Bolhuis GK, Lerk CF. Magnesium stearate susceptibility of directly compressible materials as an indication of fragmentation properties. Powder Technol 1988; 54(1):39–44.
52. Bos CE, Vromans H, Lerk CF. Lubricant sensitivity in relation to bulk density for granulations based on starch or cellulose. Int J Pharm 1991; 67(1):39–49.
53. Egermann H. Zur Beeinträchtigung der Komprimatfestigheit durch Magnesiumstearat in Abhängigkeit von der Mischdauer. 5. Mitteilung der Beiträge zum technologischen Verhalten von Gleitmitteln. Sci Pharm 1978; 46:137–138.
54. Riepma KA, Vromans H, Zuurman K. A coherent matrix model for the consolidation and compaction of an excipient with magnesium stearate. Int J Pharm 1993; 97(1–3):195–203.
55. Duberg M, Nyström C. Studies on direct compression of tablets. VI. Evaluation of methods for the estimation of particle fragmentation during compaction. Acta Pharm Suec 1982; 19(6):421–436.
56. Bossert J, Stamm A. Effect of mixing on the lubrication of crystalline lactose by magnesium stearate. Drug Dev Ind Pharm 1980; 6(6):573–589.
57. Roblot L, Puisieux F, Duchêne D. Etude de la lubrification par le stéarate de magnesium. Influence des proportions de lubrifiant et du procédé de mélange sur les charactéristiques de comprimés. Labo-Pharma Probl Techn 1983; 31(336):843–847.
58. Malmqvist K, Nyström C. Studies on direct compression of tablets. IX. The effect of scaling-up on the preparation of ordered mixtures in double-cone mixers. Acta Pharm Suec 1984; 21(1):21–30.
59. Bolhuis GK, De Jong SW, Van Kamp HV, et al. The effect on tablet crushing strength of magnesium stearate admixing in different types of labscale and production-scale mixers. Pharm Technol 1987; 11(3):36–44.
60. Strickland WA, Higuchi T, Busse LW. The physics of tablet compression X. Mechanism of Action and Evaluation of Tablet Lubricants. J Am Pharm Ass Sci Ed 1960, 49 (1), 35–40.
61. Bowden FP, Tabor DD. Friction and Lubrication of Solids. Paperback edition. Oxford: Clarendon Press, 1986.

62. Müller BW, Steffens KJ, List PH. Tribologische Gesetzmässigkeiten und Erkenntnisse in der Tablettentechnologie. 5. Mitteilung: Zur Methode der Bestimmung der tribologischen Eigenschaften von Feststoffschmiermitteln bei der Tablettierung. Pharm Ind 1982; 44(6):636–640.
63. Hölzer AW, Sjögren J. Friction coefficients of tablet masses. Int J Pharm 1981; 7(4):269–277.
64. Miller TA, York P. Pharmaceutical tablet lubrication. Int J Pharm 1988; 41(1–2):1–19.
65. Junghäni H, Bier HP, Sucker H. Die Bestimmung der minimalen Schmiermittelkonzentration von Tablettiermischungen durch Leistungsmessung an Planetenmischern. Vorläufige Mitteilung. Pharm Ind 1981; 43(10):1015–1018.
66. Hölzer AW, Sjögren J. Comparison of methods for evaluating of friction during tableting. Drug Dev Ind Pharm 1977; 3(1):23–27.
67. Kikuta J, Kitamori N. Evaluation of the die wall friction during tablet ejection. Powder Technol 1983; 35(2):195–200.
68. Kikuta J, Kitamori N. Frictional properties of tablet lubricants. Drug Dev Ind Pharm 1985; 11(4): 845–854.
69. Hölzer AW, Sjögren J. The influence of the tablet thickness on measurements of friction during tableting. Acta Pharm Suec 1978; 15(1):59–65.
70. Hölzer AW. Sodium stearyl fumarate (Pruv). A successful alternative to magnesium stearate. Acta Pharm Suec 1987; 24(2):71–72.
71. Schrank-Junghäni H, Bier HP, Sucker H. Studies in quantitative determination of lubricant properties for tableting processes. Pharm Technol 1983; 7(9):71.
72. Schrank-Junghäni H, Bier HP, Sucker H. The measurement of die-wall forces to determine the minimum concentration of lubricant needed for tablet formulations. Acta Pharm Technol 1984; 30(3):224–234.
73. Staniforth JN, Ahmed HA. Influence of ternary components on compressibility of microcrystalline cellulose following blending with magnesium stearate. J Pharm Pharmacol 1986; 38(suppl):50P.
74. Bolhuis GK, Lerk CF. Ordered mixing with lubricant and glidant in tableting mixtures. J Pharm Pharmacol 1981; 33(12):790.
75. Rowe RC. Interactions in ternary powder system microcrystalline cellulose, magnesium stearate and colloidal silica—a solubility parameter approach. Int J Pharm 1988; 45(3):259–261.
76. Staniforth JN, Ahmed HA. Influence of ternary components on lubrication of microcrystalline cellulose following blending with magnesium stearate. J Pharm Pharmacol 1987; 39(suppl):68P.
77. Staniforth JN, Cryer S, Ahmed HA, et al. Aspects of pharmaceutical tribology. Drug Dev Ind Pharm 1989; 15(14–16):2265–2294.
78. Gruber P, Gläsel VI, Klingelhöller W, et al. Presskammerbeschichtung, ein Beitrag zur Optimierung der Tablettenherstellung. Pharm Ind 1988; 50(7):839–845.
79. Hersey JA. Avoiding powder-mixing problems. Aus J Pharm Sci 1972; 1:76.
80. Lael BJ, Irvington H, Pinto PJ. Method of making a compressed tablet. US pat 3.042.531, 1962.
81. Fukumori Y, Carstensen JT. Lubricative properties of mixtures of dicalcium phosphate dihydrate and magnesium stearate. Int J Pharm Tech Prod Mfr 1983, 4 (4), 1–5.
82. Bavitz FJ, Shiromani PK. Granulation surface area as basis for magnesium stearate concentration in tablet formulations. Drug Dev Ind Pharm 1986; 12(14):2481–2492.
83. Hölzer AW. Evaluation and characterization of tablet lubricants. Acta Pharm Suec 1981; 18(2):72.
84. Stamm A, Kleinknecht A, Bobbé D. Etude de quelques lubrifiants pour compression directe. Deuxième partie. Comparaison des résultats obtenus avec différents lubrifiants. Labo-Pharma Probl Techn 1977; 25:215.
85. Staniforth JN. Use of hydrogenated vegetable oil as a tablet lubricant. Drug Dev Ind Pharm 1987; 13(7):1141–1158.
86. Lindberg NO. Evaluation of some tablet lubricants. Acta Pharm Suec 1972; 9(3):207–214.
87. Vezin WR, Khan KA, Pang HM. Adjustment of precompression force to reduce mixing-time dependence of tablet tensile strength. J Pharm Pharmacol 1983; 35(9):555–558.

10 Crystal engineering and particle design for the powder compaction process

Anant Paradkar and Peter York

INTRODUCTION

For several decades in addition to investing huge resources in the drug discovery process for identifying new chemical entities (NCEs), pharmaceutical interests have been engaged in the engineering of drug substances to overcome challenges in drug delivery linked to their pharmaceutical properties. These physicochemical and mechanical properties can be classified as primary and secondary functionalities of the particles (Fig. 1). Particle size, shape, crystal habit, crystal form, density, porosity, etc., are primary properties of solids, whereas flowability, compressibility, compactibility, consolidation, dust generation, and air entrapment during processing are secondary ones.

In the pharmaceutical industry, rapid and efficient production of small and/or substantially uniform particles is especially needed, since variables such as particle size, particle size distribution, polymorphic form, and crystal habit have been found to affect in vitro and in vivo performance of pharmaceuticals. For example, in the manufacture of tablets these properties govern performance of the crystalline powder in the die during compression. Thus, it is critically important to employ efficient and effective particle design techniques with the ability to produce small, uniform particles with desired primary and secondary properties for the development of pharmaceutical products.

CRYSTAL STRUCTURE AND POLYMORPHISM

A crystal is a homogeneous particle of a chemical element or a compound formed by solidification under favorable conditions. Boundary surfaces of crystal are planes symmetrically arranged to each other at definite angles, in a definite geometric form. The orientation and shape of the structural units, known as unit cells, are defined by the translational vectors (a, b, and c) and the individual angles between these vectors (α, β, and γ) in a crystal lattice. A finite number of symmetrical arrangements of constituent atoms are possible for a crystal lattice, termed as crystal forms (internal structure) and the arrangement of the planar faces of crystal is termed crystal habit (external appearance). The ratio between the lengths of the translational vectors for a given crystal must be constant.

Crystalline solids can exist as polymorphs and molecular adducts (solvates and hydrates). The ability of a specific compound to exhibit alternative crystal forms is known as polymorphism. Different crystalline polymorphs differ in molecular packing as well as crystal lattice energy and entropy. Based on the type of transformation of one form to another, polymorphs are further classified as enantiotropes and monotropes. Polymorphs do exhibit differences in physicochemical and mechanical properties such as hardness, powder flow, compactibility and compressibility. For example, the orthorhombic and monoclinic polymorphic forms of acetaminophen exhibit different mechanical behavior. The monoclinic form is difficult to compress directly because of a poor densification property (1). In contrast, the orthorhombic form is suitable for direct compression because of the presence of parallel sliding intermolecular planes within the crystal lattice (2). For a drug material exhibiting polymorphism, one of the polymorphs will be the most stable form due to its lowest crystal lattice energy (stable polymorph) compared with other form(s), while the other(s) form is known as a metastable polymorph that will undergo transition to the stable polymorph but over different

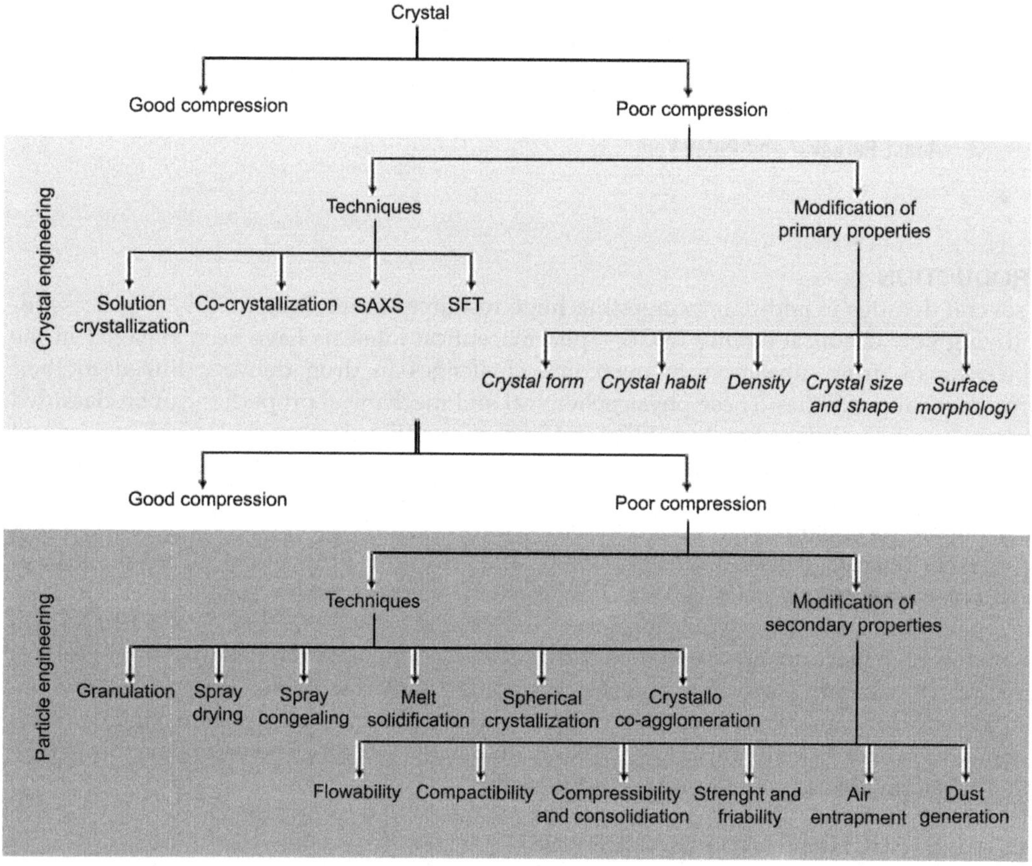

Figure 1 Primary and secondary functionalities of the particles.

time scales ranging from seconds to years. Changes in polymorphic form result from changes in processing variables like temperature, pressure, etc. (3), while crystallization of the same drug from alternative solvents may generate different polymorphs (4).

A crystalline solid with a solvent molecule(s) incorporated into the crystal structure is known as a solvate. The solvent may be a pure organic solvent or a mixture of solvents. When the solvent is water, the solvate is called a hydrate. It is estimated that about one-third of all active pharmaceutical ingredients (APIs) can form hydrates due to the small size of the water molecule and its potential to form hydrogen bonds during structural void filling (5).

Crystalline forms are considered to be more stable when compared to amorphous forms. Amorphous substances have high entropy and hence are thermodynamically unstable, and which on storage may undergo transition to corresponding crystalline forms (devitrification). Techniques like spray drying and melt quenching have been used to generate amorphous solids which do not have internal crystal structure but achieve a state of dynamic arrest for the molecules. The lack of definite crystal lattice structure may impart poor compressibility characteristics to such solids. If they are melted and cooled below the glass transition temperature (vitrification), a glass can be formed, which is tough and undergoes brittle fracture on compression. It has also been observed that amorphous forms of novobiocin, indomethacin, glibenclamide, and griseofulvin exhibit higher initial levels of aqueous solubility than their crystalline counterparts (6,7).

During tabletting other formulation additives may cause unexpected effects on polymorphic drugs. For instance, for carbamazepine granules a change in polymorph has been observed, which depends on the binder solution used during the wet granulation process, and which resulted in altered mechanical strength of tablets (8). In some cases, tabletting performance can be depicted if crystal structures of the different polymorphs of a drug are available and studied by molecular simulation techniques (9). Microindentation and nanoindentation techniques have also been used to study fracture properties of single crystals and link results to the performance of particulate solids during compression (10,11). Polymorphic materials with well-defined intermolecular slip planes are thought to exhibit good tabletting performance and the detailed investigation of slip-planes of acetaminophen has been described by Liao and Wiedmann (11). This study has also correlated crystal hardness, creep, and elastic modulus with the indentation force of acetaminophen.

CRYSTAL HABIT

Crystal habit can be expressed in terms of the aspect ratio, which is the ratio of length to width of solid particles. Aspect ratio values below 5 are generally preferred to avoid problems regarding flow property, while values approaching unity are an indication of spherical or cubic shape and are considered a desirable characteristic for pharmaceutical particulate solids. It has been observed that crystal habit can influence compaction profiles through its effect on the relative orientation of crystallites during compression (12).

Equidimensional crystals are usually preferred in the industry due to better handling and processing characteristics such as flowability, compatibility, and compressibility (13,14). In addition, crystal habit profoundly affects surface-dependent properties such as drying, dissolution, sedimentation, dispersibility, mixing, and bulk density. It has also been observed that external factors such as crystallization solvent, temperature, extent of supersaturation, agitation, impurities, and rate of crystallization can modify the resultant crystal habit. To change crystal habit, alternative solvent crystallization has been widely attempted. One example is for celecoxib where different solvent crystallization approaches and vapor deposition have been studied (15), as this material exhibits poor manufacturing and handling due to its needle-shaped crystals. The major effect of solvents on particle habit has been attributed to preferential adsorption of solvent molecule on specific crystal faces, which delays their growth (16).

A similar mechanism has been proposed for improving the powder flow properties of citric acid. A relative humidity of 69.9% was used to obtain the pure monohydrate form from the anhydrate form with nearly identical particle size and morphology but different surface properties. This transformation reduced the strength of interactions between crystal surfaces of citric acid monohydrate leading to an improvement in flow properties (17). This behavior was analyzed and demonstrated using shear cell testing, which showed a higher unconfined yield strength (f_c) for anhydrous citric acid, which was indicative of poor flow compared to the monohydrate (Fig. 2).

Crystals of the same crystal form are termed isomorphs. Mechanistic analysis of the formation of isomorphs has shown that solvent polarity controlling solute-solvent interaction at the molecular level together with other solvent properties such as viscosity and surface tension affect the crystal habit (4). A report (18) describes the change in crystal habit of phenytoin from smooth to rough surfaces due to reduction in interfacial tension favoring solute-solvent interactions. For celecoxib various crystal habits were generated by different solvent and processing conditions. Figure 3 shows acicular-shaped and lath-shaped crystal habits obtained by solvent crystallization and the vapor diffusion method. These crystal habits were found to be isomorphs of native acicular celecoxib (19). For this study, the formation of solvates was ruled out from analysis of the energy transitions seen during thermal analysis.

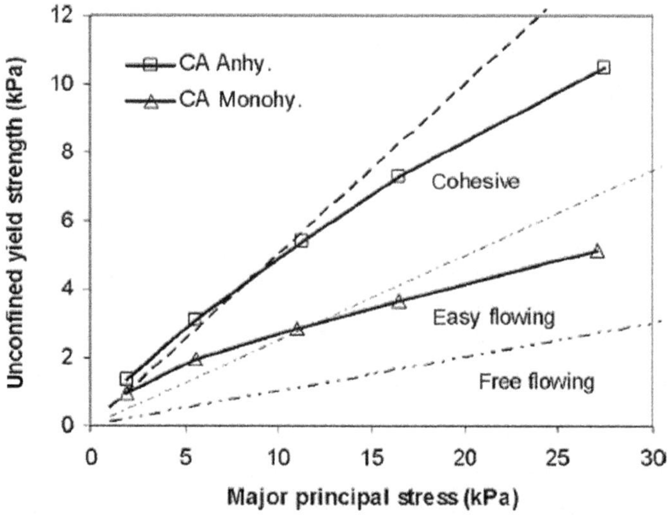

Figure 2 Flow functions of citric acid anhydrate and monohydrate. Flow properties of the monohydrate are significantly better. The Jenike flow classification criteria are shown as broken lines. *Source*: From Ref. 17.

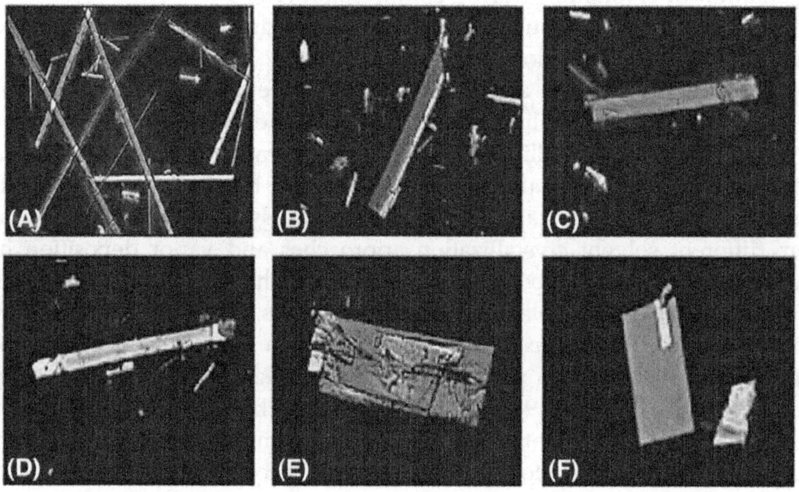

Figure 3 Celecoxib habits: polarized photomicrographs of solvent crystallized and vapor diffusion products at 63x (x10): (**A**) recrystallized from butanol, $T = 25°C$, concentration $= x$; (**B**) recrystallized from toluene, $T = 25°C$, concentration $= x$; (**C**) recrystallized from chloroform, $T = 25°C$, concentration $= x$; (**D**) recrystallized from carbon tetrachloride, $T = 25°C$, concentration $= x$; (**E**) recrystallized from toluene, $T = 60°C$, concentration $= x - 25\%$; (**F**) recrystallized from chloroform, $T = 60°C$, concentration $= x - 25\%$. *Source*: From Ref. 15.

Since crystal habit is directly related to the external morphology and topography of particulate solids, it has major importance in determining their flow properties and compressional behavior. It is also clear that changes in crystal habit will alter the size distribution and surface characteristics and hence modify the contact area and alignment of particles during powder compression. In further studies with the acicular and lath type crystals of celecoxib, an improvement in compressibility of the lath-shaped crystals was noted (15). However, the acicular form did not exhibit any marked improvement in flowability and compressibility even after addition of Aerosil 200.

More specifically it was noted that both acicular and lath crystals of celecoxib consolidated by plastic deformation (15). Modified crystals of lath habit exhibited a lower net work done than acicular particles indicating an improvement in compression behavior. The higher work done for the acicular form was attributed to rough particle surfaces and high surface area, responsible for adherence of material to the tablet punches. In the case of laths, the propensity to consolidate was accentuated due to a small volume reduction during particle packing. Thus, in addition to initial particle rearrangement, crystal habit can affect the entire compaction profile. Hence, intentional modification of crystal habit by altering crystal nucleation and growth can be carried out to obtain a crystal habit with improved processability and for that purpose parameters such as solvent polarity, degree of supersaturation, viscosity, temperature, agitation, etc., can be manipulated.

Acetaminophen can be considered as a classic example of a poorly compressible powdered solid, wherein tabletted monoclinic form I (stable) produces tablets with a high capping tendency due to rigid behavior of molecules in the crystal lattice. In contrast, orthorhombic form II shows better compression behavior (20) due to sliding intermolecular planes. However, the metastable orthorhombic form may undergo transition to monoclinic form.

To study the compression behavior of solids, the *T*-factor has been devised (21), which take into account the compressibility (factor S_{Fmax}/F_{max}) and tablettability (crushing strength/ F_{max}) of solids. In this case, F_{max} is maximum upper punch force and S_{Fmax} is displacement of upper punch at F_{max}. In addition, plastic and elastic deformation has also been considered in calculation of the *T*-factor. A higher value of *T*-factor for a material indicates good tablettability. Studies on various crystal habits of ibuprofen (habit I and II) and acetaminophen (I to V) have shown major effects of habit on the handling and processing qualities of crystals. Different crystal habits exhibited different flowability values due to difference in adhesive and cohesive forces. In the case of ibuprofen, free-flowing crystals (thin plates, habit II) were shown to have the best compaction behavior. For acetaminophen, prismatic crystals (habit III) and small plates (habit V) were preferred, with higher values of *T*-factor compared crystals with other habits (Fig. 4). When excipients were blended with the pure drug prior to compression, the values of *T*-factor rose indicating better compressibility (21,22). Thus, improvement in compaction behavior can also be achieved by crystal habit modification and addition of excipient.

Crystal habit modification by the solvent selection approach does however suffer from shortcomings such as solvent cost and potential toxicity issues, poor crystallization efficiency and the purity requirements of the final product. The incorporation of pharmaceutically accepted excipient additives during crystallization as crystal habit modifiers of drug substances is another practical alternative. For example, hydroxypropyl cellulose (HPC) has been used to modify the crystal habit of erythromycin A dihydrate for improved compaction,

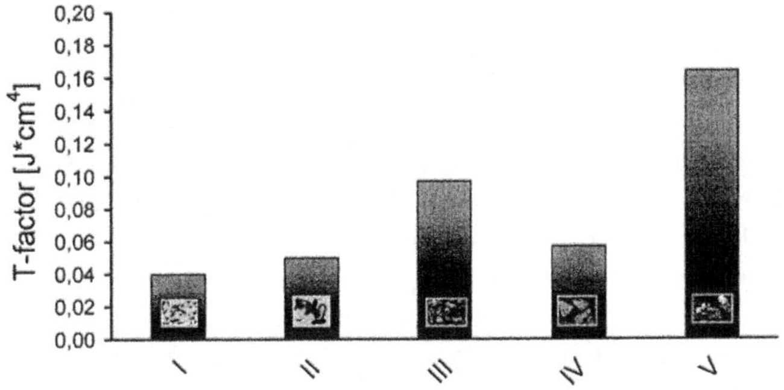

Figure 4 *T*-factor values for tabletting properties of acetaminophen. *Source*: From Ref. 21.

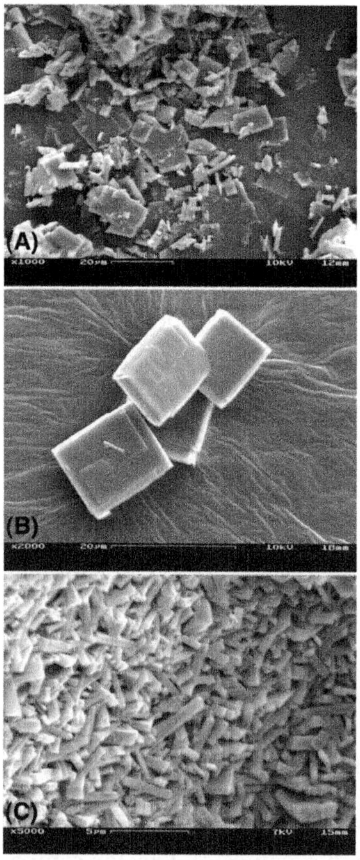

Figure 5 SEM images of erythromycin A dihydrate crystals grown in the presence of various concentrations of HPC: (A) 0 (reference), (B) 0.45 (L crystals), and (C) 4.5 weight percent (H crystals). *Source*: From Ref. 26.

as cellulose derivatives can interact with drug molecules via H-bonding (9,23–25). Change in crystal habit of erythromycin A dihydrate has been found to be HPC concentration dependent (acicular to plate like, see Fig. 5), which may be due to a change in degree of supersaturation during the crystallization process.

In addition to recrystallization of drug from alternative solvent and addition of polymeric additives, routine processing such as wet granulation has also been shown to lead to transition of anhydrous drug to hydrous forms which on subsequent drying reverts to the anhydrous state. Theophylline, upon aqueous wet granulation followed by drying, exhibited such multiple transformations (26), with theophylline anhydrate converted to monohydrate during wet granulation and reverted to anhydrate on dying. The crystallinity of processed theophylline anhydrate was lesser than that of unprocessed anhydrate which resulted in increased mechanical strength of the tablet. Simultaneously the transformation of anhydrate to monohydrate form during dissolution studies was effected primarily at the surface and reduced with increased distance from the surface.

In addition to established methods of compression requiring large specimens and careful sample preparation, fracture mechanics studies using single crystal (e.g., microindentation technique) are gaining popularity to understand how the fracture behavior affects compression behavior (27–29). This type of technique uses small single crystal samples to help in determine the indentation harness and Young's modulus simultaneously. The technique can also provide

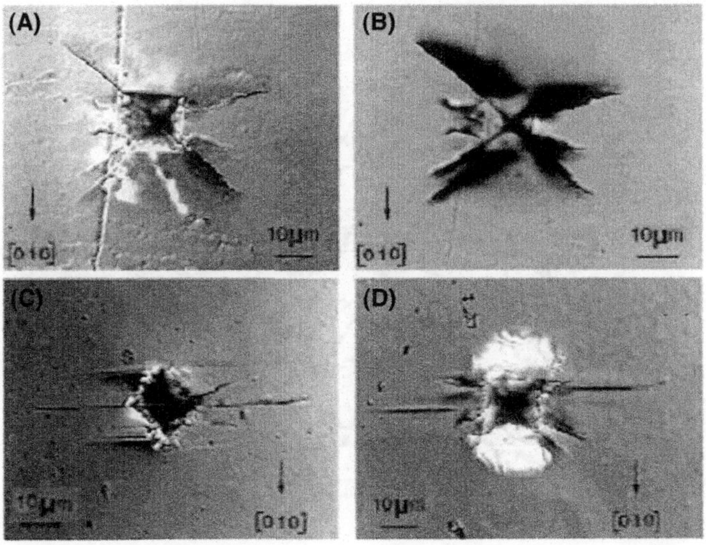

Figure 6 (**A**) Reflection and (**B**) transmission photomicrographs of a Vickers impression on the {001} face of acetaminophen showing radial cracks (applied load 4 0.147 N). Reflected light photomicrographs of a Vickers impression (0.147-N load) showing radial cracks and some slip traces on {201} surfaces. (**C**) indenter diagonal aligned parallel to [010] direction and (**D**) indenter diagonal rotated 45° to the [010] direction. *Source*: From Ref. 10.

detail of the fracture toughness, K_c, (critical stress intensity factor for tensile fracture under plane stress conditions), while the toughness data obtained by this technique is semiempirical because anisotropy in the stress field around the indentation complicates the process. Acetaminophen single crystals have been studied by the microindentation technique and anisotropic fracture behavior observed for the {001} surfaces, whereas slip traces were observed running parallel to the direction of {100} (10). From Figure 6 the formation of slip traces are clearly evident in the vicinity of indentation. The molecular packing on the ab crystallographic plane of monoclinic acetaminophen clearly demonstrates the slip planes as indicated by dotted lines in Figure 7.

Similarly, the nanoindentation technique has been used to study the slip planes of potassium chloride. Figure 8 indicates schematic diagram showing the predicted slip planes in potassium chloride and paracetamol (11).

TECHNIQUES OF PARTICLE ENGINEERING

Various new techniques have been devised or adapted from those used in other material processing industries to bring about the engineering of particulate pharmaceutical solids. Among them spray drying (SD) (30), spray congealing (SC) (31), extrusion-spheronization (ES) (32), melt solidification (MS) (33), melt sonocrystallization (MSC) (34), melt pelletization (MP) (35–37), solution atomization and crystallization by sonication (SAXS) (38), spherical crystallization (SCZ) (39,40), and crystallocoagglomeration (CCA) (41,42) are considered as promising methods. At present ES and SD have been used at an industrial level, whereas SAXS, SCZ, and CCA are being further researched and developed to explore the potentially wide array of applications.

Spray Drying

The spray drying technique often leads to the formation of totally or partially amorphous phase of a drug substance. As a result of the amorphous highly energetic form produced,

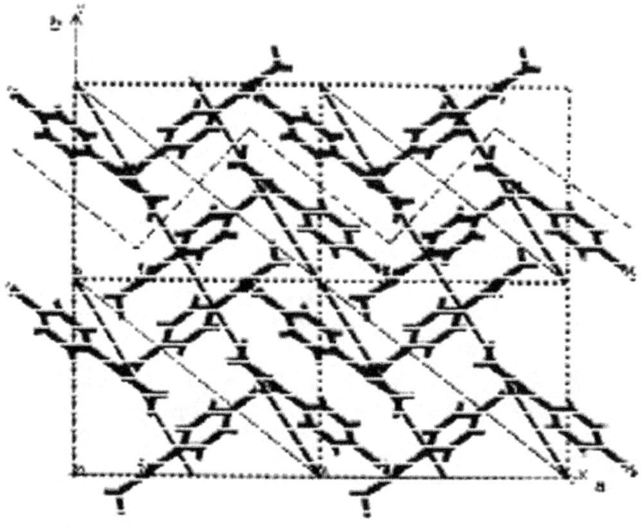

Figure 7 Molecular packing on the ab plane of monoclinic acetaminophen. The serrated dotted line indicates the position of the (010) cleavage plane. The fine and thick diagonal lines represent the (110) and (210) planes, respectively. *Source*: From Ref. 10.

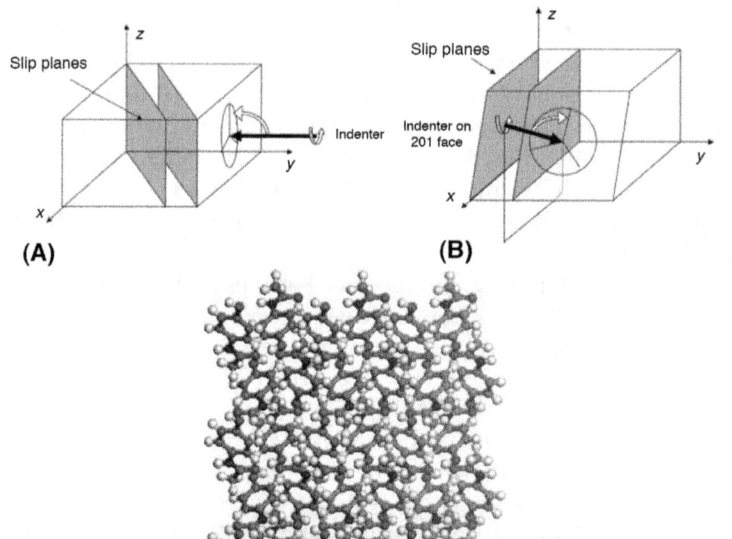

Figure 8 Schematic diagram showing the most probably slip planes in the crystal structure of (**A**) potassium chloride and (**B**) acetaminophen with respect to the unit cell dimension (*x*, *y*, and *z* axes), the indenter tip with pyramid shape, and the plane that was indented with the recrystallized samples. In addition, arrows indicate the angles of ascension and declination reflecting the range of orientation possibilities of indenting the quenched samples. (**C**) Crystal structure of acetaminophen viewed at (201) face. *Source*: From Ref. 11.

improved dissolution and bioavailability of poorly soluble drugs can be achieved. However, in a highly energetic and thermodynamically unstable form, the spray dried solid will over time return to the least energetic, more stable crystalline form. To avoid this conversion and increase the stability profile of the amorphous state of drug substance, various stabilizing agents such as polymers, lipids, and adsorbents have been used. Examples of drug substances studied by spray drying include celecoxib (43), etoricoxib (44), simvastatin (45), curcumin (46), and acetazolamide (30).

Supercritical Fluid Technology

Over the last decade many applications have been studied and developed for crystal and particle engineering using supercritical fluid assisted processing. Supercriticality is the state of a fluid above its critical point, that is, its critical temperature T_c and critical pressure P_c where the supercritical fluid has both gaseous and liquid-like properties. This state of fluid exhibits appreciable solvation power, allows higher rates of mass transfer and is highly compressible. The solvation power of the fluid in the supercritical state which is controlled by changes in temperature and/or pressure, can be modified by incorporating a small amount of volatile cosolvent such as ethanol or acetone. Carbon dioxide is commonly used for crystal and particle engineering of pharmaceuticals because of its low critical temperature (31.2°C) and pressure (7.4 MPa). It is nonflammable, nontoxic, and inexpensive. This process offers a single step green crystal and particle engineering technique with solvent recovery and recycling capability.

There are different methods used for formation of particles using supercritical fluids. The rapid expansion of supercritical solution (RESS) technique uses supercritical fluid as a solvent. The solute is dissolved in the supercritical fluid and the resulting solution is rapidly expanded on reducing pressure and due to rapid expansion its solvent power decreases suddenly causing the solute to precipitate. Although the method is simple, it is applicable only to solutes that exhibit high solubility in the supercritical fluid or in the presence of small amounts of a cosolvent (47).

Supercritical Antisolvent (SAS) is another method, which uses supercritical fluid as an antisolvent. The solute is dissolved in an organic solvent, which is brought in contact with supercritical fluid, which acts as antisolvent. The rate of migration of the antisolvent into the organic phase determines the rate of nucleation and crystal growth. The particle size distribution of the product is controlled by the ratio of solvent to supercritical fluid, rate of addition of supercritical fluid, and miscibility of solvent and supercritical fluid. SAS is suitable for crystallization of compounds poorly soluble in supercritical fluids.

Hanna and York (48) developed the solution-enhanced dispersion by supercritical fluids (SEDS) process, in which the drug solution and supercritical fluid are introduced simultaneously into a vessel maintained under controlled temperature and pressure through a coaxial nozzle with a mixing chamber. The interdiffusion of supercritical fluid and solvent from the solvent droplets causes faster nucleation and generation of smaller particles (49). SEDS is a highly controlled and reproducible process that has been scaled-up and operated under GMP manufacturing environments for many drugs including albuterol sulphate and budesonide (50).

Wet Spherical Agglomeration

This is a technique of spherical agglomeration in which agglomerates of suspended particles in liquid suspension are prepared by the addition of a bridging liquid, which plays a role in wetting of the solid surface. This technique was developed by Smith and Puddington (51) and is carried out in two ways.

Selective Agglomeration
This is a simple separation technique in which one suspended solid can be separated from other suspended solids or liquids by the introduction of a suitable bridging liquid (20% calcium chloride, ethanol + chloroform, etc.). In this case, the shape of agglomerates is not controlled.

Spherical Agglomeration
In spherical agglomeration, a suspension of fine particles of the drug in an organic liquid is agitated with the addition of a small amount of a bridging liquid, which preferentially wets the particles to be agglomerated. This process can be carried out in a cylindrical vessel equipped with a revolving agitator. The size of the agglomerate can be manipulated by various experimental factors such as amount of bridging liquid, agitation speed, and agitation time. In spherical agglomeration, when a considerable number of particles are suspended in a bridging liquid, some particles are completely or partially dissolved, and undissolved particles are converted to spherical agglomerates on continuous stirring (52–54).

Spherical Crystallization

Spherical crystallization (SC) is a particle size–enlargement technique devised by Kawashima and his coworkers (52,55,56) for simultaneous crystallization and agglomeration of solids. The mechanism of SCZ is illustrated in Figure 9. In this technique, agglomeration of crystallized solids is carried out by the use of a bridging liquid, which forms liquid bridges between crystallized particles. Various methods have been developed based on SC, to improve micromeritic, mechanical, compressional, and drug release properties of agglomerates produced. These include the solvent change method (SCM), ammonia diffusion (AD) method, quasiemulsion solvent diffusion method (QESD), and neutralization technique (NT) (57). The SC technique is associated with some limitations such as suitability for size enlargement of large dose drugs, which are directly compressible without any excipient, and this method is not useful for multidrug combinations due to difference in to the physicochemical properties of component drugs. In addition, the entire SC process requires a flame proof facility to handle

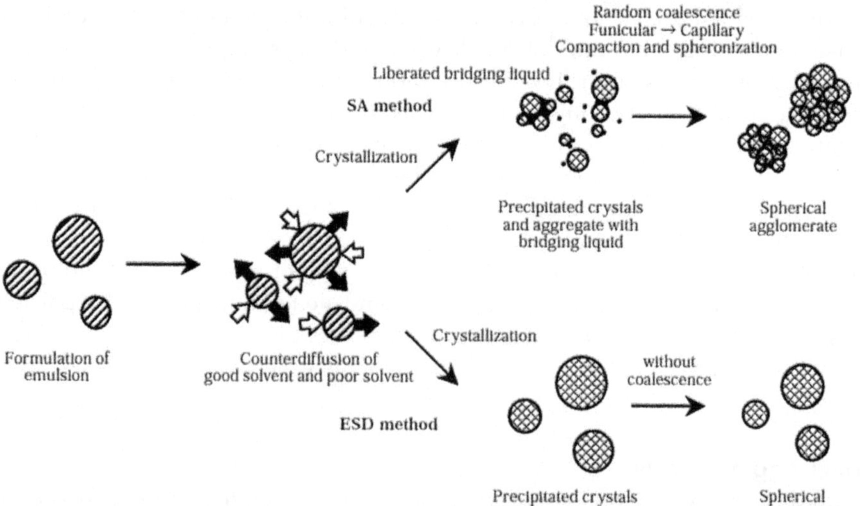

Figure 9 Schematic representation for mechanism of spherical crystallization. *Source*: Reproduced from Ref. 59.

Table 1 Drugs Agglomerated by SC

Method	Drug	Solvent system	Results	References
SCM	Salicylic acid	Ethanol, chloroform, water	Improved micromeritics	56
SCM	Naproxen	Acetone, hexanol/octanol/ toluene, water	Improved micromeritics	58
SCM	Propyphenazone	Good solvent (ethyl alcohol), nonsolvent (DW), bridging liquid (isopropyl acetate)	Improved flowability and compression behavior	59
QESD	Bucillamine	Ethanol, HPMC 1% weight/ volume in water	Improved micromeritics	60
QESD	Bucillamine	Ethanol, water, dichloromethane	Improved micromeritics	60
SCM	Aspirin	Acid buffer (pH 2.5), methanol, and chloroform	Improved flowability	61
SCM	Trimethoprim	Water-methanol-chloroform with PEG	Improved flowability and compressibility	62
SCM	Fenbufen	Tetrahydrofuran—DM water, isopropyl acetate (bridging liquid)	Improved dissolution capacity	63
SCM	Flurbiprofen	Acetone-water-hexane	Improved flowability, wettability, and compaction	64
QESD	Naproxen	Acetone, water with HPC	Improved flowability and compressibility of drugs	65
QESD	Ketoprofen	Acetone/DM water	Improved compressibility	66

Abbreviations: SC, spray congealing; SCM, solvent change method; QESD, quasiemulsion solvent diffusion method.

the organic solvents. Some examples of particulate solids reported to have been prepared by SC and SC-related methods are listed in Table 1.

Crystallo-coagglomeration

Crystallo-coagglomeration (CCA), an extension of SC, has been developed by Kadam et al. (41,42) for the SCZ and agglomeration of pharmaceuticals. While SC is applicable for particle size enlargement of large dose drugs having good compressibility, CCA is applicable for particle size enlargement for a wider range of materials; low dose, high dose, one, two, or more drugs in combination with or without diluent.

For over a decade, CCA has been explored as a technology to produce spherical agglomerates required for design of multiparticulate drug delivery systems (MUPS). The agglomerates obtained by this technique have shown improvement in micrometric, mechanical, and compressional properties (67–69). With the use of suitable excipients and polymers, modified drug release can also be achieved from the intact agglomerates or compacts thereof. Moreover, agglomerates obtained by this technique can be used as directly compressible tablet intermediates and/or spheres to be encapsulated.

In the process of CCA, crystallization of drug and agglomeration of particles takes place simultaneously leading to the formation of matrix beads having a uniform dispersion of crystallized substance/drug in the body of diluent/insoluble substance. The system design for CCA recommends the use of a good solvent to solubilize drug(s), a poor solvent to enable precipitation/crystallization of drug(s), and a bridging liquid that essentially has to be immiscible with the poor solvent to form the liquid bridges between crystallized particles and insoluble solids. In some cases, the bridging liquid also acts as a good solvent. The technique of

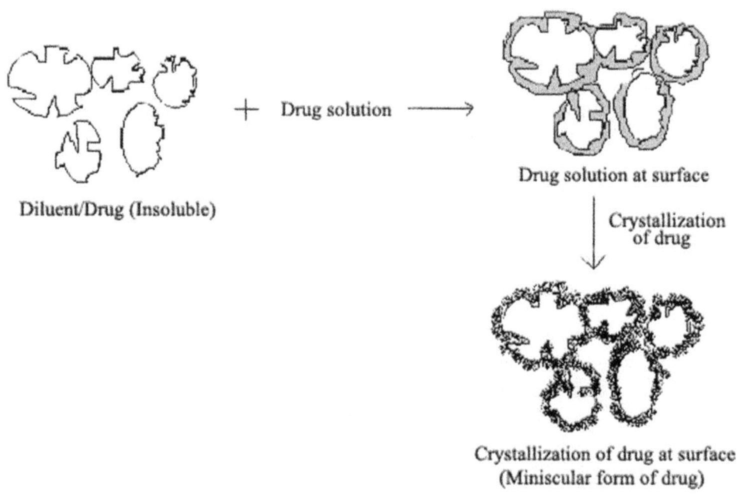

Figure 10 Schematic representation of crystallo-coagglomeration process.

CCA is schematically represented in the Figure 10, wherein the formation of the miniscular form of drug due to crystallization of drug on the diluent/insoluble drug has been illustrated.

To date, two methods have been developed for CCA. The SCM involves simultaneous crystallization and agglomeration of two or more drugs from good solvent and a bridging liquid by the addition of a poor solvent. The second method involves crystallization of drug from a system containing good solvent and bridging liquid and its simultaneous agglomeration with insoluble drug/diluent by the addition of a poor solvent (67,68). The suitability of either of these methods requires knowledge of the physicochemical properties of the drug and solvent system. Once selected, then processing can be performed in a vessel such as that described by Morishima et al. (60) for SC crystallization. The controlled agitation of the contents in the vessel is carried out using a propeller agitator until the end point of the agglomeration process when spherical agglomerates can be obtained. The end point of the agglomeration process can be assessed by the required size growth of agglomerates, clarity of supernatant, and complete vaporization of organic solvent from the system (69).

Recently, Pawar et al. (68) have applied CCA to the agglomeration of ibuprofen-talc and obtained directly compressible spherical agglomerates. It has been observed that compacted agglomerates of ibuprofen-talc show sustained zero order release. In this case, drug release retardation was attributed to the hydrophobic nature of talc, which was used as an inert diluent. The authors have also studied the effect of different polymers on the quality attributes of such agglomerates (70), and in another study the preparation of direct compressible agglomerates of ibuprofen-paracetamol as a tablet intermediates has been described (67). Preparation of directly compressible naproxen-starch and sodium starch glycolate agglomerates for direct compression has been reported by Maghsoodi et al. (65) using CCA. The processing characteristics of the agglomerates, such as flowability, packing, and compactibility were dramatically improved, resulting in successful direct tabletting without capping. The main factor in the improvement of the flowability and packing was attributed to a major reduction in interparticle friction due to the spherical shape of the tabletted particles. The dissolution rate of naproxen from the naproxen-disintegrant agglomerates was enhanced with increasing amounts of disintegrant. Sanghavi et al. (71) have reported that, large surface area available due to the miniscular form of drug particles in CCA, may have improved dissolution and ultimately bioavailability.

Extrusion/Spheronization

Extrusion spheronization (ES) is a multiple step process used to produce beads/pellets. To date, it has been practiced in the pharmaceutical industry for the preparation of blank pellets and drug loaded pellets (direct pelletization). The process of ES involves dry blending and wet granulation as initial stages. Subsequently, the wet mass is extruded through an extruder screen giving rise to cylindrical extrudates. In the next stage, the small pieces of extrudates (cylinders) are spheronized by the action of a rotating friction plate. The spherical pellets obtained are then dried and subject to further fuctionalization if required, by the application of a polymer coating.

Recently, fluidized bed rotary processors have been used to prepare compaction-ready pellets, similar to ES, but as a single step process. This technique, being single step, has curtailed the time, space, and manpower requirement, reducing the processing costs (72,73).

Solution Atomization and Crystallization by Sonication

The solution atomization and crystallization, or crystallization by sonication (SAXS) technique, consists of three interdependant processes. As an initial step, aerosol droplets of the dissolved drug substance from a carrier solvent are generated using a suitable aerosol generator. These droplets containing highly supersaturated drug are then collected in a crystallization vessel containing a nonsolvent for the drug. The application of ultrasonic waves to the crystallization vessel is done to induce homogeneous nucleation and crystal growth in a controlled manner, resulting in high purity, micron-sized crystalline particles in a single-step operation (38). The particles produced by SAXS are more reproducible in both shape and size with a relatively uniform particle size distribution than from conventional crystallization and are composed of crystalline drug.

Studies on the preparation of acetaminophen particles by SAXS have indicated the resulting particles to exhibit uniform particle size. The degree of sphericity and particle-size characteristics were manipulated by the solute concentration of the drug solution. SAXS-produced particulates were more uniform in shape with increased surface roughness when compared to micronized acetaminophen particles (38).

Melt Sonocrystallization

As mentioned earlier, ultrasound was introduced several years ago into pharmaceutical processing as a technique to improve the particle formation behavior of drug following solution preparation. In this technique, the initial nucleation stage of crystallization is influenced by the application of ultrasound, which leads to the reduction of width of the metastable zone in crystallization and ultimately nucleation starts at a lower level of supersaturation (74,75). Propagation of ultrasonic waves through the liquid medium results in the formation of cavitation bubbles and formation of a localized hot spot, along with an extremely large temperature and pressure gradient takes place when a cavitational bubble implodes. When solid particles are present in the fluid system, the cavitational event might occur symmetrically or asymmetrically, depending on the proximity and size of the solids. Particles that are close to each other move away from the cavitational event in a radial direction with high speed due to the formation of particle acceleration shock waves. The high force collisions between the particles result in particle breakage (76) generating fine particles.

Recently, MSC of ibuprofen has been attempted (34) to generate irregular agglomerates with porous surfaces. It was found that specific surface area and intrinsic dissolution rate increase with the treatment of ultrasonic energy during particle formation. A change in crystal habit of ibuprofen was confirmed by scanning electron microscopy, resulting in improved compressional properties and reduction in particle sticking to the die wall. Similarly, MSC of celecoxib formed a viscous melt when processed below its glass transition temperature. The particles obtained after sonocrystallization were porous, irregular in shape, amorphous in nature, and showed increase in apparent solubility. Amorphous particles formed by the process also exhibited higher stability in the amorphous state when compared with particles obtained by melt quenching (77).

COCRYSTALLIZATION

The range of different crystalline forms possible for a drug substance or API is shown in Figure 11. While salts of APIs have been routinely used, the area and potential of cocrystals is arousing much current interest. A cocrystal is made up of two or more components (API/guest(s)), usually in a stoichiometric ratio, each component being an atom, ionic compound or molecule held together by noncovalent forces (78). The guest molecule must be a pharmaceutically acceptable compound which may include food additives, preservatives, pharmaceutical excipients, vitamins, minerals, amino acids, neutraceuticals and other biomolecules, as well as other APIs. In forming a cocrystal, the physicochemical properties of the API can be modified while maintaining its intrinsic therapeutic activity. Cocrystal formation may not only improve solubility and bioavailability but also modify the physical properties of API including density, hygroscopicity, crystal morphology, loading volume, and compressibility. Furthermore, other properties such as toxicity, taste, physical and chemical stability, production costs, and manufacturing method may be improved by using a cocrystal rather than the API alone or as a salt (79,80). Different pharmaceutical cocrystallization techniques have been reported such as solution crystallization (81,82), grinding/solvent drop grinding (83–85), melt crystallization (86,87), and cocrystallization using ultrasound (88–90).

The importance of cocrystals in terms of polymorphism has been studied by Hickey et al. (91) and a carbamazepine/saccharin cocrystal provided advantages over carbamazepine alone with respect to polymorphism, physical and chemical stability, and improved oral bioavailability. Sun and Hou (92) reported improvement in the compressional properties of caffeine by formation of cocrystal with methyl gallate. Karki et al. (93) developed new compressible cocrystal forms of acetaminophen. In this study, cocrystals of form I of acetaminophen were formed with various conformers such as oxalic acid, phenazine, and

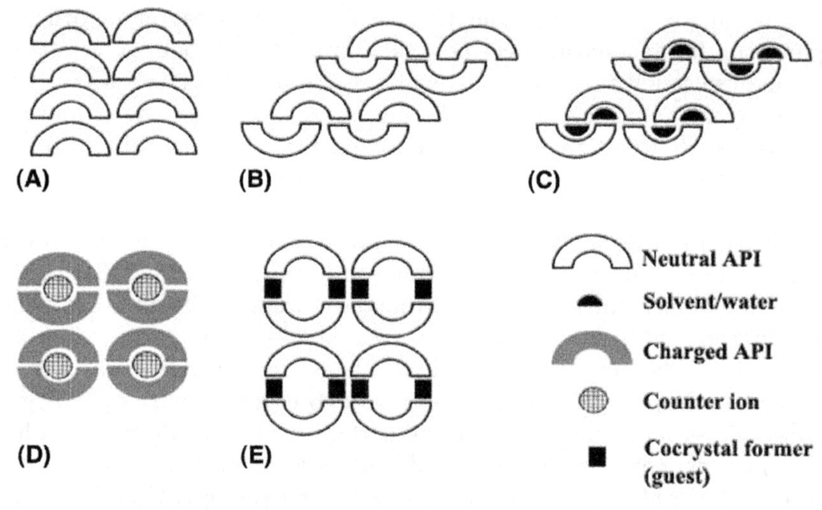

Figure 11 Schematic representation for the range of single crystalline forms that are possible for an API: (**A**) pure API, (**B**) polymorph of pure API, (**C**) hydrate/solvate of API, (**D**) salt of API, and (**E**) cocrystal. (Salts and cocrystals can also form hydrates, solvates, and polymorphs.)

naphthalene. All cocrystals showed improved tabletting properties as compared to acetaminophen form I, with oxalic acid cocrystals exhibiting superior tablettability. The improved compressional performance of the cocrystals was attributed to a layered molecular structure formed by hydrogen bonding between the API and coformer.

SUMMARY

The recently studied areas in crystal engineering focus on directed alteration of the primary properties of the crystals so as to impart improved physicochemical and compressional properties. New experimental approaches, such as cocrystallization, provide opportunities to obtain improved crystalline materials. Computational methods give complementary theoretical knowledge. For example, Beyer et al. (94) predicted mechanical properties of acetaminophen using molecular modeling, while Payne et al. (95) showed good agreement between the predicted and experimental mechanical properties of A and B forms of primidone. This type of approach will provide basic understanding to direct the crystal engineers to design directly compressible crystal forms of commercially important drug molecules. Similarly, other approach in the modeling of powder flow and compression into, the die cavity by techniques including finite element analysis will provide support in manipulating secondary particle properties (96,97).

In the area of particle engineering, the approaches are focused on alteration of primary and secondary particle properties by a single step process using techniques such as SCZ and CCA. Green processes like supercritical fluid technology may well be preferred over solvent based crystallization methods, considering stringent regulatory controls and required directed control of primary and secondary drug particle characteristics.

ACKNOWLEDGMENT

Authors wish to acknowledge the help provided by Dr N.R. Jadhav and Mr Suyog Aher in manuscript compilation.

REFERENCES

1. Deburg M, Nystrom C. Studies on direct compression of tablets. XVII. Porosity-pressure curves for the characterization of volume reduction mechanisms in powder compression. Powder Technol 1994; 46:67–75.
2. Joiris E, Martino P, Berneron C, et al. Compression behavior of orthorhombic paracetamol. Pharm Res 1998; 15(7):1122–1113.
3. Getsoian A, Lodaya RM, Blackburn AC. One-solvent polymorph screen of carbamazepine. Int J Pharm 2008; 348:3–9.
4. Gu C, Li H, Gandhi RB, et al. Grouping solvents by statistical analysis of solvent property parameters: implication to polymorph screening. Int J Pharm 2004; 283:117–125.
5. Stahl HP. The problems of drug interactions with excipients. In: Braimar DD, ed. Towards Better Safety of Drugs and Pharmaceutical Products. Amsterdam, The Netherlands: Elsevier/North-Holland Bio-medical Press, 1980:265–280.
6. Mullins JD, Macek TJ. Some pharmaceutical properties of novobiocin. J Am Pharm Assoc 1960; 49 (4):245–248.
7. Hancock BC, Parks M. What is the true solubility advantage for amorphous pharmaceuticals? Pharm Res 2000; 17(4):397–404.
8. Otsuka M, Hasegawa H, Matsuda Y. Effect of polymorphic transformation during the extrusion-granulation process on the pharmaceutical properties of carbamazepine granules. Chem Pharm Bull 1997; 45:894–898.
9. Mirza S, Miroshnyk I, Heinamaki J, et al. Crystal morphology engineering of pharmaceutical solids: tabletting performance enhancement. AAPS Pharm Sci Technol 2009; 10(1):113–119.
10. Korlakunte VRP, Sheen DB, Sherwood JN. Fracture property studies of paracetamol single crystals using microindentation techniques. Pharm Res 2001; 18(6):867–872.

11. Liao X, Wiedmann TS. Measurement of process-dependent material properties of pharmaceutical solids by nanoindentation. J Pharm Sci 2005; 94:79–92.
12. Higuchi WI, Lau PK, Higuchi T, et al. Polymorphism and drug availability. Solubility relations in the methylprednisolone system. J Pharm Sci 1963; 52:150–153.
13. Bandyopadhyay R, Grant DJ. Influence of crystal habit on the surface free energy and interparticulate bonding of L-lysine monohydrochloride dihydrate. Pharm Dev Technol 2000; 5:27–37.
14. Sun C, Grant DJW. Influence of crystal structure on the tableting properties of sulfamerazine polymorphs. Pharm Res 2001; 18(3):274–280.
15. Banga S, Chawla G, Varandani D, et al. Modification of the crystal habit of celecoxib for improved processability. J Pharm Pharmacol 2007; 59:29–39.
16. Berkovitch-Yellin Z. Toward an ab initio derivation of crystal morphology. J Am Chem Soc 1985; 107:8239–8253.
17. Sun C. Improving powder flow properties of citric acid by crystal hydration. J Pharm Sci 2009; 98(5):1744–1749.
18. Nokhodchi A, Bolourtchian N, Dinarvand R. Crystal modification of phenytoin using different solvents and crystallization conditions. Int J Pharm 2003; 250:85–97.
19. Chawla G, Gupta P, Thilagavathi R, et al. Characterization of solid-state forms of celecoxib. Eur J Pharm Sci 2003; 20:305–317.
20. Martino PD, Hermann AMG, Conflant P, et al. A new pure paracetamol for direct compression: the orthorhombic form. Int J Pharm 1996; 128:1–8.
21. Rasenack N, Muller BW. Crystal habit and tableting behavior. Int J Pharm 2002; 244:45–57.
22. Rasenack N, Muller BW. Ibuprofen crystals with optimized properties. Int J Pharm 2002; 245:9–24.
23. Katzhendler I, Azoury R, Friedman M. Crystalline properties of carbamazepine in sustained release hydrophilic matrix tablet based on hydroxypropyl methylcellulose. J Control Release 1998; 54:69–85.
24. Wen H, Morris KR, Park KJ. Hydrogen bonding interactions between adsorbed polymer molecules and crystal surface of acetaminophen. J Colloid Interface Sci 2005; 290:325–335.
25. Tian F, Sandler N, Gordon KC, et al. Visualizing the conversion of carbamazepine in aqueous suspension with and without the presence of excipients: a single crystal study using sem and raman microscopy. Eur J Pharm Biopharm 2006; 64:326–335.
26. Debnath S, Suryanarayanan R. Influence of process ing-induced phase transformations on the dissolution of theophylline tablets. AAPS PharmSciTech 2004; 5(1):1–11.
27. Lawn BR, Wilshaw TR. Indentation fracture: principles and applications. J Mater Sci 1975; 10:1049–1081.
28. Lawn BR, Swain MV. Microfracture beneath point indentations in brittle solids. J Mater Sci 1975; 10:113–122.
29. Anstis GR, Chantikul P, Lawn BR, et al. critical evaluation of indentation techniques for measuring toughness: I, direct crack measurements. J Am Ceram Soc 1981; 64:533–538.
30. Piera D, Mara S, Etienne JG. The spray drying of acetazolamide as a method to modify crystal properties and to improve compression behaviour. Int J Pharm 2001; 213:209–221.
31. Asker AF, Beeker CH. Some spray-dried formulations of sulfaethylthiadiazole for prolonged-release medication. J Pharm Sci 1966; 55:90–94.
32. Woodruff CW, Nuessle NO. Effect of processing variables on particles obtained by extrusion-spherosnisation processing. J Pharm Sci 1972; 61:787–790.
33. Paradkar AR, Maheshwari M, Ketkar AR, et al. Preparation and evaluation of Ibuprofen beads by melt solidification technique. Int J Pharm 2003; 255(1–2):33–42.
34. Maheshwari M, Harshal J, Paradkar AR. Melt sonocrystallization of ibuprofen: effect on crystal properties. Eur J Pharm Sci 2005; 25(1):41–48.
35. Wong TW, Chan LW, Heng PWS. Study of the melt pelletization process focusing on the micromeritic property of pellets. Chem Pharm Bull 2000; 48(11):1639–1643.
36. Thies R, Kleinebudde P. Melt pelletization of a hygroscopic drug in a high shear mixer. Part 3. Effects of binder variation. Chem Pharm Bull 2001; 49:140–146.
37. Thomas V, Jakob K, Torben S. Melt pelletization with polyethylene glycol in a rotary processor. Int J Pharm 2004; 275(1–2):141–153.
38. Kaerger JS, Price R. Processing of spherical crystalline particles via a novel solution atomisation and xtallisation by sonication (SAXS) technique. Pharm Res 2004; 21:372–381.
39. Kawashima Y, Okumura M, Takenaka H. Spherical crystallization: direct spherical agglomeration of salicylic acid during crystallization. Science 1982; 216:1127–1128.
40. Kawashima Y. Spherical crystallization as a novel particle design technique for oral drug delivery system. Chin Pharm J 1989; 41:163–172.

41. Kadam SS, Mahadik KR, Paradkar AR. A process for making agglomerates for use as or in a drug delivery system. Indian Patent (183036) 1997.
42. Kadam SS, Mahadik KR, Paradkar AR. A process for making agglomerates for use as or in a drug delivery system. Indian Patent (183481) 1997.
43. Paradkar AR, Chauhan B, Yamamura S, et al. Preparation and characterization of glassy celecoxib. Drug Dev Ind Pharm 2003; 29(7):739–744.
44. Chauhan B, Shimpi S, Paradkar A. Preparation and characterization of etoricoxib solid dispersions using lipid carriers by spray drying technique. AAPS Pharm Sci Technol 2005; 6(3):E405–E412.
45. Ambike A, Mahadik KR, Paradkar AR. Spray-dried amorphous solid dispersions of simvastatin, a low Tg drug: in vitro and in vivo evaluations. Pharm Res 2005; 22(6):990–998.
46. Paradkar A, Ambike AA, Jadhav BK, et al. Characterization of curcumin-PVP solid dispersion obtained by spray drying. Int J Pharm 2004; 271(1-2):281–286.
47. Pasquali I, Bettini R, Giordano F. Solid-state chemistry and particle engineering with supercritical fluids in pharmaceutics. Eur J Pharm Sci 2006; 27:299–310.
48. Hanna M, York P. Method and apparatus for the formation of particles. WO95/01221. 1995.
49. Palakodaty S, York P. Phase behavioral effects on particle formation processes using supercritical fluids. Pharm Res 1999; 16:976–985.
50. York P, Hanna M, Shekunov BY, et al. Microfine particle formation by SEDS (solution enhanced dispersion by supercritical fluids): scale up by design. Proc Resp Drug Deliv 1998; VI:169–175.
51. Smith HM, Puddington IE. Spherical agglomeration of Barium Sulphate. Can J Chem 1960; 38(10):1911–1916.
52. Kawashima Y, Takagi H, Takenaka H. Wet spherical agglomeration of binary mixtures. II. Mechanism and kinetics of agglomeration and the crushing strength of agglomerates. Chem Pharm Bull 1981; 29(5):1403–1409.
53. Vanangamudi M, Rao TC. Kinetic study of agglomerate growth in coal-oil agglomeration process. Fuel 1984; 63(6):738–743.
54. Ching-Chung H, Kono HO, The granulation of partially prewetted alumina powders—a new concept in coalescence mechanism. Powder Technol 1988; 55(1):19–34.
55. Kawashima Y, Aoki S, Takenaka H, et al. Preparation of spherically agglomerated crystals of aminophylline. J Pharm Sci 1984; 73:1407–1410.
56. Kawashima Y, Okumara M, Takenaka H, et al. Direct preparation of spherically agglomerated salicylic acid crystals during crystallization. J Pharm Sci 1984; 73:1535–1538.
57. Paradkar AR, Pawar AP, Mahadik KR, et al. Spherical crystallization a novel particle design technique. Indian Drugs 1998; 31(6):283–299.
58. Gordon MS, Chowhan LT, Manipulation of naproxen particle morphology via the spherical crystallization technique to achieve a directly compressible raw material. Drug Dev Ind Pharm 1990; 16:1279–1290.
59. Martino PD, Roberta D, Cristofaro CB, et al. Improved compression properties of propyphenazone spherical crystals. Int J Pharm 2000; 197(1–2):95–106.
60. Morishima K, Kawashima Y, Takeuchi H, et al. Micromeritic characteristics and agglomeration mechanisms in the spherical crystallisation of bucillamine by the spherical agglomeration and the emulsion solvent diffusion methods. Powder Technol 1993; 76:57–64.
61. Deshpande MC, Mahadik KR, Pawar AP, et al. Evaluation of spherical crystallization as a particle size enlargement technique for aspirin. Ind J Pharm Sci 1997; 59(1):32–34.
62. Pawar PH, Pawar AP, Mahadik KR, et al. Evaluation of tabletting properties of agglomerates obtained by spherical crystallization of trimethoprim. Ind J Pharm Sci 1998; 60(1):24–28.
63. Martino DP, Barthelemy C, Piva F, et al. Improved dissolution behavior of fenbufen by spherical crystallization. Drug Dev Ind Pharm 1999; 25:1073–1081.
64. Paradkar AR, Maheshwari M, Tyagi AK, et al. Preparation and characterization of flurbiprofen beads by melt solidification technique. AAPS Pharm Sci Technol 2003; 4(4):E62.
65. Maghsoodi M, Omid T, Martin G, et al. Particle design of naproxen-disintegrant agglomerates for direct compression by a crystallo-co-agglomeration technique. Int J Pharm 2008; 351:123–133.
66. Ribardiere A, Tchoreloff P, Couarraze G, et al. Modification of ketoprofen bead structure produced by spherical crystallization technique with a two solvent system. Int J Pharm 1996; 144(2):195–207.
67. Pawar AP, Paradkar AR, Kadam SS, et al. Crystallo-co-agglomeration: a novel process to obtain ibuprofen-paracetamol agglomerates. AAPS Pharm Sci Technol 2004; 5(3):E41, 1–8.
68. Pawar AP, Paradkar AR, Kadam SS, et al. Agglomeration of ibuprofen with talc by novel Crystallo-co-agglomeration technique. AAPS Pharm Sci Technol 2004; 5(4)E55:1–6.

69. Jadhav NR, Pawar AP, Paradkar AR. Design and evaluation of deformable talc agglomerates prepared by crystallo-co-agglomeration technique for generating heterogeneous matrix. AAPS Pharm Sci Technol 2007; 8(3):E59, 1–7.
70. Pawar A, Paradkar A, Kadam S, et al. Effect of polymers on crystallo-co-agglomeration of ibuprofen-paracetamol: factorial design. Ind J Pharm Sci 2007; 69:658–654.
71. Sanghavi NM, Sivanand R, Kotwaney HN. Dissolution pattern of miniscular sulfisoxazole. Ind J Pharm Sci 1979; 41:116–117.
72. Vertommen J, Rombaut P, Michoet A, et al. Estimation of the amount of water removed by gap and atomization air streams pelletization in a rotary processor. Pharm Dev Technol 1998; 3(1):63–72.
73. Kristensen J, Schaefer T, Kleinebudde P. Direct pelletization in a rotary processor controlled by torque measurements. II: effect of changes in the content of microcrystalline cellulose. AAPS Pharm Sci Technol 2000; 2(3):24.
74. Luque de Castro MD, Priego-Capotea F. Ultrasound-assisted crystallization (sonocrystallization). Ultrasonics Sonochem 2007; 14(6):717–724.
75. Li H, Lib H, Guob Z, et al. The application of power ultrasound to reaction crystallization. Ultrasonics Sonochem 2006; 13(4):359–363.
76. Ruecroft G, Hipkiss D, Ly T, et al. Sonocrystallization: the use of ultrasound for improved industrial crystallization. Organic Proc Res Dev 2005; 9:923–932.
77. Paradkar AR, Maheshwari M, Kamble R, et al. Design and evaluation of celecoxib porous particles using melt sonocrystallization. Pharm Res 2006; 23(6):1395–1400.
78. Aakeroy CB. Crystal engineering: strategies and architectures. Acta Cryst 1997; B53:569–586.
79. Blagden N, de Matas M, Gavan PT, et al. Crystal engineering of active pharmaceutical ingredients to improve solubility and dissolution rates. Adv Drug Deliv Rev 2007; 59:617–630.
80. Shan N, Zaworotko MJ. The role of cocrystals in pharmaceutical science. Drug Discov Today 2008; 13(9/10):440–446.
81. Fleischman SG, Kuduva SS, McMahon JA, et al. Crystal engineering of the composition of pharmaceutical phases: multiple-component crystalline solids involving carbamazepine. Cryst Growth Des 2003; 3(6):909–919.
82. Trask AV, Motherwell FWDS, Jones W. Physical stability enhancement of theophylline via cocrystallization. Int J Pharm 2006; 320:114–123.
83. Xyrofin OY. Composition comprising co-crystals methods for its manufacture and its use. WO 96/07331, 1996.
84. Trask AV, Motherwellb WDS, Jones W. Solvent-drop grinding: green polymorph control of cocrystallisation. Chem Commun 2004:890–891.
85. Jayasankar A, Somwangthanaroj A, Shao ZJ, et al. Cocrystal formation during cogrinding and storage is mediated by amorphous phase. Pharm Res 2006; 23(10):2381–2392.
86. DuRoss, James W. Melt cocrystallized sorbitol/xylitol compositions. United States Patent 5158789, 1992.
87. Berry DJ, Seaton CC, Clegg W, et al. Applying hot-stage microscopy to co-crystal screening: a study of nicotinamide with seven active pharmaceutical ingredients. Cryst Growth Des 2008; 8(5):1697–1712.
88. Childs SL, Mougin P, Stahly BC. Screening for solid forms by ultrasound crystallization and cocrystallization using ultrasound. WO/2005/089375, 2005.
89. Bucar DK, MacGillivray LR. Preparation and reactivity of nanocrystalline cocrystals formed via sonocrystallization J Am Chem 2007; 129(1).
90. Friscic T, Childs SL, Rizvic SAA, et al. The role of solvent in mechanochemical and sonochemical cocrystal formation: a solubility-based approach for predicting cocrystallisation outcome. Cryst Eng Comm 2009; 1(3):375–522.
91. Hickey A, Magali B, Matthew L, et al. Performance comparison of a co-crystal of carbamazepine with marketed product. Eur J Pharm Biopharm 2007; 67:112–119.
92. Sun CC, Hou H. Improving mechanical properties of caffeine and methyl gallate crystals by cocrystallization. Crystal Growth Design 2008; 8:1575–1579.
93. Karki S, Friscic T, Fabian L, at al. Improving mechanical properties of crystalline solid by cocrystal formation: new compressible forms of paracetamol. 2009, Advanced Materials, 21, 1–5.
94. Beyer T, Day GM, Price S. The prediction, morphology, and mechanical properties of the polymorph of paracetamol. J Am Chem Soc 2001; 123:5086–5094.
95. Payne RS, Roberts RJ, Rowe RC, et al. The mechanical properties of two forms of primidone predicted from their crystal structures. Int J Pharm 1996; 145:165–173.
96. Michrafy A, Ringenbacher D, Techoreloff P. Powder Technol 2002; 127:257–266.
97. Sinka IC, Cunnigham JC, Zavalinagos A. Powder Technol 2003; 133:33–43.

11 Compaction of combination products

Carl T. Allenspach and Erika A. Zannou

INTRODUCTION

The pharmaceutical industry, whether brand name or generic, has over the past few years been increasing its focus on developing combination products (1). These products have been a key component of very successful life cycle management strategies, mainly due to significant advantages in both therapeutic and commercial aspects, including extension of patent coverage (2). The main therapeutic advantages and rationale for developing fixed-dose combination products include better long-term patient compliance (3) by decreasing the overall number of dosage forms to be administered, whether treating one or multiple indications (4–6) and often decreasing the dose compared to monotherapies due to synergistic effects, thus potentially minimizing side effects. A wide array of combination products are already available in many therapeutic areas, including diabetes, hypertension, lipid lowering, pain management, HIV/AIDS treatments as well as against various types of bacterial infections, with the objective of maximizing safety and efficacy for the patient's benefit.

From a technical perspective, combination products are challenging. Various formulation approaches can be considered for the development of tableted combination products. The preferred approach is a standard/monolayer tablet containing two or more active ingredients along with excipients. The compaction of these combination product monolayer tablets can be difficult due to the added complexity and inhomogeneity of the formulation components compared to mono-active tablets; some critical considerations will be briefly discussed (see Section "Monolayer Combination Products"). The reader should also refer to other chapters in this book discussing theory, modeling, and simulation as well as compaction of pharmaceutical powders in general. This chapter will mainly focus on combination products formulation options and their impact on compaction behavior when a monolayer tablet is not feasible mainly due to (*i*) compaction properties, (*ii*) intended biopharmaceutical performance (e.g., dissolution/bioavailability and/or stability), or (*iii*) line extension or market differentiation. The majority of this chapter will be addressing multilayer tablet compression (bilayer being the most common) including tablet strength and layer adhesion strength measurement and prediction, formulation considerations to optimize compaction properties and manufacturing considerations during the compression unit operation. Compaction considerations during development and production of minitablets will also be discussed along with a short section on compression-coated tablets (tablet-in-tablet). While compaction principles discussed in sections I and II can be extended to combination products, each of these combination formulation approaches presents some unique advantages as well as hurdles for compaction measurement, prediction and practical application (1). In addition, it is to be highlighted that compaction is only one part of developing a tablet: other physicochemical aspects, such as formulation composition, powder flow mechanics, and chemical and physical stability have to be considered and balanced with optimal compaction properties. This can be especially challenging for combination products, and is not part of the scope of this chapter.

MULTILAYER TABLETS: INTRODUCTION

Multilayer tablets are versatile dosage forms that present significant formulation, production, and analytical challenges. During the development of multilayer tablets, the formulation scientist needs to take into consideration and balance multiple potentially critical parameters such as excipient composition/formulation of each layer, particle size distribution and flowability, relative compaction and moisture sorption properties of each layer, thickness of each layer, tablet size and shape, tablet press equipment, and various processing parameters including compression order of the layers, individual layer weight control, compression speed, and compression force on each layer and the multilayer tablet. Each parameter in this long and nonexhaustive list is interdependent and can significantly impact compaction as will be discussed in detail in the following sections.

When discussing multilayer tablets in this chapter, the authors mainly refer to the simplest and most commonly developed bilayer tablets (Fig. 1). The principles are, however, applicable to more complex systems with higher numbers of layers. The literature refers to some trilayer tablets also being developed (7–12), and tablet presses that can compress up to five layers are available in the industry. The increase in popularity of multilayer tablet compression is highlighted by the number of laboratory and pilot scale tablet presses that have become available in the past five years to meet the customer needs and demands for developing these products. Korsch, for example, has recently launched their new XL-400 FT tablet press that allows for production of single-layer, bilayer, trilayer, and compression-coated tablets on the same tablet press platform using interchangeable modules (Fig. 2).

The process of multilayer compaction involves filling the first layer of material, then slightly compressing (tamping) this layer to remove entrapped air thereby providing space in the die for the addition of the next layer. The steps of filling and tamping can be repeated to create the required number of multiple layers depending on the tablet press capabilities. This is followed by an optional precompression step to help remove entrapped air in the combined layers of the multilayer tablet and a final main compaction and ejection of the formed tablet (Fig. 3).

Similarly to monolayer tablets, multilayer tablet strength is the main attribute either measured to assess compaction and/or predicted for optimum formulation selection. In addition, for multilayer tablets, the strength of adhesion between layers is a critical attribute that must also be taken into account. Measurement is thus challenging since the techniques used for the monolayer tablets may not be producing relevant data for the heterogeneous multilayer tablets that exhibit a tendency to delaminate at the interface between layers due to inherent binding weakness. In addition, the potential differences in breaking force (sometimes referred to as the crushing strength, or hardness, of the tablet) between layers may not be detected and/ or may result in erroneous readings due to the measurement principle (see section "Multilayer Tablet Compaction, Strength, and Adhesion Strength"). Alternatively, the mechanical strength of each layer can be measured and/or predicted separately providing some valuable information to be used during development. In addition, the mechanical strength at the layer interface or adhesion strength has been measured using various techniques. Some studies have also been conducted to determine which formulation parameters are to be considered for the design of multilayer tablets of optimal mechanical strength (see section "Formulation Considerations for Optimum Compaction Properties of Multilayer Tablets"). Finally, the large scale and high-speed production of such complex and heterogeneous systems requires careful and unique settings during the compaction unit operation, highly dependent on the

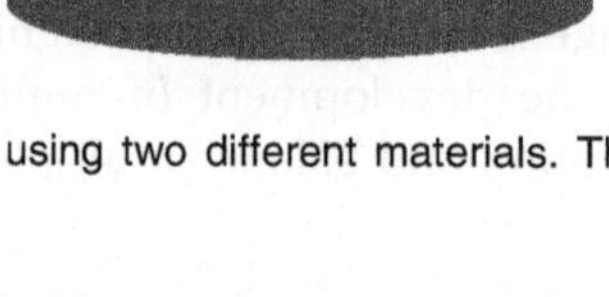

Figure 1 Bilayer tablet manufactured using two different materials. The first layer is on the bottom, and the second layer is added on top.

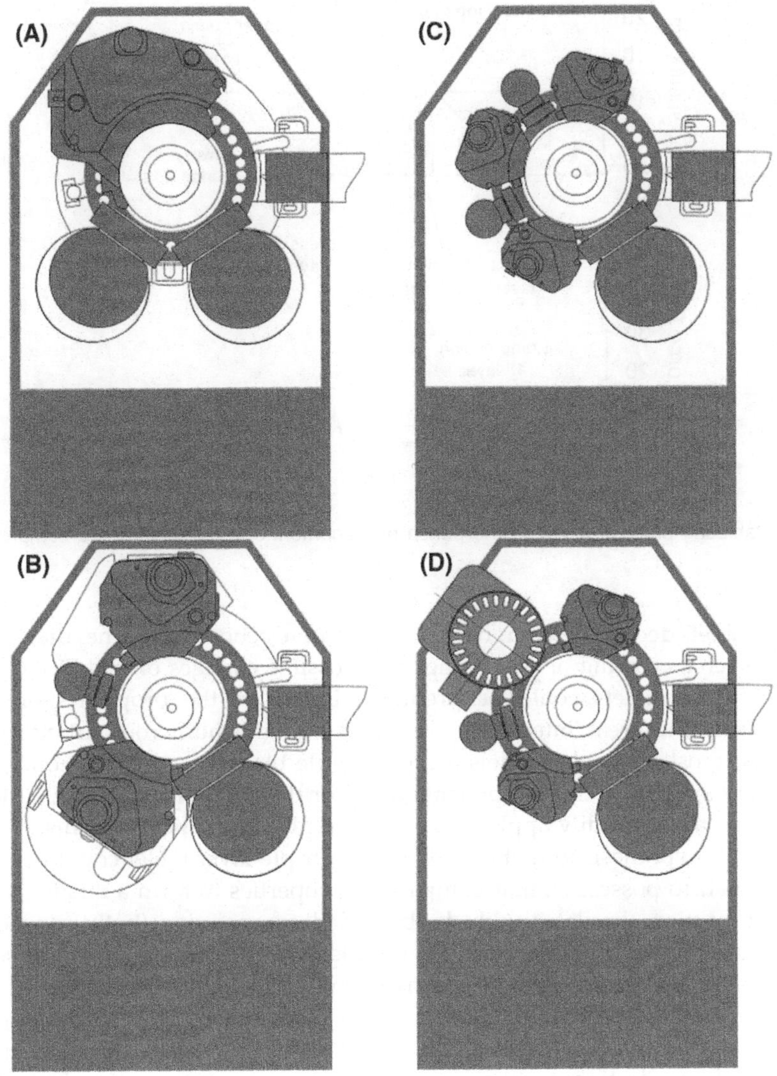

Figure 2 Alternate modular applications for Korsch XL 400FT tablet press for producing (**A**) single-layer, (**B**) bilayer, (**C**) trilayer, and (**D**) compression-coated tablets. *Source*: Courtesy of Korsch.

multilayer tablet formulation and specific tablet press used (see section "Manufacturing Considerations for Multilayer Tablet Compression Unit Operation").

MULTILAYER TABLET COMPACTION, STRENGTH, AND ADHESION STRENGTH

Compaction and Strength of Multilayer Tablets: Theory, Measurement, and Prediction

The understanding of multilayer tablet mechanical strength (both whole tablet strength and adhesion strength between layers) is a scientific challenge with significant commercial impact. Failure of the tablet to remain intact until its intended in vivo use can lead to enormous financial losses (13) and can have significant regulatory repercussions.

The compaction behavior of multilayer tablets and the resulting tablet strength are influenced by the mechanical (such as viscoelastic) properties of the formulation powder, compression process parameters (such as press speed and compression forces), resultant

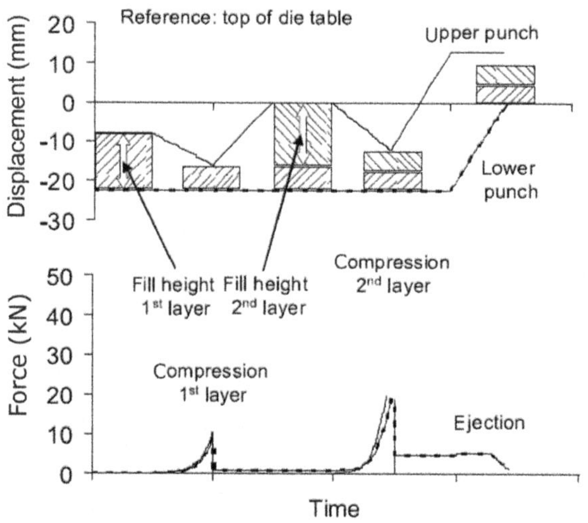

Figure 3 Bilayer tableting process following the punch displacement and forces during a compression cycle. *Source*: From Ref. 32.

stresses during tablet decompression (relaxation) and ejection from the die, subsequent processing steps such as coating or packaging, and possible changes on stability (12). Ideally, the tensile strength of tablets could be predicted based on the properties of individual components. Despite the great interest in this field and the numerous articles describing predictive models, most available models are for simple binary or ternary mono- or bilayer tablets and are very specific to the components used for the study. Much work is still needed for predictions that can be readily applied to complete pharmaceutical systems.

Yang et al. (12) hypothesized that each layer of a triple-layer tablet could be different in composition but need to possess similar compaction properties to form a quality tablet. They used a Mand compaction simulator with double-sided compression (both punches moving during compression) and flat-faced 1-cm round tooling to assess the Manesty Betapress compaction of various excipient blends of matrix formulations. They demonstrated with their model formulation that for optimal multilayer tablet compaction, each of the layers should exhibit comparable consolidation characteristics (brittle fracture or plastic deformation or a balance of both) and low elastic recovery during decompression and ejection. This is however challenging to achieve especially in the case of combination products, when the layer composition and/or manufacturing process may be significantly different for each layer to attain specific and different release profiles.

Adhesion Strength of Multilayer Tablets: Theory, Measurement, and Prediction

Despite many similarities in compaction problems shared by monolayer and multilayer tablets, one issue multilayer tablets are particularly prone to is delamination. Delamination, sometimes also referred to as lamination or layer separation, occurs when the bonding between the layers is inadequate (weak or under stress), resulting in separation of the tablet at the interface between the layers. In this chapter, the strength of the bond between the layers will be referred to as layer adhesion, also sometimes called in the literature interfacial adhesion, interlayer adhesion, layer cohesion, layer interface strength or, conversely, delamination potential.

Poor adhesion between layers can be attributed to many potential factors resulting in each layer exhibiting a different physical behavior, thus increasing stress at the interface.

Potential factors include the formulation, layer ratios, bonding surface area, compaction process (compression and relaxation), subsequent processing (coating and/or packaging), and even material changes upon stability (moisture sorption, heat expansion). These stresses can result in delamination of multilayer tablets at the layer interface. Since multilayer tablets are composed of layers of different materials they are susceptible to layer separation due to differential behavior of the layers as described above. It is obvious that even a single delaminated tablet in a commercial batch is viewed as a critical tablet defect since it could potentially result in a patient not receiving one of the intended medications. During manufacturing, an additional step should be included and validated to mechanically or visually inspect multilayer tablets to ensure that no delaminated tablets reach the patients. The strength of layer adhesion is a topic of great scientific interest; however, due to the complexity of pharmaceutical formulations and specific equipment required to manufacture multilayer tablets, it is generally studied under simplified/idealized conditions: with pure materials (one per layer), flat-faced tooling, a low compaction speed and often with single sided compaction instead of both upper and lower punches moving during compaction. These conditions do not represent the real life scenarios of manufacturing multilayer products, but are a necessary step to begin to understand the root principles of multilayer compaction. The process of delamination and the root causes are still poorly understood and can potentially result in an unacceptable product or at least significant time spent troubleshooting to improve the product.

The potential for delamination has been recognized since the first multilayer tablets went into production (14), but it was not until 1997 that a method to specifically measure the adhesion strength was developed by Dietrich et al. (15,16). A shear apparatus was designed and inserted into a standard diametrical tensile strength tester (hardness tester) to measure the shear force required to separate the two layers of a bilayer tablet (Fig. 4). The bilayer tablet is placed into the apparatus and its height adjusted so that the interface of the layers is leveled and the bottom layer is below the surface. A fixing screw is used to apply a constant torque to the top of the tablet, holding it in place. The tensile strength tester then pushes the shear block against the top layer and measures the force required to separate the layers. The resulting shear force is used as a measure of the adhesion strength of the layers. This simple method provides reproducible results with similar variability as when measuring monolayer tablets tensile strength, and allows quantification of the adhesion strength between layers. Additionally, it replicates stresses and shear that a tablet may be submitted to during coating, packaging, and/or shipping, which could result in delamination. A similar insert for the standard tensile strength tester is now commercially available from Schleuniger Pharmatron (Switzerland) for measuring the force needed to delaminate bilayer tablets.

To measure the layer adhesion strength, bilayer tablets can also be subjected to an axial tensile stress test where the tablet halves are pulled apart. In this process each side (top and

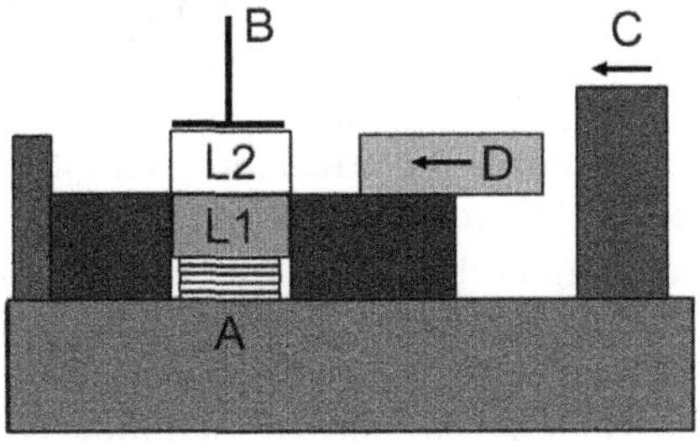

Figure 4 Shear apparatus for measurement of adhesion strength between layers. (A) Adjustment screw for height. (B) Fixing screw. (C) Moving platen of tensile strength tester. (D) Pushes on shear block causing (L2) layer 2 to separate from (L1) layer 1 of the bilayer tablet. *Source*: From Ref. 15.

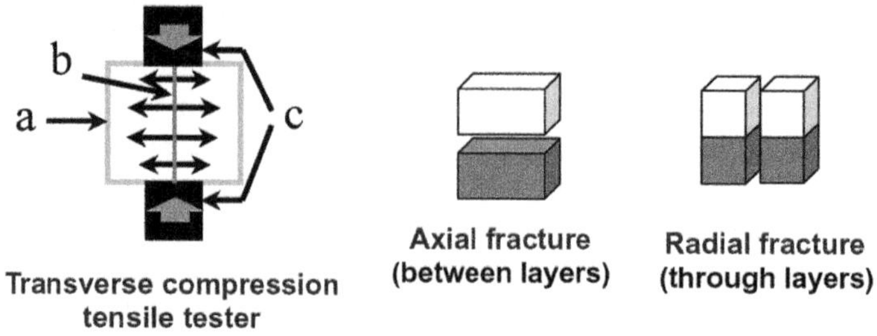

Figure 5 Compression test of compacts normal and parallel to compression axis for measurement of tensile strength. *Source*: From Ref. 18.

bottom) of the tablet is attached to adapters using adhesive. A material testing instrument such as an Instron^{TM} material tester slowly pulls the tablet at a specified rate of displacement until the two halves delaminate and a tensile fracture force is measured when the layers are separated. It should be confirmed that the fracture occurs at the layer interface and represents the adhesion strength between layers rather than fracture in the plane of the material. Inman et al. incorporated 1% of ferric oxide in one of the layers and used energy dispersive X-ray (EDX) analysis to map fractured surfaces (17). They confirmed the presence of ferric oxide at significantly lower amounts on both fracture surfaces demonstrating the fracture occurred at the interface. Had the fracture occurred within either of the layers either no ferric oxide or a much higher amount would have been detected depending in which layer the fracture occurred.

Layer adhesion strength can also be assessed by measuring the tensile strength of cubic compacts in both axial (between layers) and radial (through layers) directions (Fig. 5). An example of this was given by Mullarney and Lam who prepared monolayer and bilayer cubic compacts at different solid fractions using a triaxial press with a split die. They found that generally the measured axial tensile strength is smaller than the radial tensile strength, indicating the interface between the layers is the weakest point. The ratio represents tensile strength anisotropy and can be used to evaluate the layer adhesion strength in different materials (18).

Adhesion strength of multilayer tablets can also be assessed by indirect methods. One such method relates the adhesion strength to critical tamping force during compression (19,20). The critical tamping force is defined as the maximum compression force that can be used to compress the first layer of the tablet while still allowing for bonding of the subsequent layer. If the critical tamping force is high, then the adhesion strength is also high. In such a case the initial layer can be compressed to a high force and the next layer will still be able to adhere adequately. The method for determining the critical tamping force involves increasing the first-layer compression force until the multilayer tablet delaminates after ejection from the tablet press and then decreasing the force until no delamination is observed when the tablets are gently shaken.

Another indirect method to evaluate the adhesion strength using standard compaction and tablet testing equipment is an extended friability test (21). If the adhesion between layers is poor, the tablets will delaminate under the mechanical stress imparted during extended friability testing. While this test seems practical and the number of delaminated tablets can be counted, there is no quantification of potential differences in adhesion strength between batches where tablets stay intact. One additional shortcoming is that failure to delaminate during extended friability testing is not an assurance that the tablets will not delaminate during further processing such as coating and/or during packaging, as the stresses involved under high tablet load may be of greater magnitude than during friability testing. It is

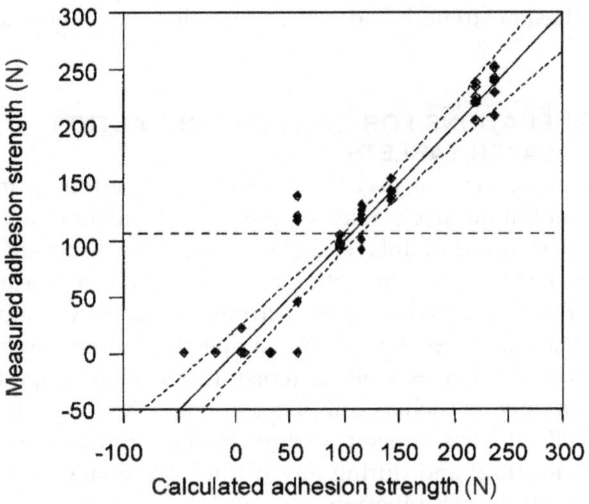

Figure 6 Measured values versus calculated values for adhesion strength. The horizontal line represents the overall mean; the fit is displayed with the 95% confidence interval. *Source*: From Ref. 15.

important to note that a standard friability test of 100 rotations or even an extended test of 400 rotations may not be rigorous enough to detect potential issues with layer adhesion, which is a common misconception.

Although adhesion strength has garnered the most attention as a critical attribute of multilayer tablets, its prediction even for specific formulations is very complex. Dietrich et al. (15) carried out a thorough design of experiments (DOE) quantifying the impact of the first-layer compression force, the multilayer tablet compression force, and the lubricant concentration in one of the layers on the adhesion strength of a specific bilayer tablet formulation. The equation listed below was obtained by fitting the experimental data, and they saw reasonable agreement between the predicted and actual values for layer adhesion strength (Fig. 6), the results and implications of which are discussed later in this chapter.

$$\begin{aligned} \mathrm{AS} &= A + B \\ A &= 152.7 - 23.1 \cdot \mathrm{CF1} + 19.6 \cdot \mathrm{CF2} - 370.4 \cdot \mathrm{Lub} + 1.6 \cdot (\mathrm{CF1})^2 \\ B &= 398.8 \cdot (\mathrm{Lub})^2 - 1.5 \cdot (\mathrm{CF1} \cdot \mathrm{CF2}) - 13.1 \cdot (\mathrm{CF1} \cdot \mathrm{Lub}) \end{aligned} \tag{1}$$

with AS = adhesion strength in Newton, CF1/2 = compression force for central-layer and multilayer tablets, and Lub = lubricant concentration in central-layer tablet.

Measuring or predicting adhesion strength between layers either directly or indirectly is critical for the optimal development and manufacturing of multilayer tablets. As described in this section, multiple methods have been attempted and published. The shear apparatus that can be inserted in standard diametrical tensile strength tester seems to be the easiest and fastest method to implement, which gives quantitative results; however, other methods are also relevant and should be compared. One of the key challenges is that tablet characteristics (besides formulation, such as layer thickness and ratio as well as tablet shape and size) have a significant impact on the adhesion strength measurement. Additional efforts in modeling and simulation of adhesion strength based on formulation properties and processing parameters as well as tablet characteristics described above need to be investigated to thoroughly understand their impact on layer adhesion strength and benefit overall development and manufacture of multilayer tablets.

FORMULATION CONSIDERATIONS FOR OPTIMUM COMPACTION PROPERTIES OF MULTILAYER TABLETS

Formulation of multilayer tablets is especially challenging: in addition to formulating each layer for desired and optimum drug release, particular attention needs to be paid to maximizing the layer adhesion and multilayer tablet strength. As discussed in detail later on in this section, layer adhesion may be dependent on the respective layer formulations, including the layer composition, thickness, particle size distribution, density, morphology, extent of lubrication, physicomechanical properties (flow, compaction and relaxation), and moisture content at the time of compaction as well as moisture sorption properties (12). It is also dependent on process parameters such as compression forces (tamping, pre and main), press type and speed as well as compression tooling design. Storage conditions following compression (heat and moisture) and during any further processing such as film-coating or packaging can also impact the layer adhesion.

Multilayer tablet formulations should be designed to minimize the tablet size (due to process limitations regarding layer thickness—section "Manufacturing Considerations for Multilayer Tablet Compression Unit Operation"), which for multilayer tablets tend to be larger than standard monolayer tablets. This is driven by the use of two separate formulations and the likely constraint not to change formulations for marketed products due to potential biopharmaceutical risk. In addition, optimization of each layer formulation should be done in the context of the multilayer tablet formulation and compaction. A change made to one layer could have significant impact on the other(s), influencing either their processability (compaction properties or layer adhesion), quality attributes (i.e., drug release), or both. It may not be possible to optimize the layers separately and expect the combined multilayer tablet to be optimized. Also, relative properties of each layer and the potential physical changes of each layer within a multilayer tablet upon various stresses could jeopardize the integrity of the layer adhesion. In summary, the particularities and challenges of multilayer tablet formulation are that all of the above mentioned variables could both have an impact on compaction properties and be impacted by the compression process. Moreover, changing one or more of these can have a significant impact on both the physical strength of the tablet, the adhesion between the layers, and its in vivo performance. With this high level of complexity, multilayer tablet formulations are usually optimized for in vivo performance but compaction optimization is not always possible through formulation alone. Therefore, a delicate balance between formulation compaction properties and optimal processing conditions should be sought.

Impact of Formulation Composition and Extent of Lubrication on Multilayer Tablet Compaction Properties and Layer Adhesion

As discussed in previous sections, the formulation composition of each of the layers can have a significant impact on multilayer tablet compaction properties and layer adhesion. The complexity of the formulations used in pharmaceutical tablets renders a universal model and specific modeling and simulation difficult.

Lubricants, such as magnesium stearate, are typically used in formulations to reduce the incidence of sticking to the tablet punch faces as well as to decrease the die wall forces during ejection of the formed tablet from the die. The use of lubricants can, however, be detrimental to the tablet strength as described elsewhere in literature (22–30). It is therefore not surprising

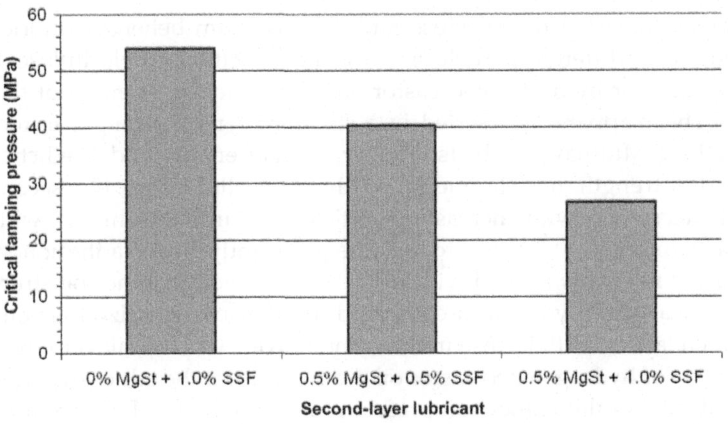

Figure 7 Critical tamping pressure as a function of second layer lubrication amount and type. *Abbreviations*: MgSt, magnesium stearate, and SSF, sodium stearyl fumarate. *Source*: From Ref. 20.

that the use of lubricants in multilayer formulations would have a similar impact on tablet strength as well as a significant impact on layer adhesion.

Birringer et al. (20) studied the impact of lubricant type and level for a wet granulated product using a Fette 3090 high-speed bilayer tablet press. The layer adhesion was evaluated using critical tamping force measurements as described in section "Adhesion Strength of Multilayer Tablets: Theory, Measurement, and Prediction." It was determined that increasing the lubricant level in the range of 1.0% to 1.5% for both the first- and second-layer formulations resulted in significantly decreased layer adhesion. Additionally, the type of lubricant seemed to impact the layer adhesion in this system. They found that the use of magnesium stearate was more detrimental to layer adhesion than sodium stearyl fumarate (Fig. 7). When formulating for multilayer tablets, it is recommended to evaluate different options for lubrication that might otherwise be considered for monolayer tablets where layer adhesion is not an issue.

Evaluation of the compaction properties using placebo formulations is not recommended since incorporation of the active ingredient(s) can impact the compaction properties of the formulation and is commonly detrimental to compaction. However, in this reported case, increasing the drug load improved the layer adhesion. The impact of drug substance on compression is highly dependent on the properties of the active ingredient as well as the method for incorporating it into the formulation, which in this case was *via* a wet granulation process (20). It should also be mentioned about this study that tablets appearing to be well adhered after compaction tended to delaminate after storage for one week at 40°C/75% RH. It is possible that the components of each layer had varying degrees of expansion after exposure to heat and moisture, which resulted in failure at the layer interface. Care should be taken when choosing materials for layers, and it is recommended to evaluate the impact of storage stability on the layer adhesion to use as a criterion for determining the feasibility of the formulation.

Impact of Formulation Particle Size Distribution on Multilayer Tablet Compaction Properties and Layer Adhesion

One key attribute of the formulation, the particle size, can significantly impact the layer adhesion strength. In an attempt to evaluate the effect of particle size on adhesion strength, Dietrich et al. (16) manually produced bilayer tablets using 10-mm round flat-faced punches, with 5-kN compression force on the first layer, followed by either 5 or 10 kN on the bilayer tablet. The first layer consisted of cellactose and 0.5% calcium behenate (lubricant) and the second layer consisted of different particle size fractions of either ethylcellulose, lactose (with 0.5% calcium behenate), or hydrogenated castor oil. The adhesion strength of the tablets was measured using a shear apparatus inserted in a standard tensile strength tester (see section "Adhesion Strength of Multilayer Tablets: Theory, Measurement, and Prediction" for more details on adhesion strength measurement). With ethylcellulose in the second layer, the adhesion strength decreased with increasing particle size and was higher with the higher bilayer compression force of 10 kN. Lactose had significantly lower adhesion strength than ethylcellulose and did not show a trend with particle size although adhesion strength with the largest particle size was sharply reduced compared to other particle size fractions. There was no adhesion between layers (with lactose in the second layer) when a bilayer force of 5 kN was used. The particle size of hydrogenated castor oil did not impact the adhesion strength possibly due to its waxy nature, but the higher compression force resulted in lower adhesion strength. Although the first-layer compression force for this study was much higher than is typically used with multilayer compression (when manufactured using standard tablet press equipment), the overall effect of material type, particle size, and bilayer tablet compression force was demonstrated and highlights the importance of evaluating this for specific formulations.

Birringer et al. (20) evaluated the impact of particle size by changing the wet granulation process of the top layer of bilayer tablets prepared using a Fette 3090. Particle size D50 values ranging from about 120 to 560 μm for a low drug load formulation and from about 100 to 390 μm for a high drug load formulation were studied. The particle size did not significantly affect the adhesion in the low drug load formulation, but increasing the particle size resulted in significantly reduced layer adhesion for the high drug load formulation indicating that the particle size effect is dependent on the material and formulation.

Impact of Formulation Mechanical Properties on Multilayer Tablet Compaction Properties and Layer Adhesion

During their evaluation of the importance of surface roughness on bonding and tensile strength of tablets, Karehill et al. (31) investigated the compaction of various directly compressible materials exhibiting different volume reduction mechanisms mainly by fragmentation or by plastic deformation. Sodium chloride, microcrystalline cellulose, partially pregelatinized starch 1500, lactose α-monohydrate, dicalcium phosphate dehydrate, and sucrose with specific particle size fractions between 212 and 250 μm were examined. Compression was performed with an eccentric Korsch EK-0™ tablet press using 1.13-cm diameter flat-faced punches that were lubricated manually prior to compaction. The first layer was compressed with a pressure between 25 and 200 MPa to vary its surface roughness. Additional powder was added on top to form a bilayer tablet and compacted at a pressure of 200 MPa. The amount of powder was varied so as to keep the first layer and bilayer thickness constant (0.3 cm for the first layer and 0.6 cm for the bilayer tablet). After compaction, the first-layer surface roughness was evaluated using SEM and the adhesion strength was measured using the axial tensile strength test to pull apart the layers (see section "Adhesion Strength of Multilayer Tablets: Theory, Measurement, and Prediction"). For all of the materials studied, increasing the first-layer compression pressure resulted in a decrease in the adhesion strength, with the plastically deforming material being more sensitive than the fragmenting material (Fig. 8). Increasing the compression pressure resulted in less surface roughness and less surface area available for further bonding. The fragmenting materials were not impacted as significantly by the increase in compaction pressure as they were able to form high surface areas available for bonding *via* fragmentation (Fig. 9). The plastically deforming materials failed at the layer interface during the axial strength test, indicating that the bonding strength

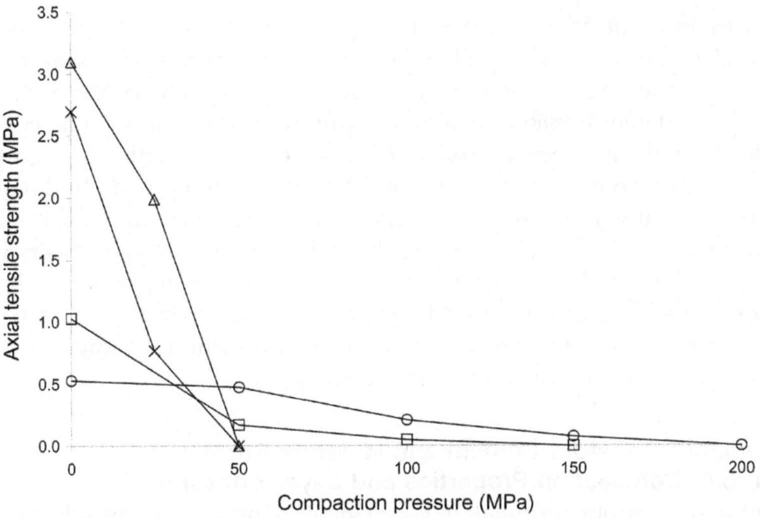

Figure 8 The effect of first-layer compaction pressure on the axial tensile strength of the bilayer compact compressed at 200 MPa for plastically deforming materials. Δ, Avicel PH101; X, Avicel PH101 < 10 μm; □, starch 1500; and ○, sodium chloride. *Source*: Modified from Ref. 31.

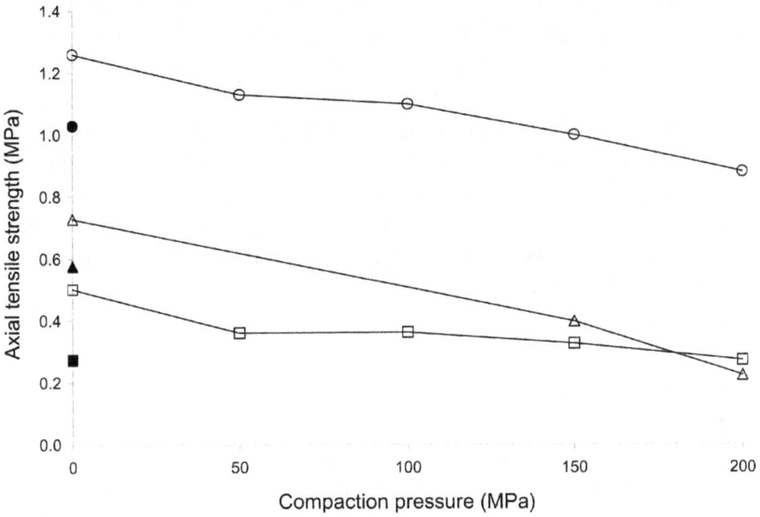

Figure 9 The effect of first-layer compaction pressure on the axial tensile strength of the double layer compact compressed at 200 MPa for mainly fragmenting materials. The strength of a single tablet compressed twice at 200 MPa is denoted by a filled symbol on the *y*-axis. ○, Emcompress; Δ, sucrose; and □, lactose monohydrate. *Source*: Modified from Ref. 31.

was lower at the interface than in the individual layers. The fragmenting materials failed in the first layer during the axial strength test, indicating that the bonding at the layer interface was stronger than the individual layers. This indicates that material fragmentation created a high surface area, which increased bonding between the layers in the compact.

Materials can exhibit different deformation mechanisms such as elastic deformation, plastic deformation, brittle fracture, and/or viscoelastic deformation to varying degrees in response to an applied pressure. These differences can impact the porosity of each layer of a multilayer tablet to different degrees. To evaluate the impact of solid fraction (porosity) and layer composition, Mullarney and Lam (18) compressed monolayer and bilayer compacts of microcrystalline cellulose, lactose, and dibasic calcium phosphate at three different solid fractions using a triaxial compression press and a split die to create a 0.375-in. cubic compact. The axial tensile strength (between layers) and radial tensile strength (through layers) was measured for the cubic compacts (see section "Adhesion Strength of Multilayer Tablets: Theory, Measurement, and Prediction"). It was determined that the tensile strength was always lower in the axial direction, indicating that the bonding between layers was weaker than the bonding within the layers. Bilayer tablets produced using material that exhibited differences in compressibility resulted in differences in the solid fraction of each layer in the bilayer compact. They showed that these differences had a significant impact on the bilayer compact strength as well as on the layer adhesion strength.

Impact of Formulation Moisture Content and Moisture Sorption on Multilayer Tablet Compaction Properties and Layer Adhesion

Moisture content of the various formulation components significantly impacts the compaction properties as discussed in other sections of this book. It similarly impacts the compaction behavior of multilayer tablets but seems to be seldom studied in this context. Moisture sorption, and its combination with exposure to heat, is a parameter more often studied as its extent can significantly weaken layer adhesion upon storage and during stability studies.

Bilayer or multilayer tableting can be even more challenging as tablets can delaminate long after compression. It is not uncommon to observe delamination upon further processing, such as coating, packaging, and/or stability testing. As described earlier in this chapter, Birringer et al. (20) assessed measurement and optimization of layer adhesion in bilayer tablets. They have defined the critical tamping force as the greatest tamping force at which both layer still adhere to each other at the time of compression and assessed whether tablets of various critical tamping forces maintain their integrity upon exposure to heat and humidity. They observed that the higher the critical tamping force, the less the tablets were susceptible to delaminate upon storage. In designing multilayer tablets, layers should not only exhibit similar compaction properties but also limited and/or similar sorption properties and expansion with heat.

MANUFACTURING CONSIDERATIONS FOR MULTILAYER TABLET COMPRESSION UNIT OPERATION

As eluded to in the previous sections of this chapter, process parameters such as compression forces (tamping, pre, and main), press type, compression zone (penetration depth) and compression speed as well as compression tooling design are critical in the compaction of multilayer tablets. Tamping forces applied after each individual layer especially need careful optimization and monitoring to achieve enough adhesion between layers and to avoid cross contamination (12). In some multilayer production presses, compression forces also control the weight of each layer, this can be of importance since it depends on the quality of the formulation (especially its flowability) and significantly impacts the tablet strength. Other manufacturers have taken slightly different approaches to the manufacturing of multilayer tablets and two of these presses will be highlighted in this section.

High-speed and high-volume manufacturing of multilayer tablets is very challenging with typically longer compression runs due to slower press speeds, low yields as a result of the very special requirement to prevent contamination of the layers, and challenges to sampling and accurate weight control of each of the layers. This is even more the case since the formulation is not, or cannot, always be fully optimized for multilayer production. Often, combination products are developed using at least one commercial product in combination with a new chemical entity or another already marketed product. In the case of a marketed product, the formulation may not be changed significantly at the risk of impacting the bioequivalence of the product depending on the biopharmaceutical classification. This section focuses on compaction aspects; however, other important characteristics are powder flow and required fill weight for each specific multilayer formulation. The compaction properties are also influenced by many other parameters such as layer thickness, tablet size and shape, number and composition of the layers. Process parameters such as compression force for each layer and the multilayer tablet compression force, precompression on the multilayer tablet, compression zone (location within the die where the tablet is compressed), layer order, as well as tablet press speed can impact the compaction process for multilayer tablets and are discussed in detail in the rest of this section.

A case study on computer modeling of density distribution in bilayer tablets highlighted that the properties at the layer interface have significant importance and contribute to the overall strength and delamination potential of bilayer tablets (32). The density was modeled using 20-mm round flat-faced punches with single-sided compression (lower punch held stationary). Each layer contained the same material: microcrystalline cellulose. A high compression force of 10 kN was used for compression of the first layer and a compression force of 40 kN was used to form the bilayer tablet. Densification of the first layer was shown to occur initially following the application of the first-layer force. When the second layer was added and a force applied to the bilayer tablet, there was no change in the first layer that was already compressed, while the second layer proceeded to become more dense. Densification of both layers occurred only after the second layer reached a density similar to that of the first (i.e., around 10 kN in this example). This confirmed that the first-layer compression force is critical as mentioned by many authors and discussed further below. It was also observed that the layers exhibit different densities at the interface, especially near the die wall (Fig. 10). In this case, the frictional effect was exaggerated by simulating unlubricated dies. A distortion of the interface with the first layer becoming concave was predicted in this simulation and confirmed by Inman et al. (study described further below). This modeling study was performed only for the application of force and did not consider changes that might occur during unloading and ejection. The first-layer compression force used in this model (10 kN) was also high compared to the range of compression forces most often used for first-layer tamping of bilayer and multilayer tablets (around 1 kN). This study, however, had the merit of predicting the behavior of a simple formulation upon bilayer compression. Such model and simulation approaches should be performed more often and have the potential of explaining or predicting the behavior of complex and inhomogeneous pharmaceutical systems.

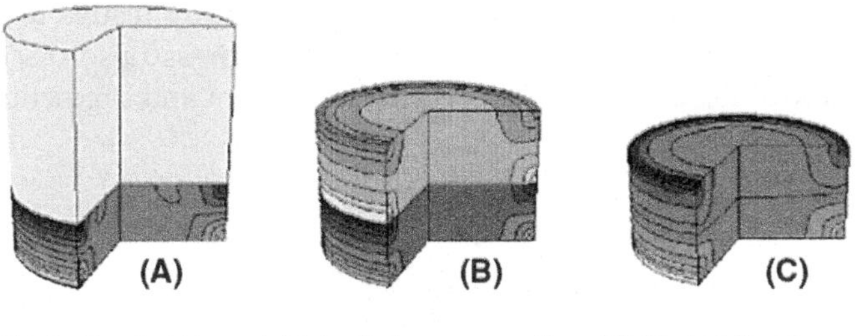

Figure 10 Density distribution in bilayer tablets during compaction. (**A**) End of first-layer compaction after the second layer was filled. (**B**) Snapshot during bilayer compression. (**C**) End of bilayer compression. *Source*: From Ref. 32.

Impact of First-Layer Compression, Precompression, and Main Compression Forces on Multilayer Tablet Compaction Properties and Layer Adhesion

The first processing factor to consider that has an influence on the layer adhesion in a multilayer tablet is the compression force used on the first layer of material, sometimes referred to as tamping force. In the limited published literature on multilayer compaction, the impact of first-layer compression force on layer adhesion has been the most widely studied aspect of multilayer tableting. Tamping of the first layer has multiple purposes. It creates sufficient volume in the die to add the next layer of material prior to the final, main compression. It also forms a smoother interface, thus preventing mixing of the materials and providing a visually appealing separation of layers. If the first layer is not compressed sufficiently, the high-speed filling and rotation of the press can lead to centrifugal forces sufficient to result in a skewed layer interface that is not straight across the tablet (Fig. 11). In some instances the layer interface can reach the top of the tablet and significant mixing of the layers can be observed, which can be a stability concern or at least an appearance defect. The tamping of the first layer also allows for automatic weight control of the first layer when a tablet press uses compression force to control the tablet weight. A sufficiently high enough force must be applied when there is a force feedback loop used to control the first-layer weight (33) as described further in section "Alternate Tablet Press Designs for Multilayer Tablet Production." A balance must however be struck between the need of a high force for the best weight control and the potential negative impact of increasing first-layer compression force on layer adhesion. More sensitive compression rolls may have to be installed on the first layer to accurately monitor potentially small compression forces allowing optimum layer adhesion as well as weight control.

Some degree of plastic deformation of the material at the interface is needed to achieve bonding of the layers. The higher the compression force used for the first layer, the less plastic deformation or fragmentation is available for bonding between the layers, which results in weaker adhesion. This weak adhesion leads to a high potential for delamination of the bilayer tablet, which should be avoided with multilayer tablets for the reasons already discussed in section "Adhesion Strength of Multilayer Tablets: Theory, Measurement and Prediction". In their evaluation of bilayer tablet machines, Li et al. (21) demonstrated that the compression force of the first layer was the most critical factor that affected layer adhesion.

Dietrich et al. (15) carried out a thorough DOE to assess the impact of compression force (and lubrication discussed in section "Impact of Formulation Composition and Extent of Lubrication on Multilayer Tablet Compaction Properties and Layer Adhesion") on the adhesion strength using single-sided compression with 10-mm flat-faced round tooling at slow compaction speed for bilayer and central layered tablets. Although the compaction force used for the first layer was also high at 5, 10, and 15 kN (they were interested in making a tablet that could be physically transferred), the authors observed that increasing compression force of the first layer resulted in decreased adhesion strength between the layers whereas increasing the main compression force for the multilayered tablet from 10 to 20 kN resulted in increased adhesion strength (Figs. 12A, B and 13 A, B). They fitted the study results to an equation that then predicted the adhesion strength on a production scale press (also discussed in section "Adhesion Strength of Multilayer Tablets: Theory, Measurement and Prediction"). The highest

Figure 11 Example of a skewed layer interface.

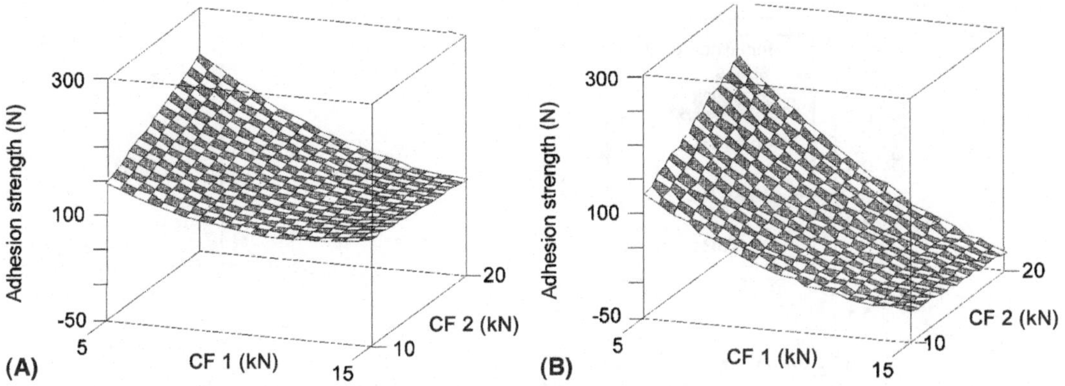

Figure 12 (**A**) Contour plot for influence of compression forces on adhesion strength *at low level of lubricant concentration.* (**B**) Contour plot for influence of compression forces on adhesion strength at *high level of lubricant concentration. Abbreviations*: CF1, compression force for central layer; and CF2, compression force for multilayer tablet. *Source*: From Ref. 15.

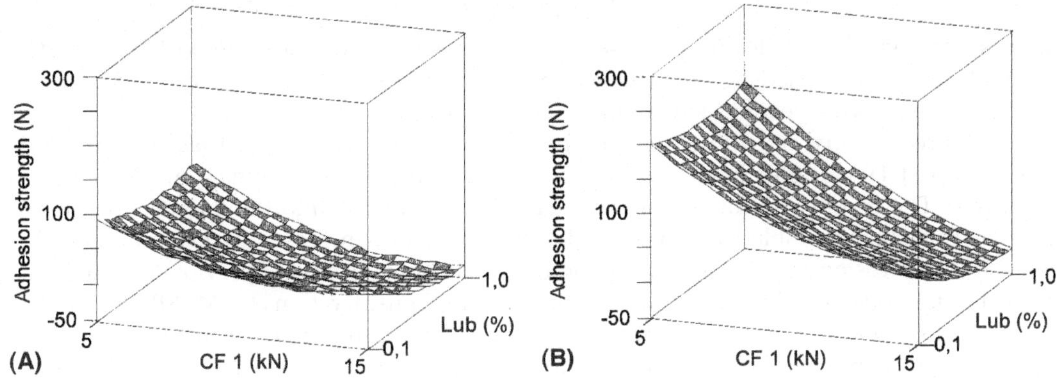

Figure 13 (**A**) Contour plot for influence of compression force and lubricant concentration on adhesion strength at *low compression forces.* (**B**) Contour plot for influence of compression force and lubricant concentration on adhesion strength at *high compression forces. Abbreviations*: CF1, compression force for central layer; Lub, lubricant concentration in central tablet. *Source*: From Ref. 15.

layer adhesion was determined to be at low lubricant concentrations, low compression forces for the first layer, and high compression forces for the multilayer tablet.

Using the method of axial tensile stress (pulling apart the tablet) to determine the fracture force (as described in section "Adhesion Strength of Multilayer Tablets: Theory, Measurement and Prediction"), bilayer tablets containing microcrystalline cellulose in each layer were produced while varying the compaction forces for the first layer from 28.6 to 95.5 MPa and for the bilayer tablet compression forces from 10 to 156 MPa (17). The tablets were flat-faced, 20-mm round diameter and manufactured with an E-Z 50^{TM} materials tester machine. The weight of each layer was varied to maintain a constant aspect ratio for each layer and a tablet thickness of 10 mm. As the first-layer force was increased, the fracture force decreased indicating reduced layer adhesion (Fig. 14). The authors concluded that the amount of plastic and elastic deformation that occurred in the initial layer during compaction influenced the adhesion strength of the layers at the interface. They performed surface roughness measurements using laser profilometry on the first layer at forces of 50.9 and 95.6 MPa. The results showed that when the first-layer compression force was increased, the surface roughness was reduced due to more extended deformation of the particles. The authors hypothesized that if there is sufficient surface

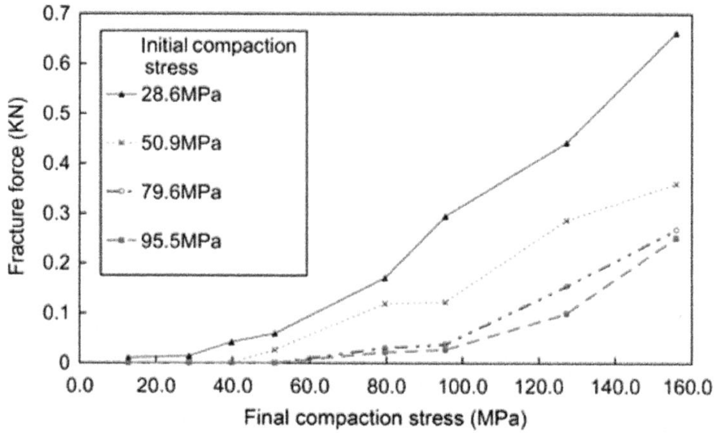

Figure 14 The fracture force of bilayer tablets of microcrystalline cellulose with different first-layer compression forces as a function of increasing final compression forces. *Source*: From Ref. 17.

roughness into which particles from the second layer can fit, mechanical interlocking has the potential to occur. This increases the contact area and possibly creates a less direct route for fracture to spread, which could increase the adhesion strength.

Laser profilometry and X-ray tomography were used to analyze the surface topography of the fractured layers after breaking and to visualize cracks within intact bilayer tablets, respectively. The results indicated that a curvature is introduced into the interfacial surface when the final bilayer tablet is compressed. This curvature increases in magnitude with increasing compression force, thus confirming the predictions discussed previously based on density distributions modeling in section "Manufacturing Considerations for Multilayer Tablet Compression Operation". The appearance of the curvature in the center of the compact was attributed to the combined effect of die wall friction (preventing movement of particles near the die wall) and of transferred force acting through the second layer material. The difference between the edges and center of the compact may indicate a reduced bonding near the edges of the tablet due to the reduced force being transferred though the tablet. As a consequence, any relaxation or release of elastic energy may not be uniform across the interface. This inhomogeneity could create cracks at the interface as observed in some of the tablets generated in the study. Cracks in the tablet at the interface of the layers could concentrate stresses and result in weaker adhesion of the layers.

While it is useful to perform evaluations with pure materials to gain fundamental insight into the underlying mechanisms, it is often not representative of pharmaceutical applications that use multicomponent formulations, potentially process the materials *via* granulation, milling, etc., and high-speed compression machines (where both upper and lower punches are displaced during compression and the compaction occurs at a high speed).

Birringer et al. used a Fette 3090^{TM} high-speed bilayer rotary press and material processed *via* wet granulation (20) to evaluate the precompression and main compression forces for the overall bilayer tablet. It has long been recognized that the use of precompression can significantly impact the performance of multilayer tablets (14). It can remove entrapped air *via* application of low pressure prior to main compression to aid in bonding and help prevent delamination. It was demonstrated that increasing the main compression force for the bilayer tablet from 250 to 300 MPa reduced the adhesion strength of the layers. This was thought to be a result of higher compression forces accentuating the effect of differing axial recovery (expansion) of the two layers after the compression force is removed and the tablet ejected from the die. It was also noted that increasing the precompression force of the overall bilayer tablets from 20 to 60 MPa resulted in improved adhesion of the layers to a much higher degree than the effect of the main compression force. This highlights the importance of evaluating first-layer tamping force, precompression and main compression for a bilayer tablet in a particular combination system.

The overall force used to compress the multilayer tablet can impact the tensile strength of the tablet as well as the layer adhesion. Tablets generally need to be soft enough to exhibit disintegration leading to dissolution for the drug to be absorbed. Tablets that are too soft tend to have low strength and unacceptable friability and layer adhesion. Increasing the compression force increases the strength of the tablet as well as the adhesion of the layers, but there tends to be a maximum force after which the tablet is over compressed and the layer adhesion is reduced. By using X-ray computed tomography, Wu and Seville (13) found that high maximum compression forces lead to increased delamination or fractures along the interface between the layers. Bilayer tablets were prepared with specific microcrystalline cellulose and lactose monohydrate particle size fractions. The first layer was compressed by hand with a low force, and increasing compression forces of 3, 9, and 18 kN were applied for bilayer compression. Increasing forces resulted in increasing fracture planes between the layers after tensile strength testing.

Impact of Other Processing Parameters Such as Compression Zone, Speed, and Layer Order on Multilayer Tablet Compaction Properties and Layer Adhesion

Altering the compression zone or location within the die where the tablet is compressed is accomplished *via* upper punch penetration depth. This is often done to compress the tablet near the top of the die so that air entrapment is reduced (less distance for the air to escape) and the distance required for ejection of the tablet is also reduced, thus decreasing ejection forces and reducing incidence of tablet capping. The compression zone is also important for the density distribution within the tablet, but this impact is poorly understood.

Ideally the upper and lower punches should have the same displacement during compression to equalize the density distribution within the layers. If either the upper or lower punch is displaced more than the other, the density distribution throughout the tablet may not be completely uniform. Wu and Seville (13) showed that the nonuniform distribution that occurs using single-ended compaction by a stationary lower punch resulted in differences in relative density of two layers. They prepared bilayer tablets using specific microcrystalline cellulose and lactose monohydrate particle size fractions and analyzed the tablets after tensile strength testing using X-ray computed tomography. The results indicated more fracture between layers when the microcrystalline cellulose was loaded first (as the bottom layer).

The effect of tableting speed on the layer adhesion has not been widely studied in the literature, but it can be extrapolated that materials that exhibit strain rate sensitivity (reduced bonding at high compaction speeds) may also exhibit reduced layer adhesion. This was demonstrated in one study of compaction of a triple-layered tablet composed of a multicomponent mixture, representative of typical formulations and using a Mand compaction simulator with 1-cm round, flat-faced punches (12). The authors observed that at the fastest compression speed of 500 mm/sec punch velocity (shortest dwell time), occasional delamination of the tablet occurred during compression of one of the layers. Although adhesion strength was not directly measured, the authors hypothesized that such short dwell times were not realistic for typical production of triple-layered tablets and their formulation was acceptable. This conclusion demonstrates another difference between multilayer tableting and monolayer tableting: the compression speeds used in multilayer tableting are often much slower than with standard tableting due to some of the manufacturing considerations that have been discussed in this section.

Figure 15 Cross-sectional view indicating difference in layer shape for (L1) first layer and (L2) second layer of a bilayer tablet produced using concave tooling.

Typically the layer that exhibits the highest weight/bulk density or has the poorest flow is used as the first layer. During filling of the first layer, suction is created *via* the pull down of the lower punch (when it is under the feeder), which aids significantly in die filling (34). This punch pull down and suction do not occur with the second layer since the first-layer tablet is already lightly compressed and still in the die. Additionally, the first layer allows for a deeper fill volume compared to the second-layer fill volume, which can only be as much as the upper punch penetrates into the die during compression of the first layer (typically around 8 mm for most commercial rotary multilayer tablet presses). Li et al. (21) utilized the compression profiles for individual materials to determine which layer to put first. An antacid formulation was fed first because of its higher tensile strength. The weaker aspirin layer was added as the second layer so that it would be compressed only once by the main compression force to optimize the adhesion of the layers, although no evaluations were reported to confirm this theory.

Another interesting and sometimes complicating aspect of bilayer tablet compaction is when using concave tooling to make tablets: the tool design and the order in which the layers are compressed can significantly impact the density of the material due to the different shapes of the layers (Fig. 15). The same material has a higher density around the edges when compressed in the first layer compared to a more homogeneous density throughout the layer when compressed in the second layer (35). The layer order can be a critical parameter in the compaction of bilayer tablets and can impact the bilayer tablet strength as well as potential layer delamination due to differences in density distribution.

Alternate Tablet Press Designs for Multilayer Tablet Production

Most commercially available multilayer presses operate in a similar manner to standard rotary presses except for not ejecting the tablet after the first layer is produced. Instead, the second layer is filled on top and the bilayer tablet compressed and then ejected. This results in a tablet with the layers compressed horizontally. However, Korsch has designed a series of multilayer presses that compress tablets vertically with examples shown in Figures 16 and 17. These have been termed longitudinally compressed tablets (LCTs). This requires the use of specially designed compression tooling and press turret that allows for deeper penetration into the die. Benefits of this method include the ability to fill more material or less dense material and to add increased number of layers. Most manufacturers stop at a maximum of two- or three-layered presses; however, Korsch offers a five-layered tablet press. While the activity of accurately controlling weight for five individual active layers may seem daunting there are also possibilities to use two adhesion layers in between three active layers or provide for a dual combination tablet with an inactive center layer to allow for accurate splitting of the tablet (Fig. 18) termed Accubreak™. There have been no articles to date evaluating this mode of compression compared to the standard mode, the layer adhesion could be even more critical as the surface area available for bonding is smaller but the ability to utilize non-active bonding layers may alleviate the issue.

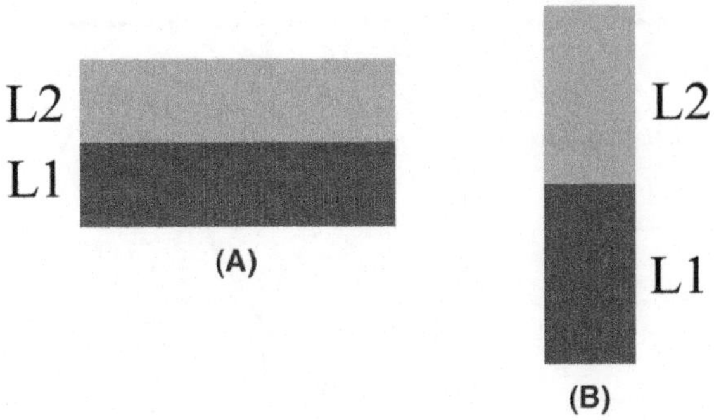

Figure 16 Bilayer tablet (**A**) compressed horizontally and (**B**) compressed vertically to create a longitudinally compressed tablet (LCT).

Figure 17 Ejection of longitudinally compressed tablets from a tablet press. *Source*: Courtesy of Korsch.

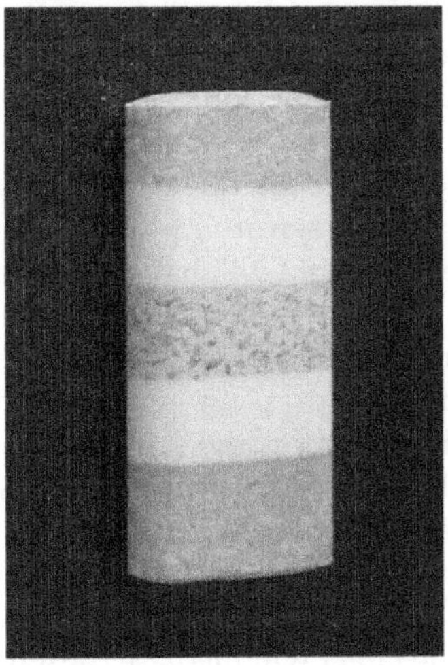

Figure 18 An example of a five layer tablet with three active layers and two bonding layers compressed vertically (longitudinally compressed tablet). *Source*: Courtesy of Korsch.

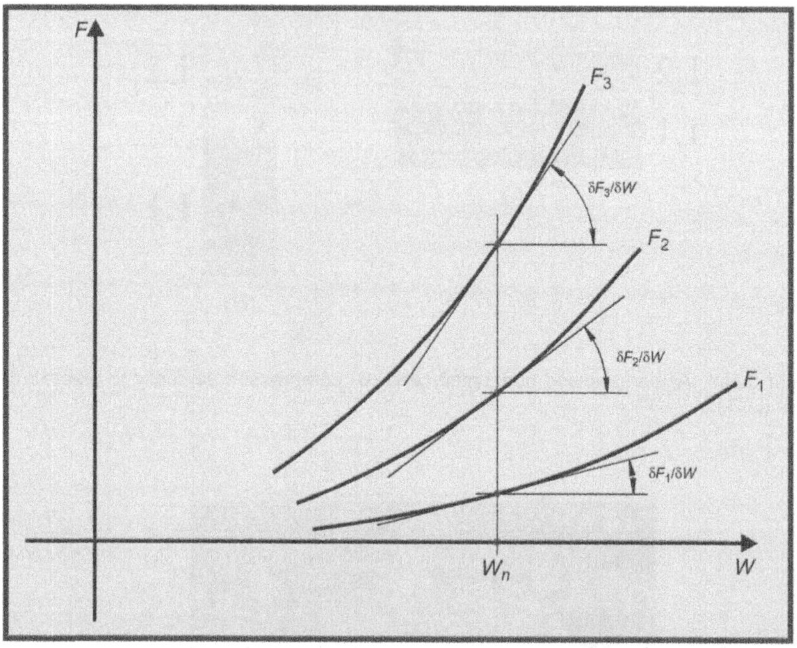

Figure 19 Force *versus* weight sensitivity at different compression force levels demonstrating reduced sensitivity at low compression forces. *Source*: From Ref. 36.

While most tablet press manufacturers measure a force variation at main compression to monitor and control tablet weight, Courtoy has taken a different approach and measures a variation of tablet height at precompression, using the precompression roll displacement as an alternative method (36). Typically, a press is designed to compress tablets at a defined thickness since the compression rollers stay at a constant distance apart to achieve a desired compression force. Changes in weight will lead to changes in the force applied to the powders at a constant thickness but force differences can also arise from having punch lengths that are variable. Compression force monitoring works well in most situations, but has limitations when dealing with multilayered tableting. Multilayer tablets require a very low compression force for the initial layers to maximize the adhesion between the layers during multilayer tablet compaction as described above. As many papers have indicated, high compression force on the initial layers leads to poor bonding of subsequent layers. This low compression force is a problem because the force *versus* weight sensitivity curve decreases as the force decreases (Fig. 19). This indicates that changes in weight will have less impact on the force when it is low and results in less sensitive control over the weight. Additionally, the compression rollers and sensors in most multilayer tablet presses are designed to be used with a high force to allow the press to also be used for double-sided compression—compression of a single monolayer tablet on each half of the tablet press, which reduces the sensitivity at low forces.

Courtoy's approach is to apply a constant force on the tablet at precompression using the "air compensator." The precompression roller is fixed to a piston, which can move inside a cylinder that is filled with compressed air, and the air pressure is maintained constant using a regulating valve and expansion chamber. F = surface of piston x air pressure. During compression, the force exerted on the tablet will increase until it equalizes, the precompression roller will then be moved up by the punch to allow the punch to travel under the roller. During the upward and downward movement of the compression roller, the force exerted on the

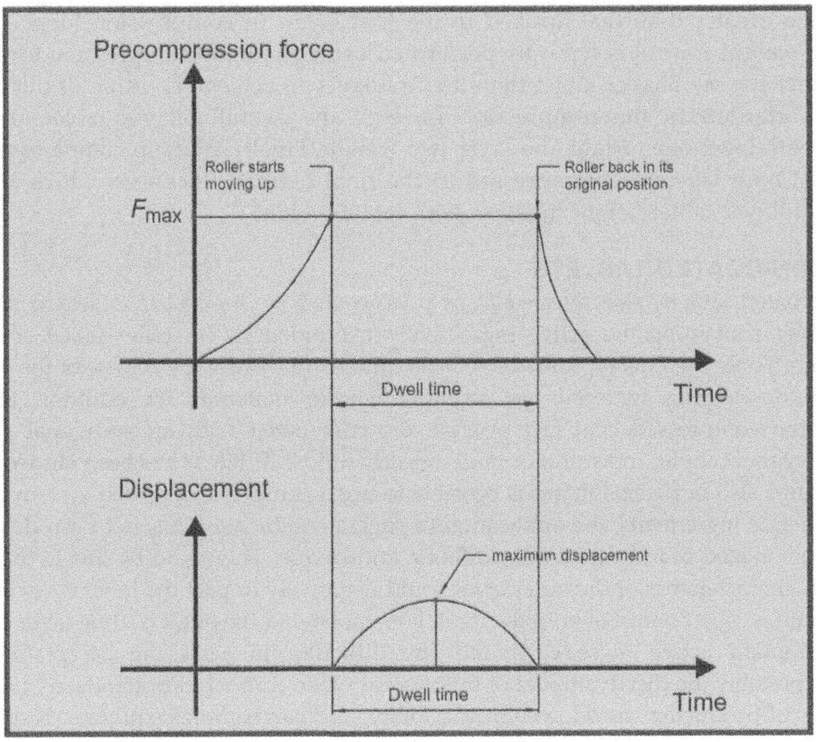

Figure 20 Force and displacement profiles during precompression using air compensation. *Source*: From Ref. 36.

tablet remains constant and equal to the set force in the control. Therefore, a tablet with more weight or mass will create more movement or displacement than a tablet with less weight or mass resulting in variable precompressed tablet height. While the tablets at precompression have a variable height, at the main compression station all tablets are compressed to the same thickness to ensure a correct volume of fill for the second layer.

The advantage of this technique is that the sensitivity of the system is not reduced at low forces and the dwell times (time at which the compression force is within 90% of the maximum) are increased at precompression. Increased dwell times are a result of the pneumatic compensator that allows the roll assembly to move after a set compression force is reached enabling the tablet to have the maximum force applied for an increased amount of time (Fig. 20). The advantages of increased dwell times at precompression are increased deaeration and powder rearrangement, which can help improve the tablet breaking force and reduce the possibility of capping due to air entrapment. Additionally, for materials that exhibit a high strain rate sensitivity, the additional dwell time allows for increased tablet strength.

Besides the high sensitivity at low forces, the system also allows for an independent weight control of individual layers. Once the first layer is compressed to its final thickness, the second layer is filled. The precompression force of the second layer will be lower than the main compression force of the first layer. As a result, the first layer will not be compressed while measuring the tablet weight of the second layer. This leads to an individual weight control of the second layer and tablet weight variations of the first layer will not be carried over to the second layer.

This was also described above in the modeling of in-die density changes of a bilayer tablet during compaction where the first layer did not begin to deform until the bilayer tablet reached a force greater than that applied to the first layer. In compression force controlled machines, the weight control is typically performed using the main compression force, which is always higher for the bilayer tablet than the first layer to achieve bonding of the layers. In these cases, variations in the compression force of the overall bilayer tablet are due to variations in both layer one weight and layer two weight. Finally, at main compression for the Courtoy press, both layers are compressed to the final defined thickness, which allows for control of the bilayer tablet height together with tablet weight.

COMPRESSION-COATED TABLETS

Compression-coated tablets, also referred to as press-coated or dry-coated tablets, consist of an inner core tablet containing the active ingredient surrounded by an outer tablet compressed around the core. These systems typically have been utilized to modify the release of the inner core tablet *via* enteric coating materials or swelling/gelling materials. In addition to release modification, the compressed coat can provide the core tablet with an additional protection from the environment (light, moisture) or body (gastric fluids). While it has been eluded to in the literature (37) and also in patents that it is possible to use a compression-coated system to deliver two different active ingredients, the application of such a combination has not been described in the literature or utilized in industry to the authors' knowledge. This could be due to the fact that the dissolution characteristics of the outer layer could negatively impact the inner core release, the need for accurate weight control of both the inner core tablet and the outer coating layer since they would both contain active ingredients and the difficulty in achieving acceptable content uniformity, or possibly the disadvantages of this system when compared to standard multilayered tablets in terms of production speed, weight variability, and less complex equipment and process.

Compression-coated tablets are slower and costlier to manufacture as they require the separate manufacture of the central core tablets and subsequent transfer into the die of another tablet press. Improper placement of the core tablet can result in multiple problems such as off-center cores, lack of cores and multiple cores (38). Many of these challenges have been addressed with the recently developed one-step dry-coating system (OSDRC®) where the coating and core tablets are manufactured in one single process on the same press using double-structured punches and a specially designed tablet press (38–42). Although currently under evaluation, the applicability of this system to combination products has not yet been demonstrated to the authors' knowledge. For combination products the weight accuracy and variability of both the inner core material and the outer coating layer is critical, whereas in most standard compression coating systems the outer coating layer does not contain an active ingredient and does not need to be controlled as rigorously. In the OSDRC process a bottom layer is produced followed by the central core layer with compression after each fill. Finally the top and sides are filled and the final tablet compressed and ejected (Fig. 21). One significant advantage of this process is that the central core layer can consist of uncompressed material such as multiparticulates, which is not possible with standard compression coating press designs where an intact tablet needs to be transferred. Other benefits include improved accuracy in the placement of the core, the ability to have thinner outer walls and the possibility to have multiple cores in one tablet.

MONOLAYER COMBINATION PRODUCTS

Several chapters have discussed in detail the theory, modeling and simulation, as well as practice of compaction of binary mixtures and directly compressible materials. These principles can be extended to monolayer combination products, which is the preferred approach when physical, chemical and biopharmaceutical properties allow multiple actives to be formulated and compacted in a single matrix. Some additional considerations however should be taken into account for combination products and are highlighted in this section.

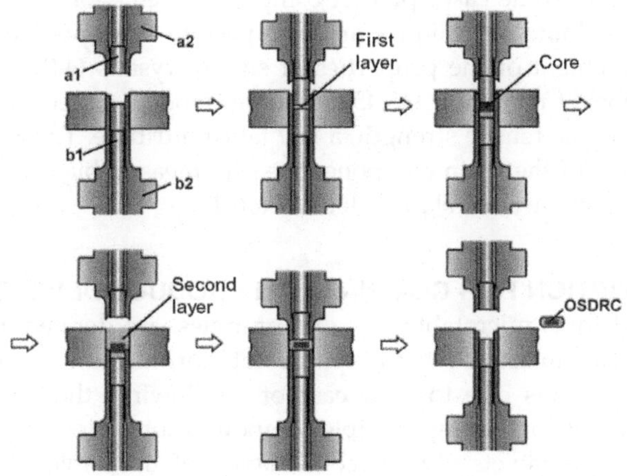

Figure 21 Schematic of OSDRC® (One-Step Dry-Coating) manufacturing method. (a1, b1) center punches; (a2, b2) outer punches. *Source*: From Ref. 38.

Despite the wealth of work and knowledge in the area of compaction of pharmaceutical powders, predicting mechanical behavior of tablets is very challenging due to the complexity and diversity of pharmaceutical formulations: monolayer tablet combination products are compressed from one formulation combining two or more active ingredients and multiple excipients. The formulation can be prepared as a dry blend if directly compressible or as a dry, wet, or melt granulation or extrusion to improve among other properties the packing density of the heterogeneous mixture. The "pharmaceutical powder" can also be composed of a mixture of multiparticulates of different compositions: each multiparticulate section prepared separately, potentially by a different method with different excipients, containing one or more different actives, and mixed in specific ratios for compression. The multiparticulates can potentially be large granules, beads prepared by spheronization, and/or beads coated with active ingredients. Many options are possible but often limited by drug-loading considerations and tablet size: a significant amount of excipients is usually necessary to ensure compressibility and adequate tablet strength, which in turn has a deleterious impact on drug loading and may render this formulation approach unsuitable for combinations containing high-dose compounds. Moreover, the desired release pattern for each of the actives of the combination monolayer tablet, whether immediate or controlled, may be affected by the complex formulation and functionality of each component. These limitations render multilayer approaches most attractive despite the compaction challenges discussed in previous sections of this chapter.

Depending on the formulation and/or process, one could potentially predict the tablet characteristics based on the properties of the main components or on the behavior of the binary mixture of the dominant components. Alternatively, data analysis techniques can be used to characterize the complex pharmaceutical powder or main components and predict the compaction behavior of the combination products as monolayer tablets. This approach would allow the formulator to further tailor the formulation for optimum compaction properties, mitigating the impact of other key tablet properties, which could affect the desired in vivo performance.

The main criterion for the successful compaction of a tablet is adequate bonding (tensile strength) combined with nonfracture (balance between elastic, plastic and brittle behaviors) (43). Since these mechanical properties are a priori nonpredictable, physical measurements are necessary to quantify and in some cases predict compaction behavior of complex pharmaceutical powders into tablets. Many techniques and compaction concepts have been developed, ranging from the measurement of the properties of single crystals (44), binary mixtures and well-defined multicomponent compacts (43,45). In many models tablet mechanical properties are assessed by measuring the tensile strength, a key tablet attribute. Tensile strength has been directly correlated to some of the main components easily measurable critical properties, such as their particle size and porosity or relative density (46,47).

MINITABLETS AS AN OPTION FOR COMBINATION PRODUCT DEVELOPMENT

Minitablets, also referred to as microtablets or oral granules, are defined as compacts of 1 to 5 mm in diameter. They are an attractive multiparticulate formulation option for pediatric (48) and geriatric oral dosage forms due to their ease of swallowing, the dosing flexibility they provide and the possibility of combining multiple compounds or various delivery rates through a mixture of monolithic minitablets (49). A combination of minitablets containing different actives can be filled into capsule shells or sachets, or compacted into a large tablet (50). From a clinical perspective, as with other multiparticulate systems, minitablets usually provide consistent in vivo drug release, minimize the risk of dose dumping in the case of controlled release and can alleviate gastrointestinal irritation due to high local concentrations sometimes observed with standard size tablets (49,51). Bioadhesive bioerodible minitablets have also been used for ocular administration with potential increased bioavailability and local safety compared to eye drops (52). From a technical perspective, minitablets have the significant advantage to use the existing tableting technology, with a different punch design. They can also be readily coated and thus possess significant technical advantages compared to other multiparticulate dosage forms such as granules, beads, or pellets in regards to ease of manufacture.

The number of tips on a multiple tooling used to compress minitablets can range from about 2 to 20. The requirements for the tooling are more stringent than for standard size tablets: the tooling must fit very precisely in the die and must be of high mechanical strength, while being flexible enough to minimize punch stiffness due to higher friction with the die wall (53). Associated with minitablet compaction are two key formulation requirements: the formulation must exhibit adequate flow for good weight control on fast production tablet presses, and its upper particle size must be controlled to alleviate obstruction of the die opening (54).

Compacting smaller tablets seems to provide in most cases some technical advantages in terms of tablet strength and capping tendency, as demonstrated in the literature with poorly compressible, highly brittle actives such as paracetamol (53) or pancreatin (55). In the case of paracetamol, the capping tendency upon compression stems from residual stress in the tablet after compression, released through elastic recovery. The capping tendency seems to be reduced at high-compaction pressure in a smaller diameter tablet (53). Tablet tensile strength has also been demonstrated to increase with decrease of tablet size for minitablets (53,56,57). Multiple hypothesis have been discussed in the literature to explain this size dependence, such as structural differences due to different level of friction, different stress distribution due to differences in tablet formation or size, increase in mechanical activation with decrease in tablet diameter, greater pressure on smaller tablets due to smaller tablet height. The latter was however disproved by Flemming (58) as he demonstrated that for minitablets of less than 3 mm in diameter, the increased force transmission is counteracted by higher friction compared to the compaction of larger tablets. Lennartz and Mielck (53) studied tablet strength and capping tendency of minitablets: tablets containing 0%, 30%, and 50% of paracetamol and ranging from 1.5 to 5 mm in diameter were compacted to the same relative densities using a reciprocating tablet press. The authors demonstrated that the weakening effect of paracetamol (which negatively impacted the capping tendency and final tablet strength of the minitablets) was reduced with a decrease in tablet diameter. As the minitablets got smaller there was less capping and the tensile strength increased. They argued that during compression, a protective shell is formed around the tablet due to shear stress upon contact with punch and die surfaces. This shear stress creates binding sites locally, and is higher as the tablet size decreases due to the increased ratio between tablet outer surface and volume.

Minitablets, due to their low porosity, seem to be more amenable to compression into a standard size tablet than pellets and granules. This provides significant advantages in reproducibility and flexibility for multiphasic drug release and/or for combining chemically or physically incompatible drug substances. Lopes et al. (50) successfully developed a biphasic delivery system platform composed of slow-release film-coated minitablets compressed into a standard size tablet, outer phase of which provided immediate drug release. They observed that the minitablets integrity was mostly maintained upon standard size tablet compression: minitablets showed slight deformation but no fragmentation, with minimum impact on their dissolution profile. They hypothesized that due to the minitablets low porosity, particles could not rearrange significantly upon additional compression into a standard size tablet, thus resulting in minimum deformation. In addition, during the uniaxial compression, minitablets are submitted to stress in all directions, thus rendering fragmentation difficult.

CONCLUSIONS

Combination products are an attractive and elegant formulation answer to the rising clinical need to combine drug substances for synergistic effects. They also result in better patient compliance and provide significant marketing advantages. A combination product can potentially be formulated as a monolayer or multilayer tablet, a compression-coated tablet or minitablets, depending on the combined drug substance properties and the desired dosage form performance and market image. Compaction of these combination products is driven by the mechanical properties of the pharmaceutical mixtures, with similar characterization techniques as for typical monolayer, standard tablets containing only one active. Modeling and simulation of mechanical properties and behavior upon compaction can also be useful for the optimum design of combination products, and additional research is needed in this field especially for multilayer tablets. These combination dosage forms exhibit significant challenges for formulation, compaction, and overall processing. Additional aspects for formulation, compaction, and equipment as well as for characterization thus need to be carefully taken into consideration depending on the combination product dosage form being developed, especially when compacting multilayer tablets, compression-coated tablets, and minitablets.

ACKNOWLEDGMENTS

We would like to thank Chris Balducci, Jay Lakshman, Richard Nause, and Colleen Ruegger for providing critical reviews; Rich Ferriera and Johan Van Evelghem for their discussions and contributions.

REFERENCES

1. Miller FH. Combination Oral Products: The Time is Now! Available at: http://www.ondrugdelivery.com. Accessed Jan 2009, 12–15.
2. Zannou EA, Li P, Tong W-Q. Product life-cycle management (LCM). In: Qiu Y, Chen Y, Zhang GGZ, *et al.*, eds. Developing Solid Dosage Forms: Pharmaceutical Theory and Practice. 1st ed. Elsevier, 2009:911–920.
3. Bangalore S, Kamalakkannan G, Parkar S, et al. Fixed-dose combinations improve medication compliance: A meta-analysis. Am J Med 2007; 120(8):713–719.
4. Wald NJ, Law MR. A strategy to decrease cardiovascular disease by more than 80%. Br J Med 2003; 326(7404):1419.

5. Waeber B. Fixed low-dose combination therapy for hypertension. Curr Hypertens Rep 2002; 4(4): 298–306.
6. Bailey C. Whence and whither the fixed-dose combination? Diab Vasc Dis Res 2005; 2(2):51–53.
7. Streubel A, Siepmann J, Peppas NA, et al. Bimodal drug release achieved with multi-layer matrix tablets: transport mechanisms and device design. J Control Release 2000; 69(3):455–468.
8. Siahi MR, Barzegar-Jalali M, Monajjemzadeh F, et al. Design and evaluation of 1-and 3-layer matrices of verapamil hydrochloride for sustaining its release. Aaps PharmSciTech 2005; 6(4):E626–E632.
9. Narendra C, Srinath MS, Rao BP. Development of three layered buccal compact containing metoprolol tartrate by statistical optimization technique. Int J Pharm 2005; 304(1–2):102–114.
10. Krishnaiah YSR, Karthikeyan RS, Sankar VG, et al. Three-layer guar gum matrix tablet formulations for oral controlled delivery of highly soluble trimetazidine dihydrochloride. J Control Release 2002; 81(1–2):45–56.
11. Krishnaiah YSR, Karthikeyan RS, Satyanarayana V. A three-layer guar gum matrix tablet for oral controlled delivery of highly soluble metoprolol tartrate. Int J Pharm 2002; 241(2):353–366.
12. Yang LB, Venkatesh G, Fassihi R. Compaction simulator study of a novel triple-layer tablet matrix for industrial tableting. Int J Pharm 1997; 152(1):45–52.
13. Wu CY, Seville JPK. A comparative study of compaction properties of binary and bilayer tablets. Powder Technol 2009; 189(2):285–294.
14. Multi-Layer Tablet Production. Drug Cosmet Ind 1968; 102(1):84–85.
15. Dietrich P, Bauer-Brandl A, Schubert R. Influence of tableting forces and lubricant concentration on the adhesion strength in complex layer tablets. Drug Dev Ind Pharm 2000; 26(7):745–754.
16. Dietrich P, Cremer K, Bauer-Brandl A, et al. Adhesion strength in two layer tablets. Pharm Res 1997; 14(11):S-420.
17. Inman SJ, Briscoe BJ, Pitt KG. Topographic characterization of cellulose bilayered tablets interfaces. Chem Eng Res Des 2007; 85(A7):1005–1012.
18. Mullarney MP, Lam DS, eds. Mechanical Properties of Bilayer Pharmaceutical Compacts. American Association of Pharmaceutical Scientists (AAPS) Annual Meeting; 2006.
19. Plank R, ed. Challenges of Multi-layer Compression for Combination Products. AAPS Workshop on Fixed-Dose Combination Products, 2006.
20. Birringer N, Shoemaker S, Gilman C, et al. eds. Measurement and Optimization of Layer Adhesion in Bilayer Tablet. American Association of Pharmaceutical Scientists (AAPS) Annual Meeting, 2005.
21. Li SP, Karth MG, Feld KM, et al. Evaluation of bilayer tablet machines—a case-study. Drug Dev Ind Pharm 1995; 21(5):571–590.
22. He XR, Secreast PJ, Amidon GE. Mechanistic study of the effect of roller compaction and lubricant on tablet mechanical strength. J Pharm Sci 2007; 96(5):1342–1355.
23. Bastos MD, Friedrich RB, Beck RCR. Effects of Filler-Binders and Lubricants on Physicochemical Properties of Tablets Obtained by Direct Compression: A 2(2) Factorial Design. Latin American Journal of Pharmacy 2008; 27(4):578–583.
24. Almaya A, Aburub A. Effect of particle size on compaction of materials with different deformation mechanisms with and without lubricants. AAPS PharmSciTech 2008; 9(2):414–418.
25. Patel S, Kaushal AM, Bansal AK. Compression physics in the formulation development of tablets. Crit Rev Therap Drug Carrier Syst 2006; 23(1):1–65.
26. Kato H, Kimura K, Izumi S, et al. The effect of magnesium stearate particle size on tablet properties and tableting characteristics of granules prepared with standard formulation. J Drug Deliv Sci Technol 2005; 15(6):475–480.
27. Kikuta J, Kitamori N. Effect of mixing time on the lubricating properties of magnesium stearate and the final characteristics of the compressed tablets. Drug Dev Ind Pharm 1994; 20(3):343–355.
28. Sheikhsalem M, Alkaysi H, Fell JT. The tensile-strength of tablets of binary-mixtures lubricated with magnesium stearate. Drug Dev Ind Pharm 1988; 14(7):895–903.
29. van Veen B, Bolhuis GK, Wu YS, et al. Compaction mechanism and tablet strength of unlubricated and lubricated (silicified) microcrystalline cellulose. Eur J Pharm Biopharm 2005; 59(1):133–138.
30. Otsuka M, Yamane I, Matsuda Y. Effects of lubricant mixing on compression properties of various kinds of direct compression excipients and physical properties of the tablets. Adv Powder Technol 2004; 15(4):477–493.
31. Karehill PG, Glazer M, Nystrom C. Studies on direct compression of tablets XXIII: the Importance of Surface-Roughness for the Compactibility of Some Directly compressible materials with different bonding and volume reduction properties. Int J Pharm 1990; 64(1):35–43.
32. Sinka IC, Cocks ACF. Modelling die compaction in the pharmaceutical industry. In: Brewin PR, Coube O, Doremus P, et al. eds. Modelling of Powder Die Compaction. Springer-Verlag, London, 2008:235–241.

33. Ebey GC. Bilayer tablet weight control theory. Pharm Technol 1996; Yearbook:54–57.
34. Schneider LCR, Sinka IC, Cocks ACF. Characterisation of the flow behaviour of pharmaceutical powders using a model die-shoe filling system. Powder Technol 2007; 173(1):59–71.
35. Ohmori S, Makino T. Sustained-release phenylpropanolamine hydrochloride bilayer caplets containing the hydroxypropylmethylcellulose 2208 matrix. II. Effects of filling order in bilayer compression and manufacturing method of the prolonged-release layer on compactibility of bilayer caplets. Chem Pharm Bull 2000; 48(5):678–682.
36. Vogeleer J, De Smet P. Bi-layer tablets—why special technology is required. Pharm Process 2002; 30–36.
37. Bose S, Bogner RH. Solventless pharmaceutical coating processes: a review. Pharm Dev Technol 2007; 12(2):115–131.
38. Ozeki Y, Watanabe Y, Inoue S, et al. Comparison of the compression characteristics between new one-step dry-coated tablets (OSDRC) and dry-coated tablets (DC). Int J Pharm 2003; 259(1–2):69–77.
39. Shimizu K, Ando M, Nakayama Y. Characterization of tableting using the OSDRC system. Pharm Res 2007; 24(10):1902–1909.
40. Ozeki Y, Watanabe Y, Okamoto H, et al. Development of dividable one-step dry-coated tablets (Dividable-OSDRC) and their evaluation as a new platform for controlled drug release. Pharm Res 2004; 21(7):1177–1183.
41. Ozeki Y, Ando M, Watanabe Y, et al. Evaluation of novel one-step dry-coated tablets as a platform for delayed-release tablets. J Control Release 2004; 95(1):51–60.
42. Ozeki Y, Watanabe Y, Inoue S, et al. Evaluation of the compression characteristics and physical properties of the newly invented one-step dry-coated tablets. Int J Pharm 2003; 267(1–2):69–78.
43. Hiestand EN. Mechanical properties of compacts and particles that control tableting success. J Pharm Sci 1997; 86(9):985–990.
44. Duncanhewitt WC. Uniaxial compaction modeled using the properties of single-crystals. Drug Dev Ind Pharm 1993; 19(17–18):2197–2240.
45. Wu CY, Best SM, Bentham AC, et al. Predicting the tensile strength of compacted multi-component mixtures of pharmaceutical powders. Pharm Res 2006; 23(8):1898–1905.
46. Kuentz M, Leuenberger H. A new theoretical approach to tablet strength of a binary mixture consisting of a well and a poorly compactable substance. Eur J Pharm Biopharm 2000; 49(2):151–159.
47. Wu CY, Best SM, Bentham AC, et al. A simple predictive model for the tensile strength of binary tablets. Eur J Pharm Sci 2005; 25(2–3):331–336.
48. Thomson SA, Tuleu C, Wong ICK, et al. Minitablets: new modality to deliver medicines to preschool-aged children. Pediatrics 2009; 123(2):E235–E238.
49. Ishida M, Abe K, Hashizume M, et al. A novel approach to sustained pseudoephedrine release: differentially coated mini-tablets in HPMC capsules. Int J Pharm 2008; 359(1–2):46–52.
50. Lopes CM, Lobo JMS, Pinto JF, et al. Compressed mini-tablets as a biphasic delivery system. Int J Pharm 2006; 323(1–2):93–100.
51. Vuong H, Levina M, Rajabi-Siahboomi AR, eds. Evaluation of the Enteric Performance of Lansoprazole Mini-tabs Coated in a Perforated Pan. American Association Of Pharmaceutical Scientists (AAPS) Annual Meeting, 2008.
52. Weyenberg W, Vermeire A, Remon JP, et al. Characterization and in vivo evaluation of ocular bioadhesive minitablets compressed at different forces. J Control Release 2003; 89(2):329–340.
53. Lennartz P, Mielck JB. Minitabletting: improving the compactability of paracetamol powder mixtures. Int J Pharm 1998; 173(1–2):75–85.
54. Flemming J, Mielck JB. Requirements for the production of microtablets—suitability of direct-compression excipients estimated from powder characteristics and flow-rates. Drug Dev Ind Pharm 1995; 21(19):2239–2251.
55. Pich CH, Moest T. BASF Aktiengesellschaft, assignee. Cylindrical Microtablets. Patent 4797287, Jan 10, 1989.
56. Kennerley J, Newton J, Stanley P. The influence of dimensions on the strength and strength variability of tablets. J Pharm Pharmacol 1977; 29:39.
57. Stanley P, Newton JM. Variability in the strength of powder compacts. J Powder Bulk Solids Technol 1977; 1:13–19.
58. Flemming J. Experimentelle Mikrotablettierung: Fließeigenschaften von Hilfsstoffen und Tablettierverhalten von Cellactose® bei Herstellung von Tabletten mit einem Durchmesser von 5 bis 1.5 mm. University of Hamburg, 1998.

12 Quality by design and compression

Chris Balducci, Rosario LoBrutto, Alan Royce, James Cheney, Dong Xiang, Olivia Darmuzey, and Hedinn Valthorsson

OVERVIEW: QUALITY BY DESIGN AND PRODUCT DEVELOPMENT

Quality by design (QbD) is not a new term. In 1985, R. Kacker stated that "consideration of quality in manufacturing should begin before manufacturing starts, even before any capital commitments are made and this is quality by design" (1). The experimentation at the manufacturing stage is orders of magnitude more costly than experimentation at the R&D stage (2), and it is the designs of both the product and the manufacturing process that play crucial roles in determining the degree of performance variation and the overall manufacturing cost (1). Then in 1986, M. Walton (3), said that the manufacturing group (the consumer) should receive from R&D (the producer) a process that has inherent good quality characteristics, and in particular, R&D should develop—in collaboration with manufacturing—a process that is rugged with respect to anticipated manufacturing variables.

However in the current state, QbD is a new pharmaceutical development philosophy being defined and proposed by several regulatory authorities in conjunction with the pharmaceutical industry. The regulatory agencies involved include the Food and Drug Administration of the United States (FDA), European Medicines Agency of the European Union (EMEA), and the Pharmaceuticals and Medical Devices Agency of Japan (PMDA). The goal of the QbD development process is to define a drug product manufacturing process that is flexible enough to allow mitigation of inherent variability in the process that may arise from raw materials or equipment fluctuations. Additionally, the process is open to continual improvement.

Traditional manufacturing processes, in contrast, are considered too rigid to account for inherent variability of the process and provide no opportunity for continual improvements. In traditional development and validation processes, it was desired to obtain a robust process that could absorb the inherent variability of the process without impacting the product quality. It is the authors' experience that the new knowledge obtained from development according to the QbD process allows one to better define a robust process, and batch-to-batch adjustments in the design space would not typically be required.

This comparison of the two development processes is shown in Figure 1.

In summary, QbD is a more systematic, science-based approach to development. Prior knowledge, proactive team-based risk assessments, multivariate experiments, use of process analytical technologies (PATs), and continual improvements highlight the QbD techniques that are used to develop product understanding to a level necessary to establish a design space for a drug product (Fig. 2). The design space reflects the product understanding as a multivariate area where the impact on product quality is understood and controlled based on variability of processing parameters and/or material attributes within a known range. In essence, the multivariate design space replaces the proven acceptable ranges (PARs) already specified in the past for processing parameters during prevalidation and validation batches (e.g., mixing times, air volume, etc.). However, PARs lack the capability to provide understanding of potential interactions among other parameters throughout the drug product manufacturing process. These interactions (either synergistic or antagonistic) can potentially affect a process and may not necessarily be determined if studied one at a time in a univariate manner. Essentially, the manufacturing process runs in a multivariate environment, therefore the parameters of that

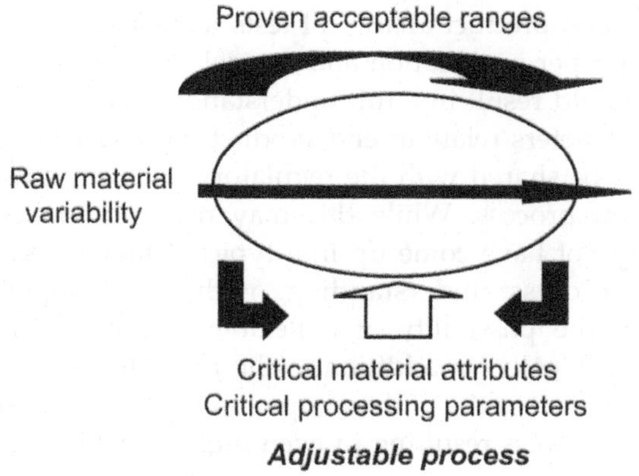

Figure 1 Proven acceptable ranges versus a QbD process.

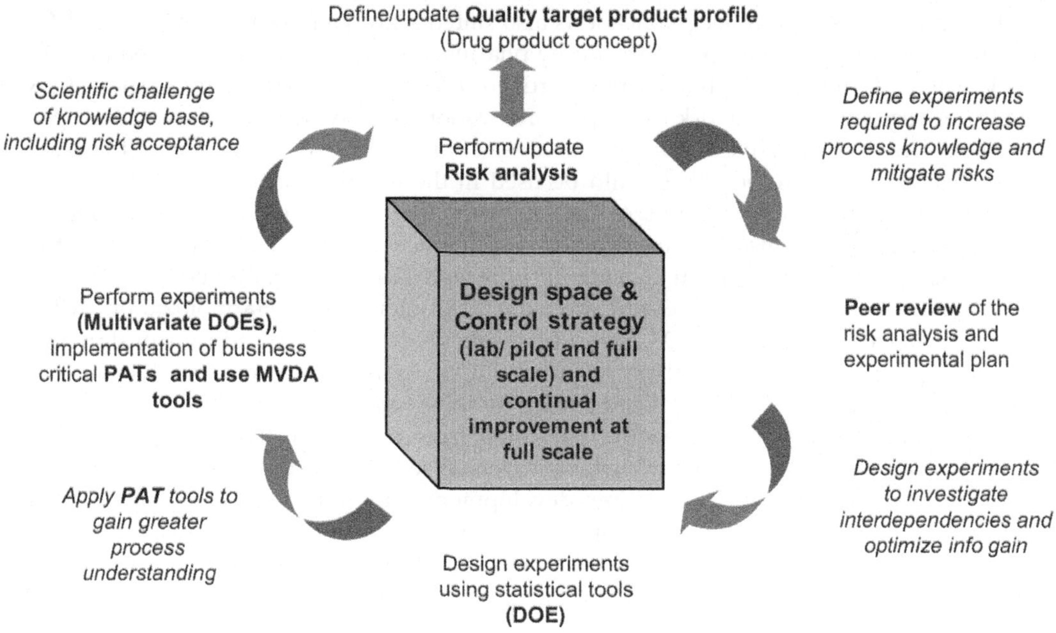

Figure 2 Quality-by-design life cycle.

process should be studied accordingly in order to fully understand their affects (main effects of each process parameter and their interactions with other process parameters).

The design space is held within specification via a control strategy. QbD requires the establishment of a control strategy to ensure compliance of the drug product processing parameters and material attributes remain within the design space. Typically, the control strategy includes items such as additional in-process testing, use of PAT tools/probes, IT/ software programs (i.e., feedback loops), well defined batch records, and specifications. Schematically, this process is depicted in Figure 2.

QbD development leads to submissions where flexibility in critical processing parameters is allowed within the approved design space(s), and real-time release may replace many of the drug specific, end-product analytical tests (e.g., assay, content uniformity (CU), dissolution, etc.) with the proper justification and control strategies in place.

Utilization of QbD should result in a full understanding of how material attributes and manufacturing process parameters relate to end product performance. QbD information and conclusions are expected to be shared with the regulatory authorities to a greater extent than the traditional development process. While this may raise additional questions from the dossier reviewers that may not have come up in a typical/standard submission, it provides the advantage of better process understanding, with more opportunities for control. Furthermore, it allows for the possibility of a flexible manufacturing process that could reduce costly batch failures. Finally, in addition to the regulatory benefits of change control flexibility and real-time release, it is the authors' opinion that QbD development could be the most efficient development process resulting in even higher quality product.

QUALITY RISK MANAGEMENT AND ICH GUIDELINES

In August 2002, the FDA outlined an initiative on Pharmaceutical cGMPs for the 21st century (4), and then in 2003 further defined the "desired state" for pharmaceutical manufacturing in the 21st century (5). Several other publications also stated a growing awareness of a need for improvement in manufacturing processes in the pharmaceutical industry (6–9). Then in 2003, the FDA issued additional guidelines with regards to risk-based inspections, PAT, and 21CFR Part 11 (5). A regulatory framework to support innovation and PAT guidance was developed by the FDA. The PAT initiative is part of a broader effort of the cGMPs for the 21st century efforts, which are essential tools that could be used in the manufacturing process to further build quality into the product. The FDA has sought international scientific collaboration on PAT from its inception and has been engaged in very interactive discussions with the industry, academia, and health agencies from Canada, Europe, and Japan. Several other key guidance documents have been prepared through International Conference on Harmonization (ICH): Q8 (Pharmaceutical Development), Q8 Annex (focusing on design space and control strategy), Q9 (Quality Risk Management), and proposed Q10 (Quality Management) (10). Elements of Q8, Q8 Annex, Q9, and Q10 are discussed further in this chapter.

The advent of QbD has shifted the paradigm from a retrospective to a prospective, systematic, and risk-based approach to developing enhanced process understanding (11). The introduction of QbD toward formulation development and optimization of process for particular unit operations, such as compression allows for the movement away from the existing check-the-box paradigm to one of knowledge sharing and continual improvement. QbD utilizes risk-based assessments to guide process development activities with an aim at enhancing process robustness.

Therefore, there is an inherent need to understand the main effects that critical process parameters (CPPs), material attributes, and their mutual interactions have on the critical quality attributes (CQAs) defined for a drug product. Moreover, since there are many variables a scientist must consider when developing or validating a robust process, standardized risk assessments coupled with statistical experimental design and analysis allow for an efficient and effective means for execution. This will lay the basis for defining the design space boundaries and ultimately a control strategy to ensure quality is built into the process and product. The designation of critical parameters and the critical material attributes allows for a systematic approach on demonstration of quality that is commensurate with reducing the impact to safety and efficacy of the product for the patient.

Moreover, risk management is key to identifying the criticality of process parameters and material attributes within a process. The integrity of the risk assessment is predicated on a robust quality system that is in alignment with the principles in ICHQ9 and Q10 (10).

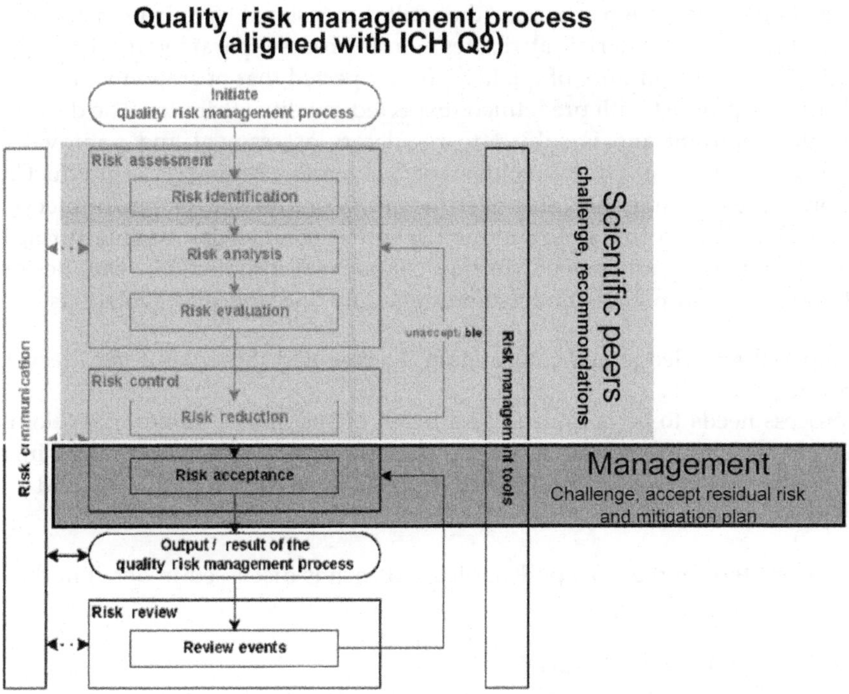

Figure 3 Quality risk management process.

The elements of ICHQ9, which include risk identification, risk analysis, risk evaluation, and risk control, are integral to the risk management process and are depicted in Figure 3. Examples of risk identification, risk analysis, risk evaluation and risk control include risk matrices/fishbones, failure mode and effects analysis (FMEA), design of experiments (DoEs), and control strategy/multivariate statistical process control (MSPC), respectively.

Scientific peer reviews ensure input from senior scientific experts and customers on key technical development activities. Internal experts and customer inputs strengthens the science and hence the quality of the development work. In scope of these reviews are all technical activities such as drug substance synthesis, drug substance finishing steps (crystallization, filtration, milling, drying), drug product manufacture, analytical method development/validation, and specification setting. Associated with all of these technical activities are risks. The initial risk analysis and the FMEA are performed by the project subteam and cross-functional experts during the risk identification and risk analysis phase. Proposals for how to mitigate the risks, including a DoE plan, are also discussed. The summary of this risk assessment phase are then presented at a scientific peer review before performing the experimental work. After experimental work is conducted, generally using a DoE approach and the CPPs and material attributes are identified, a control strategy would be proposed and further discussed at a subsequent scientific peer review meeting.

Quality peer reviews provide decision-making management the opportunity to review and acknowledge the risks associated with development projects at key project milestones. This aspect of management oversight (risk acceptance) is consistent with ICHQ9 guideline, during the risk control phase as depicted in Figure 3.

As previously discussed, the design space for a manufacturing process provides the mechanism at which defined CPPs and material attributes deliver a drug product's defined

CQAs. ICHQ8 defines the design space as "the multidimensional combination and interaction of input variables (e.g., material attributes) and process parameters that have been demonstrated to provide assurance of quality." It is expected that operation within the design space will result in a product with predefined (expected) quality attributes. The design space is proposed by the applicant and is subject to regulatory assessment and approval (ICHQ8, November 2005) (10). Design space development is dynamic, beginning at product development inception and continuing to evolve over the entire product development lifecycle (from small scale development to full scale production at the commercial manufacturing facility) allowing for continual improvement during commercial production. The design space represents the best overall process understanding at any one time and evolves as

- additional knowledge and information is generated throughout the lifecycle of the product,
- a process needs to be optimized as a result of successive, iterative developments of the physical properties of the drug substance/active pharmaceutical ingredients (APIs), and as new stability data for drug product become available, and
- upon scale-up of process.

ICH defines seven steps in the QbD philosophy and each is discussed in detail in this chapter (Fig. 4).

DEFINE TARGET PRODUCT QUALITY

A key aspect of QbD is the forward looking focus of the drug product development. In essence, a quality target product profile (qTPP) needs to be predetermined, establishing the critical attributes of the drug product that impact the product's quality as realized by the customer and/or patient. These attributes are referred to as the CQAs, and the development process is designed to understand and provide control over these CQAs to ensure product quality. Hence, the qTPP is the collection of the product specific CQAs. As an example, the qTPP for a tablet would include the dosage strength and dosing regimen that achieves a predefined

Figure 4 Seven steps of QbD.

clinical result, including established efficacy and safety profiles. The CQAs that define the qTPP could include the drug product assay, dissolution specifications (surrogate for in-vivo performance; i.e., biorelevant), stability, impurity profile, microbial burden, physical appearance, etc. Presently, CQAs are similar to the drug product release, in-process controls (IPCs), and stability specifications.

As a forward looking process, QbD can be applied very early in development, including the drug discovery process where new molecular entities are selected for further clinical development. Indeed, there are many advantages for using a QbD approach early in development, the primary reason being the increased ability to affect the quality of the final drug product (building quality into the product). Early in development where QbD process could be used, the main focus is on the API characteristics that could impact the biopharmaceutical properties of the drug product. Additionally, the CQAs are influenced by multivariate parameters, which may interact with other parameters throughout the development process. Hence, all of the parameters identified need to be carried forward to the final drug product as received by the patient. Therefore, as the development process unfolds, new knowledge may be uncovered that challenges some of the conclusions drawn earlier in development.

Additionally, it has to be expected that the qTPP will evolve during the development process. As an example, typically the target dose will change as the clinical program advances and therefore the CQAs will also need to adapt to these changes. To balance these inherent changes, the QbD process may be defined in stages. As an example, the attributes of the drug substance such as the salt counter ion, may be defined before the formulation of the drug product, and the formulation may be defined before the final drug product process. In each case, the drug substance form, formulation and final drug product manufacturing process that are investigated thoroughly using QbD techniques, are those that are represented within the product dispensed to the patient. Similar to the current development process, QbD development can be aligned with the attrition curve of new products. In essence, traditionally very little processing optimization is done for a phase I compound, and for QbD the focus would be on the drug substance form and required biopharmaceutical characteristics (e.g., absorption, distribution, metabolism, etc.).

Theoretical identification of compression CQAs (i.e., those CQAs established during traditional validation) are physical, chemical, biological, or microbiological properties or characteristics that should be within an appropriate limit, range, or distribution and must be controlled directly or indirectly to ensure the desired product quality (10). CQAs of tablets are characteristics affecting product purity, dosage strength, drug release, and stability for which a target profile has been established as knowledge was gained during early phase development.

Typical CQAs of tablets include hardness, tensile strength, friability, disintegration time (DT), dissolution, dissolution profile, thickness, density, porosity, moisture content, visual attributes, CU, assay, weight, identity, purity, and stability.

This list is not exhaustive. CQAs are identified as the ones relevant to the qTPP and from prior knowledge. They can change as more process understanding is gained during the development phases. Risk management tools can be used to prioritize relevant CQAs, as described in ICHQ9.

A specific, actual example of applying QbD principles will be discussed throughout this chapter. The first step the Novartis team initiated in developing this product using the QbD philosophy was defining the target product quality. This step focused on identification of those attributes that were critical to ensure quality of a product. The CQAs for this example were determined based on traditional development conclusions. Because the candidate selected for these QbD activities was a legacy product (i.e., a commercial product released to market), the team decided to keep the same CQAs as originally developed, where the target CQAs are adequate and did not need to be altered. These CQAs include routine traditional end product quality attributes.

- Dissolution
- Assay
- CU
- Weight
- Hardness
- Friability
- Disintegration time
- Appearance
- Purity

These CQAs were used as response variables when conducting DoE trials at lab and full scale. Impact to any of these attributes rendered a process parameter as critical.

IDENTIFY KNOWLEDGE BASELINE

Before a robust process can be developed and then further optimized, one must first understand the state of the current process. That is to say, all current and previous knowledge of a process needs to be identified and collected in order to establish a baseline process assessment. To identify the current knowledge baseline, a cross-functional team of experts from manufacturing, quality assurance, and development is needed. This results in a cross-functional group that should be able to provide the team a broad level of experience for the given process. When starting the identification of the knowledge baseline, it is imperative to have as much relevant expertise on a team as possible. These experts should then provide all available documented experience and knowledge that exists for the process in question, in this case, the compression unit operation. Based on this data, the group then needs to focus on each aspect of the compression process and the potential impact on product CQAs.

A fishbone diagram (Fig. 5) was the initial step in identifying possible product/process risk for the Novartis QbD example, helping the team to discuss the different compression

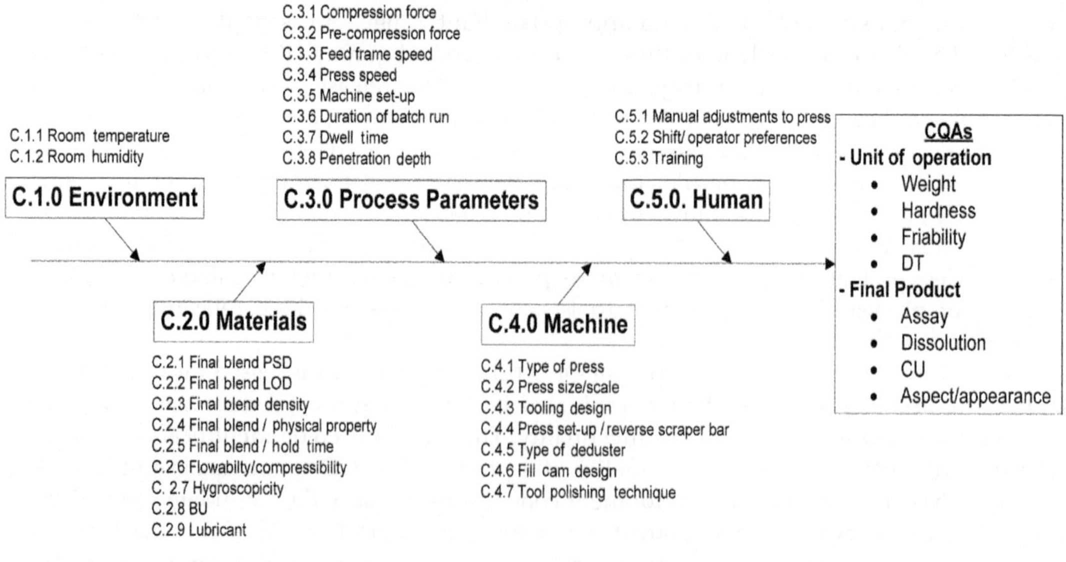

Figure 5 Fishbone diagram (compression).

variables in an organized manner. In this example, the compression process was mapped out in a way so that all incoming variables, process parameters, and CQAs were identified.

An Ishikawa, or fishbone diagram consists of all the components encompassing a particular process. Typically, this includes the five "M's" *M*an (human operators/personnel), *M*ethod (process parameters/controls), *M*achine (equipment), *M*aterials (incoming and raw materials), and *M*other nature (environment), but can be customized to suit a specific exercise. These segments (identified as C.1.0–C.5.0) and subsegments (C.1.x–C.5.x) are numbered to allow for traceability throughout the assessment process. The example above is not exhaustive, and it is encouraged that a fishbone be tailored to each individual/specific process. While there may be some "lessons learned" that can be applied to all future products using the same process, the goal is identifying all aspects that can impact an individual process/product. For example, if there is a known CQA deficiency (i.e., if picking/sticking or dissolution is identified as a specific risk or actual process shortcoming at baseline process assessment, the fishbone can be tailored to focus on parameters affecting these quality attributes).

Once the fishbone diagram for compression was completed, the resultant list of possible influential parameters was assessed for potential/actual risk. This was accomplished using an FMEA. Each parameter was evaluated in a similar manner as outlined in Figure 6. This FMEA is where all the baseline knowledge is documented, and when this is performed the items with high risk priority numbers (RPNs) are listed as the high risk/knowledge gaps.

When assessing the current knowledge baseline of a process variable, it should be referenced from where this knowledge/experience came from or where it is documented. The assessment of the risk or criticality of each variable and process parameter based on the current knowledge should again be documented and linked to the previous process map (e.g., fishbone). This way the current knowledge of the process (knowledge baseline) is laid out and most importantly, it highlights the knowledge gaps.

CQAs and the possible CPPs that were studied in the DoEs were documented through an iterative process of risk assessment reports.

The fishbone/FMEA process used as suggested is a proactive and very systematic method for documenting and communicating process knowledge. This is a method that is accepted by both the U.S. FDA and European EMEA. The fishbone and FMEA process has been used in many industries over the years, but mainly in retrospect for trouble shooting in the pharmaceutical industry.

The risk associated with each process parameter was determined based on three criteria: *probability* (*P*), *impact* (*I*), and *detectability* (*D*). A resultant RPN ($P \times I \times D$ = RPN) is calculated from evaluation of a parameter's likelihood of deviating from the current process, and the impact on final product attributes this deviation would cause and what measures are embedded in the process to detect and correct this deviation.

Fishbone diagrams and consequent FMEAs were performed by a cross-disciplined team of scientists (comprising statisticians, formulators, chemists, project managers, etc.) resulting in the identification of the following parameters as high risk because of their associated calculated RPNs above the acceptable threshold limit and/or high theoretical impact where risk mitigation would be warranted. Even if an RPN is not above the critical threshold limit, but the impact is high or the impact not known, the parameter is deemed critical and would warrant further investigation.

The identified potentially critical parameters were

- compression force,
- compression speed,
- feeder speed, and
- press type/scale-up.

No	Variable Name/ Input	Current Setting	Ref No.	Potential Setting Variability	P	(Potential) Impact on Output Quality	I	Detection in Current Process	D	Comment	RPN	DoE	Follow up, Control/ Mitigation Strategy
C 3.1	Compression Force	1–5 kN	1	>5 kN Machine mal-function In-consistent granulation flowability.	2	Weight Adjustments to main compression force via fill depth parameter could affect tablet weight.	4	MCF is set at beginning of compression process based on batch record instruction. Adjustments throughout run are driven by IPC results (weight, thickness, hardness, DT, friability)	2	Flowability of blend is considered to be good. DT for one strength high at upper end of hardness range.	16	Y	Relation of tablet hardness, friability DT and dissolution will be evaluated in a DOE.
					6	Hardness High MCF lead to harder tablet production.	6				72		
					4	Dissolution Based on relationship of tablet hardness to dissolution, harder tablets can retard tablet dissolution.	6				48		
					4	Friability Tablets produced at low MCF produce softer tablets and can increase tablet friability.	4				32		
					6	DT Evidence of increased tablet DT at upper hardness range could be attributed to high MCF.	6				72		

Figure 6 Failure mode and effects analysis (FMEA).

DEVELOP PRODUCT UNDERSTANDING

After the fishbone and FMEA processes are completed and the knowledge gaps identified, the next step is to develop a plan for acquiring data/information for the purpose of bridging these gaps. In an effort to establish in-depth product knowledge, a series of experiments can be planned in order to develop product understanding. These experiments should be set up according to a DoEs in

Table 1 Plackett Burman Design

Experiment	X1	X2	X3	X4	X5	X6	X7	X8
1	+1	+1	+1	+1	+1	+1	+1	+1
2	−1	+1	−1	+1	+1	+1	−1	−1
3	−1	−1	+1	−1	+1	+1	+1	−1
4	+1	−1	−1	+1	−1	+1	+1	+1
5	−1	+1	−1	−1	+1	−1	+1	+1
6	−1	−1	+1	−1	−1	+1	−1	+1
7	−1	−1	−1	+1	−1	−1	+1	−1
8	+1	−1	−1	−1	+1	−1	−1	+1
9	+1	+1	−1	−1	−1	+1	−1	−1
10	+1	+1	+1	−1	−1	−1	+1	−1
11	−1	+1	+1	+1	−1	−1	−1	+1
12	+1	−1	+1	+1	+1	−1	−1	−1

order to get the most information possible. Incorporation of DoEs into the development strategy, leads to increased knowledge/process understanding with minimal amount of experiments compared to a one variable at a time approach where less information is obtained per same unit number of experiments. DoEs are statistically based experiments aimed at establishing correlations/relationships between process parameters and outcomes or quality attributes.

The first round of DoEs focus on the process variables that had an RPN score greater than the critical value identified in the FMEA process. This type of DoE is called a "main effect screening" DoE. With this type of DoE, any process variable that has an impact on the defined quality attributes experimentally determined will be identified.

An example of main effect screening DoEs is Plackett Burman designs. These experiments can estimate main effects between two level factorials, but cannot estimate two-way interactions (two way interactions are moderately correlated with main effects). A popular DoE is a 12-run design that can estimate a maximum of eight factors (parameters) each at two levels (e.g., high and low) and leave three degrees of freedom for error. Table 1 illustrates an example of a Plackett Burman design for eight factors in 12 runs for factors X1 to X8 all at two levels (−1 and 1). For example, X1 can be compression force, with the +1 designated as high force and the −1 designated as low force. X2 can be press speed, with the +1 designated as high speed and the −1 designated as low speed, and so on.

The next stage of DoEs is designed to determine the interaction and the magnitude of the interaction between the different CPPs. An example is a "Full Factorial" DoE. Full factorial designs will also have each variable run at two or more levels. The number of levels is denoted by "X" and the number of variables is denoted by "y," X^y. For example, three CPPs each set at two levels (low and high) would result in a 2^3 full factorial, which would require eight experiments, and if there were four CPPs this would result in a 2^4 full factorial, which would require 16 runs. A complete full factorial will allow one to independently estimate all main effects of each variable, and all possible interactions between all of the variables, which include all two-way interactions, three-way, up to the k-way interaction for k variables in the design. One can use fractional factorials for designs (for two level factorials, they are often used for five or more variables) to reduce the total number of experiments and still be able to estimate all of the main effects and some or all of the two-way and higher interactions.

There is no preset number or predetermined rule as to the number/amount of DoEs or number of batches that comprise a design. The goal is that at the end of the statistical evaluation all interactions and effects should be known and documented. At the end of this process, a "design space" encompassing the impact that the CPPs have on the predetermined

CQAs should be developed for lab or pilot scale. This proposed "design space" will be verified at full scale through the execution of a confirmation DoE.

The Novartis example was conducted using a slightly different DoE strategy than the one previously described. In the following example, the compression study focused on assessing the impact of compression conditions and granulation LOD on the manufacturing and final tablet properties of a drug product.

The DoE study was composed of one granulation batch manufactured at target moisture (target LOD), one granulation batch manufactured at "high" moisture (above target LOD) and one granulation batch manufactured at "low" moisture (below target LOD), see Figure 7. These batches were then compressed varying compression parameters detailed by the following statistical study:

- Compression speed
- Tablet hardness
- Tablet press type/size
- Feeder speed

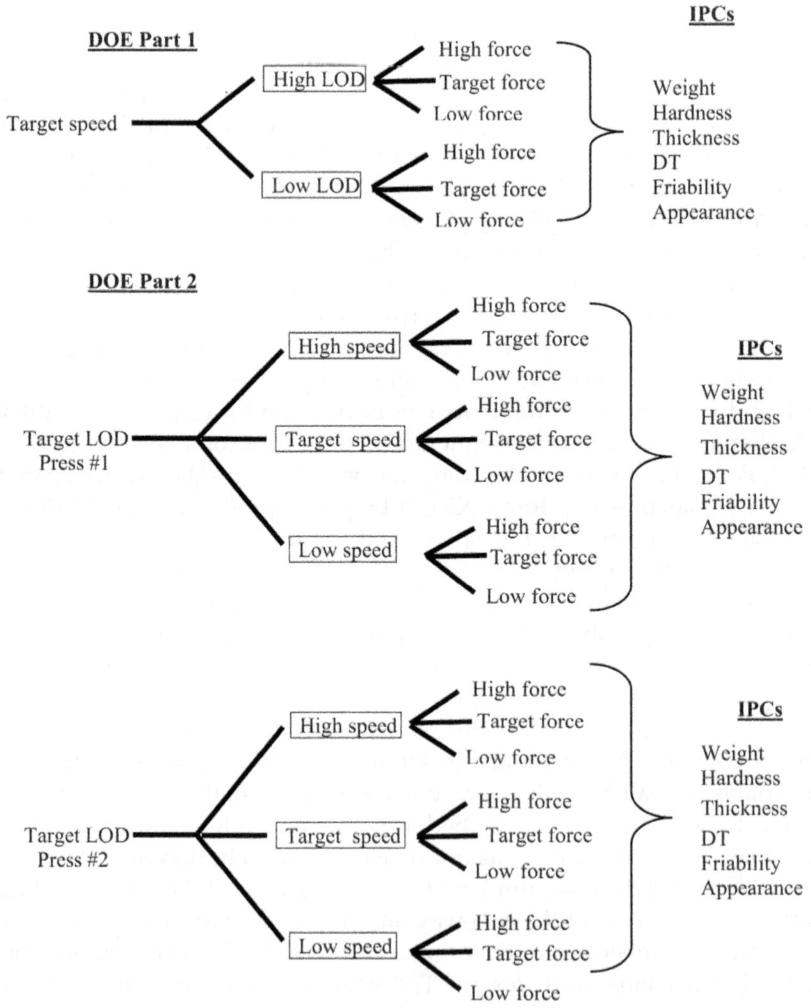

Figure 7 Experimental design.

Compression speed and tablet hardness were evaluated simultaneously on different tablet press scales. Feeder speed was only evaluated on one press type, as the feeders were identical and deemed scale independent.

The aim of the study was to evaluate the effect of LOD on IPCs, such as tablet weight, thickness, hardness, disintegration, friability and appearance, and compression parameters. For each batch, samples were taken and tested after drying, sizing, final blending, and compression to allow for a comprehensive mapping of the process. The testing included IPC, blend uniformity (BU) by NIR, CU by NIR, assay, and dissolution on selected tablets.

DESIGN SPACE

The design space defines the multivariate range of the critical processing parameters and material attributes that are known to comply with the product specific CQAs and hence qTPP. Implicit in the design space is the criticality of the identified parameters and material attributes, since any parameter can become critical if it is allowed to fluctuate widely. Criticality can be defined by comparing the magnitude of the impact on the CQAs versus the required change in the range of the parameter needed to create this impact, and often is also compared to the inherent variability of the parameter based on historical data (if available) or machine operating tolerances. An approach adopted by the authors for process criticality is where criticality is correlated to the risk prioritization number assigned through the quantitative risk assessment (i.e., FMEA) conducted for each unit operation or processing step. Hence, all parameters with a calculated RPN above an established threshold value after multiplying the probability, impact and detectability scores, are considered critical. Additionally, those parameters that are assessed high solely regarding product *impact* alone are also considered critical.

The parameters that are demonstrated as not critical via the risk assessment mechanism are nevertheless identified and documented in the risk assessment in the same manner as the critical parameters. Their ranges, although demonstrated as noncritical, are controlled within these tested ranges during drug product manufacture. Since these parameters are not critical to CQAs, they comprise the knowledge space of a process. That is to say that scientific understanding/knowledge has been attained through risk assessments, DoEs, etc., and are understood within this defined range as noncritical.

Typical compression process parameters that potentially could comprise a design space include compression mode (constant force or thickness), speed, dwell time, main compression force, precompression force, feed frame type and speed, hopper design, height and vibration, fill depth, punch penetration depth, ejection force, and lubrication type (internal or external) (12,13).

Attributes of incoming raw materials can have a major impact on the tablet CQAs and should be included in the risk assessment and potentially be investigated in DoE studies. Although it might be difficult to acquire raw materials that exhibit the range of properties of interest, it is possible to manipulate material to achieve desired characteristics (i.e., humidify to increase LOD or take sieve cuts for particle size effects). Although raw material attributes are not considered "parameters," their attributes are considered for potential risk. Examples of raw material attributes that can influence the compression process and subsequently final product's CQAs are excipients and API particle size distribution, water content (LOD), true and bulk densities, polymorphs and amorphous content, purity, API mean yield pressure (viscoelastic properties of a material), and strain rate sensitivity (material classification based on how tablet strength is affected by punch speed). Attributes of the materials generated from previous process steps should also be considered for their potential impact on subsequent process steps as well as final tablet CQAs (i.e., incoming granulation impact on blending and compression processes).

Parameters comprising the design space should be identified as relevant according to baseline knowledge, developed product understanding (i.e., DoE results), expertise from development scientists and prioritized using risk assessment tools. The effect of these parameters on the tablets CQAs is studied over predefined ranges together with the interaction effects of these parameters and a design space is determined for each investigated parameter ensuring the production of tablets will yield the predetermined quality. If failure is reached with the studied parameter ranges for the CQAs, it might be useful or needed to further study those parameters in subsequent DoE(s) to determine the edge of failure of a design space (10).

In this example, two responses were demonstrated as a result of the compression study.

- The granules dried to the low LOD endpoint required approximately 5-kN additional compression force when compared to the high LOD endpoint to achieve equivalent tablet hardness.
- The granules dried to the low LOD endpoint resulted in significant higher (approximately 30–40%) tablet DTs when compressed at the mid to high end of the allowable hardness range.

The results of the DoE were not surprising; batches produced with higher LOD (moisture) required less compression force than batches with lower LOD. However, an interesting discovery was made with regard to one aspect of the QbD product. Through the DoE trials, it was learned that the tablet compression-hardness–DT relationship was a function of force per unit area of the compression, and hence the main compression force and hardness ranges used in production were adjusted accordingly within the given design space.

As a result of this finding, a suitable "compromise" between tablet hardness, DT and friability was established, using main compression force as the CPP for determining all three quality attributes.

CONTROL STRATEGY

ICHQ10 defines a control strategy as "a planned set of controls derived from current product and process understanding that assures process performance and product quality" (10). It is the mechanism(s) put in place to ensure that the critical processing parameters and material attributes comply with the design space during drug product manufacture. Essentially, it is any measure implemented to keep a process operating within its predetermined settings; its design space. The control strategy may include familiar items such as incoming material specifications/testing, in-process testing, statistical process control charts, batch records, sampling protocols, or more sophisticated measures such as PAT probes, and recipe driven processes. Essentially, the entire quality system of the manufacturer contributes to the product control strategy.

Traditional approaches to control strategy mainly include the testing of intermediate and end product, an enhanced risk-based QbD approach will provide controls that are shifted upstream and may allow real-time release or reduced end product testing (14).

Typical controls for a compression step include the following:

- Raw material specifications such as for particle size distribution, water content, polymorph and amorphous content, bulk and tap density, and LOD
- Intermediate material specifications such as for granule particle size distribution, granule bulk and tap density, flow, BU, and granule or blend LOD
- IPCs such as tablet mean weight, tablet hardness or tensile strength, tablet DT, tablet thickness, friability, and appearance
- Online and at-line testing such as for CU or assay of tablets by NIR

- In process measurements such as main and precompression forces or punch displacement to monitor tablet weight and porosity with a feedback control system (15)

In a QbD approach, the control strategy is closely linked to criticality and design space: material and process sources of variability have been identified through risk assessment and must be included in the control strategy (16). The ideal scenario allows for changes to be made to the critical processing parameters and/or material attributes within the design space to account for inherent variability in the process. These adjustments are part of the filed operating parameters, and do not require any submission of a variance, however internal change control processes would be applied. Typically, such variability can occur because of fluctuations in excipient quality, supplier and/or other uncontrolled parameters. In its most valuable state, a control strategy built around prior knowledge of inherent process variability can be used to adjust process parameters before it is used (feedforward mechanism). In-process data can be used to adjust an ongoing process in real time (feedback). PATs are useful to provide real-time adjustments to an ongoing process via in-line, at-line, or online testing. However, it is the authors' experience that PATs are best suited for those unit operations where the process is controlled via the online technology, such as blending where the online BU data is used to stop the process when uniformity is achieved.

For all of the approaches to define a control strategy, a company's existing data collection system or IT infrastructure must be sufficient to collect, store, retrieve and process all relevant data. Deficiencies in any of these areas of the IT infrastructure will negate the ability to determine if the control strategy was successful in maintaining the process within the defined design space for all of the parameters and attributes defined. In the authors' experience, systems capable of multivariate data analysis (MVDA) may prove useful for monitoring all of the critical parameters defined in the design space through use of MSPC models, and these models may also provide real-time information to aid batch control in a manner similar to online PATs.

In our example, DT analysis at lab and full scale was found to be related to tablet hardness showing that compression performance is consistent with the theoretical relationship expected. However, the window for suitable compression force to ensure that hardness values results in DTs within specification for one strength of the product is significantly smaller than for the other two product strengths. The impact of compression force on tablet DT (DT) as previously discussed was further evaluated at full scale. Tablets produced during full scale DoE at the low end of the hardness/disintegration profile resulted in acceptable DT time and demonstrated no friability issues. On the basis of hardness profiles, resulting DTs and friability data for these DoE batches, the hardness specification for this strength product was changed.

QUALITY MONITORING

Quality monitoring involves strategically placed sensors/probes at critical location points of a process. The objective is to monitor the status of identified critical parameters for the purpose of early fault detection. Since these parameters have been identified as critical, their performance during a production run has direct impact on the finished products CQAs. One of the methods used for monitoring the compression process is MVDA. This is a very powerful tool for overall process monitoring and has been used in other industries (chemical, bulk paper and computer chip manufacturing) for many years. The pharmaceutical industry has started adapting these tools for QbD. MVDA is considered a "soft" PAT tool.

Traditionally, a tabletting process can be monitored in two ways.

- Compression force control systems allow the monitoring of the tablet main compression force and adjust the fill depth to maintain a constant force (e.g., Fette). However, some companies/operators still perform "retroactive" adjustments to compression outputs (i.e., tablet weight, thickness, and hardness).
- Punch displacement control system involves the compression at a variable force and a constant precompression height (e.g., Courtoy).

For both of the above systems a relationship between the monitored parameter and the tablet weight is established.

MVDA software allows one to establish MSPC plots. These plots are generated from normalized process parameter data converting each unit of measure into one trajectory graph. This graph then represents how a process is proceeding in real time compared to a "normal or golden" batch.

CONTINUAL IMPROVEMENT

Continual improvement of a manufacturing process is not unique to QbD. Health Authority regulations mandate that processes be updated through revalidation when it becomes necessary (17). The QbD approach views every batch as a validation batch, thereby continually focusing on process/product improvement. Based on the process understanding gained through risk assessment and/or experimental exercises, it is possible to demonstrate the true CPPs and material attributes that impact the final products CQAs. If, at any time, new knowledge is gained, it can be used to update the design space and/or move within the design space to a more optimal range/set point.

Traditionally, changes in formulation or manufacturing processes for established drug products required variances granted by the health authorities, some requiring prior approval. These variances prove to be a deterrent to continual improvements being made to the drug product because of the associated financial, time, and overall resource requirements necessary to implement such changes. QbD employs a new concept where changes within the established design space do not require any regulatory variances, and improvements to the drug product are expected based on the production experience.

To facilitate changes, the established procedures of annual product review (APR), which is a regulatory requirement, provide an opportunity to review the quality of the product and identify any trends that suggest improvements. Typically, the analyses from these established procedures, such as APRs, deviations, out of specification (OOS), production control charts, and batch record reviews, are discussed in a quality review meeting, which are also a common attribute of modern quality systems.

Within the framework of the QbD, several new opportunities are available for detecting potential product improvements. Foremost in this process is the updating of the knowledge base of the drug product with any new information obtained during commercial production. In the authors' experience, this is associated with an update of the risk assessment (FMEA) developed and modified for the product prior to launch. This approach is akin to the current device regulations, imposed to improve the quality of medical devices. Another potential source of detecting improvements is the use of MVDA models that reflect many of the critical parameters used for each batch manufacture, and can correlate changes in the output of the established process with changes to the parameters of the model. However, these correlations need further confirmation since noncausal correlations are possible with MVDA analysis. The use of this approach requires an adequate IT infrastructure to collect and process the data in a flexible manner.

All improvements identified via these mechanisms can be implemented without regulatory filing if they are within the established process design space. These changes to parameters/values within the design space will be handled through internal change control and quality systems. Modifications to the design space itself can also be made, however this would involve regulatory oversight.

Continual improvement of a compression process could be attained through knowledge gained from MVDA models using CPPs and CQAs (raw material inputs, compression force, lubrication spray rate, at-line NIR, tablet weight, etc.). Through the establishment of MSPC models, the CPPs responsible for the entire tabletting process can be illustrated. This will provide the manufacturer the ability to

- gain better understanding of the overall tabletting process and
- provide CQA predictions (e.g., dissolution) when the process trajectories are correlated with other data sets (such as material inputs or end product characteristics).

FULL-SCALE DoE/DESIGN SPACE VERIFICATION AT FULL SCALE

After conducting lab/pilot scale DoEs on all identified knowledge gaps for all process steps, the CPPs should be identified and their individual impact on final product quality understood. For an understanding of any potential interactions, these CPPs need to be combined in one final DoE, which could be called a critical interaction DoE. This particular DoE evaluates all demonstrated CPPs throughout the process as factors. After this DoE is executed, the process as a whole should be understood and all CPPs and any interactions identified. This knowledge can then be used for establishing the lab scale design space.

Having a lab/pilot scale design space can cause some challenges in expanding to full scale. There are different options that can be followed, such as conducting a full factorial DoE with all CPP and possibly nCPPs and/or using engineering principals of scale-up, such as dimensional analysis (geometric scaling relationships, especially if the same type of equipment is used in small scale and production scale operations). It is highly recommended to perform a DoE around the CPPs that were confirmed in the critical interaction lab scale DoE. Depending on the number of CPPs, the number of batches needed for the design increases accordingly. Therefore the number of batches needed for a full factorial design might get very high and not realistic on full scale. In those cases a DoE with a lower degree of resolution might be chosen, or bracketing used performing only the cornerstones of the design space to confirm that these still fulfill the CQAs. This exercise can also be used to confirm the nCPPs by including one or two of them along with the identified CPPs in the DoE study. This improves the scale-up understanding of this process step and variable as well as provides connectivity to lab scale (i.e., those parameters identified as nCPPs at lab scale, demonstrated to be not critical at full scale). This approach might prove useful for potentially scale-dependant parameters.

PAT TOOLS AND TABLETTING

PAT has been used in various industries for more than 70 years (18). Early applications focused on the problem solving associated with chemical process/reaction monitoring as well as quality control involving physical property characterization and chemical composition determination of the desired products in a process. As PAT is evolving, there has been increasing realization that real-time measurement provides more useful information for process understanding and optimization during manufacturing.

FDA recently defined PAT as "a system for designing, analyzing, and controlling manufacturing through timely measurements (i.e., during processing) of critical performance attributes, of raw and in-process materials and processes with the goal of ensuring final product quality" (19). The use of PAT tools allow for the ability to obtain real-time information on identified CPPs during a specific unit operation or across the span of unit operations used to manufacture a product. PAT tools could also be used at-line or off-line to obtain additional information about the product, that could not otherwise be possible with conventional analytical methodologies.

If the impact of the CPPs on the CQAs is known, then the PAT tool can be used as part of the control strategy for the manufacturing process to ensure quality is built into the product and may alleviate the need for end product testing. Moreover, the blue sky vision entails measuring, in real-time, the progress of CPPs for a particular unit operation and knowing the impact of the process deeply enough to decide when and how to intervene if it starts to deviate and what modifications may need to be made to parameter settings in the current/subsequent unit operation to ensure final quality in the drug product. The boundaries of the allowable parameter settings in the subsequent unit operation are based on the knowledge gained during development and the design space that has been constructed for the final drug product. Contrast this with control charting of the final product quality where the process is already completed and it is too late to intervene.

Because of unprecedented advances in instrumentation and automation during the past 10 years, the pharmaceutical industry is embracing an increasing opportunity to implement a breadth of PAT tools for process monitoring, which include process analyzers, process control tools, multivariate tools for experimental design and data analysis, and data management tools for knowledge sharing and continual improvement. The availability of those PAT tools enables better process understanding for scientific, risk-managed pharmaceutical development, manufacture, and quality assurance.

Near-infrared (NIR) based diffuse reflection or transmission spectroscopy is one of the most developed spectroscopic tools that has been widely used for in-situ measurement in a variety of pharmaceutical unit operation, such as granulation, drying, BU, and tabletting. NIR offers the advantages of probing samples rapidly and nondestructively over the traditional methods such as chromatography. Other emerging PAT techniques in the pharmaceutical industry for the analysis of tablets include chemical imaging (CI), far-infrared or terahertz spectroscopy, and laser induced breakdown spectrocopy.

CI is usually associated with NIR, but it is also applicable to Raman and MIR spectroscopy, which have been used for pharmaceutical evaluations (20). Standard spectroscopy measures the "average" properties of a sample; however, CI uses multiple pixel imaging to generate a chemical and spatial image of the sample. The spectra of individual points on a surface are captured in a two-dimensional matrix and this "hyperspectral cube" of data can then be analyzed for many statistical properties: spatial distribution of chemical moieties (APIs, API polymorphs, and excipients), statistical distribution of particle sizes, homogeneity, and so forth.

Another off-line technique is far-infrared and terahertz spectroscopy (21), which can be used to look at interfaces within complex dosage forms such as bilayer tablets and coated tablets (coating thickness homogenity). The rotations of atoms are mainly affected by intramolecular forces and since the source of terahertz absorption is intermolecular, the interfaces between surfaces can be readily studied. Interfaces or layer boundaries of coated tablets can be studied, for example, air–tablet surface interface, coating 1–coating 2 interface, and coating 2–core tablet interface. Laser-induced breakdown spectroscopy (LIBS) is another off-line tool that could be used especially for formulation troubleshooting and for tablet compression trials during the formulation optimization stage or upon scale-up. This technique is a destructive technique, but is an excellent tool to show the homogeneity of both a tablet matrix (APIs and excipients) and its coating thickness, and homogenity. It has also been used as a photodegradation-predictive tool, whereby, the increasing levels of ferric oxide determined by the technique, provided a new understanding of the photoprotection mechanism in coated formulations (22).

LIBS uses a finely-focused laser that strikes, for example, a tablet core or coated tablet, causing vaporization at the point of contact. The time at which the light is measured gives elemental or molecular information. The light emitted can be used to determine metals contained within the structure of the API and/or excipient as well as a number of other elements, such as nitrogen and halides. The emission spectra gathered over successive pulses at a particular point provides a depth-profile of various elements. By repeating the operation at adjacent points on the sample, the analyst generates a three-dimensional picture of the distribution of an excipient/API containing a particular element, for instance, the magnesium stearate in the tablet or titanium dioxide in the coating.

NEAR-INFRARED SPECTROSCOPY FOR CONTENT UNIFORMITY DETERMINATION

Tablet CU test is one of the critical quality control tests required for release of tablets to patients. It assures the presence of accurate and uniform APIs content in the dosage unit. Conventional analytical methods for CU involve the laborious sample preparation steps and determination of concentration using high-performance liquid chromatography (HPLC). The sample preparation typically involves dissolving, extracting, and diluting APIs into a solution of appropriate concentration that can be accurately detected by HPLC. This procedure generally accounts for most of the time spent on the analysis. For example, it can easily take several hours to prepare 30 tablets for a CU test. Multiple steps in the preparation also increase the likelihood of occurrence of laboratory errors. In addition, like most of wet chemistry–based analytical methods, considerable amounts of chemicals and waste (either from extraction solvents or active ingredient usage) have to be handled and disposed of cautiously during and after the analysis.

Use of NIR spectroscopic techniques to determine the tablet CU is one of the applications of the ongoing PAT activities (23–30). NIR offers the advantages of probing samples rapidly and nondestructively over the traditional methods based on chromatography. Successes on this front can lead to decreased analysis time and cost as well as reduced handling of and exposure of analysts to hazardous chemicals. More importantly, implementation of this approach allows considerably larger sample size to be inspected within the same time frame with less resources, as opposed to conventional chromatographic method, thus furnishing improved quality control of the final product (31).

METHOD FEASIBILITY

Before NIR method development, a feasibility study should be performed to ensure that the instrumentation furnishes the sufficient selectivity and sensitivity for desired quantification measurement. This feasibility study should involve identification of unique NIR fingerprints of the active ingredient(s) from the placebo matrix, modality selection (if multiple modalities are furnished on the instrumentation), and development of a quantification model (32).

Chemical selectivity of a NIR model is largely determined by spectral discrimination of the APIs from excipients. It would be extremely challenging to achieve chemically selective measurement if the NIR signature of API is overwhelmingly overlapped with the bands from the excipients. Furthermore, at low API drug loads in the tablet for an API that has a weak NIR signature, this may make quantification even more challenging.

NIR CU measurement can be achieved either by diffuse transmission or by diffuse reflection mode. Comparison of the two modalities has been conducted for the tablet content measurement (24,33,34). Compared to the reflection mode, the transmission mode has the advantage of sampling a large portion of the tablet thus being less sensitive to the nonhomogeneity within the tablets. The strong absorption associated with the transmission measurement could cause the loss of spectral information in the combination and first overtone regions (associated with C-H stretching), and also attenuate the light reaching the detector for thick tablets. However, because of the fact that analytical signal-to-noise ratio is a compromise of the optical path length and noise, this loss may be compensated by the deep penetration of the sample in transmission measurement. The major disadvantage associated with the reflection modality is the sensitivity to the nonhomogeneity and matrix changes within the tablets because of the small penetration depth of the measurement. This could present a major robustness problem for an analytical method developed using this modality.

It is recommended that a quantification model be built using the minimum size of samples required by the multivariate calibration. Sensitivity and selectivity can be roughly estimated from the accuracy of calibration because prediction accuracy of external validation samples theoretically could never exceed that of the calibration. If it is determined that the required accuracy will not be achieved using the proposed NIR method, an alternative analytical method should be explored.

A feasibility study can help analysts make a quick decision upfront on whether the proposed NIR method would be suitable for the intended analytical application or not, saving considerable resources by avoiding the execution of a full validation.

METHOD DEVELOPMENT METHODOLOGY

The general methodology used for CU by NIR technique starts with a set of calibration tablets with varied API content spanning the range of analysis (Fig. 8). The NIR spectra are then collected with a NIR spectrometer, and the true API contents are established afterward by a fully validated reference method, HPLC, for example. The calibration model or curve is then developed by correlating the NIR spectra with API content using multivariate regression techniques. One particularly effective, and hence popular, algorithm is partial least-squares (PLS) regression. In multivariate PLS regression, high spectral dimension is reduced to a few

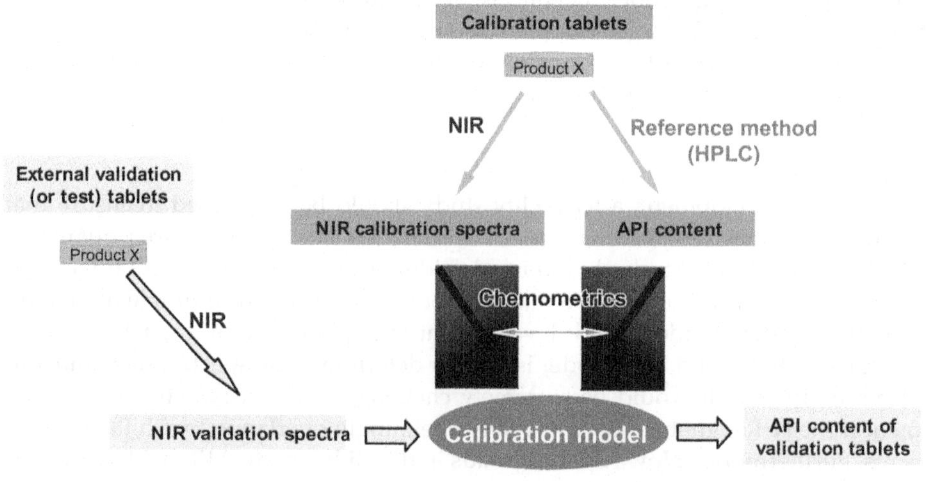

Figure 8 Calibration model development.

orthogonal latent variables, or factors. These factors are selected to maximize the correlation between spectral variance and concentration. Once the model is developed, the accuracy and robustness of the model will be assessed by the external validation tablets that do not belong to the set of the calibration tablets.

There are some key considerations in the calibration design for developing a multivariate PLS model. First of all, the calibration tablets must be both chemically and physically representative, exhibiting the full range of variations expected for the future unknown. Secondly, an appropriate and reasonable sample size should be defined. Ideally, a larger sample size is preferred for multivariate analysis. However, from a practical point of view, the sample size should be reasonable as long as they are sufficiently representative. One way is to use the American Society of Testing and Materials (ASTM) recommendation, (35) where the sample size can be estimated by multiplying the number of principal components or latent variables by a factor of 6. Thirdly, also most importantly, any step that could potentially introduce chance correlations with API concentration into the spectra should be avoided (36). Special care should be taken with the calibration design, sample selection, data acquisition and evaluation. NIR quantification heavily relies on MVDA tools because of the weak and overlapping features of NIR bands. Therefore, a robust calibration design and careful implementation of chemometric techniques are integral to assure reliable NIR methods are developed (36).

METHOD VALIDATION

Implementation of the NIR technique in the drug product CU release test has been encountering tremendous hesitancy from the pharmaceutical community. Few quantitative NIR methods are used in pharmaceutical development and quality control laboratories. The main challenge is developing a robust NIR method that meets the stringent measurement accuracy required by this highly regulated industry. Validation of the method must be aligned with the existing standard operational procedure (SOP) established for chromatographic method validation, but should also embrace the uniqueness of NIR technology.

NIR methods fundamentally differ from chromatographic methods in many aspects for quantification analysis. One key distinction between the two techniques would be the methodology that each technique uses for selective measurement. Selectivity of a chromatographic method is achieved by chemical separation of the analyte of interest from synthetic by-products, degradation products, and system peaks. For an NIR method, spectral discrimination is a prerequisite for successful quantification, however, selectivity requires utilization of mathematical multivariate techniques. Since the analyte is detected after the separation in chromatography, a univariate calibration method built on reference standards is generally sufficient for quantification analysis of the unknown samples. For an NIR method where samples are analyzed intact and spectra are characterized by broad and overlapping features, multivariate calibration methods are required for selective extraction of analyte specific information from the matrices.

Fundamental difference in the two calibration methods suggests that different figures of merits should be utilized for assessment of an NIR method. For HPLC, calibration model is mainly characterized by univariate correlation coefficient and intercept. Accuracy of a chromatographic CU method is determined by recovery studies that involve analyzing the drug substance content in spiked placebo matrix. The experiment focuses on the matrix effect in the sample preparation on measurement accuracy. For a NIR CU method, the calibration model is characterized by a multivariate correlation-coefficient and root-mean-squared standard error of calibration (RMSEC, eq. 1). The model accuracy is assessed by comparing NIR predictions to the HPLC reference values and calculating root-mean-squared standard errors of prediction (RMSEP, eq. 2).

$$\text{RMSEC} = \sqrt{\frac{\sum_{i=1}^{n_{\text{cal}}} \left(C_{i,\text{HPLC}} - C_{i,\text{NIR}}\right)^2}{n_{\text{cal}} - 1 - k}} \tag{1}$$

$$\text{RMSEP} = \sqrt{\frac{\sum_{i=1}^{n_{\text{pred}}} \left(C_{i,\text{HPLC}} - C_{i,\text{NIR}}\right)^2}{n_{\text{pred}}}} \tag{2}$$

where $C_{i,\text{HPLC}}$ and $C_{i,\text{NIR}}$ represent content of APIs in percent (relative to label claim) determined by HPLC and NIR, respectively; n_{cal} represents the number of determinations in the calibration; k represents the number of latent variables or factors used in the calibration; n_{pred} represents the number of determinations in the prediction (or external validation). The two figures of merit (RMSEC and RMSEP) both include computation of the residuals between the reference HPLC and NIR methods.

According to ICH guidelines for validation of analytical methods, the following parameters should be evaluated for the analysis of a major component in the samples: linearity, accuracy, range, precision, repeatability, specificity, intermediate precision, reproducibility, and robustness. Linearity of the NIR methods should be partially assessed in the calibration model development, and be further assessed based on the external validation data sets. Validation of the other parameters can be performed by submitting the NIR spectra associated with the external validation tablets to the calibration models and evaluating the output. The dual sensitivity of NIR technique to chemical and physical properties of the samples suggests that robustness of a multivariate NIR method should include an evaluation of physical property impact of samples on the measurements such as hardness, moisture content, particle size of powder, and excipient vendor.

As an adaptation to the conventional method validation, the specialized chemometric parameter RMSEP and bias should be adopted to assess the method accuracy. Bias has been conventionally used to measure the degree of equivalency of two analytical methods (eq. 3).

$$\text{Bias} = \frac{\sum_{i=1}^{n} \left(C_{i,\text{HPLC}} - C_{i,\text{NIR}}\right)}{n} \tag{3}$$

METHOD MAINTENANCE AND UPDATE

As per USP 1119, calibration model of a NIR method should be subject to ongoing performance evaluation and corrective action would be necessary if unacceptable performance is indicated. It is recommended that the model performance should be monitored at least on a monthly basis on a set of designated tablets to ensure that no instrument drift occurs over time. Prediction accuracy of the model should also be periodically checked against the reference method in case there is any change in the tablet characteristic induced by the process variability.

If it is determined that the NIR analysis no longer meets the model suitability check criteria, and it is determined that the cause of this situation is a change in the properties of the tablets, then an update of the calibration curve or recalibration may be required. The use of another preprocessing algorithm should be evaluated first. If this approach is successful, then the method would require validation before use. If not, then freshly acquired tablet spectra may be needed to replace some of the original spectra. The modified curve would require validation by checking of the bias and RMSEP at points along the curve. If this curve cannot meet the acceptance criteria, then a completely new curve would need to be established with batches produced using the appropriate excipients and drug substance, taking into consideration any change in manufacturing process involved. The new curve would then require validation prior to use. If variations to the process are implemented within the approved design space, the model should be monitored and updated if necessary. Any changes to the existing calibration method should be addressed by the analytical expert(s) in spectroscopy and chemometrics under the proper change control/quality system in place.

CONCLUSIONS

When QbD is fully implemented into the development strategy, the critical sources of process and material attribute variability can successfully be identified, measured and understood so they can be controlled with the appropriate control strategy. Statistical design provides an economical use of resources, especially when many process parameters exist and provides a greater chance of finding optimum conditions from a large amount of data generated from a limited number of experiments. Also, statistical design facilitates an in-depth understanding of the process and justifies the choice of ranges for parameters and finds a robust (optimum) region for the final process to ensure final product quality. Moreover, statistical optimization experimentation and analysis provide strong assurances to regulatory agencies regarding superior process quality. The robustness of the process is truly challenged, and the resulting business benefits are significant, which include the following:

- Established structured approach that is aligned with a risk and science-based approach for formulation development, development of design space at small (lab/pilot) scale and verification of design space at full scale
- Reduced failures at validation of process at full scale/commercial production
- Lower operating costs from fewer failures and deviation investigations
- More effective dialogue between industry and regulator during the application review process
- Regulator to more readily see the connection between design space, criticality, and control strategy

These benefits could translate into significant reductions in working capital requirements, resource costs, and nonvalue added time (i.e., because of replications, poor knowledge preservation, knowledge exchange, and transfer), as well as reduced regulatory burden (i.e., freedom to move within the filed and approved design space).

ACKNOWLEDGMENTS

We especially thank Carl Allenspach, Colleen Ruegger, and Robert Wagner whom reviewed this manuscript and Georgia Floropoulos for technical editing of the manuscript.

REFERENCES

1. Kackar RN. J Qual Technol 1985; 17:176–209.
2. Deming SM. Quality by Design, Part 1. Chemtech, September 1988, 560–566.
3. Walton M. The Deming Management Method. New York: Dodd, Mead, 1986:131–157.
4. "Pharmaceutical cGMPs for the 21st Century: A Risk-Based Approach." Available at: www.fda.gov/oc/guidance/gmp.html. Accessed May 3, 2006.

5. PQRI/FDA Report on the Workshop "A Drug Quality System for the 21st Century," April 22–24, Washington, D.C. (prepared June 16, 2003).
6. Staff Reporter. "Trouble in the Making: Pharmaceuticals" The Economist(US), (August 31, 2002).
7. Carey J, Arndt M. Making Pills the Smart Way. Business Week Online, (May 3, 2004).
8. Abboud L, Hensyle S. New prescription for drug makers: update the plant. Wall Street Journal. September 3, 2003.
9. Pisano GP. The development factory: unlocking the potential of process innovation. Boston, Massachusetts: Harvard Business School Publishing, 1996.
10. ICHQ8, Q8 annex, Q9, Q10. Available at: http://www.fda.gov/cder/guidance/6419fnl.htm.
11. Nosal R, Schultz T. PQLI definition of criticality. J Pharm Innov 2008; 3:69–78.
12. Levin M. Changing Tableting Machines in Scale-Up and Production: Ramifications for SUPAC, background notes for FDA CDER DPQR Seminar, 2000.
13. Yu LX. Pharmaceutical quality by design: product and process development, understanding, and control. Pharm Res 2008; 25(4).
14. Davis B, Lundsberg L, Cook G. PQLI control strategy model and concepts. J Pharm Innov 2008; 3: 95–104.
15. Van Evelghem J. Improving tablet quality with compression to equal force technology. Pharm Technol 2008.
16. Garcia T, Cook G, Nosal R. PQLI key topics – criticality, design space and control strategy. J Pharm Innov 2008; 3:60–68.
17. Process Validation: General Principles and Practices, DRAFT GUIDANCE, U.S. Department of Health and Human Services Food and Drug Administration Center for Drug Evaluation and Research (CDER) Center for Biologics Evaluation and Research (CBER) Center for Veterinary Medicine (CVM), November 2008 Current Good Manufacturing Practices (CGMP).
18. Workman J, Koch M, Lavine B, et al. Anal Chem 2009; 81:4623–4643.
19. "Guidance for Industry. PAT—A Framework for Innovative Pharmaceutical Development, Manufacturing, and Quality Assurance," September 2004. Available at: http://www.fda.gov/cder/guidance/ 6419fnl.htm.
20. Jean-Claude Wolff, Warrack JK. GlaxoSmithKline, and Linda Kidder & E. Neil Lewis, Pharmaceutical Manufacturing, May 2008.
21. J Pharm Sci 2005; 94(no. 1).
22. Cecilia Madamba M, Mullett WM, Debnath S, et al. AAPS PharmSciTech 2007; 8(4) Article 103. Available at: http://www.aapspharmscitech.org.
23. Broad NW, Jee RD, Moffat AC, et al. Application of transmission near-infrared spectroscopy to uniformity of content testing of intact steroid tablets. Analyst 2001; 126:2207–2211.
24. Corti P, Ceramelli G, Dreassi E, et al. Near infrared transmittance analysis for the assay of solid pharmaceutical dosage forms. Analyst 1999; 124:755–758.
25. Kirsch JD, Drennen JK. Appl Spectrosc Rev 1995; 30:139–174.
26. Wang Q, DeJesus S. J Near Infrared Spectrosc 1998; 6:A223–A226.
27. Williams RL, Adams WP, Poochikian G, et al. Pharm Res 2002; V19:359–366.
28. Trafford AD, Jee RD, Moffat AC, et al. A rapid quantitative assay of intact paracetamol tablets by reflectance near-infrared spectroscopy. Analyst 1999; 124:163–167.
29. Ritchie GE, Roller RW, Ciurczak EW, et al. J Pharm Biomed Anal 2002; 29:159–171.
30. Xiang D, Konigsberger M, Wabuyele B, et al. Development of robust quantitative methods by near-infrared spectroscopy for rapid pharmaceutical determination of content uniformity in complex tablet matrix. Analyst 2009; 134:1405–1415.
31. Sandell D, Diener M, Vukovinsky K, et al. Drug Inf J 2006.
32. Xiang D, LoBrutto R, Cheney J, et al. Appl Spectrosc 2009; 63:33–47.
33. Merckle P, Kovar KA. J Pharm Biomed Anal 1998; 17:365–374.
34. Cogdill RP, Anderson CA, Delgado-Lopez M, et al. AAPS PharmSciTech 2005; 6.
35. Annual Book of ASTM Standards, Designation: E1655-94.
36. Xiang D, Berry J, Buntz S, et al. J Pharm Sci 2009; 98:1155–116.

13 Expert systems and their use in pharmaceutical applications

Metin Çelik

INTRODUCTION

The pharmaceutical industry has entered the twenty-first century, a new era that will be far more scientific, technologic, and sophisticated than anyone would have imagined just a quarter of a century ago when it was still a tradition to develop formulations and processes mostly based on trial and error. The future success in all areas of pharmaceutical science will depend entirely on how fast pharmaceutical scientists will adapt to the rapidly changing technology. The regulatory agencies seem to enforce such changes as evidenced by the fact that the goal of FDA's recent Process Analytical Technology (PAT) initiative is basically to understand and control the manufacturing process and to realize that quality cannot be tested into products but it should be built-in or should be by design.

Pharmaceutical scientists will gradually enjoy the availability of the harmonized and fingerprinted (in terms of functionality testing) excipients. Also, the awareness of and the use of artificial intelligence (AI)-based expert systems (ESs) [rule-based systems, fuzzy logic, genetic algorithm (GA), artificial neural networks (ANNs), simulations, etc.] in the areas of preformulation, formulation and process development, regulatory affairs, new drug delivery system development, and project management will increase dramatically (1,2).

ESs have been defined in various ways, but all the definitions share a common thread suggesting that ESs are artificial means to emulate the way in which human (domain) experts solve problems.

A definition of such systems that may be appropriate for the applications in pharmaceutical science would be: "ES is a computer program capable of making recommendations, decisions or predictions based on knowledge gathered from the experts and/or experimental data obtained in the field."

ESs are designed to work in a narrow field of focus (such as compaction, spray drying, film coating) and have distinct architecture of components outlined in the following sections. Functional areas of ESs include, but are not limited to, control, design, diagnosis, instruction, interpretation, monitoring, planning, prediction, prescriptions, selection, and simulation. ESs are being used in many disciplines such as agriculture, business, chemistry, communications, computers, education, electronics, engineering, environment, geology, image, information, law, manufacturing, mathematics, medicine, meteorology, military, science, space, and transformations. The literature reported less than 50 ESs in use in 1985; this number increased to more than 12,000 in about seven years (3). However, although problems in the pharmaceutical industry are not necessarily more complicated than some of the problems encountered in the above-listed fields, the number of ESs used in pharmaceutical science is still negligibly low. One of the reasons for the insignificant use of ESs in pharmaceutical applications is the challenge facing ES developers in terms of their verification and validation (V&V) processes, in part because of FDA's interest in the V&V of all types of software. However, the main reason is that pharmaceutical scientists prefer to use well-established concepts. Many scientists will let somebody else try a new concept first, and if it works, then join the crowd. It is a safe approach to use an established system, but it does not provide us with the immediate benefits of being on the technologic edge. On the other hand, it is always risky to try a new concept, even though the outcome may prove to be rewarding for both the scientist(s) and the company.

The history of ESs has played an integral part in the development of its structure and components. ESs did not begin as a known program with defined components and relationships. Instead, the ES was preceded by the general development of AI.

The Dartmouth Summer Research Conference on AI was considered to give birth to the field of AI, The conference occurred in 1956 at then at Dartmouth College and formally proposed by McCarthy et al. (4). Their proposal stated "We propose that a two-month, 10-men study of AI be carried out during the summer of 1956 at Dartmouth College in Hanover, New Hampshire. The study is to proceed on the basis of the conjecture that every aspect of learning or any other feature of intelligence can in principle be so precisely described that a machine can be made to simulate it" marks the debut of the term "AI" (4).

BUILDING AN EXPERT SYSTEM

To build an ES, the full participation of a domain expert, knowledge engineer, and user is essential. A domain expert possesses the knowledge and skill to solve a specific problem in a manner superior to the others. This expert's highly specialized knowledge is stored in the knowledge-base component of an AI-based program by the knowledge engineer. The user also can help define the interface specifications.

WHY BUILD AN EXPERT SYSTEM?

In general, the reasons for the development of an ES can be listed as follows although every company may have different motivations.

Improved productivity: The system is expected to be capable of improving the quality of decisions, to reduce the time to reach a decision and/or to provide expertise to locations within the organization where this capability is lacking.

Lower costs: The system is expected to improve the use of materials during manufacturing and/or to reduce labor costs by allowing a time-consuming task to be completed quickly or acts in place of a highly paid expert.

Improved quality: The system is expected to improve the quality of the final product or the services supplied by the organization and/or to provide training to personnel that improve their work activities.

Improved image: The system is expected to improve the organizations image as a leader and innovator.

The above given list also explains the advantages of the ESs over the human experts. The advantages of the ESs are related to knowledge, decisions, safety, and cost. The knowledge of an ES is permanent and easy to transfer while the human expert's knowledge is perishable and difficult to transfer from one worker to another. The decisions made by human experts can be unpredictable and difficult to document. In contrast, the ES decisions process is consistent and easy to document. In an unsafe or hostile environment, an ES is replaceable while the human expert is definitely irreplaceable. When cost is an issue, the ES services are often more affordable than those of the human expert.

When compared with human experts, ESs have the following advantages: An ES's knowledge is permanent and can be easily transferrable. The decision process is fast and consistent, therefore predictable, and it is easily documented. Despite these advantages, ESs are not intended to take the place of formulation scientists. They must be considered as vital tools to be used by formulators for the rapid, cost-effective, and scientifically sound development of a dosage form as well as useful for training inexperienced scientists.

However, there are some factors that favor the human expert as opposed to an ES. These factors are much more difficult to quantify, but can often be important to a project. A human expert is creative and adaptive and uses sensory experiences. The computer expert system is uninspired, needs to be directed, and uses only symbolic input. A human expert has a broad focus and may be able to use knowledge from another field or experience to aid in the problem solving whereas an ES has a narrow focus constrained to the domain knowledge. Lastly, the human expert has the ability to use common sense knowledge while the ES can only use technical knowledge.

PHASES OF AN ES DEVELOPMENT PROCESS

There are many textbooks addressing the strategies and/or tools employed in building ESs in depth (3,5–8). The following should be considered only a general overview of the phases involved in the development of an ES.

Feasibility Study

A project team assesses whether an ES can or should be developed for a specific problem or project. The team evaluates the motivation for the development of the ES in terms of improving productivity, quality, and image as well as cost reduction. The team also must consider the problem and the people-related feasibility issues very carefully. Some of the important questions that must be answered positively are as follows:

Is the problem solvable?
Are the problem-solving steps definable?
Is the problem stable, well focused, and its complexity reasonable?
Is the management supportive of the project, receptive to change, not skeptical, and does it have reasonable expectations?

If all the answers to these questions are in the affirmative, then the project team should continue to evaluate the other problem—the deployment-related issues concerning the development of the ES for that particular problem or project.

If and when a decision is made in favor of the development of the ES, then the project team defines the features and specifications of each component of the ES and develops flow charts for each specific problem.

ACQUISITION OF THE KNOWLEDGE

The objective of knowledge acquisition is to compile a body of knowledge on the problem of interest that can then be encoded into the ES. There are different type of knowledge and different methods of obtaining them. Some of these types of knowledge are as follows:

1. Procedural (e.g., rules, strategies)
2. Declarative (e.g., facts, objects)
3. Heuristic (rule of thumb)
4. Structural (e.g., rule sets, concept relationship)

The major difficulties with the knowledge gathering from the human experts lie in the facts that some domain expert may be unaware of or unable to verbalize the knowledge or may provide irrelevant, incomplete or inconsistent knowledge.

Design of the ES

The knowledge engineer determines which software to use to transform the acquired knowledge into a coded program for the development of the ES. Some of the AI tools (knowledge representation techniques) used alone or in combinations in the development of an ES include decision trees, object-attribute-value (OAV) triplets, rules (if-then-because statements) with forward and/or backward chaining, fuzzy logic, GA, case-based reasoning, and ANNs. A successful ES usually is developed by combining more than one AI technique.

Testing the Modules and Development of the Prototype

Case studies with known results are used to test the ability of the rules, databases, and programming to perform properly.

Implementation, Testing, and Troubleshooting of the Final Program

Case studies as well as untested materials and parameters are used to verify the proper operation of the program and to troubleshoot any additional problems identified.

Training of Users

A user acceptance questionnaire is used during the implementation of the program.

Maintenance and Upgrade of the Program

Depending on the availability of the new knowledge and/or the data in the field of a particular ES, and upgrade may be needed to ensure that the ES will evolve continuously to overcome new challenges concerning that specific project or problems.

Critical Issues Concerning the Verification and Validation of an ES

Verification of an ES determines whether the system is developed according to its specifications. Validation of an ES determines whether the system meets the purpose for which it was intended. Very critical differences exist between an ES and conventional systems in terms of V&V of an ES. An ES is both a piece of software and a domain model, and there may not be a unique, correct answer to a problem given to an ES. An ES can adapt itself by modifying its behavior in relation to changes in its internal representation of the environment.

An ES should be considered correct when it is complete, consistent, and satisfies the requirements that express expert knowledge about how the system should behave. If a system has hundreds of rules, however, it may require thousands of distinct decision paths, and this makes the aspect of correctness hard to establish. This is not, of course, a problem in a conventional programming technique.

These differences between the AI and conventional programming tools provide flexibility and special capabilities to an ES, but these differences also make the use of traditional V&V of an ES difficult. This is one of the problems slowing the development and acceptance of ESs. Experts do not agree on how to accomplish the V&V of ESs. One of the impediments to a successful V&V effort for ESs is the nature of ESs themselves. They are often used for working with incomplete and uncertain information or ill-structured situations. Because the ES specifications often do not provide precise criteria against which to test, there is a problem in verifying and validation them according to the definitions. This is unavoidable. If there are precise enough specifications for a system, there would not be any need to use an AI tool to develop the system, and the conventional programming language would be sufficient for the development of a piece of software for that system.

In reality, the first part of V&V, that is, verification of an ES, is not so difficult to establish because it is possible, and also highly recommended, to build small modules (sub-ESs) for each problem within a system. This is a significant help to the verification process of the whole system. This is true even if the ES is developed by combining mere than one system.

The main problem is the second part of V&V, that is, validation. ESs will make a recommendation based on the domain knowledge. If the domain knowledge is junk, then the recommendation of the ES naturally will be junk. How can someone validate the correctness of knowledge provided by domain expert, or if two domain experts have conflicting views over a problem-solving process, who will decide which is correct?

FDA's requirements for the submission of the software code can also add additional burden to the software validation of an ES. This is a serious obstacle because only a few AI tools providers and ES developers will be willing to share the code. As some of the AI tools may cost more tens of thousands of dollars, who could blame the software providers if they do not wish to share the code.

EXPERT SYSTEM COMPONENTS

An ES contains three basic components, a knowledge base, a working memory, and an inference engine as described below. These components can be found in many types of AI programs including decision trees, ANNs, GAs, and fuzzy logic. In addition, many of the AI tools have an explanation facility providing the reason for the decision or recommendation of the ES.

Knowledge Base

The knowledge base contains the domain knowledge, the information pertinent to the field or problem. The domain knowledge can be acquired from literature and/or experts in the field and is in an electronic form that can be searched and updated easily. The knowledge base is similar to a human's long-term memory or experience.

One typical way of representing the knowledge in an ES is the rules. In its very basic form, a rule is an IF/THEN structure that logically relates information contained in the IF part to the other information contained in the THEN part. Some derivations of such a structure could also include ELSE and/or BECAUSE parts as well.

For example, the following rule represents the knowledge for the selection of the plasticizer polyethylene glycol 400 (PEG 400) with the film forming polymer hydroxypropylmethylcellulose (HPMC):

IF: The selected polymer is HPMC only
AND: There is no regulatory restriction for the use of PEG 400 in that country
THEN: Recommend PEG 400
BECAUSE: PEG 400 is compatible with HPMC and it is efficient in its functionality.

WORKING MEMORY

The working memory contains the facts about the problem discovered during the problem-solving session. This component is similar to a human's short-term memory or current experience. Knowledge in the working memory can be inferred by the system or it can be obtained by user input. Knowledge inferred by the system is obtained by matching user input with knowledge in the knowledge base to produce new facts.

INFERENCE ENGINE

The inference engine is the component that models the human reasoning process. It matches facts in the working memory with domain knowledge in the knowledge base and draws a conclusion. It works by searching the database for a match between its contents and the information in the working memory. If a match is found, the conclusion from the match is added to the working memory and the inference engine continues to scan the database for additional matches.

EXPLANATION FACILITY

A unique feature of an ES is its ability to explain the reasoning used to reach a conclusion. The following part of the example rule given above represents the explanation facility of the ES.

BECAUSE: PEG 400 is compatible with HPMC and it is efficient in its functionality.

Because an ES can explain why user input was requested or how a conclusion was reached, the system developer can use this component to uncover errors in the system's knowledge and the user can benefit from the transparency provided into the system's reasoning.

KNOWLEDGE REPRESENTATION

There are a number of techniques that represent the knowledge (3,5–9) including but not limited to the following:

OAV triplets
Semantic networks
Frames
Rule-based systems
Fuzzy logic
ANNs
GA
Others: decision trees, hybrid systems (e.g., neuro-fuzzy systems), case-based reasoning, etc.

These techniques will be briefly described in the following sections. The most successful ES applications integrate more than one technique. For example, the rule-based system are very good in providing the reasoning for how and why they reach and decision but they are not best in automated learning (without being updating its knowledge base) or recognizing patterns in large amount of data. This gap can be filled by integrating ANNs are very powerful in automated learning although they lack in providing the justification for their predictions. Therefore, combining these two techniques can bring the strength of both approaches while eliminating their weaknesses.

Object-Attribute-Value Triplets

OAV triplets provide a particularly convenient way in which to represent certain facts within a knowledge base. Each OAV triplet is concerned with some specific (conceptual) entity or (physical) object. For example, our object of interest may be a granule (particle). Associated with every object is a set of attributes. Using the granule as an example (i.e., object), some of the attributes include the following: Particle size, particle shape, particle density, particle porosity, surface roughness, moisture content, etc.

For each attribute, there is an associate value, or set of values. For example, in the granule example, the particle size attribute can have the values of large, small, fine, etc. Please note that values could be numerical as well.

Most of OAV triplet systems have also confidence factor associated with each specific triplet. Confidence factors, or certainty factors, refer to a numerical weight given to a fact or a relationship to indicate the confidence one has in that fact or relationship (Fig. 1). There are two kinds of confidence: "expert confidence" (the confidence that an expert feels when suggesting a rule) and "user confidence" (the confidence that a user feels when answering a question).

In a typical ES programming language, there are several ways of handling the uncertain data such as confirmatory (Yes/No) system, numerical range (−1 to 1, 0 to 10, −100 to 100,

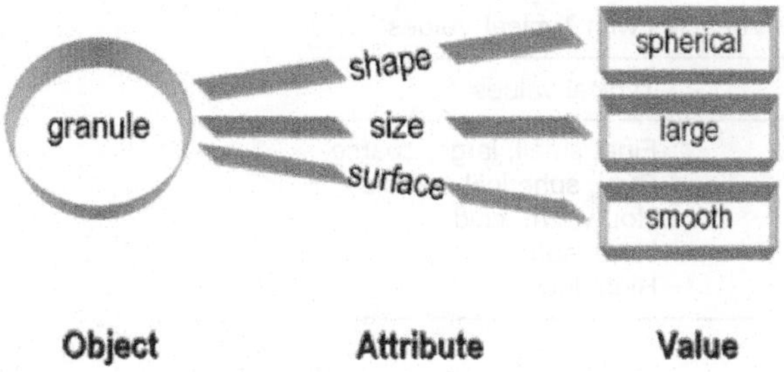

Figure 1 An example for a object-attribute-value triplet (fuzzy variables: shape, size, surface).

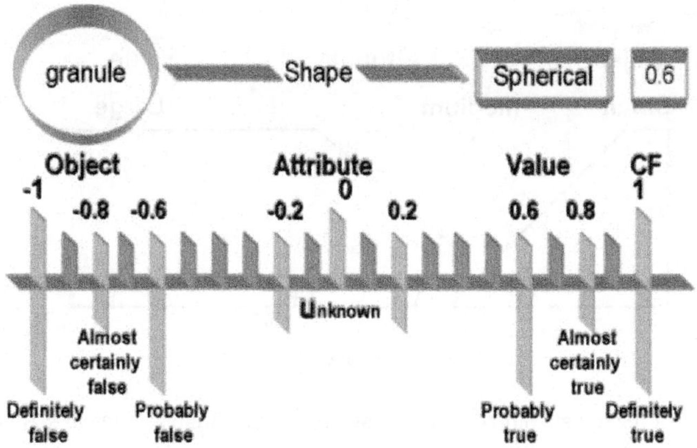

Figure 2 An example for an object-attribute-value triplet with confidence factors.

etc.), systems, increment/decrement system, custom formula systems, and fuzzy logic. In many instances, the user may have to answer a question to determine the confidence factor. In the numerical approach, this is achieved by asking the trueness (definitely false, almost definitely false, probably false, unknown, probably true, almost definitely true, and definitely true) or sureness of a fact or value (Fig. 2). The ES inference engine then converts the answer to a numerical value that computers understand.

Semantic Networks

Semantic networks consist may be thought of as a network that is composed of multiple OAV triplets in network and characterizes their interrelationships. An advantage of this method is its flexibility new objects whenever needed.

Frames

A frame contains an object plus slots for any and all information related to the object. The contents of slots are typically the attributes, and the attribute values, of a particular object. Therefore, a frame is a natural extension of the semantic networks.

Fuzzy Logic

Fuzzy logic is mainly concerned with quantifying and reasoning about vague or fuzzy terms that appear in our daily lives. In fuzzy logic, these terms are referred to as linguistic variables

Table 1 Examples of Fuzzy Variables with Typical Values

Fuzzy variable	Typical values
Size	Fine, small, large, coarse
Shape	Oval, spherical, needle
Temperature	Hot, warm, cold
Tablet strength	Hard, soft
Pressure	High, low

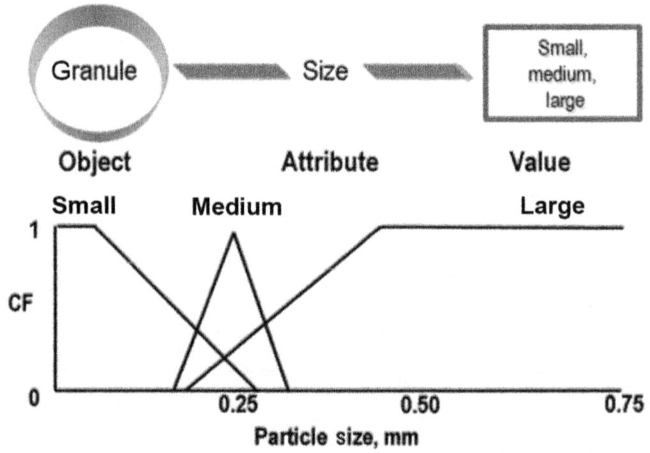

Figure 3 An example for a fuzzy set with confidence factors (fuzzy variable: particle size; fuzzy values: small, medium, large).

or fuzzy variables. Some examples of fuzzy variables that are encountered in pharmaceutical applications are given in Table 1.

Fuzzy Sets

Classical set theory establishes systematic relation among objects with asset as well as between elements of various sets. A set is a collection of any number of definite, well distinguished objects, called the elements of the set that share common properties. Thus, an object may either belong to the set or be completely excluded. In other words, if A is a set and x is an element to the set, then x belongs to A if and only x satisfies all the membership requirements if A, otherwise x does not belong to A.

Fuzzy set theory differs from classical set theory in one critical aspect. An element can belong to the fuzzy set, be completely excluded from the fuzzy set, or it can belong to the fuzzy set to any intermediate degree between these two extremes. The extent to which an element belongs to a given fuzzy set is called grade of membership or degree of membership. It can be said, therefore, that classical set theory is a special case of fuzzy sets.

Fuzzy sets can be obtained to reflect the general opinion of the scientists or experts in the fields. For example, in Figure 3, fuzzy sets are shown in a piecewise linear form for the issues of three different categories (small, medium, and large) of the size of granule(s). In this fuzzy subset, a granule particle with the size of 0.25 mm is a member of medium size with a membership value of about 1, and at the same time a member of small and large sizes with a value of about 0.15 and 0.25, respectively.

Rule-Based Systems

The most common way of representing knowledge is found in rule-based systems that employ rules to represent the experts' knowledge. Such rules are typically of IF-THEN variety. However in some instances this is extended to include: IF-THEN-ELSE or IF-THEN-ELSE-BECAUSE type or rules. In rule-based system, the uncertainty of the knowledge is handled using the method of confidence factors as described above in the OAV triplets.

In the rule-based ESs, there are different ways of executing the rules. Backward chaining is by far the most common strategy used in the simple rule-based systems and it is a term used to describe running the rules in a "goal-driven" way. A "goal" is an attribute for which the ES tries to establish a value. In backward chaining, if a piece of information is needed, the program will automatically check all the rules to see if there is a rule that could provide the needed information. The program (inference engine) will then "chain" to this rule before completing the first rule. This new rule may require information that can be found in yet another rule. The program will then again automatically test this new rule. The logic of why the information is needed goes backward through the chain of rules.

Forward chaining is a "data-driven" way to run the rules. In backward chaining, there is always a goal to be satisfied and a specific reason why rules are tested. In pure chaining, rules are simply tested in order they occur based on available data. If information is needed, other rules are NOT invoked. Instead, the user is asked for information. Consequently, forward chaining systems are dependent on the order of the rules. However, since time is not spent determining whether the information can be derived, forward chaining is much faster.

In a control by hybrid backward and forward chaining, the basic approach is data driven but information needed by rules derived through backward chaining.

Another technique is to divide an ES to subsets of rules and run some in forward chaining and some in backward chaining.

Artificial Neural Networks

ANNs can be defined as machine-based computational techniques that attempt to simulate some of the neurologic processing ability of the human brain. In the human brain, neurons are the information carriers. In the same way, ANNs are composed of interconnected simulated neurons capable of pattern recognition or data analysis. Processing of the data using pattern recognition produces classification of the data while data analysis produces numerical output. One of the most powerful characteristics of ANNs is the ability to find complex and latent patterns in the information being processed. Unlike most statistical experimental design, analysis of data using ANN does not require a specific number of experiments. Also, neural networks can generate hypotheses that can be tested by other scientific methods and the outputs of one network can become the inputs to a subsequent network.

ANN Elements

ANNs can be represented by a neuron model like the one found in Figure 4.

As seen in this model, an ANN is composed of interconnected processing elements (PE) or neurons. The interconnections represent weights or weighing factors applied to the input values of the neuron as the information is passed forward through the network. These weights are sometimes referred to as synaptic weights since the interconnections are similar to the synapses of the human brain. Output values from each neuron are passed forward to the next layer through its interconnections or used as part of the final output of the network. The architecture of a network is defined by the number of layers in the network, the number of neurons in each layer, the configuration of their interconnections, and the way in which the weights of the interconnections are calculated.

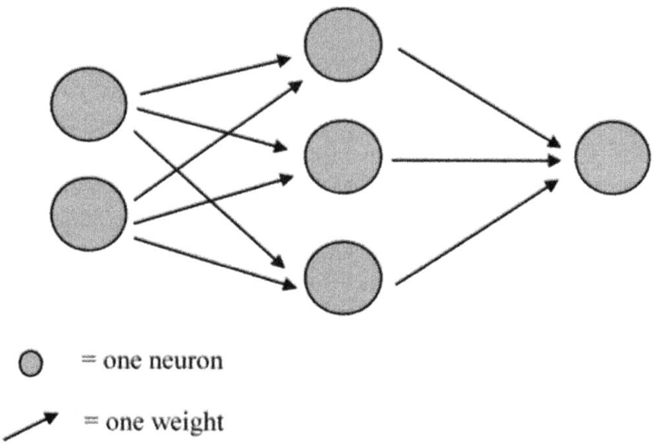

Figure 4 Neuron model.

Network Types

Generally, there are three basic types of neural networks, feedforward, feedback, and self-organizing. The network type to be utilized depends on the task to be accomplished. The following paragraphs will describe the network architectures and types of input data suitable for each network type.

Feedforward Networks

Feedforward networks (Fig. 5), also called error backpropagation or backprop networks contain the basic network components described in the neuron model. A feedforward network is designed by defining its number of layers and the number of neurons in each layer. The number of neurons in the input and output layers is equal to the number of independent and dependent variables, respectively. The input layer serves as a distribution point for the data to the first hidden layer and can only scale the data, not calculate weighting factors. The purpose of a scaling the data is to normalize it to a constant numerical range, such as 0 to 1 or −1 to +1. Scaling can be performed using linear or nonlinear scaling functions. The number of hidden layers is based on personal preference and rules of thumb. The purpose of the hidden layers is to provide a balance between network accuracy and network generalization. A higher number

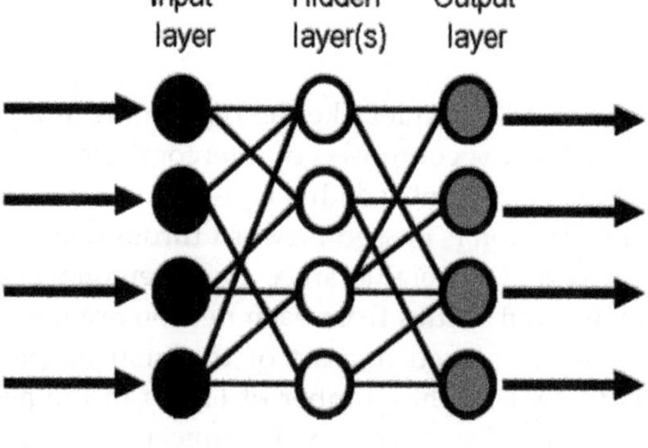

Figure 5 An example for feedforward networks.

of hidden layers lead to a narrow, accurate network with a decreased ability to predict outside the boundaries of its original data. Fewer hidden layers will produce a more generalized robust network, but may smooth the curve between the data points too much. The balance required between network accuracy and generalization depends on the purpose of the network.

Once the network is designed, it is ready to be trained. Training is the process of tuning the synaptic weights to minimize the difference between the actual output and the network output values. The next step is the error back propagation step or learning step. Learning in this context does not imply the human qualities of understanding, consciousness, or intelligence. Instead, it simply implies the use of data for tuning a set of parameters or, in this case, the tuning of the synaptic weights. Once the network training is complete, the network does not store or refer to the training data. Instead, the trained network is an independent summary of the data. With the weights established by training, the network is capable of producing outputs for input data not originally contained in the training data set. The use of a data set to train a network is called supervised learning and requires that the output data corresponding to the input data be available during training.

Classification Networks

The third kind of network is the classification network or self-organizing network. This network type is able to separate the data into a specified number of categories. It is always unsupervised, which means that the network has the ability to learn without being shown correct outputs in sample patterns. The network architecture contains only two layers, input and output. The number of neurons in the input layer is defined by the data and output layer has one neuron for each possible output category. During training, the data is presented to the input layer, propagated to the output layer resulting in one neuron providing an active response or being a "winner." The network adjusts the weights for the neurons in a "neighborhood" around the winning neuron based on a two-dimensional feature map whose cells form a rectangular grid. During training, the locations of the responses become ordered as if some meaningful coordinate system for different input features were being created over the network. The neighborhood size is variable. It starts large and decreases with learning until the neighborhood approaches zero and only the winning neuron's weights are changed. The training process is repeated for all patterns for a number of predetermined epochs. At the end of the training, each neighborhood becomes an output classification.

Genetic Algorithms

GAs are mathematical tools that solve optimization problems. This type of problem is usually composed of a number of variables that control a process or outcome, and a formula or algorithm that combines these variables to fully model the process. The goal of the problem is then to find the values of the variables that optimize the model in some way, usually by minimizing or maximizing one of the dependent variables. While there are many mathematical methods that can solve optimization problems, these traditional methods tend to break down when the problem is more complex. Examples of complex problems include combinatorial problems or problems where the fitness function is not a smooth, continuous mathematical formula, such as a neural network function.

GAs optimize these complex problems using the methods of evolution, specifically survival of the fittest. Much of the terminology used to describe GAs is partially based on concepts from biology; however, some terms may have different names depending on the author. In this case, "survival of the fittest" means that the GA solves the problem by allowing the less fit individuals in the population to die and selectively breeding the most fit individuals, that is, those that solve the problem best.

The use of GA in combination with neural networks for the optimization of process parameters has been investigated by Cook et al. (10). In this example, a neural network model was developed to predict the effect of several process operating parameters and conditions on the internal bond strength of particleboard. A GA was applied to this neural network model to determine the process parameters that would result in the optimal strength for a given set of operating conditions. This ANN GA system was successful in predicting the process parameters, which allowed the manufacturer to achieve optimal levels of board strength based on the current, variable operating conditions. The ANN portion was used to model the process parameters while the GA utilized this model to obtain the optimal processing parameters under actual manufacturing conditions.

OTHER METHODS OF KNOWLEDGE REPRESENTATION

Decision Trees

A decision tree takes as input an object or situation described by a set of properties, and outputs a yes/no decision. Decision trees therefore represent Boolean functions. Functions with a larger range of outputs can also be represented. Decision trees are considered to be auxiliary tools in ES development and are usually incorporated with other systems.

Neuro-Fuzzy Logic

In the field of AI, neuro-fuzzy refers to combinations of ANNs and fuzzy logic. Neuro-fuzzy hybridization results in a hybrid intelligent system that synergizes these two techniques by combining the human-like reasoning style of fuzzy systems with the learning and connectionist structure of neural networks.

Case-based Reasoning

In case-based reasoning, to solve a problem, the inference engine searches and find a similar problem solved in the past and adapts the old solution to solve the new problem. The systems retrieves, reuses, revises, and retains the solutions and provides the basis for almost limitless applications in domain where there are many exceptions to rules and where the problems are not fully understood, but there is a database of past examples.

Cause-and-Effect Diagrams

Another useful tool is a cause-and-effect diagram (also known as an Ishikawa diagram or a fishbone diagram) (11). Such a diagram clearly illustrates the various causes affecting product quality by sorting and relating the causes. Therefore although there is no set format, a good cause-and-effect diagram is one that fits the purpose. It is useful because it

i. serves as a guide for discussion,
ii. actively seeks causes,
iii. allows data collection,
iv. shows the level of technology,
v. provides education even via its preparation, and
vi. can be used for almost any problem.

Figure 6 shows a cause-and-effect diagram for a direct compression process (12). The relationship between the variables and the responses can also be expressed in a so-called "influence-matrix" format, which categorizes the impact of any cause on the response(s) as

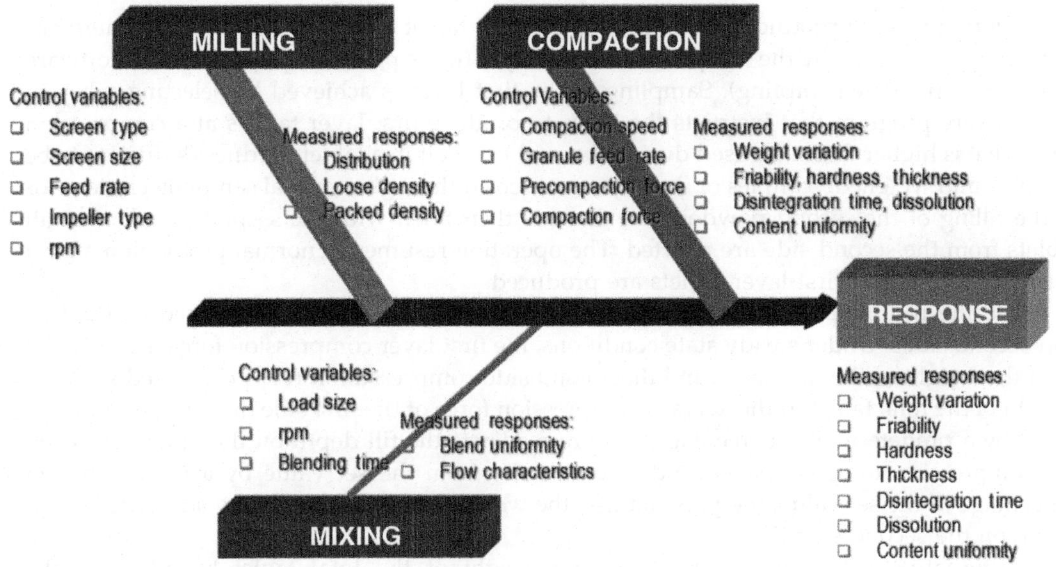

Figure 6 A cause-and-effect diagram for a direct compression process. *Source*: From Ref. 12.

none, weak, moderate, or strong. Cause-and-effect diagrams and, especially, associated influence-matrix analysis provide vital knowledge for understanding a given process.

AN EXAMPLE TO EXPERT SYSTEMS—BILAYERex: A BILAYER TABLETTING ES

The example ES, namely BILAYERex, was developed by PTI, Inc. (13) in collaboration with the Department of Industrial and System Engineering at Rutgers University to accurately simulate the operation of a bilayer tablet press. Broadly, it allows a realistic simulation and visual presentation of monolayer and bilayer tablet production as per the details discussed below.

Before describing the BILAYERex, it is appropriate to explain briefly the bilayer tablet presses, although the process of bilayer compression tablets has been addressed in depth in chapter 9 as part of the compaction of combinations drug products. The following discussion is applicable for most of the advanced bilayer tablet presses although the discussion is based on Fette Bilayer Tablet Press Model 3090.

Advanced bilayer tablet presses are usually equipped with a computer controlled software program to regulate production operation. The software generally utilizes compression force measurement of the first layer and total tablet to regulate the weight of the first layer and that of the total (two-layer) tablet. At the outset of a typical production operation, the weight of the first layer is adjusted through the use of an appropriate fill cam and adjustment of the filling depth followed by scrape-off of the excess powder. A soft compact of the first layer is formed at a relatively low compression forces. The penetration position of the upper punch into the die cavity determines the maximum volume fill available for the second layer. This volume is totally filled with the second layer powder as the die passes through the second feed frame. The weight of the total tablet is regulated by use of an appropriate fill cam (on the second side) and adjustment of the filling depth (on the second side). Excess second layer powder is again scraped off. The weight of the first layer is adjusted to the target weight followed by that of the total tablet target weight within an operational and a tolerance limit. The difference between the two layers gives the weight of the second layer.

During routine production, samples of the total tablet are obtained at regular intervals, immediately followed by the samples of the first layer (if the press is equipped with a software program to regulate sampling). Sampling of the first layer is achieved by selecting a key in the software program that instructs the press to produce first layer tablets at a compression force (that is higher than that used during the run) to result in a tablet hardness sufficient to be sampled and weighed. Samples of first layer are ejected through a special sampling chute prior to the filling of the second powder. The dies are then filled with the second powder but all tablets from the second side are rejected. The operation resumes to normal production after a sufficient number of first layer tablets are produced.

During routine production, the press measures the compression force used on the first and second sides. Under steady state conditions, the first layer compression force is correlated with the weight of the first layer and the second side compression force is correlated with the weight of the total tablet. If the average compression force of the first side deviates from the set value by a predetermined percentage, the press adjusts the fill depth on the first side. When the compression force on the second side deviates from the set value by a predetermined percentage of the set value, the press adjusts the weight of the total tablet by adjusting the fill depth on the second side.

Alternatively, the press can adjust the weight of the total tablet by adjusting the penetration depth of the upper punch that would alter the fill volume of the second powder. By controlling the weight of the first layer and total tablet within limits, the press controls the weight of the second layer.

The BILAYERex software was developed using *Arena* (1), which is an object-oriented general purpose simulation tool, extensively used in modeling industrial scenarios in connection with a user interface written in Visual Basic.

Arena is one of the powerful simulation tools in use today. Its simulation environment consists of module nested in templates, built around SIMAN language constructs and other facilities, and augmented by a visual front end. It implements the discrete-event simulation (DES) paradigm. In the DES paradigm, the simulation model possesses a *state S* (possibly vector-valued) at any point in time. A system state is a set of data that captures the salient variables of the system and allows us to describe system evolution over time. In the simulation model, the state is stored in one or more program variables that represent various data structures (e.g., the number of customers in a queue, or their exact sequence in the queue). Thus, the state can be defined in various ways, depending on particular modeling needs, and the level of detail incorporated into a model.

The state trajectory in time is abstracted as a step function, whose jumps are triggered by discrete events, which induce changes in the system state at particular points in time. Although computer implementation of events varies among DES simulators, they are all conceptually similar: An event is a data structure that always has a field containing its time of occurrence, and any number of other fields. Furthermore, the "occurrence" of an event in a DES simulator is implemented as the execution of a corresponding procedure (computer code) at the scheduled event occurrence time. When that procedure is run, the event is considered to be *processed* or *executed*.

The evolution of any DES model is governed by a *clock* and a chronologically ordered *event list*. That is, events are linked in the event list according to their scheduled order of occurrence. That is, as time passes, events are scheduled and placed into the event list and they are executed when their occurrence times are reached. The execution of an event can change state variables and possibly schedule other events in the event list. As state variable changes, the simulation model collects observations and summarizes it when the simulation run is completed.

FEATURES OF BILAYERex

User Interface

The following information is entered by the user (as shown in Figure 7A–C) before a simulation run is executed.

1. Project identification (Fig. 7A).
2. Press, model, and turret type (i.e., number of stations) (Fig. 7A): The user can choose a specific press from a library.
3. Press mode (Fig. 7A): The user can select either a monolayer or bilayer operation for the simulation.
4. Operative punches (Fig. 7A): User can choose all or any number and combination of the operative punches for a selected turret type.
5. Punch selection (Fig. 7A): Punch parameters provide precise tool geometry for realistic calculation of fill volumes.
6. Bulk density parameters (Fig. 7A): The user enters minimum and maximum bulk density values and the density distribution mode for each layer. These specific input parameters are used to calculate weights corresponding to the fill volumes. Bulk density parameters also include noise and draft factors.

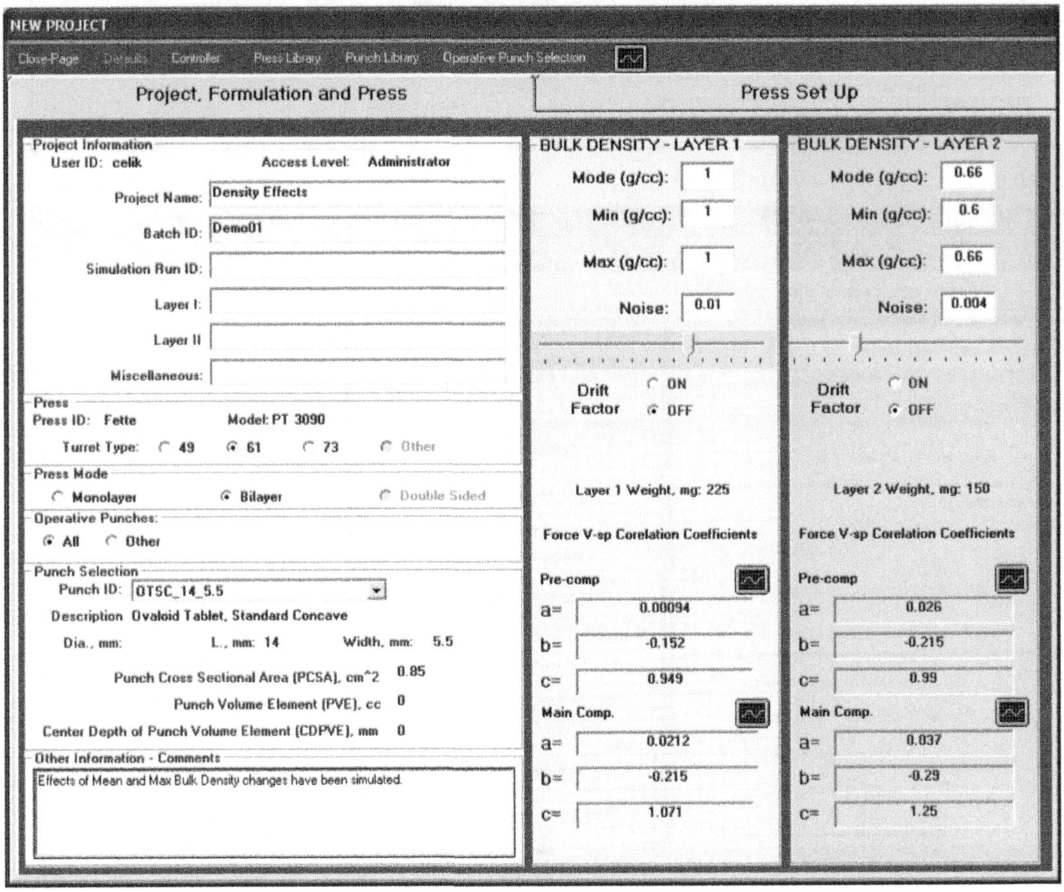

(A)

Figure 7 (**A**) BILAYERex user interface module: formulation and press parameters (continued).

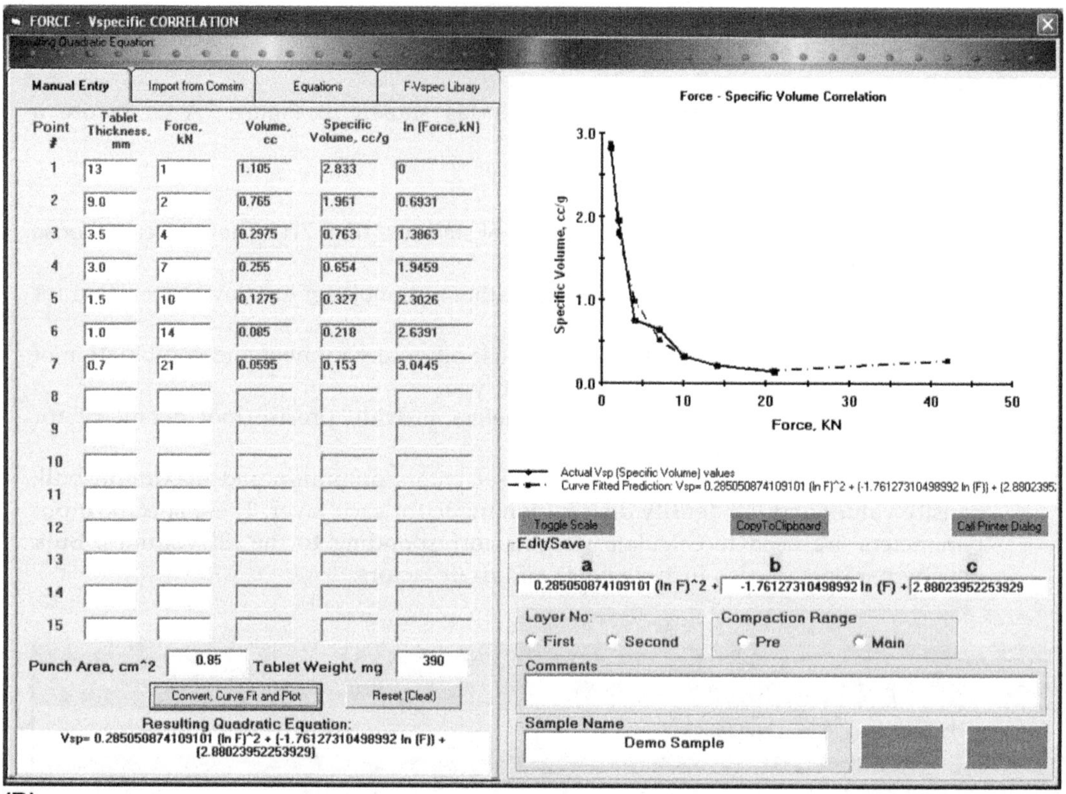

(B)

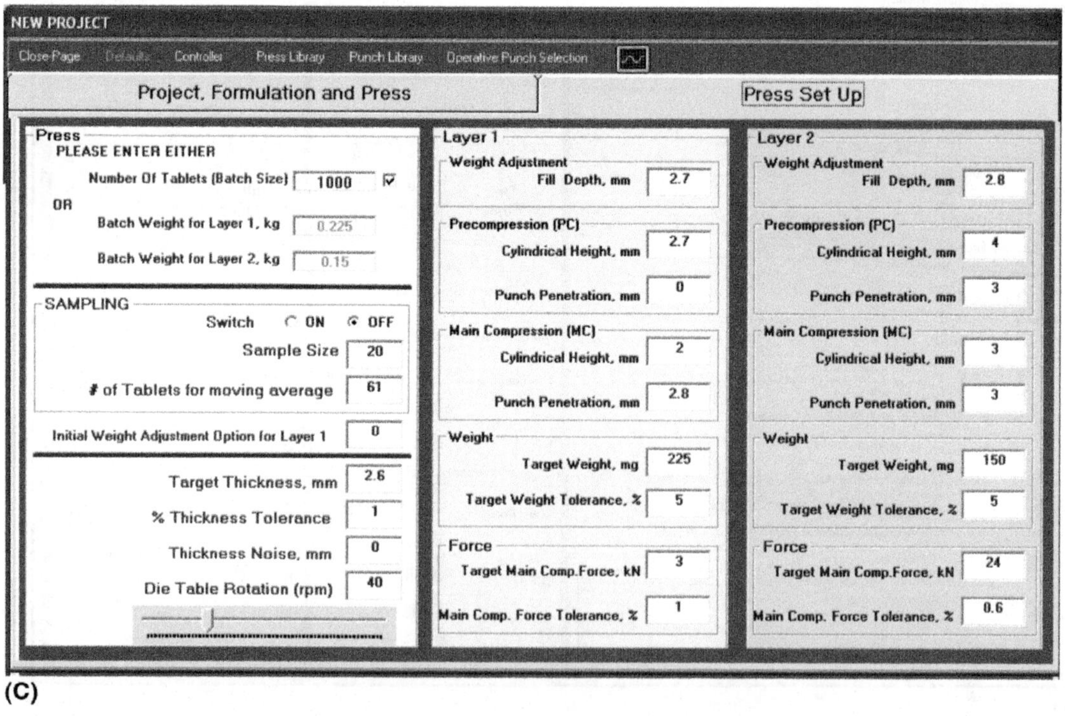

(C)

Figure 7 (**B**) BILAYERex user interface module: force-specific volume correlation parameters. (**C**) BILAYERex user interface module: press set up parameters.

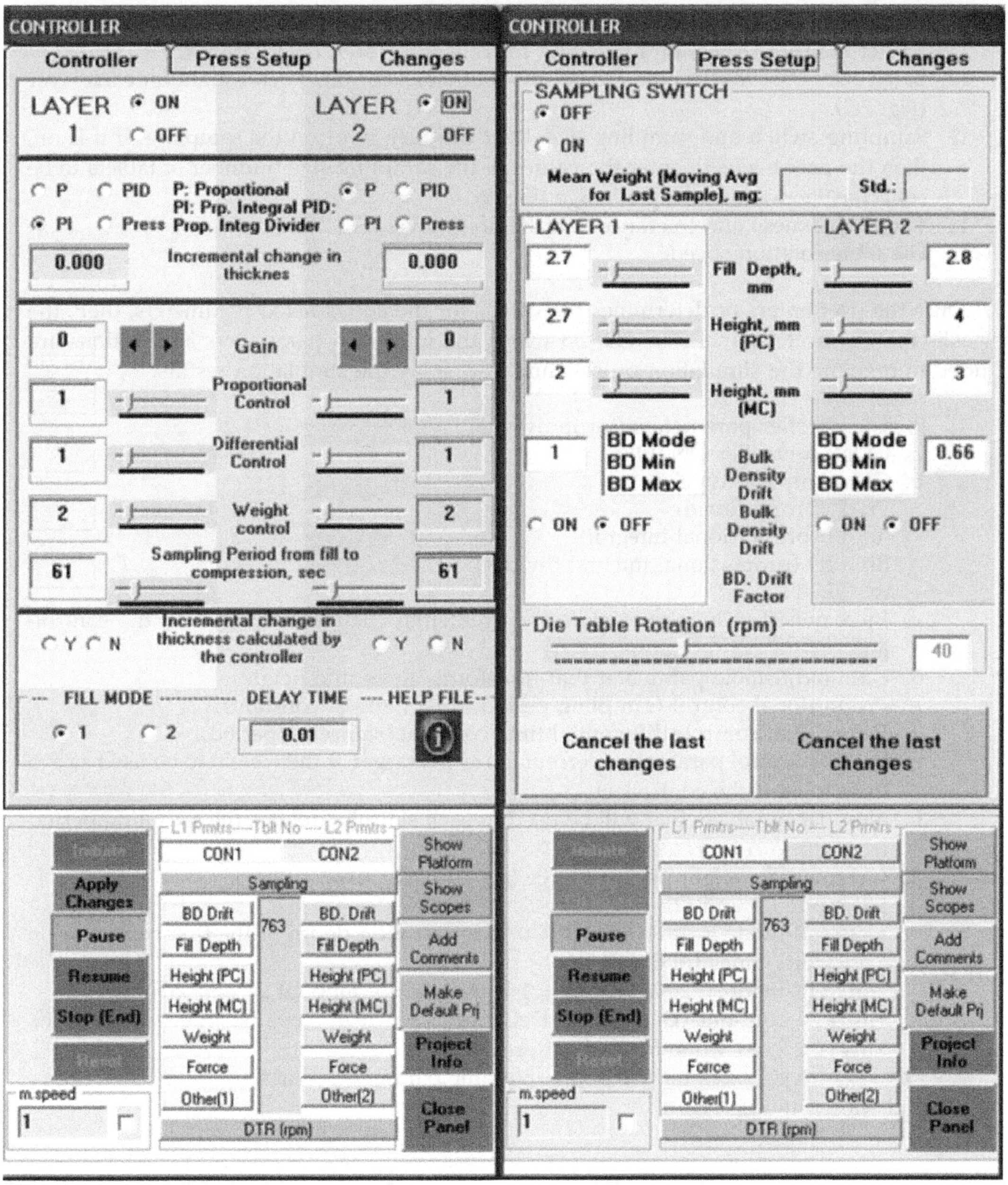

(D)

Figure 7 (**D**) BILAYERex user interface module: controller panel.

7. Force-specific volume correlation coefficient parameters (Fig. 7B): The user enters the values for these parameters for each layer for use in monitoring the densities of the layers while in the die cavity during compaction. The user determines these parameters by utilizing a sub function of the software shown in Figure 7B.
8. Fill depth in weight adjustment stations for each layer (Fig. 7C).
9. Cylindrical height and magnitude of punch penetration for each layer at the precompression and main compression stations (Fig. 7C).

10. Target weight and target weight tolerance for each layer (Fig. 7C).
11. Target main compression force (and tolerance) for each layer (Fig. 7C).
12. Batch size: User inputs either the number of tablets or the batch weight for each layer (Fig. 7C).
13. Sampling switch and sampling size: User can turn of or on the sampling. If it is on, then the user needs to enter the values of the sampling size (number of tablets to be collected) and the moving average (Fig. 7C).
14. Target thickness and tolerance and the noise level (Fig. 7C).
15. Die table rotation speed.

Once the user enters or determines the values for the above listed parameters, then, the user calls the Controller panel (Fig. 7D) on which the following parameters and options are included to regulate the simulation process and to display the simulation results.

1. Press controller parameters (for individual layers).
   a. CON: controller ON/OFF.
   b. CN: controller type.
      i. P (proportional).
      ii. PI (proportional integral).
      iii. PID (proportional integral divider).
      iv. FETTE.
   c. Incremental change in thickness (incremental change calculated by the controller).
   d. Gain (adjustable parameter that transforms force into height).
   e. Proportional control (sampling period/time integral constant).
   f. Differential control (differential time constant/sampling period).
   g. Weight control parameter (percentage of the target of the weight to be used as a ± limit to start the controller).
   h. Number of tablets that will be used in each sample (defines the sampling period from fill to compression).
   i. Incremental weight per unit force (for display purpose only).
   j. Fill mode (f1 and f2):
      i. Fill mode = 1: f1 = 0, f2 = 0 Þ. Penetration depth of the first compression determines fill depth.
      ii. Fill mode = 2: f1 = 0, f2 = 1 Þ. Adjust fill depth at second layer fill.
   k. Fill time constant (DLYTIME): Delay time between steps (stepper motor time). This is a fixed value for a given press.
2. Press set-up parameters: (These are the same parameters that have been listed above and shown in Fig. 7C).

Arena Module and Communication Between the Arena Module and User Interface

Once the input entries necessary to run a simulation even is completed, pressing "Initiate" button loads the Arena simulation module in the background and the values of these input parameters are passed to the Arena environment via communication step shown in Figure 8A and the simulation module stands by for triggering. When the user is ready to run the simulation, pressing the "Resume" button triggers the simulation event.

The Arena module of BILAYERex has the operational logic from filling to compression of the bilayer tablet press and a PID controller. As shown in Figure 8B–G) The operational logic consists of a number of steps modeling the filling, weight adjustment, pre-compression and compression stages as stations and the tablets as entities that move from station to station of

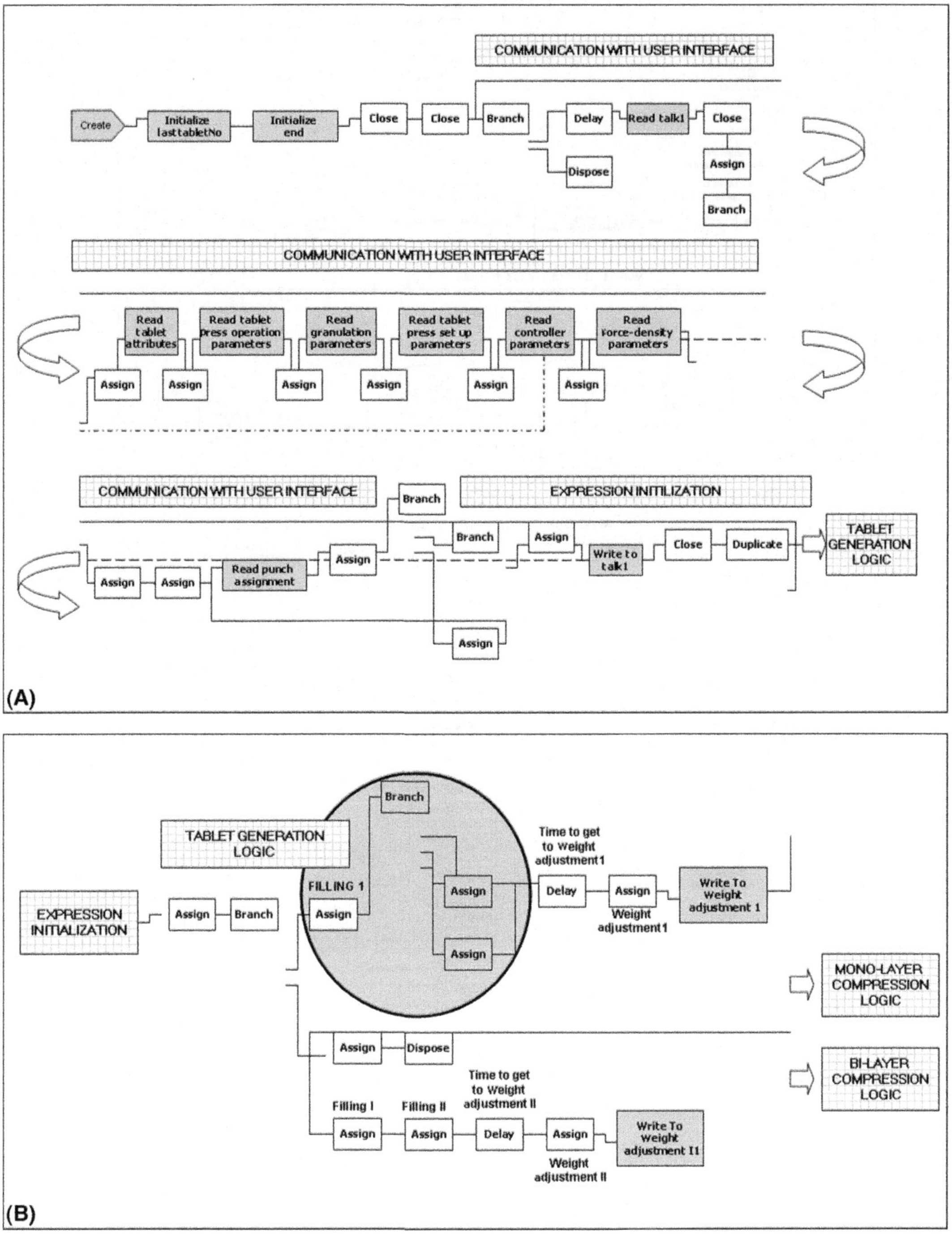

Figure 8 (**A**) BILAYER arena module: communication interface. (**B**) BILAYER arena module: filling and weight adjustment stages (continued).

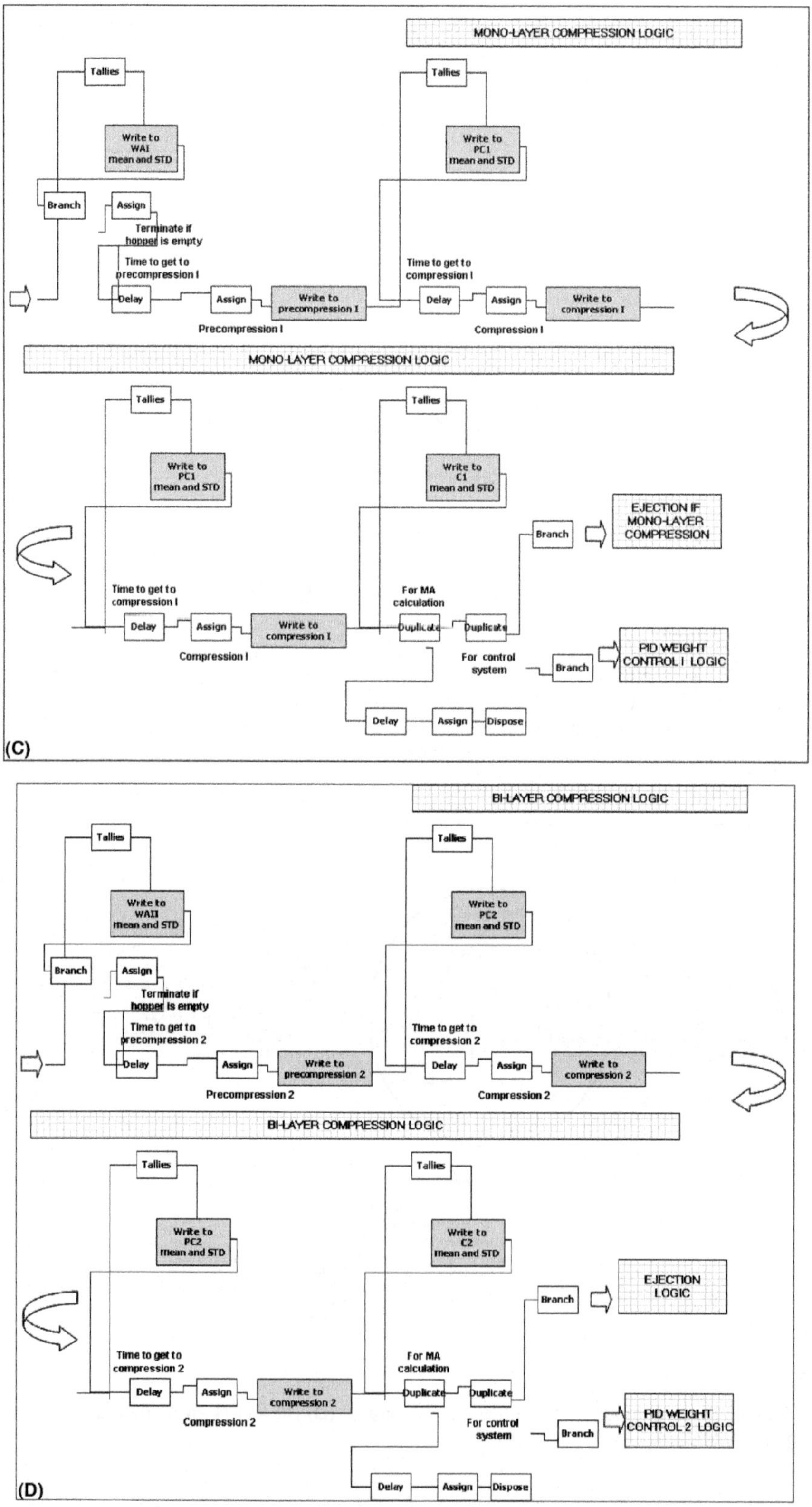

Figure 8 (**C**) BILAYER arena module: first layer (or monolayer) logic. (**D**) BILAYER arena module: second layer (bilayer) logic.

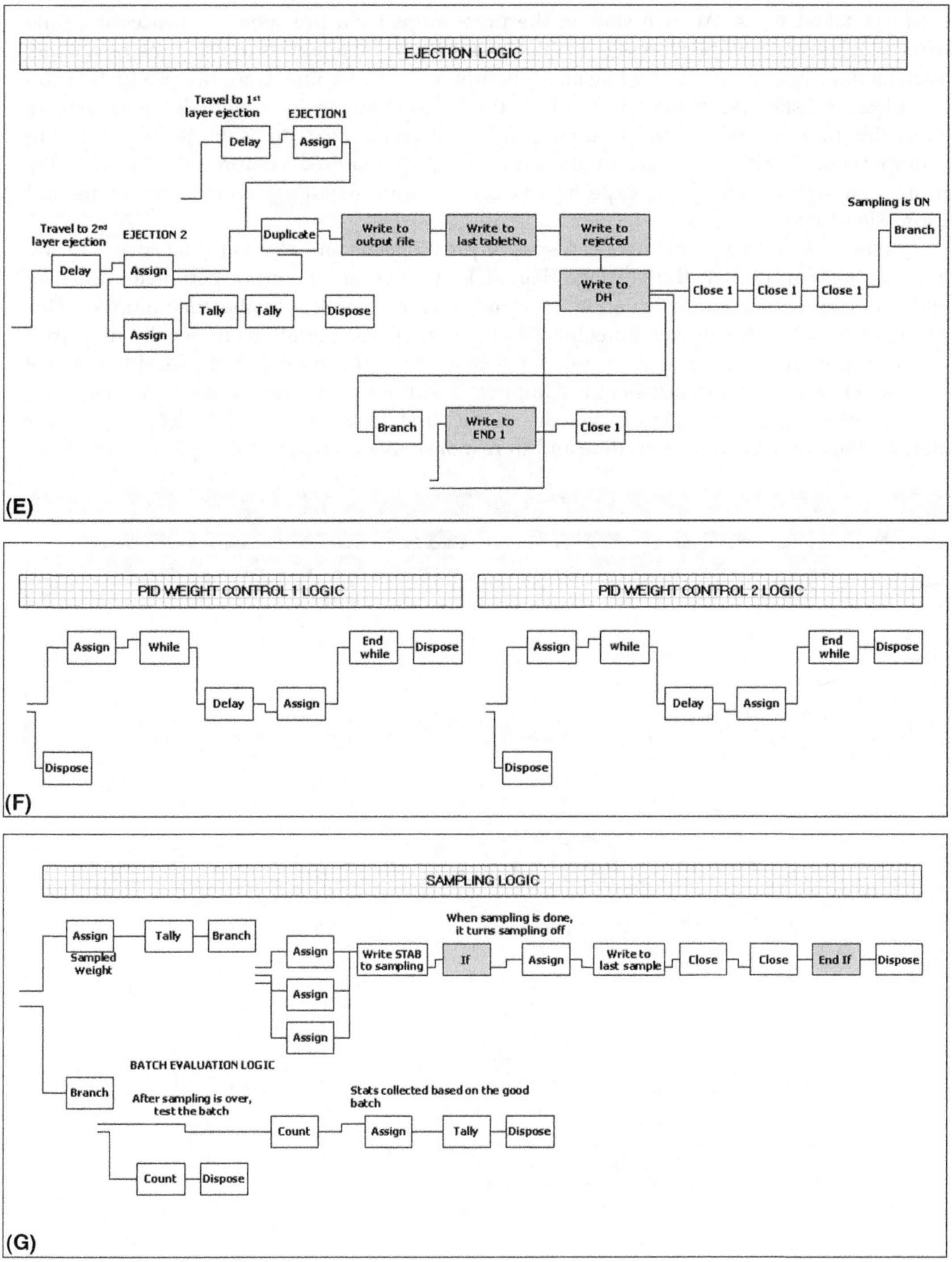

Figure 8 (**E**) BILAYER arena module: ejection stage. (**F**) BILAYER arena module: PID control logic. (**G**) BILAYER arena module: sampling logic.

the bilayer tablet press. At each station, the press stops for a pre specified processing time during which every station processes a tablet and then the table moves to bring the next tablet to each station. The system works like an asynchronous transfer line. Once the production run is complete, statistics are flushed out to show the system performance. Note that bulk density is typically randomized at the beginning and this creates variations in tablet weight in precompression, which later creates the corresponding required compression force in the compression station. This way, experiments can be done regarding weight variations and compression forces.

As mentioned earlier, the user can specify press speeds and also vary the speed during the simulation run itself within a range (Fig. 8C). Product specific input parameters such as powder densities help the calculation of weights corresponding to the fill volumes. This accommodates the relevant equations for weight calculations. Variations in density may arise from poor powder flow. The user will be able to vary the bulk density to simulate the variations in the powder densities at the hoppers and the powder feeder mechanisms.

Output parameters included weight of the first layer and total tablet weight and variations therein as a function of time and in response to a change in input parameters. Each

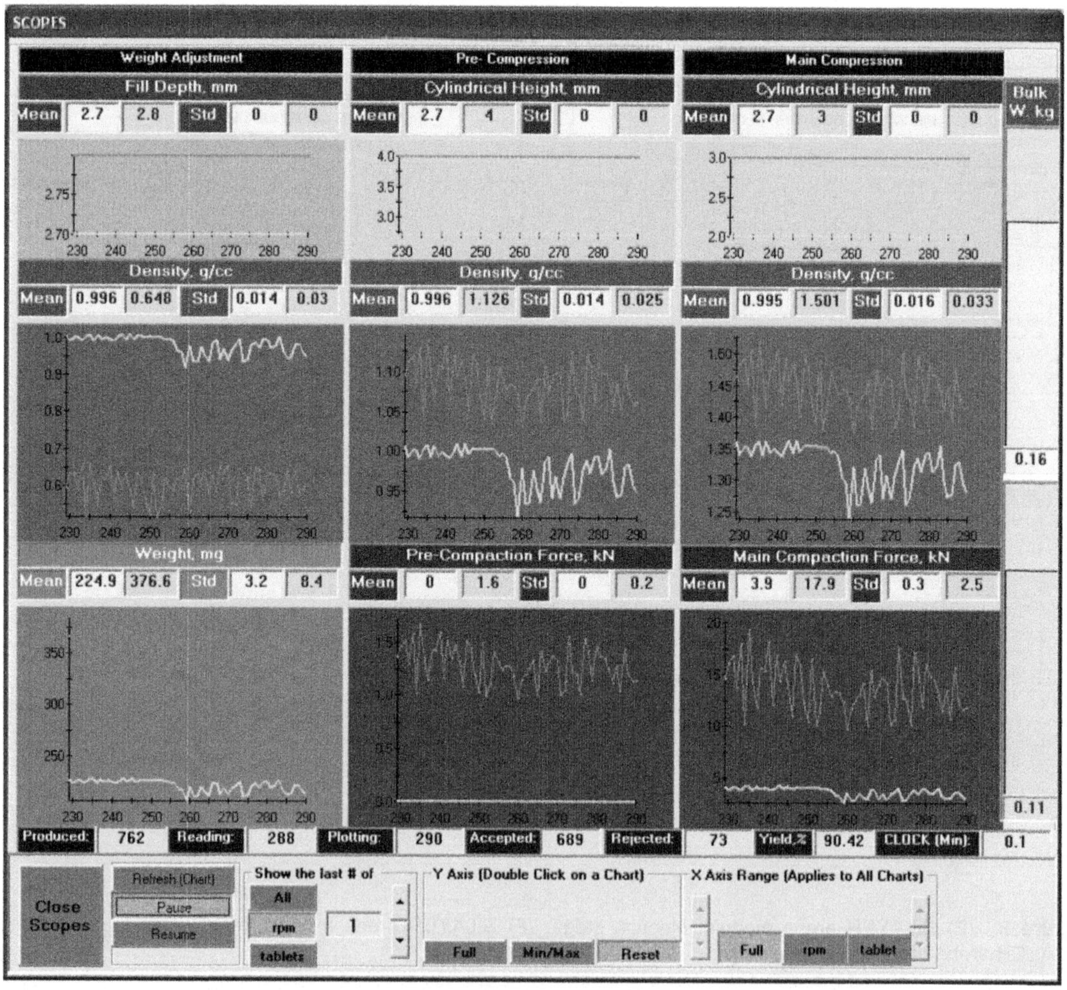

Figure 9 BILAYER-user interface: simulation scope.

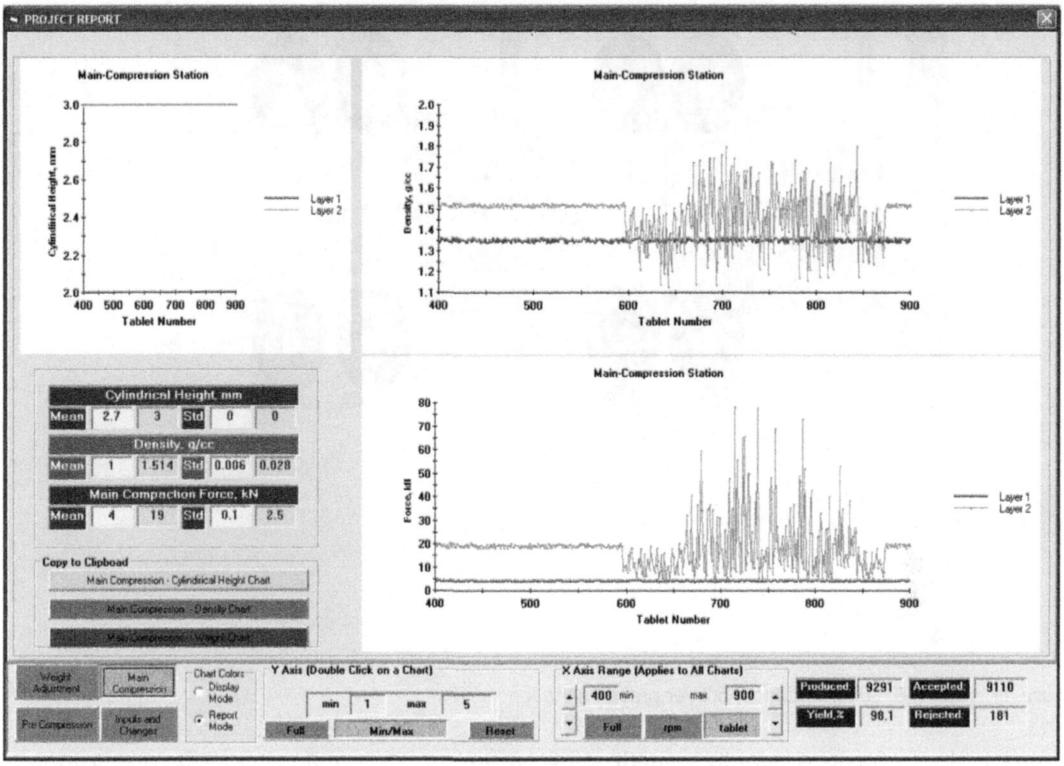

Figure 10 BILAYER-user interface: example for a report display.

single run of the model is called a replication. The replication length is controlled by the total number of tablets to be produced. At the end of each replication, a scope consisting of a number of plots continuously display the process details using end-of-replication calculations. These variables include total produced, accepted, rejected (based on weight), weight, volume, peak force, density, tablet thickness at different stations and throughput among other key parameters. Some of the key parameters are plotted to show behavior over time (Fig. 9). The input parameters (including the changes in a given parameter) as well as the output results are stored at the end of simulation and can be displayed later both in a summarized report form and in the form of a replay of the scope mode (Fig. 10).

The model also has an animation component consisting of higher-level view showing the entire tablet operation from filling granules to tablet exit, as well as the single-punch detail (modeled using visual basic component) from filling to compaction and ejection (Fig. 11).

PHARMACEUTICAL APPLICATIONS OF EXPERT SYSTEMS

The applicability of ESs to the pharmaceutical industry has been reviewed by Klinger (14). The review contains definitions and explanations of AI and ESs as well as information about the components and available programming languages. Possible applications for the pharmaceutical industry outlined include pathologic evaluation, molecular modeling, biologic activity screening, statistical design/analysis/interpretation, manufacturing process/control, automated QA monitoring, drug interaction predictions, production scheduling, and marketing/sales plans.

The specific application of ESs to manufacturing process and control was addressed in more detail by Murray (15). The article begins with an outline for choosing processes *which*

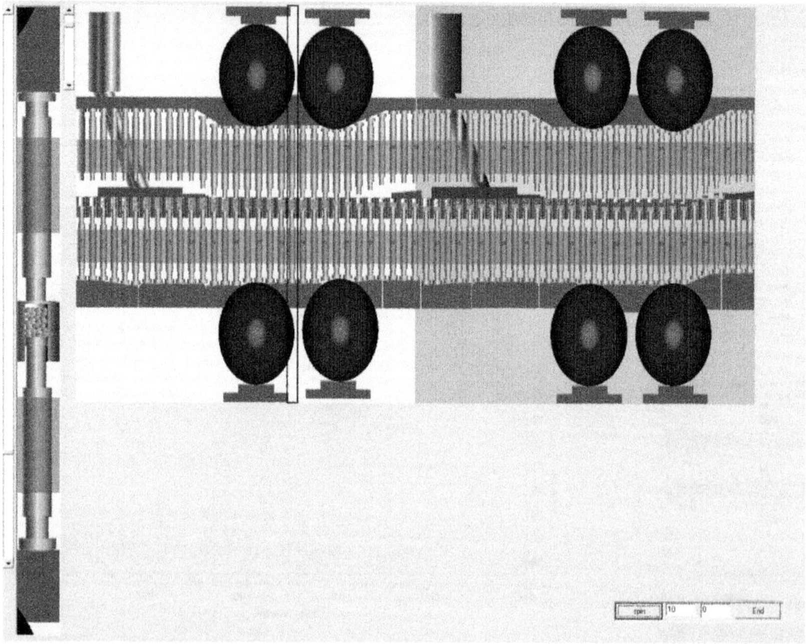

Figure 11 BILAYER-user interface: bilayer press animation.

manufacturing processes would benefit most from an ES application. The application described in additional detail is a rule-based ES for the troubleshooting and diagnostics of a high-speed tablet press that was in the process of being developed and some of the experiences resulting from this development.

Another formulation ES was described in the literature by Rowe et al. (16) This ES was based on a decision tree and was used for the development of parenteral formulations. The decision trees utilized by the system were described in detail. Additional detail about the software used and the advantages for this formulation tool were also included.

Bateman et al. (17) described an ES for the development of powder formulations for hard gelatin capsules. A team process incorporating formulators and software engineers was utilized for the acquisition of the information for the knowledge base. From this process the rules for the knowledge base were discovered and evolved using a process of "iterative refinement." The system also required an excipient database containing excipient physical properties.

A comprehensive review of the commercially available software for use in developing intelligent systems was provided by Rowe (18). Rowe divided the software into five types describing the applications, advantages, and disadvantages of each type as well as diagramming the operation processes. Software tool names and supplier information are also provided.

Within the pharmaceutical literature, ANNs have been applied to several areas. These include clinical pharmacy, drug design (QSAR), product development and optimization, protein drug delivery, biopharmaceutics, and pharmacokinetics.

Hussain et al. describe an ES for the prediction of the in-vitro drug release profile from hydrophilic matrix tablets (19). The ES is based on ANN software that is defined as the main component of CAFD (computer-aided formulation design). The purposes outlined for CAFD include the prediction of formulation/process conditions, the simulation of studies, the storage of information for training purposes, and the reduction of time and cost in the product development process. The specific ES described in this work was built using data from the release profiles of eleven active ingredients and three polymer grades of hydroxypropyl cellulose combined at several drug to polymer ratios. The developed system was able to differentiate between the active ingredient salt types, the polymer grades, and the drug to polymer ratios and successfully predicted the release profiles of most drugs within the ranges of the training sets. Additional components such as additional formulation variables, process conditions, and performance tests were recommended to make the "CAFD" a useful system.

Neural networks have also been applied to the process of fluidized bed granulation. Watano et al. (20) have specifically applied neural networks to fluid bed granulation scale-up. A three layer, backpropagation network was used with the input variables being vessel diameter, moisture content, fluidization air, and agitator rotational speed. The number of neurons in the output layer was also four and generated the following outputs, granule mass median diameter, geometric standard deviation, apparent density, and shape factor. Various numbers of middle layer units were tested to determine the optimal number based on the behavior of the error convergence during learning. Evaluation of the final error after 1000 epochs showed the optimal number of middle layer units to be four. The data used to train the network was obtained from three sizes of laboratory scale granulators. The trained network was used to predict the granule characteristics of material produced using commercial scale equipment. These granulations were produced and the actual granule data was compared to the predicted values and an excellent correlation was observed. Additional networks using the same architecture were also trained by the authors using fewer data points in the training set. From this investigation, it was shown that the training data could be decreased while retaining good accuracy. However, the authors noted the when the number of training sets was less than 13, the accuracy of the predictions decreased.

Murtoniemi et al. have also used ANN to model the fluid bed granulation process (21). In their work, three input variables, inlet air temperature, atomizing air pressure, and binder solution amount, were varied at three levels. The output variables, mean granule size, and granule friability were measured. This training data was processed using a modified backpropagation algorithm in a basic feedforward architecture containing one or two hidden layers. The number of neurons in each hidden layer was varied from 3 to 15. In all, 36 networks were trained. Evaluation of the training data revealed that the number of hidden neurons did not greatly affect the average error except when the networks were small and contained only three or four hidden neurons. The data produced by the optimal network was also compared to data calculated using a regression model. For both outputs, the ANN data were closer to the experimental values than the regression data. In a second article by the same authors, the topology and the training end point of this network were investigated further (22). The purpose of this study was to optimize the ability of the ANN to generalize by varying the number of hidden layer neurons and the training end point. The results again showed that the number of hidden neurons did not affect the ability of the network to generalize. However, the training end point had a significant effect on generalization and on the number of iteration epochs required.

The prediction of in vitro dissolution as a function of formulation variables was also the goal of work performed by Ebube et al. (23). This study demonstrated the importance of optimizing the number of hidden layers and the number of iterations or epochs. The developed network and two inputs, the level of polymers 1 and 2, and one output, the percent dissolved in one hour. Optimization of the network resulted in three neurons for the hidden layer and an optimal number of iterations, which varied from 81 to 671, depending on the number of formulations in the training, set. The authors also found that the network predicted data outside the training set less accurately than data bounded by the training set. However, the predictive capability of the network was improved using replicate input and output data.

Two reviews of neural network computing are published in the pharmaceutical literature. In a 1993 article, Erb comprehensively describes the backpropagation architecture citing much of the original neural network literature as well as additional helpful books (24).

CONCLUSION

It is a highly complicated process to develop an ES to the full satisfaction of the users, domain experts, company, FDA, etc. However, none of these obstacles should discourage pharmaceutical scientists. On the contrary, despite all of these problems, the overwhelming advantages of ESs must encourage pharmaceutical scientists to learn more about them. Despite these advantages, ESs are not intended to take the place of formulation scientists. They must be considered as vital tools to be used by formulators for the rapid, cost-effective, and scientifically sound development of a dosage form as well as useful for training inexperienced scientists. In the same way that we cannot do much without computers today, we will not be able to do much without ESs in the future. Sooner or later, all of us will be happily using them. Those who use them sooner will enjoy being the pioneers in their fields. They also will have the personal satisfaction of contributing to pharmaceutical science by catching up with the rest of the world in the application of such useful tools.

ACKNOWLEDGMENT

The author of this chapter would like to thank Drs Susan Wendel, Tayfur Altiok, and Nicholas G. Lordi for their invaluable contributions at varying degrees to both the theory and practice of expert systems described in this chapter.

REFERENCES

1. Çelik M. Catching up with expert systems. Pharm Technol 2001; 25(7):122–124.
2. Çelik M. The past, present and future of tableting technology. Drug Dev Ind Pharm 1996; 22(1):1–10.
3. Durkin J. Introductions to Expert Systems. Expert Systems Design and Development. Eaglewood Cliffs, New Jersey: Prentice Hall, 1994:1–25.
4. McCarthy J, Minsky M, Rochester N, et al. A Proposal for the Dartmouth summer research project on AI. 1955. Available at: http://www-formal.stanford.edu/jmc/history/dartmouth/dartmouth.html.
5. Harmon P, Sawyer B. Creating Expert Systems for Business and Industry. New York: Wiley, Inc., 1990.
6. Ignizio JP. Introduction to Expert Systems: The Development and Implementation of Rule-Based Expert Systems. New York: Mc Graw-Hill, Inc., 1991.
7. Schneider M, Kandl A, Langholz G, et al. Fuzzy Expert System Tools. Chichester, U.K.: Wiley, 1996.
8. Swingler K. Applying Neural Networks: A Practical Guide. London, U.K.: Academic Press, 1996.
9. Rowe C, Roberts RJ. Intelligent Software for Product Formulation. London, U.K.: Taylor & Francis, 1998.
10. Cook DF, Ragsdale CT, Major RL. Combining a neural network with a genetic algorithm for process parameter optimization. Eng Appl Artif Intell 2000; 13:391–396.
11. Ishikawa KA. Guide to Quality Control. 2nd ed. Asian Productivity Organization, 1982.
12. Çelik M. Eye on excipients: tools for risk-based product development. Tablets Capsules 2009; 7(6):55–59.
13. BILAYERex. Available at: http://www.pt-int.com/Bilayerex.html.
14. Klinger DE. Expert systems in the pharmaceutical industry. Drug Inf J 1988; 22:249–258.
15. Murray FJ. The application of expert systems to pharmaceutical processing equipment. Pharm Technol 1989; 13(3):100–110.
16. Rowe RC, Wakerly MG, Roberts RJ, et al. Expert systems for parenteral development. PDA J Pharm Sci Technol 1995; 49:257–261.
17. Bateman SD, Verlin J, Russo M, et al. The development and validation of a capsule formulation knowledge-based system. Pharm Technol 1996; 20(3):174–184.

18. Rowe RC. Intelligent software systems for pharmaceutical product formulation. Pharm Technol 1997; 21(3):178–188.
19. Hussain AS, Shivanand P, Johnson RD. Application of neural computing in pharmaceutical development: computer aided formulation design. Drug Dev Ind Pharm 1994; 20(10):1739–1752.
20. Watano S, Takashima H, Miyanami K. Scale-up of agitation fluidized bed granulation by neural network. Chem Pharm Bull 1997; 45:1193–1197.
21. Murtoniemi E, Yliruusi J, Kinnunen P, et al. The advantages by the use of NN in modeling the fluidised bed granulation process. Int J Pharm 1994; 108:155–164.
22. Murtoniemi E, Merkku P, Kinnunen P, et al. Effect of NN topology and training end-point in modeling the fluidised bed granulation process. Int J Pharm 1994; 110:101–108.
23. Ebube NK, McCall T, Chen Y, et al. Relating formulation variables to in vitro dissolution using an ANN. Pharm Dev Technol 1997; 2(3):225–232.
24. Erb RJ. Introduction to backpropagation neural network computation. Pharm Res 1993; 10:165–170.

Index